The Theory of Cubature Formulas

Mathematics and Its Applications

Volume 415

The Theory of Cubature Formulas

by

S. L. Sobolev †

and

V. L. Vaskevich

Sobolev Institute of Mathematics,
Siberian Division of the Russian Academy of Sciences,
Novosibirsk, Russia

Springer-Science+Business Media, B.V.

A C.I.P. Catalogue record for this book is available from the Library of Congress.

ISBN 978-90-481-4875-2 ISBN 978-94-015-8913-0 (eBook)
DOI 10.1007/978-94-015-8913-0

This is a revised and updated translation
of the original Russian work of the same title.

Translated by S. S. Kutateladze.

Printed on acid-free paper

Contents

Foreword to the English Translation

Academician *Sergeĭ L'vovich Sobolev* (1908–1989), a great Russian scholar and the founder of the Institute of Mathematics of the Siberian Division of the Russian Academy of Sciences at Novosibirsk which is now named after him, is world renowned for his contribution to distribution theory, sharing the fame of its propounding with L. Schwartz. S. L. Sobolev successfully applied his new functional-analytical technique not only to partial differential equations but also to computational mathematics, changing the layout of the field of numerical integration.

The present edition is a translation of the posthumous monograph finished by V. L. Vaskevich, the last pupil of S. L. Sobolev, and published in Russian by the Sobolev Institute Press in 1996. The book contains all contributions of S. L. Sobolev to numerical integration as well as results of his students and followers on cubature formulas, thus granting an updated definitive source of this direction in modern mathematics. For the first time, the book includes recent data about invariant cubature formulas exact for spherical harmonics up to a given degree and Sobolev's research on optimal cubature formulas and Euler polynomials.

This edition was typeset using $\mathcal{A}\mathcal{M}\mathcal{S}$-TeX, the American Mathematical Society's TeX system.

S. L. Sobolev was one of my inspired teachers in mathematics and life although I have never plunged into numerical integration before. Translating the book was thus an imperative but onerous duty since it covers the field of research deeply rooted in classical mathematics as well as in the brand-new applied sections of functional analysis. One technicality needs explanation, namely, absence of the term *distribution* which is in common parlance in the West. The competing term *generalized function* proliferates in Russia and is reverently retained as coined by the inventor as far back as in the thirties of this century.

It gives me a blended feeling of great pleasure and deep sorrow to invite the reader to get acquaintance with this masterpiece which is a part of the memory of S. L. Sobolev to whom I am greatly indebted.

S. Kutateladze

Preface

A noticeable place in the scientific legacy of S. L. Sobolev is occupied by his contributions to the theory of approximate integration which were accomplished during his stay of twenty-five years in Novosibirsk. His first article in this direction was published in 1961 and the last in 1985. Their total number exceeds three dozens, including the definitive monograph "Introduction to the Theory of Cubature Formulas" [250]. Several chapters of the monograph, completely revised and updated, form the basis for the present book.

The object of further research is formulas for approximate calculation of multiple integrals. The domain of integration Ω has a piecewise-smooth boundary of finite area. It is arbitrary in other aspects. We include the one-dimensional case in which integration is carried out over an interval of the numerical axis. To calculate an integral over Ω approximately, we propose to use a *cubature formula*, i.e., an approximate identity of the shape

$$\int_{\Omega} \varphi(x)dx - \sum_{k=1}^{N} c_k \varphi(x_k) \cong 0.$$

The parameters x_k are the *nodes* or *points* and c_k are the *weights* or *coefficients* of a cubature formula. Disposition of x_k in Ω may be arbitrary in principle. However, most results of the book relate to formulas with a parallelepipedic lattice of nodes. In this event the nodes x_β are enumerated with the multi-index $\beta = (\beta_1, \dots, \beta_n)$ with integer entries. Each of the nodes may be found by the formula $x_\beta = hH\beta$, with H a matrix of size $n \times n$, $\det H = 1$, and h a positive parameter called the *lattice mesh-size*.

The nodes and weights of all formulas dealt with in the monograph are given either explicitly or as a result of a rather definite succession of operations. To each formula we assign the set of functions for which it is exact. Most often this set is the finite-dimensional space $\mathbf{P}_{m-1}$ comprising polynomials of degree less than m. If a cubature formula is exact for the members of $\mathbf{P}_{m-1}$ and the number m is maximal among those with the same property, then m is referred to as the ***order*** of a cubature formula. The number $m-1$ is usually called the ***polynomial degree*** (or simply ***degree***) of a cubature formula. The authors thus deal with cubature formulas of order m or degree $m-1$.

In the book we mainly pursue a functional-analytical approach. This implies that, first, the integrands are combined in a Banach space and, second, the difference between the integral and the approximative combination of the values of the integrand is treated as the result of applying some linear functional. This functional, called the ***error*** of a cubature formula, is usually continuous. Knowledge of the value of its norm allows us to derive guaranteed estimates for the accuracy of the cubature formula under study at the elements of the chosen space. This feature incarnates the essential advantage of the functional-analytical approach over its purely algebraic counterpart.

The algebraic and functional-analytical approaches lead to different criteria for quality of a cubature formula. In the first case, the best is a formula exact for possibly more polynomials. In the second, we prefer a formula whose error has a lesser norm. The book explicates the interrelation of the two criteria in detail. We prove that the norm of an error may tend to infinity as the number of exactly integrated polynomials increases indefinitely. In other words, that of two errors may have a lesser norm which is exact for essentially fewer polynomials.

In addition to describing the construction of the formulas under consideration, i.e., indicating their nodes and weights or algorithms for their determination, the functional-analytical approach implies the study of the norms of the respective errors in a chosen Banach space. In particular, we derive two-sided estimates for these norms or their asymptotic expansions with respect to the mesh-size of the lattice of integration and the order of a cubature process.

Functional-analytical methods became widespread in the theory of approximate integration since the articles by S. M. Nikol'skiĭ and the first edition of his book [136]. The results of [136] relate mostly to quadrature formulas over the function classes $W^m_{L_p}$. The origination of the theory of cubature formulas is attributed to S. L. Sobolev's research and his monograph [250]. The function spaces belonging to this theory are characterized by the parameters determining the smoothness of integrands. The choice of these spaces is arbitrary in other aspects. In the present edition we consider cubature formulas in the Sobolev spaces $W_2^{(m)}(\Omega)$ and their subspaces. Among the latter we in particular treat various subspaces of $W_2^{(m)}(\Omega)$ which are composed of infinitely differentiable functions.

In the theory of cubature formulas, the term coined by S. L. Sobolev, four principal directions are specified. All are exposed in the present edition.

The first in chronological order of the directions consists in studying the cubature formulas in 3-dimensional space which possess high order and are invariant under the action of the rotation group of some regular polyhedron [224]. The requirement that a cubature formula with fixed nodes be exact for polynomials up to a certain degree reduces the problem of constructing the weights of the formula to solution of a system of linear equations. The higher the desired order is and the larger we have the number of nodes, the greater becomes the size of this system. However, in the case when the integration domain possesses some symmetry and we use an invariant cubature formula for approximate integration, it is possible to diminish substantially the size of the system to be solved. Chapter 2 addresses the question of how to achieve this. In the same chapter an algorithm is suggested for constructing the nodes of an invariant cubature formula on the 3-dimensional sphere. The research by S. L. Sobolev in invariant formulas of approximate integration was substantially developed in the articles by V. I. Lebedev [89, 105–112], I. P. Mysovskikh [133], G. N. Salikhov [192–200], and other mathematicians.

The second direction in the theory, which seems to be most advanced, consists in studying asymptotically optimal cubature formulas over a lattice on the spaces of functions of finite smoothness. In this respect S. L. Sobolev himself considered the Hilbert $L_2^{(m)}$ spaces. The construction of a regular boundary layer which he proposed makes it possible to find the weights of a cubature formula with arbitrary many nodes by solving just a few standard systems of linear equations of size depending only on the order m. The central place in this direction is occupied by derivation of an asymptotic expansion of the $L_2^{(m)*}$ norm of an error with regular boundary layer. The expansion contains two summands. The first is written explicitly via the so-called *generalized Bernoulli numbers*, whereas the second is negligible as compared with the first provided that the small mesh-size h of the lattice of integration is sufficiently small. The expansion implies that the norm of an error with regular boundary layer decreases like h^m as $h \to 0$. It is a rather deep analytical fact enabling us to give not algebraic but rather functional-analytical definition of the order of a cubature formula on some function class [264].

The expansion of the $L_2^{(m)*}$ norm of an error with regular boundary layer gives solid grounds for choosing a numerical integration formula with nodes comprising a parallelepipedic lattice. Indeed, given N nodes, we may pose the problem of finding a cubature formula whose error has $L_2^{(m)*}$ norm minimal, with the minimum taken over not only the weights but also the nodes of the formula. However, the ratio of the $L_2^{(m)*}$ norm of the error of such an optimal formula to the $L_2^{(m)*}$ norm of the error with regular boundary layer and the same number N of nodes is bounded from below by a positive quantity independent of N. This is immediate from the Bakhvalov Theorem proven in Chapter 4. Increasing N, we could hardly expect however large gain from using formulas with arbitrary disposition of nodes instead of those with nodes comprising a parallelepipedic lattice. Moreover, to optimize a formula over nodes is a difficult problem involving solution of simultaneous nonlinear equations of high order. This is in sharp contrast to the formulas with regular boundary layer whose nodes are explicit and need no calculation at all.

Note that the theory of formulas with regular boundary layer actually presents the function summation problem pertinent to the calculus of finite differences [74]. From this point of view, every cubature formula with regular boundary layer is a multidimensional analog of the classical quadrature formula of Gregory.

Constructing such a cubature formula, we thus take account of the behavior of an integrand near to the boundary of the integration domain by especially selecting the weights of the formula at the nodes belonging to some boundary layer. All remaining weights coincide.

Remarkable is the method proposed by S. L. Sobolev for finding the norm of an error $l(x)$ and his use of the concept of extremal function $u(x)$. Such function is considered as a weak solution to the many-dimensional ***polyharmonic equation*** with a special right side

$$\Delta^m u(x) = (-1)^m l(x).$$

A solution to this equation on the real axis is a piecewise-polynomial function of the class $W_2^{(m)}$, i.e., a ***univariate spline***. In many dimensions, this approach enabled S. L. Sobolev to apply the methods he invented in the theory of partial differential equations [266] to study of the classical problem of analysis. The $L_2^{(m)}$-extremal function for the error $l(x)$ turns out to take the shape

$$\psi_l(x) = P_{m-1}(x) + G(x) * l(x),$$

with $P_{m-1}(x)$ some polynomial of degree $m-1$ and $(-1)^m G(x)$ the fundamental solution to the polyharmonic equation in n-dimensional space. The ***convolution*** in the right side of the last equality is defined as

$$G(x) * l(x) = \int G(x-y)l(y)dy = \int_\Omega G(x-y)dy - \sum_{k=1}^{N} c_k G(x-x_k).$$

The extremal functions of all possible errors span an affine variety in $L_2^{(m)}$. We may describe this affine variety in terms of approximation theory. Namely, fixing the nodes $x_1, \dots, x_N$ and taking the error $l_*(x)$ in $L_2^{(m)}$ with weights $c_1^*, \dots, c_N^*$, we write the difference of extremal functions as

$$\psi_l(x) - \psi_{l_*}(x) = Q_{m-1}(x) + \sum_{k=1}^{N} d_k G(x-x_k).$$

Here $d_k = c_k^* - c_k$, and $Q_{m-1}(x)$ is a polynomial of degree $m-1$. A function of the shape is called a ***natural multivariate spline*** [84] or a D^m-spline in the whole space [2, 60]. The set of all $L_2^{(m)}$-extremal functions results thus by translating the linear space of natural multivariate splines by the member $\psi_{l_*}(x)$ of $L_2^{(m)}$. This observation allows us to claim that the theory of cubature formulas in $L_2^{(m)}$ is dual to the relevant section of the theory of multivariate splines.

The problems of the theory of asymptotically optimal formulas are addressed in Chapters 3–5. Some particularities of constructing formulas with boundary layer arise in the case when the integration domain is a rational polyhedron. In this case, some construction of a *formal boundary layer* is suggested which made it possible to invent a simple algorithm for calculating the weights of a lattice cubature formula. The relevant results occupy Chapter 3.

Research by S. L. Sobolev on asymptotically optimal cubature formulas was continued by other mathematicians. Among the latter are M. D. Ramazanov [170–181] who wrote Chapter 6 of the second Russian edition of the present monograph, and V. I. Polovinkin [142–167] who generalized the theory of asymptotically optimal formulas to the $L_p^{(m)}(\Omega)$ spaces for $1 < p < \infty$ and to the case of weighted integration.

In [25–28] O. V. Besov derived asymptotic expansions for the norms of *refinements* of a given error. Such refinement is a set of errors acting in some spaces more general than $L_p^{(m)}(\Omega)$. The latter include the classes of functions with generalized derivatives of order m belonging to the Marcinkiewicz space $M_\psi(\Omega)$ (or the Lorentz space $\Lambda_\psi(\Omega)$, or the Orlicz space $L_M^*(\Omega)$).

Cubature formulas in $W_p^{(m)}$ and in anisotropic Sobolev spaces were considered by Ts. B. Shoĭnzhurov [213–217].

Cubature formulas with regular or formal boundary layer were studied by N. I. Blinov [30–35], L. V. Voĭtishek [291–295], and F. Ya. Zagirova [299]. Asymptotic estimates for the norms of errors acting in the anisotropic classes $W_p^{(r)}M$ were found by M. D. Ramazanov [174] and V. N. Temlyakov. The latter deduced asymptotic expansions of the norms of errors over the anisotropic Nikol'skiĭ spaces $NH_q^R([0, 2\pi]^n)$ [272, 273].

The third direction of the theory comprises the S. L. Sobolev contribution to cubature formulas on the classes of infinitely differentiable functions. As such he considered the spaces of periodic functions of many variables with prescribed behavior of the integral norms in the $L_2^{(m)}$ spaces as m tends to infinity. The classification he proposed embraces the conventional spaces of entire functions of given type and order, spaces of analytic functions and the Gevrey classes containing quasianalytic functions. Considering the action on this space of the error of a lattice formula with equal weights, S. L. Sobolev obtained an asymptotic expansion of the logarithm of the norm of the error. In exact analogy with the case of the spaces of finite smoothness, the respective formula comprises two summands. One of them is explicitly expressed through the parameters of the initial class, whereas the other is negligible as compared with the first at a small mesh-size h. These results generalize the article [50] by P. Davis which treats cubature formulas in spaces of periodic analytic functions of a single variable. However, the attempts fail at obtaining analogous statements for spaces of aperiodic infinitely differentiable functions of many variables. The case of dimension 1 is the only exception.

In [17] N. S. Bakhvalov estimated the rate of convergence of the Gauss process in the space of analytic functions in a neighborhood about the unit interval. Considering the same space, V. I. Krylov proved convergence of the interpolatory quadrature process in [103]. In [278–282] V. L. Vaskevich obtained asymptotic expansions of the norms of the errors of Gregory and Euler–Maclaurin quadrature formulas in the spaces of infinitely differentiable functions which are one-dimensional realizations of the corresponding Sobolev spaces of [254]. It turns out that using formulas of moderate order at a given h leads to no decrease in the norm of the error at all; on the contrary, this may even entail a quite opposite aftermath. In [6] K. I. Babenko surveyed the properties of the formulas for approximate integration in the spaces of infinitely differentiable functions.

This research demonstrated in particular that a noteworthy effect accompanies the transition from functions of finite smoothness to infinitely differentiable functions. Namely, the norm of the error of a cubature formula, decreasing not faster than some power of the lattice mesh-size in the first case, decreases exponentially in the second case. S. L. Sobolev suggested that in the second case the order of a cubature formula be assumed *infinite*. More exactly, a *cubature formula* possesses *infinite* order in a Banach space provided that the norm of the corresponding error in the dual space vanishes faster that any degree of the mesh-size of the integration lattice. By now only a few cubature formulas are known which possess this property. S. L. Sobolev exhibited one example of the sort in the case of many dimensions in [254]. Another example of a cubature formula of infinite order is the celebrated Mean Value Theorem for harmonic functions. Hermitian cubature formulas of infinite order in spaces of harmonic and polyharmonic functions are addressed in the articles [289] and [53], respectively.

Results on cubature formulas over infinitely differentiable functions lay a basis for Chapter 7.

Finally, the fourth direction of the theory exposed in the present book comprises the S. L. Sobolev research in $L_2^{(m)}$-optimal lattice cubature formulas. A central place is occupied here by description of some analytic algorithm for determining weights of such formulas. To this end, some spaces of functions of a many-dimensional discrete argument are introduced analogous to the Sobolev spaces $W_p^{(m)}$ and $L_p^{(m)}$, and the properties of the former are studied. Chapter 8 exposes this material.

Further, S. L. Sobolev defined and studied a special finite-difference operator whose action on a function of a discrete argument may be written as convolution with a special kernel in analogy with the action of the polyharmonic operator Δ^m on an continuously differentiable function.

The problem of calculating the convolution kernel for an arbitrary m turns out rather involved. It was partly solved in the one-dimensional case: here a formula is available expressing the desired values through the roots of the Euler polynomials of degree $2m$. The weights of optimal formulas are conveniently treated as the values at the appropriate points of some compactly-supported function of a many-dimensional discrete argument. This function happens to satisfy a linear finite-difference equation with a special right side. Applying to this right side a discrete convolution analog of the polyharmonic operator, S. L. Sobolev obtained an analytical formula for the sought weights. To use it in the one-dimensional case, he revealed many properties of the roots of the Euler polynomials. In particular, he obtained asymptotic formulas for the roots of these polynomials. The results by S. L. Sobolev on the weights of optimal cubature formulas generalized some results by A. Sard, I. Meyers, I. Schoenberg and S. Silliman exposed in the articles [126], [201], [203, 204] and derived by the methods of splines. Results on $L_2^{(m)}$-optimal lattice formulas and on the Euler polynomials form the bulk of Chapter 9. Implementation of the algorithm by S. L. Sobolev was the subject of the articles by Z. Zh. Zhamalov [302, 303], F. Ya. Zagirova [300, 301], and Kh. M. Shadimetov [207].

The method of this book deserves some especial emphasis. From the first pages it becomes clear that this method is deeply rooted in such fields of theoretical mathematics as mathematical analysis, the theory of differential equations and functional analysis. At the same time, the subject of research, a formula for approximate integration, is traditionally ascribed to numerical analysis which the modern computational mathematics stems from. As a result, the theory has emerged which is undeniable import to applications. This order of events seems by far not random but rather inevitable phenomenon of modern mathematics. S. L. Sobolev expressed his view of the question in a very emphatic manner in the preface to the monograph [250]; an excerpt is cited in the present book below.

Some chapters of the present book formed the bases for optional and compulsory courses delivered for the students of Novosibirsk State University in various years. We thus hope that the material below will be of service in composing lecture courses on related topics of functional analysis, partial differential equations, and numerical mathematics.

The agreement on enumeration is traditional for the literature on physics and mathematics. The formulas in each chapter are denoted by two digits: the first is the index of a current section; and the second, following a period, is the index of the formula within the section. To refer to a formula external to a chapter we use a triple index: the first numeral indicates the chapter in which the formula is presented, while the consecutive numerals separated with a period have the same meaning as before. We refer to corollaries by analogy. Theorems and lemmas are labeled with a double index: the first numeral is the index of a current chapter and the second numeral after a period is the index of the item in the chapter.

We are to remark on the constants appearing in the book. As a rule, they are independent of the varying quantities (functions, vectors, etc.) involved in respective inequalities. Furthermore, these constants are not enumerated, and so the same letter may stand for different quantities. This agreement seems to be convenient as diminishing notation.

Several mathematicians participated in compiling the final version of this book. As the first among them we must rank M. D. Ramazanov who wrote Chapter 6 and § 1–§ 2 of Chapter 4 of the present book and whose active participation as the editor of the Russian edition undoubtedly improved the initial exposition. We acknowledge the influence of N. N. Chentsov on the composition of the present monograph. V. I. Lebedev substantially participated in writing Chapter 2, supplementing it with his own results from the articles [105–112]. The results by F. Ya. Zagirova essentially supplemented the exposition of the theory of optimal quadrature formulas in § 8 of Chapter 9. An invaluable help in preparation of the manuscript in Russian was lent by G. A. Shmyrëv. In compilation of the book also participated G. V. Demidenko, V. G. Perepëlkin, and I. I. Matveeva. Several opportune remarks were made by V. I. Polovinkin and M. V. Noskov.

The completion of this book becomes possible due to encouraging cooperation with Academician O. A. Oleĭnik and Professor A. D. Soboleva.

The English translation of the book contains numerous improvements and additions stemming from apt observations and sound advice of Professor S. S. Kutateladze who undertook the burden of translation. The authors are very grateful to each of the contributors mentioned above.

V. L. Vaskevich

Sobolev Institute of Mathematics
Siberian Division of the
Russian Academy of Sciences
Novosibirsk, 630090
RUSSIA
e-mail address: vask@math.nsc.ru

From the Preface to "Introduction to the Theory of Cubature Formulas"

We all are well aware of the sweeping changes of today in the classical branches of mathematics which are evoked by the surge of new ideas and methods related chiefly to functional analysis. These ideas first enthrall the theory of differential equations, ordinary and partial. Next captive of the realm of novel mathematics became the theory of computations to conceive which without Banach spaces is now as impossible as without electronic computers.

The methods for approximating functions, which are the heart of computation technique, inevitably became methods of approximation in function spaces. Often such are the spaces of functions with derivatives up to a given order l are locally or totally integrable to the pth power. The properties of these spaces and embedding relations between them are beneficial in addressing very diverse problems: they allow us to study solutions to boundary value problems for partial differential equations; using them we construct error estimates in numerical methods.

Calculations on lattices are treated as calculations with local errors. In the modern understanding the latter are integrals involving generalized functions.

The concept of generalized function of long implicit standing and of explicit use in theoretical physics, entered the apparatus of modern mathematics, particularly the theory of partial differential equations, in its definite shape in the thirties. At present, generalized functions are established tools of numerical analysis. Together with generalized functions, the arsenal of computational mathematics now incorporates the Fourier transform of a generalized function and other modern facilities of analysis.

Precisely to this circle of ideas belongs the present monograph treating the classical problem of finding formulas for approximate integration of a function. For the functions of one real variable, the appearance of such formulas is attributed as far back as to the eighteenth century. The simplest of them, the ***trapezoidal rule*** and the *Simpson formula*, give good approximation to integrals of smooth functions with bounded derivatives of order 2 or 4. More involved formulas, those of Gregory, Gauss, and Chebyshev, use boundedness of higher order derivatives for obtaining a more accurate result. In carrying out large numerical calculations, it becomes reasonable to optimize the processes of approximate calculation of an integral and, if the calculation is implemented with the aid of some formulas for approximate integration, the "formulas of mechanical quadratures," to optimize these formulas themselves. This problem is of particular importance for functions of many variables. For instance, in the space of dimension 10 in which the bisecting of the lattice mesh-size increases the number of necessary nodes in thousand times, saving the number of nodes becomes important.

The problem of optimizing an integration formula in a modern understanding is that of minimizing the norm of the error of the formula. It is this question that the present monograph addresses.

A few words are in order about the title of the book. By analogy with quadrature formulas which allow us to solve approximately problem of *quadrature* of a plane figure, i.e. that of calculating its area, the author calls a formula for calculating the volumes of bodies in many-dimensional space a *cubature formula*. This term seems natural since in n-dimensional space, $n \geq 3$, the fundamental domain of unit volume, defined analytically by the inequalities $0 \leq x_j < 1$ $(j = 1, 2, \ldots, n)$, is customarily called the *n-dimensional cube*. In a greater part of the book the author considers the minimization problem for the error of a cubature formula with fixed nodes (the minimum is taken over weights) in the $L_2^{(m)*}$ space of functionals on $L_2^{(m)}$ with the norm defined through the squares of the derivatives of l.

The monograph is based on the author's research. In terms of $L_2^{(m)}$ the problem of minimizing an error is stated as that of approximating a given functional

$$(\chi_\Omega, \varphi) = \int \chi_\Omega(x)\varphi(x)\, dx,$$

with χ_Ω the *characteristic function* or the *indicator* of a domain Ω, by means of a linear combination of generalized functions $\sum_k c_k \delta(x - x^{(k)})$. In this event the *error* has the form

$$l(x) = \chi_\Omega(x) - \sum_k c_k \delta(x - x^{(k)}).$$

It turns out that the calculation of the norm of this error treated as a linear functional reduces to a certain problem for the polyharmonic equation. To study this norm thus becomes a problem of the theory of partial differential equations.

The main result of our research is the discovery of the greatest lower bound of the error of a cubature formula in the case when the nodes of the formula are those of a regular lattice. The author establishes the principal term in the expansion of a minimal norm which has the form $\|l\| = Ah^m$, with A a constant expressible exactly through the number-theoretic zeta function of Epstein which depends on the form of the lattice. This estimate determines the choice of the fittest lattice in the case when the norm in $L_2^{(m)}$ is taken to be invariant under orthogonal space transformations.

The author further indicates some formulas of a special type, the "formulas with regular boundary layer," possessing an asymptotically minimal norm of the error

$$\|l\| = Ah^m + O(h^{m+1}).$$

The algorithm for determining the weights of a formula with regular boundary layer is of local character. These weights depend only on the form of a neighborhood close to a given node of the integration domain. Therefore, for instance, the number of distinct weights in a domain like a rational polyhedron is bounded and we may suggest a simple method for their computation.

Strictly optimal formulas the author studies separately. Also, separately we consider the convergence of a sequence of cubature formulas of increasing degree at individual infinitely differentiable functions.

S. L. Sobolev

Chapter 1

Problems and Results of the Theory of Cubature Formulas

This chapter is introductory. Here we state the problems under study, outline the principal ideas behind our theory and make an overview of the most important results. The intended conciseness of exposition results in making some proofs schematic. The reader interested in more details may find them in the sequel.

§1. Exact Formulas

The main problem of numerical integration consists in approximating the integral

$$I(\varphi) = \int_{\Omega} \varphi(x)\, dx = \int \chi_{\Omega}(x)\varphi(x)\, dx. \tag{1.1}$$

Here x is an n-dimensional coordinate vector, and $\chi_{\Omega}(x)$ is the *indicator* of a connected domain Ω with sufficiently smooth boundary. We seek for an approximant by taking a linear combination of the values of $\varphi(x)$ at the N points

$$x^{(1)}, x^{(2)}, \dots, x^{(N)} \tag{1.2}$$

called *nodes*, namely,

$$I^*(\varphi) = \sum_{k=1}^{N} c_k \varphi(x^{(k)}) = \int \sum_{k=1}^{N} c_k \delta(x - x^{(k)})\varphi(x)\, dx, \tag{1.3}$$

where $\delta(x)$ is the conventional *Dirac delta function.* We call (1.3) a *cubature formula* by analogy with a *quadrature formula* in the one-dimensional case.

The theory of cubature formulas consists mainly of the three branches dealing with exact formulas, formulas based on functional-analytical methods, and formulas based on probabilistic methods. Here we abstain from pursuing the probabilistic approach, addressing the reader to the articles [61, 72, 73, 298].

The main topic of the present book is the theory of formulas whose approximation properties lean on the technique of functional analysis.

To each cubature formula (1.3) we assign the *error*

$$(l,\varphi) = I(\varphi) - I^*(\varphi) = \int \Big\{ \chi_\Omega(x) - \sum_{k=1}^{N} c_k \delta(x - x^{(k)}) \Big\} \varphi(x)\, dx. \tag{1.4}$$

The error is a linear functional, therefore also referred to as *error functional*, since we require that the rules for choosing the nodes $x^{(k)}$ and the weights c_k be independent of specifying an integrable function. Integrable functions are assumed to be members of some Banach space $\mathbf{B}$ embedded into the *space of continuous functions*; i.e.,

$$\mathbf{B} \to C(\Omega). \tag{1.5}$$

This assumption guarantees that (1.3) is defined at every function in $\mathbf{B}$. We assess (1.3) by estimating

$$\sup_{\|\varphi|\mathbf{B}\|=1} |(l,\varphi)|.$$

It is natural to consider sequences of cubature formulas with *errors* $l^{(N)}$ as the *number of nodes* N, i.e., the *size* of the set of nodes, increases indefinitely. Convergence of $l^{(N)}$ to zero may be strong or weak.

We refer to a cubature formula (1.3) with the exact value of the integral (1.1) at each member of a given set of functions

$$\psi_1, \psi_2, \ldots, \psi_M \tag{1.6}$$

as *exact* for $\psi_1, \psi_2, \ldots, \psi_M$. We denote the integral of ψ_j by b_j, i.e.,

$$b_j = \int_\Omega \psi_j\, dx,$$

further arranging b_j in the column-vector b. The column-vector c is formed by collecting the weights c_j of the cubature formula (1.3). The *exact integration condition* for (1.6) is the equalities

$$\sum_{k=1}^{N} c_k \psi_j(x^{(k)}) = b_j, \quad j = 1, 2, \ldots, M.$$

These may be rewritten in matrix form

$$Sc = b \tag{1.7}$$

with

$$S = \begin{bmatrix} \psi_1(x^{(1)}) & \dots & \psi_1(x^{(N)}) \\ \dots\dots & \dots & \dots\dots \\ \psi_M(x^{(1)}) & \dots & \psi_M(x^{(N)}) \end{bmatrix}. \tag{1.8}$$

The problem of determining the weights c of an exact formula is dual to the ***interpolation problem*** of finding a linear combination of the functions (1.6) taking preassigned values at the points (1.2). We now prove the preceding claim.

Let

$$\psi = a_1\psi_1 + a_2\psi_2 + \dots + a_M\psi_M \tag{1.9}$$

be a sought linear combination with its values at the points (1.2) making the row-vector $\vec{d} = (d^{(1)}, \dots, d^{(N)})$. Also, let the coefficients of the linear combination (1.9) form the row-vector $a = (a_1, \dots, a_M)$. To determine a, we have the system of equations

$$aS = \vec{d} \tag{1.10}$$

which is adjoint to (1.7); i.e., the two problems are dual to one another.

Most usual is the case in which (1.6) consists of polynomials ordered by increasing degree. The classical statement of the problem reads: the number of nodes in (1.2) equals M, the number of monomials in n variables of degree at most m for which (1.3) is exact. Clearly, $M = (n+m)!/(n!m!)$. In this case the interpolation problem is as follows.

Find a polynomial $P(x)$ of degree at most m agreeing with $\varphi(x)$ at the given points,

$$P(x^{(k)}) = \varphi(x^{(k)}), \quad k = 1, 2, \dots, N.$$

The polynomial $P(x)$ may be written as $P(x) = ax^{\alpha}$, with a a row-vector of length M and x^{α} a column-vector of the same height

$$x^{\alpha} = \begin{pmatrix} x^{\alpha^{(1)}} \\ \vdots \\ x^{\alpha^{(M)}} \end{pmatrix}.$$

Here $\alpha^{(1)}, \alpha^{(2)}, \dots, \alpha^{(M)}$ are all integer vectors with nonnegative entries satisfying the condition

$$|\alpha^{(j)}| = \alpha_1^{(j)} + \alpha_2^{(j)} + \dots + \alpha_n^{(j)} \le m.$$

To each function $\varphi(x)$ we assign the row-vector

$$\vec{\varphi} = (\varphi(x^{(1)}), \dots, \varphi(x^{(N)}))$$

of the values of $\varphi(x)$ at the points (1.2). The values of the polynomial $P(x)$ at the points (1.2) are written down on using the *Vandermonde matrix*

$$S = \begin{bmatrix} x^{(1)^{\alpha^{(1)}}} & \dots & x^{(N)^{\alpha^{(1)}}} \\ \dots\dots & \dots\dots & \dots\dots \\ x^{(1)^{\alpha^{(M)}}} & \dots & x^{(N)^{\alpha^{(M)}}} \end{bmatrix}. \tag{1.11}$$

We have the equality

$$\vec{P} = aS.$$

The interpolation problem is uniquely solvable provided that the matrix S is nondegenerate. The solution, the row of the coefficients of $P(x)$, is determined by the formula

$$a = \vec{\varphi} S^{-1}; \tag{1.12}$$

whereas the polynomial itself, by the equality

$$P(x) = \vec{\varphi} S^{-1} x^{\alpha}. \tag{1.13}$$

The formulas (1.12) and (1.13) are called *interpolation formulas.*

The solution to the dual problem is the cubature formula (1.3) exact for every polynomial of degree m. For this formula,

$$c = S^{-1} b, \quad b_j = \int_{\Omega} x^{\alpha^{(j)}} \, dx, \quad j = 1, 2, \dots, M. \tag{1.14}$$

Observe that, by linearity, a cubature formula exact for polynomials of degree m is also exact for polynomials of degree at most m.

Considering a nonsquare matrix S, with $M \neq N$, introduce the operators S_r^{-1} and S_l^{-1}, a *right inverse* and a *left inverse* of S; surely, if such inverses exist. Now the interpolation formulas (1.12) and (1.13) become

$$a = \vec{\varphi} S_r^{-1}, \quad P(x) = \vec{\varphi} S_r^{-1} x^{\alpha}, \tag{1.15}$$

and the weights in the cubature formula (1.14) are determined from the equality

$$c = S_l^{-1} b. \tag{1.16}$$

The right sides of the cubature formula (1.3) and the error (1.4) may be defined not only for numerical functions but also for ***abstract functions*** with domain Ω of the dependent variable $x = (x_1, x_2, \ldots, x_n)$ and codomain a Banach space $\mathbf{X}$. We agree to assume that the functions f from $\mathbb{R}^n$ to $\mathbf{X}$ comprise some Banach space $\mathbf{B}$. The space $\mathbf{B}$ may be composed of functions continuous in Ω. Instead of the error functional $l^{(N)}$ we now speak about the ***error operator*** $l^{(N)}$ mapping $\mathbf{B}$ to $\mathbf{X}$. The quality of a cubature formula is again characterized in terms of the rate with which the sequence $(l^{(N)}, f)$ vanishes. We give an example of pursuing this approach.

Assume that the Banach space $\mathbf{X}$ consists of sequences $a = (a_1, a_2, \ldots, a_n, \ldots)$ and the values of $f(x)$ belong to $\mathbf{X}$. Here and elsewhere we habitually abuse notation by signifying a function f of a variable x simply as $f(x)$, despite the latter symbol also standing for the value of f at a point x of the domain of definition of f. We may thus view f as countably many ordinary numerical functions. In the example under consideration, each numerical function $f_k(x)$, expressing the kth ***component*** of f, belongs to a finite-dimensional subspace of dimension n_k. Integration of an abstract function f reduces therefore to that of each of the functions f_k. We also assume that the spaces $\mathbf{X}$ and $\mathbf{B}$ are endowed with some Banach norms.

Consider a sequence of error operators $l^{(N)}$. Say that this sequence ***vanishes uniformly*** provided that the norms of $l^{(N)}$ tend to 0 as N increases indefinitely. In this event the sequence is also referred to as ***strongly convergent*** to zero. To an arbitrary positive real ε we may further assign an integer $N(\varepsilon)$ such that, for all $N \geq N(\varepsilon)$ and every f with $\|f\|_{\mathbf{B}} \leq 1$, the inequality holds

$$\|(l^{(N)}, f)\|_{\mathbf{X}} \leq \varepsilon. \tag{1.17}$$

By linearity of $l^{(N)}$, instead of $\|f\|_{\mathbf{B}} \leq 1$ we may require the condition $\|f\|_{\mathbf{B}} \leq C$ with some C. Obviously, $N(\varepsilon)$ is then replaced with $N_1(\varepsilon)$ equal to $N(\varepsilon/C)$.

It is thus possible, given an ε-neighborhood of zero in $\mathbf{X}$ and a set bounded in $\mathbf{B}$, to find an $N(\varepsilon)$ making valid (1.17) with $N \geq N(\varepsilon)$.

Endow the space $\mathbf{X}$ with some special topology called the ***T-topology***. Assume that a neighborhood of zero in $\mathbf{X}$ consists of

$$a = (0, \ldots, 0, a_{\beta+1}, a_{\beta+2}, \ldots).$$

Varying β, find the whole system of neighborhoods in the T-topology. We call a sequence of error operators $l^{(N)}$ ***convergent in proximity order*** if to every β there is an index $N(\beta)$ such that, for all $N \geq N(\beta)$ and $f \in \mathbf{B}$, the member $(l^{(N)}, f)$ of $\mathbf{X}$ belongs to the neighborhood in the T-topology with index β. This means that, for N sufficiently large, any arbitrarily prescribed number of the components of the sequence $(l^{(N)}, f)$ vanish. We thus see that convergence in proximity order of the error operators $l^{(N)}$ amounts to the fact that all sequences $(l^{(N)}, f)$ converge uniformly in f in the T-topology.

If the norms of the error operators $l^{(N)}$ acting from $\mathbf{B}$ to $\mathbf{X} = l_2$ are bounded uniformly in N and the sequence $l^{(N)}$ vanishes in proximity order then this sequence is obviously weakly convergent.

Consider the set B of analytic functions f in two variables x and y, we also denote them by $f(x, y)$, which are defined in the unit disk and admit the convergent *Maclaurin series*

$$f(x,y) = a_0 + a_{10}x + a_{01}y + a_{20}x^2 + a_{11}xy + a_{02}y^2 + \dots. \tag{1.18}$$

We identify $f(x, y)$ with the sequence

$$\bar{f} = (a_0, a_{10}x + a_{01}y, a_{20}x^2 + a_{11}xy + a_{02}y^2, \dots)$$

whose kth entry presents a homogeneous polynomial of degree $k-1$ in the expansion (1.18). Each component $f_j(x, y)$ of the vector $\bar{f}$ is thus a member of the finite-dimensional space of homogeneous polynomials of a given degree. Introduce the norms in $\mathbf{X}$ and $\mathbf{B}$ by the formulas

$$\|a\|_{\mathbf{X}} = \sup_k |a_k|, \quad \|\bar{f}\|_{\mathbf{B}} = \max_{x^2+y^2\leq 1} \|\bar{f}(x,y)\|_{\mathbf{X}}.$$

In this event a sequence of cubature formulas converges in proximity order if for every natural m there is an index $N_0(m)$ such that

$$(l^{(N)}, P) = 0 \tag{1.19}$$

with $N \geq N_0(m)$ for every polynomial P of degree m. Under the condition that the sum of the moduli of the weights of the cubature formulas under study is bounded by some constant L, the sequence of $l^{(N)}$ converges weakly.

Indeed, using the convergence of the power series (1.18) in δ, choose β_1 so as to ensure validity of the following inequality

$$\max_{x^2+y^2\leq 1} |f_j(x,y)| \leq \frac{\delta}{L}, \quad j \geq \beta_1.$$

Let m equal β_1 and $N \geq N_0(m)$. Then from (1.19) we obtain the equality

$$(l^{(N)}, \bar{f}) = (0, 0, \dots, 0, a_{m+1}, a_{m+2}, \dots).$$

The coefficients a_k are readily written down as

$$a_j = -\sum_{k=1}^{N} c_k f_j(x^{(k)}, y^{(k)}).$$

Therefore,

$$\|(l^{(N)}, \bar{f})\|_{\mathbf{X}} \le \sup_{j\ge m} |a_j| \le \sup_{j\ge m} \sum_{k=1}^{N} |c_k||f_j(x^{(k)}, y^{(k)})|$$

$$\le \left(\sum_{k=1}^{N} |c_k|\right) \sup_{j\ge \beta_1, x^2+y^2\le 1} |f_j(x,y)| \le \delta.$$

This just means that the sequence of $l^{(N)}$ converges weakly.

Assume that the original problem consists in approximate calculation of the integral

$$\iint_{x^2+y^2\le 1} f(x,y)\,dx dy$$

by means of a sequence of cubature formulas with increasing order. The last expression means that the successive formulas exactly integrate polynomials of degree increasing with the index of the sequence. The above scheme reduces the problem to constructing a sequence of error operators which vanishes uniformly in the T-topology. The algebraic approach to the problem of approximate integration is thus replaced with the functional-analytical approach.

§2. Functional-Analytical Statement of the Problem

Specifying a cubature formula (1.3) amounts to specifying the respective error

$$l(x) = \chi_\Omega(x) - \sum_{k=1}^{N} c_k \delta(x - x^{(k)}).$$

The linear combination of Dirac delta functions in the above equality is called the ***discrete component*** of $l(x)$. We assume that the integrand $\varphi(x)$ is a member of some Banach space $\mathbf{B}$ and $l(x)$ is a bounded linear functional on $\mathbf{B}$, i.e., a member of $\mathbf{B}^*$. Consider the following two possibilities.

THE $L_2^{(m)}$ SPACE. Let the function $\varphi(x)$ with domain $\mathbb{R}^n$ have all derivatives up to order m locally integrable and

$$\|\varphi \mid L_2^{(m)}\| = \left\{\int \sum_{|\alpha|=m} \frac{m!}{\alpha!} |D^\alpha \varphi|^2\,dx\right\}^{1/2} < +\infty. \tag{2.1}$$

The integral here spreads over the whole $\mathbb{R}^n$, and summation is taken over some multi-indices $\alpha = (\alpha_1, \alpha_2, \dots, \alpha_n)$ with integer coefficients,

$$\alpha! = \alpha_1!\alpha_2!\dots\alpha_n!, \quad |\alpha| = \sum_{j=1}^{n} \alpha_j, \quad D^\alpha\varphi = \frac{\partial^m \varphi}{\partial x_1^{\alpha_1}\partial x_2^{\alpha_2}\dots\partial x_n^{\alpha_n}}.$$

The members of $L_2^{(m)}$ are equivalence classes composed of functions differing from one another by a polynomial of degree less than m and having the integral of (2.1) finite. Clearly, (2.1) defines a complete norm in $L_2^{(m)}$.

For the error l to make sense on $L_2^{(m)}$, the following conditions are necessary

$$(l, x^\alpha) = 0, \quad |\alpha| < m. \tag{2.2}$$

If in addition $2m > n$, i.e., the hypothesis is satisfied of the *First Embedding Theorem* [265]; then $l(x)$ is bounded in $L_2^{(m)}$.

By analogy, we define the $L_2^{(m)}(\Omega)$ *space* whose norm is given by the equality

$$\|\varphi \mid L_2^{(m)}(\Omega)\| = \left\{ \int_\Omega \sum_{|\alpha|=m} \frac{m!}{\alpha!} |D^\alpha \varphi|^2 \, dx \right\}^{1/2}. \tag{2.3}$$

THE $\widetilde{L}_2^{(m)}$ SPACE. Assume that a function $\varphi(x)$ has all derivatives up to order m locally integrable in $\mathbb{R}^n$ with the integral (2.3) finite for every bounded domain Ω. Also assume that $2m > n$, and $\varphi(x)$ is a *periodic function* with *period matrix* some matrix H, i.e.,

$$\varphi(x + H\gamma) = \varphi(x), \quad x \in \mathbb{R}^n, \tag{2.4}$$

where γ is an arbitrary *integer* column-vector (i.e., having integer entries). To the matrix H we assign its *fundamental parallelepiped* Ω_0 by putting

$$\Omega_0 = \{x \in \mathbb{R}^n : x = Hy, \ 0 \le y_j < 1, \ j = 1, 2, \ldots, n\}.$$

As a rule, we assume that H is an orthogonal matrix. The members of $\widetilde{L}_2^{(m)}$ are equivalence classes composed of functions differing from one another by a constant summand. The norm of $\widetilde{L}_2^{(m)}$ takes the form

$$\|\varphi \mid \widetilde{L}_2^{(m)}\| = \left\{ \int_{\Omega_0} \sum_{|\alpha|=m} \frac{m!}{\alpha!} |D^\alpha \varphi|^2 \, dx \right\}^{1/2}.$$

To distinguish the spaces that correspond to different matrices H, we sometimes use the notation $\widetilde{L}_2^{(m)}(H)$. Observe that $\widetilde{L}_2^{(m)}$ is a complete space.

The norm of $L_2^{(m)}$ is invariant under arbitrary orthogonal transformations of coordinates. In other words, if $x = Ay$, with A an orthogonal matrix, and $\psi(y) = \varphi(Ay)$ then

$$\|\varphi \mid L_2^{(m)}\| = \|\psi \mid L_2^{(m)}\|.$$

The reader can state analogous propositions for $\widetilde{L}_2^{(m)}$.

The varying parameters of a cubature formula are the nodes $x^{(k)}$ and the weights c_k. A cubature formula whose error has a minimal norm in $\mathbf{B}^*$ subject to the condition that the number of nodes N is fixed is called *optimal* in the space $\mathbf{B}$ or $\mathbf{B}$-*optimal* or simply *optimal*. We do not study the general problem with arbitrary nodes, confining exposition to the case in which the nodes comprise a *lattice*; i.e., the nodes are the points $x^{(\gamma)} = hH\gamma$. Here h is a small positive parameter called the *lattice mesh-size* of the lattice of nodes $hH\gamma$; the letter H is an $n \times n$-matrix with unit determinant referred to as the *lattice matrix*; and γ is an integer column-vector. In this event, we call a cubature formula a *lattice cubature formula*.

The error $l(x)$ of a lattice cubature formula is the generalized function

$$l(x) = \chi_\Omega(x) - \sum_{hH\gamma\in\Omega} h^n c[\gamma]\delta(x - hH\gamma). \tag{2.5}$$

In the case of the space of periodic functions $\widetilde{L}_2^{(m)}$ the domain Ω in (2.5) is replaced with the fundamental parallelepiped Ω_0. We also assume that the period matrix is a multiple of the lattice matrix in this case.

Consider the behavior of the error at small h. To this end, we need the concept of *Epstein zeta function* whose arguments are a matrix A and a natural s. The Epstein zeta function is

$$\zeta(A \mid s) = \sum_{\beta\neq 0} \frac{1}{r_\beta^s},$$

with r_β standing for the distance from the point $A\beta$ to the coordinate origin,

$$r_\beta = |A\beta|^2 = (A\beta)^*(A\beta) = \sum_{j=1}^{n} (A\beta)_j^2.$$

If A is the identity matrix then here and everywhere in the sequel the symbol $|\beta| = |A\beta|$ denotes the *Euclidean norm* of a vector $\beta \in \mathbb{Z}^n$. Only on a few occasions the symbol $|\beta|$ stands for the *order* of the multi-index $|\beta|$, i.e.,

$$|\beta| = \beta_1 + \cdots + \beta_n.$$

The reader will always see what notation is implied from the context. We have the following

Theorem 1.1. *All weights $c[\gamma]$ of the error (2.5) corresponding to an $\widetilde{L}_2^{(m)}$-optimal cubature formula are equal, and the norm of the error is*

$$\| l \mid \widetilde{L}_2^{(m)*} \| = B_{n,m} \| \chi_{\Omega_0} \mid L_2 \|^{1/2} h^m, \tag{2.6}$$

with

$$B_{n,m} = \left(\frac{1}{2\pi}\right)^m \sqrt{\zeta(H^{-1*} \mid 2m)}.$$

Observe that for $n = 1$ the quantity $\zeta(1 \mid 2m)$ is expressed through the conventional *Bernoulli numbers* B_{2m} by the formula

$$\zeta(1 \mid 2m) = (-1)^{m-1} \frac{(2\pi)^{2m}}{(2m)!} B_{2m}.$$

It is thus reasonable to call

$$B_{2m}(H) = (-1)^{m-1} \frac{(2m)!}{(2\pi)^{2m}} \zeta(H \mid 2m)$$

the *generalized Bernoulli number with matrix* H. Hence, (2.6) may equivalently be rewritten as

$$\|l \mid \widetilde{L}_2^{(m)*}\| = h^m \left| \frac{B_{2m}(H^{-1*})}{(2m)!} |\Omega_0| \right|^{1/2}.$$

PROOF OF THEOREM 1.1. Use the *strict convexity property* of the unit ball in $\widetilde{L}_2^{(m)*}$. We recall the property. Let $l_1, l_2, \dots, l_k$ be some functionals belonging to the unit sphere of $\widetilde{L}_2^{(m)*}$, and let $\mu_1, \mu_2, \dots, \mu_k$ be nonnegative reals with sum 1, namely,

$$\mu_j \geq 0, \quad \sum_{j=1}^{k} \mu_j = 1.$$

Then the respective linear combination lies in the *unit ball* of $\widetilde{L}_2^{(m)*}$, i.e.,

$$\left\| \sum_{j=1}^{k} \mu_j l_j \mid \widetilde{L}_2^{(m)*} \right\| \leq 1. \tag{2.7}$$

Equality holds here if and only if all l_j coincide.

From strict convexity it is immediate that there is a unique optimal error in $\widetilde{L}_2^{(m)*}$. For, were it otherwise, we would find at least two errors with the same minimal norm. As follows from (2.7), their half-sum would then have the norm less than each of them; a contradiction.

Now let the error $l(x)$ in the shape of (2.5) with $\Omega = \Omega_0$ have a minimal norm for a fixed lattice of nodes $hH\gamma$. Then, every $l(x - hH\gamma)$ with $hH\gamma \in \Omega_0$ clearly has the same minimal norm as $l(x)$. The arithmetic mean of these errors also has a minimal norm by (2.7) and may be written as (2.5) with $\Omega = \Omega_0$. The weights

$c[\gamma]$ of the arithmetic mean coincide, and the error (2.5) itself is exact for every constant function. Consequently, $c[\gamma] = |\Omega_0|$ for all γ. Optimality of $l(x)$ implies that this error coincides with the arithmetic mean, which proves the first part of the theorem.

Demonstrate (2.6). Observe that $\widetilde{L}_2^{(m)}$ is a Hilbert space whose *inner product* is given by the formula

$$[\varphi, \psi]_m = \int_{\Omega_0} \sum_{|\alpha|=m} \frac{m!}{\alpha!} D^\alpha \varphi D^\alpha \psi \, dx.$$

By the *Riesz Theorem* every bounded linear functional l on a Hilbert space may be written as inner product

$$(l, \varphi) = [\varphi, \psi_l]_m, \quad \varphi \in \widetilde{L}_2^{(m)}. \tag{2.8}$$

Here $\psi_l(x)$ is a uniquely determined member of $\widetilde{L}_2^{(m)}$ called the *extremal function* of $l(x)$ on, more verbosely, $\widetilde{L}_2^{(m)}$*-extremal function.* Integrating by parts the expression on the right side of (2.8) and using periodicity of φ and ψ_l, derive the equality

$$(l, \varphi) = (-1)^m \int_{\Omega_0} \Delta^m \psi_l(x) \varphi(x) \, dx.$$

So, the function $\psi_l(x)$ is a weak solution to the equation

$$\Delta^m \psi_l(x) = (-1)^m l(x). \tag{2.9}$$

If $l(x)$ has a minimal norm in $\widetilde{L}_2^{(m)}$ then it is easy to construe a solution to (2.9) as a Fourier series. Denoting the former by $\widetilde{\psi}(x)$, we have

$$\widetilde{\psi}(x) = -\left(\frac{h}{2\pi}\right)^{2m} \sum_{\beta \neq 0} \frac{1}{|H^{-1*}\beta|^{2m}} e^{i2\pi\beta h^{-1} H^{-1} x}. \tag{2.10}$$

From (2.8) it follows that

$$\left\| l \mid \widetilde{L}_2^{(m)*} \right\|^2 = (l, \widetilde{\psi}). \tag{2.11}$$

Inserting the expansion of (2.10) in (2.8), arrive at (2.6).

The proof of Theorem 1.1 is complete.

Assume that the closure of Ω lies in the interior of Ω_0, the fundamental parallelepiped of H. This agreement remains effective in what follows. Assume also that the boundary of Ω is *piecewise-smooth.*

We now explicate the last concept. Let $\Gamma^{(0)}$ be a closed part of the boundary Γ of Ω. Consider the set $\Omega_h^{(0)}$, $h > 0$, comprising the points of $\overline{\Omega}$ at a distance less than h from $\Gamma^{(0)}$. Assume existent a one-to-one change of variables $y = f(x)$ and $x = g(y)$ with continuous derivatives up to some order which sends $\Omega_h^{(0)}$ to a part of a polyhedra without selfintersections so that $\Gamma^{(0)}$ transforms into a part of the boundary of the polyhedron. Clearly, if such mapping $y = f(x)$ is available near to $\Gamma^{(0)}$ at a given h, then we may use the mapping for all smaller values of h. In this event we say that the domain Ω is *bounded near to the part* $\Gamma^{(0)}$.

Consider the indicator $\chi_\Omega(x)$ of Ω. This indicator decomposes in infinitely many ways into the finite sum

$$\chi_\Omega(x) = \sum_{j=1}^{N} \chi_j(x),$$

whose every term has range the interval $[0, 1]$ and is infinitely differentiable in the closed domain $\overline{\Omega}$. Each of the decompositions of the indicator of Ω is called an (infinitely differentiable) *partition of unity*.

Consider the supports $F_j = \operatorname{supp} \chi_j(x)$ of the entries of a partition of unity. The summand $\chi_j(x)$ is called *bounding* provided that $F_j \cap \Gamma \neq \varnothing$, with Γ standing for the boundary of Ω.

Put $\Gamma_j = F_j \cap \Gamma$. This set is the part of the boundary of F_j which is included in the boundary of Ω. Enumerate the members of the partition of unity under consideration. Denote by $\chi_0(x)$ the sum of all "inner" terms and by $\chi_1(x), \dots, \chi_N(x)$ the bounding entries

$$\chi_\Omega(x) = \chi_0(x) + \sum_{j=1}^{N} \chi_j(x).$$

Suppose that Ω possesses a partition of unity such that the support F_j of each member is bounded near to the part Γ_j for all $j = 1, 2, \dots, N$. In this case Ω is called a *domain with piecewise-smooth boundary*.

Theorem 1.2. *The norm of an error of the shape (2.5) in $L_2^{(m)*}$ satisfies the inequality*

$$\| l \mid L_2^{(m)*} \| \geq B_{n,m} |\Omega|^{1/2} h^m + O(h^{m+1}), \tag{2.12}$$

with $B_{n,m}$ the same constant as in (2.6).

We just sketch the proof, leaving exposition in more detail to Chapter 5. The function $\widetilde{\psi}(x)$, given by (2.10), depends on the lattice mesh-size h as follows

$$\widetilde{\psi}(x) = h^{2m} u\left(\frac{x}{h}\right),$$

with $u(x)$ standing for the function (2.10) taken at $h = 1$. Put

$$v(x) = u(x) - u(0).$$

Now the function $h^{2m}v(x/h)$ is extremal for the optimal error on $\widetilde{L}_2^{(m)}$. This function is 0 at every node. Define the function

$$\varphi(x, h \mid \Omega) = h^{2m} v\left(\frac{x}{h}\right) \eta(x, h \mid \Omega).$$

Here $\eta(x, h \mid \Omega)$ is a *truncator* for the domain Ω; the concurrent terms are a *cut-off* function or a *bump function.* A truncator $\eta(x, h \mid \Omega)$ is a function possessing the following properties:

(a) $\eta(x, h \mid \Omega) = 1$ at every point x in Ω at a distance greater than $2h$ from the boundary;

(b) $\eta(x, h \mid \Omega) = 0$ at every point x outside Ω and also at the points of Ω lying at a distance less than $h/4$ from the boundary of Ω;

(c) for all multi-indices α such that $|\alpha| \leq m$ and all sufficiently small h the estimates are valid

$$|D^{\alpha}\eta(x, h \mid \Omega)| \leq K_{\alpha} h^{-|\alpha|}.$$

As a truncator we may for instance take the integral

$$\int \chi_{\Omega'_h}(y)\sigma(x, y \mid h)\, dy,$$

where Ω'_h comprises the points of Ω whose distance to the boundary of Ω is greater than h, and the function $\sigma(x, y \mid h)$ is the *standard averaging kernel* [265, 266].

Calculation shows that

$$\left\| h^{2m} v\left(\frac{x}{h}\right) - \varphi(x, h \mid \Omega) \mid L_2^{(m)}(\Omega) \right\| \leq K h^{m+1},$$

with K independent of h. Therefore,

$$\|\varphi(x, h \mid \Omega) \mid L_2^{(m)}\| \leq B_{n,m} h^m \sqrt{|\Omega|} + O(h^{m+1}).$$

Calculate the value of the initial error $l(x)$ at the truncated function $\varphi(x, h \mid \Omega)$. Observe that the nodal values of $\varphi(x, h \mid \Omega)$ are all 0. Therefore, the value of $l(x)$ at $\varphi(x, h \mid \Omega)$ does not depend on the choice of weights. Hence, it coincides with the integral over Ω of the function under study. Basing on this, we may show that

$$(l, \varphi(x, h \mid \Omega)) = B_{n,m}^2 h^{2m} |\Omega| (1 + O(h)).$$

By definition, the norm of $l(x)$ is not less that the ratio

$$\frac{(l, \varphi(x, h \mid \Omega))}{\left\| \varphi(x, h \mid \Omega) \mid L_2^{(m)} \right\|} = B_{n,m} h^m |\Omega|^{1/2} + O(h^{m+1})$$

The proof of Theorem 1.2 is complete.

Fix two positive reals K_0 and L. Consider a family $l_\gamma(x)$ of functionals depending on an integer vector γ and satisfying the conditions

$$\|l_\gamma(x) \mid C^*\| \le K_0, \quad \operatorname{supp} l_\gamma(x) \subset \{x : |x| \le L\}. \tag{2.13}$$

A functional of the shape $l_\gamma(x/h - H\gamma)$ is called an *elementary error*. Say that an elementary error has *order* s, if it is exact for every polynomial of degree less than s, namely,

$$\left(l_\gamma \left(\frac{x}{h} - H\gamma \right), x^\alpha \right) = 0, \quad |\alpha| < s.$$

Denote by B_L the set of γ such that the distance from the point $hH\gamma$ to the set Ω is at most Lh. Let the error in the shape of (2.5) be expressed as

$$l(x) = \sum_{\gamma \in B_L} l_\gamma \left(\frac{x}{h} - H\gamma \right) = \sum_{\gamma \in B_L^{(1)}} l_\gamma \left(\frac{x}{h} - H\gamma \right) + \sum_{\gamma \in B_L^{(2)}} l_\gamma \left(\frac{x}{h} - H\gamma \right). \tag{2.14}$$

Here $B_L^{(1)} \cup B_L^{(2)} = B_L$, and every elementary error $l_\gamma(x/h - H\gamma)$ has order $m+1$ except possibly those with γ belonging to $B_L^{(2)}$. In this event the functional corresponding to $\gamma \in B_L^{(2)}$ has order m, and the total number of these l_γ as $h \to 0$ does not exceed $K_0 h^{1-n}$, with K_0 a constant independent of h. We call (2.14) an *equidistributed error*.

Consider an equidistributed error with the set $B_L^{(2)}$ consisting of γ such that the points $hH\gamma$ are at a distance at most Lh from the boundary of the domain Ω. The set $B_L^{(1)}$ comprises the points γ of B_2 such that $hH\gamma$ lie in Ω at a distance greater than Lh from the boundary of Ω. Redenote $B_L^{(1)}$ and $B_L^{(2)}$ by $B_L^{(i)}$ and $B_L^{(l)}$, respectively. Assume that all errors $l_\gamma(x)$ with $\gamma \in B_L^{(i)}$ are the same. Then the initial functional may be written as

$$l(x) = \sum_{\gamma \in B_L^{(i)}} l_0 \left(\frac{x}{h} - H\gamma \right) + \sum_{\gamma \in B_L^{(l)}} l_\gamma \left(\frac{x}{h} - H\gamma \right). \tag{2.15}$$

We call it an *error with regular boundary layer*.

Observe an important property of an error with regular boundary layer: Unity, the number 1, is the common value of all its weights $c[\gamma]$ corresponding to γ such that the point $hH\gamma$ lies at a distance greater than $2Lh$ from the boundary of Ω. Indeed, by (2.13) each of these weights $c[\gamma]$ appears from summing the same finite number of the elementary errors $l_0\,(x/h - H\gamma)$. Since $l_0(x)$ is exact for a constant function, all $c[\gamma]$ under consideration equal 1.

The way in which a given error $l(x)$ with regular boundary layer decomposes in the shape of (2.15) is not unique. We may take a decomposition (2.15) such that $l_0(x)$ is an arbitrary preassigned functional of order at least $m+1$ satisfying (2.13). Note that in this event the constants K_0 and L depend on the order $m+1$ and tend to infinity as the order increases indefinitely.

We now describe a constructive method for making functionals with regular boundary layer. Partition a given domain Ω into elementary parts Ω_γ each of which results from intersecting Ω with one of the *meshes* of the lattice with lattice matrix hH, the latter presenting the translates of the fundamental parallelepiped of hH. The indicator of Ω decomposes into the sum

$$\chi_\Omega(x) = \sum_\gamma \chi_{\Omega_\gamma}(x), \quad \chi_{\Omega_\gamma}(x) = \chi_\Omega(x)\chi_0\left(\frac{x}{h} - H\gamma\right).$$

Here $\chi_0(y)$ is the indicator of the fundamental parallelepiped of H. Arrange the cubature formulas for integration over each part Ω_γ and sum them.

We now specify the way of constructing cubature formulas for integration over Ω_γ. First, find some error

$$l_0(y) = \chi_0(y) - \sum_{|\gamma'|\le L} c[\gamma']\delta(y - H\gamma')$$

of order $2m+2$ for the fundamental parallelepiped Ω_0. This is possible to accomplish at sufficiently large L. More precisely, L, being at least $(n+2m+2)!/n!(2m+2)!$, will do. Further, for all γ to which there corresponds the elementary error $l_0\,(x/h - H\gamma)$ with support in Ω, put

$$l_\gamma\left(\frac{x}{h} - H\gamma\right) = l_0\left(\frac{x}{h} - H\gamma\right).$$

We agree that the set of such γ comprises $B_L^{(i)}$. For the remaining nonempty meshes Ω_γ we take the error in the shape

$$l_\gamma\left(\frac{x}{h} - H\gamma\right) = \chi_{\Omega_\gamma}(x) - \sum_{\substack{|\gamma'-\gamma|\le L \\ hH\gamma'\in\Omega}} h^n c_\gamma[\gamma']\delta(x - hH(\gamma - \gamma')).$$

The weights $c_\gamma[\gamma']$ should be taken so that $l_\gamma(x)$ has order m.

Summing the elementary errors over all $\gamma \in B_L^{(i)} \cup B_L^{(l)}$, we clearly obtain the error of a cubature formula with regular boundary layer.

Theorem 1.3. *The norm of each equidistributed error is bounded from above by $K_1 h^m$, namely,*

$$\| l \mid L_2^{(m)*} \| \leq K_1 h^m, \tag{2.16}$$

with K_1 a constant depending on Ω, m, n, and l but not on h.

Theorem 1.4. *The norm of every error with regular boundary layer satisfies the inequality*

$$\| l \mid L_2^{(m)*} \| \leq B_{n,m} h^m |\Omega|^{1/2} + K h^{m+1}. \tag{2.17}$$

Here $B_{n,m}$ is the constant of (2.6) and K is independent of h.

Comparing Theorems 1.2 and 1.4, we see that an error with regular boundary layer given a lattice of nodes $hH\gamma$ and a sufficiently small mesh-size h has the norm different from its lower bound by a higher order infinitesimal. We call such a cubature formula *asymptotically optimal.*

We now sketch the proof of Theorems 1.3 and 1.4. Our argument grounds on a detailed study of (2.9). The solution to (2.9) is written down as convolution of the right side with the *fundamental solution $G_{m,n}(x)$ to the polyharmonic equation.* In odd-dimensional space, $G_{m,n}(x) = \varkappa_{m,n} |x|^{2m-n}$, where

$$\varkappa_{m,n} = \frac{(-1)^m \Gamma(n/2 - m)}{\Gamma(m) 2^{2m} \pi^{n/2}} = \frac{(-1)^{(n-1)/2}}{\Gamma(m)\Gamma(m - n/2 + 1) 2^{2m} \pi^{n/2-1}}.$$

These descriptions for the constant $\varkappa_{m,n}$ involve the familiar rule

$$\Gamma(n/2 - m)\Gamma(m - n/2 + 1) = (-1)^{(n-1)/2+m} \pi.$$

In even-dimensional space, there are two different expressions for $G_{m,n}(x)$. For $2m < n$,

$$G_{m,n}(x) = \varkappa_{m,n} |x|^{2m-n}, \tag{2.18}$$

with

$$\varkappa_{m,n} = \frac{(-1)^m \Gamma(n/2 - m)}{\Gamma(m) 2^{2m} \pi^{n/2}},$$

and, for $2m \geq n$,

$$G_{m,n}(x) = \varkappa_{m,n} |x|^{2m-n} \log |x| \tag{2.19}$$

with the constant

$$\varkappa_{m,n} = \frac{(-1)^{(n-2)/2}}{\Gamma(m)\Gamma(m - n/2 + 1) 2^{2m-1} \pi^{n/2}}.$$

In these formulas, it is surely assumed that $|x|^2 = \sum_{j=1}^{n} x_j^2$. Each of the two forms of expressing the constant $\varkappa_{m,n}$ in (2.18) and (2.19) is applicable only if $2m \geq n$ or $2m < n$, respectively. For n odd or for $|\alpha| > 2m - n$ the derivative of order α of $G_{m,n}(x)$ satisfies the estimate

$$|D^\alpha G_{m,n}(x)| \leq K|x|^{2m-n-|\alpha|},$$

with K a constant independent of x. If n is even and $|\alpha| \leq 2m - n$, then we have the inequality

$$|D^\alpha G_{m,n}(x)| \leq K|x|^{2m-n-|\alpha|}\big|\log|x|\big|.$$

Further estimation of possible solutions to (2.9) is based on the following

Lemma 1.1. *The convolution of $G_{m,n}(x)$ with an elementary error $l(x)$ of compact support and order $s > 2m - n$ admits the estimate*

$$|G_{m,n} * l(x)| \leq K \frac{1}{(1+|x|^2)^{(n-2m+s)/2}}. \tag{2.20}$$

At sufficiently large $|x|$, the lth order derivative of the convolution under study is $O(|x|^{2m-n-l-s})$ if $l+s > 2m-n$ or n is odd; it is $O(|x|^{2m-n-l-s}\log|x|)$ otherwise.

To check the validity of Lemma 1.1, it suffices to expand $G_{m,n}(x-y)$ in the Taylor series in the powers of y with a remainder and insert this expansion in the definition of *convolution*

$$G_{m,n}(x) * l(x) = \int G_{m,n}(x-y)l(y)\,dy.$$

Further estimates are given in Chapter 5.

Write a solution $\psi_l(x)$ to (2.9) as

$$\psi_l(x) = (-1)^m G_{m,n}(x) * l(x) = (-1)^m \int G_{m,n}(x-y)l(y)\,dy.$$

Inserting (2.14) or (2.15), find that

$$\psi_l(x) = \sum_{\gamma \subset B_L} \psi_\gamma(x - hH\gamma), \tag{2.21}$$

where $\psi_\gamma(x) = (-1)^m G_{m,n}(x) * l_\gamma(x/h)$. We dominate every term $\psi_\gamma(x)$ of the series (2.21) by Lemma 1.1. Then by the integral test for series we estimate the norm of the total sum $\psi_l(x)$. Recalling that $\psi_l(x)$ is the extremal function of $l(x)$

and so its $L_2^{(m)}$ norm coincides with the $L_2^{(m)*}$ norm of $l(x)$, obtain the sought estimate (2.16) of Theorem 1.3.

The proof of Theorem 1.4 is more involved. Let $l(x)$ be an error with regular boundary layer in Ω expanded as in (2.15). Compare its $L_2^{(m)*}$ norm with the value of the *simplest* $\widetilde{L}_2^{(m)}$*-optimal error*

$$l_\infty(x) = \chi_{\Omega_0}(x) - \sum_{hH\gamma\in\Omega_0} h^n\delta(x - hH\gamma) \tag{2.22}$$

at the truncated extremal function $\chi_\Omega(x)(\widetilde{\psi}(x) - \widetilde{\psi}(0))$, where $\widetilde{\psi}(x)$ is determined from (2.10). The extremal function $\psi_l(x)$ of $l(x)$ is a solution to equation (2.9) and

$$\left\| l \mid L_2^{(m)*} \right\|^2 = (l(x), \psi_l(x)).$$

Write down the right side of this equality as

$$(l, \psi_l) = \int\limits_\Omega l_\infty(x)\widetilde{v}(x)\, dx + \int\limits_\Omega (l(x) - l_\infty(x))\widetilde{v}(x)\, dx - \int\limits_\Omega l(x)[\widetilde{v}(x) - \psi_l(x)]\, dx, \tag{2.23}$$

with $\widetilde{v}(x)$ standing for the difference $\widetilde{\psi}(x) - \widetilde{\psi}(0)$. It may be shown that the first summand in (2.23) has the form

$$\int\limits_\Omega l_\infty(x)\widetilde{v}(x)\, dx = B_{n,m}^2 h^{2m}|\Omega| + O(h^{2m+1}).$$

The second summand in (2.23) is small since the support of the difference $l_\infty(x) - l(x)$ intersects the domain Ω by a thin layer. More exactly, we have the following

Lemma 1.2. *The difference of $l_\infty(x)$ and $l(x)$ is an error with regular boundary layer for the domain $\Omega_0 \setminus \Omega$.*

Check the validity of Lemma 1.2. The value of the simplest functional at each function $\varphi(x)$ in $\widetilde{L}_2^{(m)}(H)$ may be written as

$$(l_\infty, \varphi) = \sum_{hH\gamma\in\Omega_0} \left(l_0\left(\frac{x}{h} - H\gamma\right), \varphi(x) \right),$$

with $l_0(x)$ the error in the expansion (2.15). Therefore, in the difference

$$l^*(x) = l_\infty(x) - l(x)$$

all summands $l_0(x/h - H\gamma)$ are cancelled out when $\gamma \in B_L^{(i)}$; whereas the elementary errors $l_0(x/h - H\gamma)$ remain unchanged which correspond to γ such that the points $hH\gamma$ lie in Ω_0 at a distance greater than Lh from the boundary of Ω. For all $\gamma \in B_L^{(l)}$ the elementary errors in the expansion of $l^*(x)$ take the form

$$l_0\left(\frac{x}{h} - H\gamma\right) - l_\gamma\left(\frac{x}{h} - H\gamma\right).$$

The proof of Lemma 1.2 is complete.

Using Lemma 1.2, it is easy to estimate the second summand in (2.23)

$$\left|\int_\Omega (l_\infty(x) - l(x))\,\widetilde{v}(x)\,dx\right| \leq Kh^{2m+1}.$$

The third summand in (2.23) is also $O(h^{2m+1})$. The remaining estimates are given in Chapter 5. Combining the estimates of the three summands on the right side of (2.23), arrive at (2.17) and so obtain Theorem 1.4.

We have demonstrated how to construct a family of errors with regular boundary layer. It turns out that, with a fixed matrix H and the vanishing lattice meshsize h, these errors are asymptotically optimal. In this event the constant $B_{n,m}$, occurring in the asymptotic expansion of the norm of an error with regular boundary layer, varies with H. The definition of $B_{n,m}$ implies that it stands to reason to choose the lattice matrix H so that at a given m the corresponding Epstein zeta function attain its minimal value.

The problem of finding such H is a hard problem of number theory. At m large, the Epstein zeta function coincides asymptotically with the ratio $K/r_{\min}^{2m}$, where K is a constant independent of H and $r_{\min}$ is a minimal distance from the points $H^{-1*}\gamma$, $\gamma \neq 0$, to the coordinate origin. This ratio becomes a minimum at the choice of the matrix H^{-1*} maximizing the value of $r_{\min}$. The nodes of the corresponding lattice are the centers of the spheres comprising a closest packing in n-dimensional space [191]. Consequently, the *best matrix* H is the inverse of the transpose of the lattice matrix corresponding to the lattice of the closest packing of spheres. G. F. Voronoĭ proposed an algorithm for finding such packings in space of arbitrary dimension. His algorithm is, however, hard to implement. In fact, these lattices were discovered by various scientists in dimension $n \leq 8$. The reader may find an overview of relevant results in [294].

In closing, we note that to determine the norm of an error by (2.6) given n, m, H, and h it is necessary to be able to calculate the values of the Epstein zeta function. In this connection, we refer to the articles [86, 208, 293] proposing appropriate algorithms.

§3. The Order of a Cubature Formula on Infinitely Differentiable Functions

Until now we study function spaces, not assuming that their members have derivatives of arbitrary order. Now we consider cubature formulas in the classes of infinitely differentiable functions assumed periodic for the sake of simplicity.

Say that a *function* $\varphi(x)$ *belongs to the class* $\widetilde{C}(\beta \mid A)$, provided that, first, it is *periodic with period matrix* H, i.e.,

$$\varphi(x + H\gamma) = \varphi(x), \quad x \in \mathbb{R}^n, \quad \gamma \in \mathbb{Z}^n,$$

and, second, the sequence of its derivatives has definite growth with increasing order of differentiation

$$\sup_{x\in\Omega_0} |D^\alpha \varphi(x)| \leq K\alpha! A^{|\alpha|} |\alpha|^{(\beta+1/2)|\alpha|} \alpha^{-\alpha/2} |\alpha|^\mu, \quad |\alpha| \geq 0. \tag{3.1}$$

The constant K can depend on φ but not on α. The class $\widetilde{C}(\beta \mid A)$ is a specification of the *Gevrey classes* [211, 274].

A linear change of variables, in particular, a *scaling*, sends the class $\widetilde{C}(\beta \mid A)$, to a generally new class $\widetilde{C}(\beta \mid A_1)$. Moreover, the number β, called the *order* of every member, remains unchanged and the number A, called the *type* of every member, may change. The class $\widetilde{C}(\beta \mid A)$ is, however, given so that the sequence of estimates (3.1) remains unchanged under arbitrary orthogonal transformations of the independent variable x. In other words, orthogonal transformations send this class into itself; i.e., we have

Theorem 1.5. *Let T be an orthogonal square matrix. If a function $\varphi(x)$ belongs to $\widetilde{C}(\beta \mid A)$, then the function $\psi(y) = \varphi(Ty)$ belongs to the same class.*

Certainly, $\psi(y)$ is a periodic function with period matrix the product T^*H rather than the period matrix H.

The proof of the theorem leans upon estimation of the shape

$$\Phi_m(x) = \sum_{|\alpha|=m} \frac{D^\alpha \varphi(0)}{\alpha!} x^\alpha$$

at every natural m.

Lemma 1.3. *If the derivatives of a function $\varphi(x)$ satisfies the inequalities (3.1), then $\Phi_m(x)$ admits the estimate*

$$|\Phi_m(x)| \leq KA^m m^{\beta m} m^{\lambda+n} |x|^m \tag{3.2}$$

for all complex x and $\lambda = \mu$.

Lemma 1.4. *If $\varphi(x)$ is a periodic function and for all x_0 in the fundamental parallelepiped Ω_0 the estimates (3.2) hold for $\Phi_m(x - x_0)$, then $\varphi(x)$ belongs to the class $\widetilde{C}(\beta \mid A)$ and the inequality (3.1) holds for $\mu = \lambda + n$.*

An orthogonal change of coordinates transforms $\Phi_m(x)$ in the Taylor series for a function $\varphi(x)$ into the same form in new variables. Using this and successively applying Lemmas 1.3 and 1.4, we derive Theorem 1.5.

The functions of the classes $\widetilde{C}(\beta \mid A)$ at different β and A behave differently. Let $\varkappa$ equal $\beta + 1$ and consider the following five cases.

1. $\varkappa < 0$. The class $\widetilde{C}(\beta \mid A)$ comprises constant functions.

2. $\varkappa = 0$. Each member of $\widetilde{C}(\beta \mid A)$ is a *trigonometric polynomial*, that is a (finite) trigonometric series

$$\sum_{|H^{-1*}\gamma| \le KA} a[\gamma] e^{i2\pi\gamma H^{-1}x}.$$

The order of such polynomial is finite and depends on A. Namely, as A increases, it grows.

3. $0 < \varkappa < 1$. The class $\widetilde{C}(\beta \mid A)$ contains the entire analytic functions meeting the periodicity condition and having order ρ and type σ, where $\rho = -1/\beta$ and $\sigma = A^\rho/(\rho e)$.

4. $\varkappa = 1$. The class $\widetilde{C}(0 \mid A)$ consists of analytic functions whose power series expansion at every point of Ω_0 has the radius of convergence bounded below by the ratio $1/A$.

5. $\varkappa > 1$. The class $\widetilde{C}(\beta \mid A)$ contains periodic functions that in general are not analytic, in particular, various quasianalytic functions.

The spaces $\widetilde{C}(\varkappa, A, \lambda)$, $\widetilde{C}(\varkappa, A)$, and $\widetilde{C}(\varkappa)$ analogous to $\widetilde{C}(\beta \mid A)$ are considered in Chapter 7.

Estimate the value of the error $l(x)$ of a given cubature formula at some function $\varphi(x) \in \widetilde{C}(\beta \mid A)$ to obtain

$$|(l, \varphi)| \le \|l \mid \widetilde{L}_2^{(m)*}\| \, \|\varphi \mid \widetilde{L}_2^{(m)}\|. \tag{3.3}$$

This estimate holds for an arbitrary m, and so we may choose the fittest value of m. Such is the number $m_0 = m_0(h)$ at which the right side of (3.3) attains a minimal value provided surely that the respective minimum exists.

Consider a particular example in which the expression on the right side of (3.3) admits a dominant in the shape of an explicit function of the arguments m and h.

Theorem 1.6. *The norm in $\widetilde{L}_2^{(m)}$ of an arbitrary function $\varphi(x) \in \widetilde{C}(\beta \mid A)$ admits the estimate*

$$\|\varphi \mid \widetilde{L}_2^{(m)}\| \le KA^m m^{(\beta+1)m} m^\lambda, \tag{3.4}$$

with K and λ constants independent of m.

The inequality (3.4) is easy to derive from the definition of the norm in $\widetilde{L}_2^{(m)}$ on using the series of estimates (3.1) and the *Stirling formula.*

Set in (3.3) the error $l(x)$ equal to $l_\infty(x)$ and use the formula (2.6) for the norm of $l_\infty(x)$ in $\widetilde{L}_2^{(m)*}$. Then, considering (3.4), infer that

$$|(l_\infty, \varphi)| \leq K B_{n,m} |\Omega_0|^{1/2} h^m A^m m^{\varkappa m+\lambda}. \tag{3.5}$$

The function $\zeta(H^{-1*} \mid 2m)$ obviously decreases and vanishes as m increases indefinitely. At m large this function admits an asymptotically exact expansion

$$\zeta(H^{-1*} \mid 2m) = \frac{K}{r_{\min}^{2m}} \left(1 + O\left(\frac{1}{m}\right)\right),$$

with $r_{\min}$ the minimal distance between the nodes $H^{-1*}\gamma$ and K the number of nodes at the minimal distance from the coordinate origin. At m large, (3.5) thus implies that

$$|(l_\infty, \varphi)| \leq K \left(\frac{Ah}{2\pi r_{\min}}\right)^m m^{\varkappa m+\lambda} \equiv \psi(m,h).$$

A minimum of $\psi(m,h)$, with h fixed and m varying, is easy to find. As a result, arrive at the relation

$$|(l_\infty, \varphi)| \leq K h^{-\mu} e^{-s(1/h)^{1/\varkappa}},$$

where

$$s = \frac{\varkappa}{e} \left(\frac{2\pi e r_{\min}}{A}\right)^{1/\varkappa}.$$

Consequently, for m however great, the accuracy of the cubature formula is $O(h^m)$ uniformly in all φ satisfying (3.1) with a fixed K. In this event we say that the corresponding error has *infinite order.*

The number m_0 at which $\psi(m,h)$ attains a minimum given h coincides with the integer closest to

$$\frac{1}{e} \left(\frac{2\pi e r_{\min}}{hA}\right)^{1/\varkappa}.$$

Observe that, although the cubature formula with equal weights has infinite order for periodic functions, this is of no avail for any finite $h > 0$ with the exception of the case $\varkappa \leq 0$ in which every function in $\widetilde{C}(\varkappa \mid A)$ each of which integrates exactly starting from some h.

Experimenting with numerical integration of functions by means of formulas of various order leads to results rather reminiscent of those described just away.

Namely, increasing the order of a formula at a fixed h leads to improving the accuracy of results only up to some m_0. Further increase in order involves no perceptible changes. The smaller h is, the higher order becomes beneficial.

We estimated the rate of convergence of cubature formulas on periodic functions because we knew the expression for the norm of errors as an explicit function of the order m and the mesh-size h. In the case of aperiodic functions, we unfortunately lack similar information. However, in Chapter 7 we obtain a needed expansion for the error of *Gregory quadrature formulas.*

§4. Errors in $W_2^{(m)}$

In practice it is important that calculation of an integral uses as few operations as possible while providing preassigned accuracy. Therefore, it is convenient to pose the problem as follows: at the first step, fixing the cost of calculations, i.e., the total amount of operations, provide possibly maximal accuracy; at the second step, choosing the scheme of operations as shown best at a fixed cost, find the actual number of operations.

To construct a sought approximation to a function φ or to the integral $I(\varphi)$, some job should be carried out. We assume that the cost of calculations is proportional to the number of points at which we are to know φ. The problem posed above reduces to studying the approximation methods that provide the best accuracy at a given number of nodes N. We treat these formulas as *optimal.*

Considering a cubature formula (1.3) with N fixed, we are free to choose the nodes $x^{(k)}$ and the weights c_k. Choosing weights at given nodes is a linear problem which we inspect completely. On the other hand, the choice of nodes is by far a harder problem. We do not address it in full generality, confining exposition to the case in which the system of nodes K is a parallelepiped lattice and only the parameters of this lattice vary.

In a functional-analytical setting, the quality of a cubature formula is characterized by the $\mathbf{B}^*$ norm of the error. The choice of the initial space $\mathbf{B}$, as is most often in the theory of computations, is arbitrary to some extend and depends in practical situations on the intuition and taste of the user. Unfortunately, this choice is partially dictated by a wish to have a problem solvable by the very method that the author invented since the most natural statement may seem to be too hard. We state here only two conditions that in our opinion should be imposed on the space $\mathbf{B}$ in general.

1. The value of the error

$$l(x) = \chi_\Omega(x) - \sum_{k=1}^{N} c_k \delta(x - x^k) \tag{4.1}$$

is determined at every function $\varphi(x)$ in $\mathbf{B}$ only in the case when φ is continuous. We therefore naturally require that $\mathbf{B}$ be embedded in $C(\Omega)$.

2. The presence of an embedding of $\mathbf{B}$ in $C(\Omega)$ is insufficient if only we want to have a possibility of controlling the quality of a cubature formula by increasing the number of nodes N. Indeed, if $\mathbf{B}$ coincides with $C(\Omega)$ then the norm of (4.1) is given by the expression

$$\|l \mid C^*\| = |\Omega| + \sum_{k=1}^{N} |c_k|.$$

This quantity is bounded from below by a positive real, i.e., it does not vanish as N increases indefinitely. To obviate similar nuisances, we require that the embedding of $\mathbf{B}$ to $C(\Omega)$ be compact.

Lemma 1.5. *Let the embedding operator of* $\mathbf{B}$ *to* $C(\Omega)$ *be compact. Then there is a sequence of errors like* (4.1) *whose norms may become however small as* N *tends to infinity.*

PROOF. By hypothesis, the closed unit ball of $\mathbf{B}$ becomes a subset of a compact set in $C(\Omega)$. As is known, in this case the ball consists of equicontinuous functions. In other words, to an arbitrary positive ε there is a $\delta = \delta(\varepsilon)$ such that, for every function φ in $\mathbf{B}$ with norm at most 1, the inequality holds

$$|\varphi(x) - \varphi(y)| \le \varepsilon \quad \text{if } |x - y| < \delta(\varepsilon). \tag{4.2}$$

Using such $\delta(\varepsilon)$, find N points $x^{(1)}, \dots, x^{(N)}$ in Ω such that the union of the balls $\{y : |y - x^{(k)}| < \delta(\varepsilon)\}$ over all k from 1 to N covers the whole domain Ω. These points we take as the nodes of the sought error. Further, the balls centered at $x^{(j)}$ of radius $\delta(\varepsilon)$ obviously induce a partition of Ω into some elementary portions Ω_j. We let the weight c_j equal the volume of Ω_j. Then for all functions φ in the unit ball B we obtain from (4.2) that

$$\left| \int_{\Omega} \varphi\, dx - \sum_{k=1}^{N} |\Omega_k| |\varphi(x^{(k)})| \right| \le \sum_{k=1}^{N} \int_{\Omega_k} |\varphi(y) - \varphi(x^{(k)})|\, dx \le \varepsilon |\Omega|.$$

Hence, $|(l, \varphi)| \le \varepsilon |\Omega|$ uniformly in φ belonging to the unit ball, i.e., $\|l \mid \mathbf{B}^*\| \le \varepsilon |\Omega|$. The prove of Lemma 1.5 is complete.

To find the explicit $\mathbf{B}^*$ norm of the error l_N we use the concept of extremal function. A function u in $\mathbf{B}$ is $\mathbf{B}$-*extremal for a given error* l_N provided that the equality holds

$$(l_N, u) = \|l_N \mid \mathbf{B}^*\| \|u \mid \mathbf{B}\|.$$

Choosing an extremal function u with norm 1, for $\|l_N \mid \mathbf{B}^*\|$ we obtain the following expression

$$\|l_N \mid \mathbf{B}^*\| = (l_N, u).$$

Existence of an extremal function should be proven in general. Most convenient for this purpose are the Hilbert spaces. In these spaces, each extremal function is expressible through the given functional by virtue of the Riesz Theorem on the general form of a bounded linear functional.

Along with the spaces $L_2^{(m)}$, $\widetilde{L}_2^{(m)}$, and $L_2^{(m)}(\Omega)$ introduced in § 2, we are also interested in the spaces $W_2^{(m)}$, $\widetilde{W}_2^{(m)}$, and $W_2^{(m)}(\Omega)$. Recall the definitions of the latter.

The space $\widetilde{W}_2^{(m)}$ is the direct sum of $\widetilde{L}_2^{(m)}$ and $\mathbb{R}^1$, since in the set of periodic functions each element with $\widetilde{L}_2^{(m)}$ norm 0 is a constant function. We introduce the norm in $\widetilde{W}_2^{(m)}$ by the equality

$$\|\varphi \mid \widetilde{W}_2^{(m)}\| = \left\{|\mu(\varphi)|^2 + \|\varphi \mid \widetilde{L}_2^{(m)}\|^2\right\}^{1/2},$$

where μ is an arbitrary norm on $\widetilde{L}_2$. We may for instance put

$$\mu(\varphi) = \left\{\frac{1}{|\Omega_0|}\int\limits_{\Omega_0} |\varphi(x)|^2\, dx\right\}^{1/2}.$$

In this case the norm of an element φ in $\widetilde{W_2}^{(m)}$ is easy to express through the Fourier coefficients of φ. Assume that

$$\varphi(x) = \sum_\beta c_\varphi[\beta] e^{i2\pi\beta H^{-1}x}. \tag{4.3}$$

Then

$$\|\varphi \mid \widetilde{W}_2^{(m)}\|^2 = \sum_\beta |c_\varphi[\beta]|^2 (1 + |2\pi H^{-1*}\beta|^{2m}). \tag{4.4}$$

The space $W_2^{(m)}(\Omega)$ is the direct sum of $L_2^{(m)}(\Omega)$ and the space $\mathbf{P}_{m-1}$ of polynomials of degree less than m. In the case when Ω coincides with $\mathbb{R}^n$, we simply write $W_2^{(m)}$ instead of $W_2^{(m)}(\mathbb{R}^n)$. Assume that there is an operator Π projecting $W_2^{(m)}(\Omega)$ to the subspace $\mathbf{P}_{m-1}$. Then the norm of an arbitrary function φ in $W_2^{(m)}(\Omega)$ may be written down as

$$\|\varphi \mid W_2^{(m)}(\Omega)\| = \left\{\|\Pi\varphi \mid \mathbf{P}_{m-1}\|^2 + \|\varphi \mid L_2^{(m)}(\Omega)\|^2\right\}^{1/2}. \tag{4.5}$$

Interpolation operators are natural projections of $W_2^{(m)}(\Omega)$ to $\mathbf{P}_m$. We constructed one of such operators in § 1, by assigning to each continuous function $\varphi(x)$ a polynomial $P(x)$ of degree m taking the values $\vec{\varphi} = (\varphi(x^{(1)}), \dots, \varphi(x^{(N)}))$ at the given nodes $x^{(1)}, \dots, x^{(N)}$. Recall that

$$P(x) = \vec{\varphi} S^{-1} x^{\alpha}. \tag{4.6}$$

This polynomial is defined and solves the interpolation problem if the dimension M of the space $\mathbf{P}_m$ equals the number of nodes N, and S, the *Vandermonde matrix* (1.11), is nondegenerate.

If the number of points N in (1.2) is not less than M then, generally speaking, given the system of the values $\vec{\varphi}$ of a function φ at the nodes of (1.2), it is possible to construct infinitely many operators mapping $W_2^{(m)}(\Omega)$ somehow to the space $\mathbf{P}_m$ of polynomials of degree m. All these operators have the form

$$P(x) = \vec{\varphi} Y x^{\alpha}, \tag{4.7}$$

where Y is a matrix of size $N \times M$. Each of these operators is called an *interpolation operator*; and (4.7), an *interpolation formula*.

Assume that we are to find an interpolation operator that serves as the identity operator in the space $\mathbf{P}_m$ of polynomials of degree m. Clearly, this operator recovers each polynomial ax^{α} of degree m from the given nodal values at (1.2). This condition may be written as

$$aSYx^{\alpha} = ax^{\alpha}, \quad x \in \mathbb{R}^n,$$

or, in equivalent form, as

$$aSY = a.$$

Since a is arbitrary, this is possible only if the matrix Y is a right inverse of S, which in turn happens only when $N \leq M$. If $N < M$ then the matrix Y is nonunique in general. In this event, (4.7) is written as

$$P(x) = \vec{\varphi} S_r^{-1} x^{\alpha} = \Pi\varphi. \tag{4.8}$$

When $M = N$ the polynomial (4.8) coincides with the given formula (4.6).

We exhibit an example of the system of points such that the Vandermonde matrix S has a right inverse, implying that (4.8) gives a projection of $W_2^{(m)}(\Omega)$ to $\mathbf{P}_m$. We need some elementary facts from interpolation theory which we expose below.

In the case of one variable t, given an integer $k \geq 0$, we call the *Newtonian power* $[k]$ the function $t^{[k]}$ defined by the equality

$$t^{[0]} = 1; \quad t^{[k]} = t(t-1)\dots(t-k+1), \quad k \geq 1.$$

In the case of n variables, the Newtonian power y^{α} is defined as

$$y^{[\alpha]} = y_1^{[\alpha_1]} y_2^{[\alpha_2]} \dots y_n^{[\alpha_n]}.$$

Denote by Δ the *finite difference* with mesh-size 1 for a function of one variable, and by Δ^k, the taking of the difference k times, i.e.,

$$\Delta f(t) = f(t+1) - f(t), \quad \Delta^{k+1} f = \Delta(\Delta^k f).$$

Then in the case of n variables, put

$$\Delta^{\alpha} f(x) = \Delta^{\alpha_1} \Delta^{\alpha_2} \dots \Delta^{\alpha_n} f(x_1, x_2, \dots, x_n).$$

We have the following

Lemma 1.6. *Each polynomial $P(x)$ expands uniquely in Newtonian powers*

$$P(x) = \sum_{|\alpha| \leq m} p_{\alpha} (x - x^{(0)})^{[\alpha]}. \tag{4.9}$$

In this event, the coefficients p_{α} are defined from the formulas

$$p_{\alpha} = \frac{\Delta^{\alpha} P(x_0)}{\alpha!}, \quad |\alpha| \leq m. \tag{4.10}$$

PROOF. For a function of one variable we have

$$t^{[k]} = t^k + \sum_{j=1}^{k-1} a_j t^j.$$

Therefore, the transition matrix from the system $\{t^k\}_{k=0}^{\alpha}$ to the system $\{t^{[k]}\}_{k=0}^{\alpha}$ is triangular with 1s on the main diagonal and nondegenerate. Consequently, $t^{\alpha} = \sum_{k=0}^{\alpha} l_k t^{[k]}$. In the case of n variables, we have the equalities

$$y^{\alpha} = \prod_{j=1}^{n} y_j^{\alpha_j} = \prod_{j=1}^{n} \sum_{k_j=1}^{\alpha_j} b_{k_j, j} y_j^{[k_j]} = \sum_{0 \leq \beta \leq \alpha} l_{\beta} y^{[\beta]}$$

which immediately entail (4.9).

We establish uniqueness by proving the formula (4.10) for coefficients. Applying to both sides of (4.9) the finite difference Δ^α at the point $x^{(0)}$, observe that $\Delta^\alpha y^{[\beta]}|_{y=0} = \alpha!\delta_\alpha^\beta$ where δ_α^β is the *Kronecker delta*. The proof of Lemma 1.6 is complete.

We call *Newtonian* the system of points

$$x^{(k)} = x^{(0)} + y^{(k)}, \tag{4.11}$$

where the entries of $y^{(k)}$ are integers satisfying the inequalities

$$0 \le y_j^{(k)} \le m, \quad \sum_{j=1}^{n} y_j^{(k)} \le m.$$

Show that the *Vandermonde matrix* S *corresponding to a Newtonian system is a nonsingular square matrix.*

To begin with, the number of points $x^{(k)}$ in (4.11) clearly equals M, since it coincides with the number of integer vectors α subject to the condition $|\alpha| \le m$. Further, every polynomial x^α is uniquely determined from its values at the points (4.11). Indeed, instead of the values of a function at the nodes of a Newtonian system we may introduce the finite differences that are connected with these values via linear transformations and expand each polynomial $P(x)$ in Newtonian powers

$$P(x) = \sum_{|\alpha| \le m} \frac{(x - x^{(0)})^{[\alpha]}}{\alpha!} \Delta^{[\alpha]} P(x)\Big|_{x = x^{(0)}}.$$

This expansion is clearly unique. Therefore, S is actually a nondegenerate square matrix.

If the system of nodes K includes a Newtonian subsystem then there is always a right inverse S_r^{-1} to S. We may by analogy consider a system of the shape

$$x^{(0)} + hy^{(k)} \tag{4.12}$$

and any other that results from (4.12) by an arbitrary affine transformation. Observe in particular that the matrix S has a right inverse if the system K consists of all nodes of the lattice $x^{(0)} + hH\beta$ which lie in the ball of radius $L > mh$ with center $x^{(0)}$.

Return to the space $W_2^{(m)}(\Omega)$ and define a projection Π of it to $\mathbf{P}_{m-1}$ by the equality (4.8). An arbitrary functional $l(x) \in W_2^{(m)*}(\Omega)$ may be written as

$$(l, \varphi) = (l, \Pi\varphi) + (l, \varphi - \Pi\varphi) = (l_1, \varphi) + (l_2, \varphi),$$

where l_1 is the composition of Π and l and l_2 is the composition of $I - \Pi$ and l. Clearly, l_2 vanishes at all polynomials in $\mathbf{P}_{m-1}$. Hence, l_2 belongs to $L_2^{(m)*}(\Omega)$. The norm of l in $W_2^{(m)*}(\Omega)$ may be obtained from the formula

$$\|l \mid W_2^{(m)*}(\Omega)\| = \left\{\|l_1 \mid \mathbf{P}_{m-1}^*\|^2 + \|l_2 \mid L_2^{(m)*}(\Omega)\|^2\right\}^{1/2}. \tag{4.13}$$

This equality follows from a more general proposition, Lemma 1.7.

Lemma 1.7. *Assume that a Banach space* $\mathbf{X}$ *is the direct sum of its subspaces* $\mathbf{X}_1, \mathbf{X}_2, \ldots, \mathbf{X}_\sigma$, *namely,*

$$\mathbf{X} = \mathbf{X}_1 \oplus \mathbf{X}_2 \oplus \cdots \oplus \mathbf{X}_\sigma.$$

Let the spaces $\mathbf{X}_j$, $1 \le j \le \sigma$, *be endowed with the norms* $\|\cdot \mid \mathbf{X}_j\|$ *so that*

$$\|\varphi \mid \mathbf{X}\| = g(\|\varphi_1 \mid \mathbf{X}_1\|, \|\varphi_2 \mid \mathbf{X}_2\|, \ldots, \|\varphi_\sigma \mid \mathbf{X}_\sigma\|)$$

where the elements $\varphi_j \in \mathbf{X}_j$, $j = 1, \ldots, \sigma$, *appear in the unique decomposition of* φ *into the sum* $\varphi_1 + \varphi_2 + \cdots + \varphi_\sigma$, *and* g *stands for a norm of the* σ*-dimensional Euclidean space. Then the space* $\mathbf{X}^*$ *is the direct sum of* $\mathbf{X}_1^*, \mathbf{X}_2^*, \ldots, \mathbf{X}_\sigma^*$, *namely,*

$$\mathbf{X}^* = \mathbf{X}_1^* \oplus \mathbf{X}_2^* \oplus \cdots \oplus \mathbf{X}_\sigma^*, \tag{4.14}$$

with

$$\|l \mid \mathbf{X}^*\| = g^*\left(\|l_1 \mid \mathbf{X}_1^*\|, \|l_2 \mid \mathbf{X}_2^*\|, \ldots, \|l_\sigma \mid \mathbf{X}_\sigma^*\|\right), \tag{4.15}$$

where l_j *is the restriction of* l *to* $\mathbf{X}_j$, *and* g^* *is the* **dual norm** *of* g, *i.e.,*

$$g^*(\beta_1, \beta_2, \ldots, \beta_\sigma) = \sup_{g(\alpha_1, \alpha_2, \ldots, \alpha_\sigma)=1} (\alpha_1\beta_1 + \alpha_2\beta_2 + \cdots + \alpha_\sigma\beta_\sigma).$$

PROOF. Relation (4.14) is obvious. Prove (4.15). Assume for simplicity that the functional l possesses an extremal function φ, $\|\varphi \mid \mathbf{X}\| = 1$. (The general case is settled by the limit argument and we leave the relevant calculations to the reader.)

Consider the decomposition

$$\varphi = \varphi_1 + \varphi_2 + \cdots + \varphi_\sigma, \quad \varphi_j \in \mathbf{X}_j.$$

By the definition of the norm of a functional

$$\begin{aligned}\|l \mid \mathbf{X}^*\| &= \sup_{\alpha_1, \alpha_2, \ldots, \alpha_\sigma} \frac{(l, \alpha_1\varphi_1 + \cdots + \alpha_\sigma\varphi_\sigma)}{\|\alpha_1\varphi_1 + \cdots + \alpha_\sigma\varphi_\sigma \mid \mathbf{X}\|} \\ &= \sup_{\alpha_1, \alpha_2, \ldots, \alpha_\sigma} \frac{\alpha_1(l_1, \varphi_1) + \cdots + \alpha_\sigma(l_\sigma, \varphi_\sigma)}{g(\alpha_1\|\varphi_1\|, \ldots, \alpha_\sigma\|\varphi_\sigma\|)}.\end{aligned}$$

It is easy to show that each element φ_j is an $\mathbf{X}_j$-extremal function for the functional l_j, i.e., $(l_j, \varphi_j) = \|l_j\|\|\varphi_j\|$. Consequently,

$$\|l \mid \mathbf{X}^*\| = \sup_{\alpha_1,\ldots,\alpha_\sigma} \frac{\alpha_1\|l_1\|\|\varphi_1\| + \cdots + \alpha_\sigma\|l_\sigma\|\|\varphi_\sigma\|}{g(\alpha_1\|\varphi_1\|,\ldots,\alpha_\sigma\|\varphi_\sigma\|)}$$

$$= \sup_{\alpha_1,\ldots,\alpha_\sigma} \frac{\alpha_1\|l_1\| + \cdots + \alpha_\sigma\|l_\sigma\|}{g(\alpha_1,\ldots,\alpha_\sigma)} = g^*(\|l_1\|,\ldots,\|l_\sigma\|).$$

The proof of Lemma 1.7 is complete.

Putting $\sigma = 2$ and $g(\alpha_1, \alpha_2) = \sqrt{\alpha_1^2 + \alpha_2^2}$ in the hypotheses of Lemma 1.7, obtain (4.13).

So, the space $W_2^{(m)*}(\Omega)$ splits into the direct sum of subspaces

$$W_2^{(m)*}(\Omega) = \mathbf{P}_{m-1}^* \oplus L_2^{(m)*}(\Omega)$$

in much the same way as $W_2^{(m)}(\Omega)$ before. Consequently, every element l in $W_2^{(m)*}(\Omega)$ may be written as $l = l_1 + l_2$ where l_1 in $\mathbf{P}_{m-1}^*$ and l_2 in $L_2^{(m)*}(\Omega)$, with (4.13) holding.

Using the above decomposition, we may simplify the problem of minimizing the norm of an error. We assume that the number of nodes N is sufficiently large so that there exist a matrix S_r^{-1}. Considering that the projection $\Pi\varphi$ is chosen to be the interpolation operator constructed from the nodes of the cubature formula (1.3), derive the equality

$$l_1 = \sum_{k=1}^{N} a_k \delta(x - x^{(k)}).$$

We presume this in what follows.

Minimizing the $W_2^{(m)*}$ norm of the error $l(x)$, we may separately minimize the norm of each of the corresponding functionals $l\Pi$ and $l(I - \Pi)$. If the number of nodes in (1.2) is sufficiently large then we may nullify the norm of the first of the functionals, by taking as Π the interpolation operator constructed at the nodes of the cubature formula (1.3).

Reformulate the result in a slightly different fashion. We proved that to each cubature formula in $W_2^{(m)}$ there corresponds a cubature formula of the same shape which is exact for all polynomials of degree less than m. In this event the $W_2^{(m)*}$ norm of the error of a formula in $L_2^{(m)*}$ does not exceed the $W_2^{(m)*}$ norm of the error corresponding to the initial cubature formula. This circumstance allows us to search optimal formulas in $L_2^{(m)}$.

Consider $W_2^{(m)}$ and inspect the norm of an error $l(x)$ in this space.

Assume first of all that the number of nodes N in the cubature formula in question is sufficiently large for the matrix S to have a right inverse. As we have seen, this is always so if the nodes of the lattice in the simplex (4.12) are included in the whole system of nodes.

Using (4.8), construct a projection Π of $W_2^{(m)}$ to $\mathbf{P}_{m-1}$. Clearly, $\Pi\varphi = \varphi$ for all $\varphi \in \mathbf{P}_{m-1}$.

The norm of $\mathbf{P}_{m-1}$ may be given arbitrarily. According to (4.13), this may change only that part of the norm which needs no approximation. Put

$$\|P \mid \mathbf{P}_{m-1}\|^2 = \sum_{k=1}^{N} |P(x^{(k)})|^2.$$

Then the norm of a function $\varphi(x)$ in $W_2^{(m)}$ is expressed as

$$\|\varphi \mid W_2^{(m)}\|^2 = (\vec{\varphi} S_r^{-1} S)(\vec{\varphi} S_r^{-1} S)^* + \int \sum_{|\alpha|=m} \frac{m!}{\alpha!} |D^\alpha \varphi|^2 \, dx. \tag{4.16}$$

This norm makes $W_2^{(m)}$ into a Hilbert space. The inner product on $W_2^{(m)}$ is given by the formula

$$\{\varphi, \psi\}_{W_2^{(m)}} = \vec{\varphi} S_r^{-1} S S^* S_r^{-1*} \vec{\psi}^* + \int \sum_{|\alpha|=m} \frac{m!}{\alpha!} D^\alpha \varphi D^\alpha \psi \, dx. \tag{4.17}$$

By the main property of Hilbert space, $W_2^{(m)*}$ may be identified with $W_2^{(m)}$, and, by the Riesz Theorem, every bounded linear functional l on $W_2^{(m)}$ may be written by means of the inner product (4.17) as follows

$$(l, \varphi) = \{\varphi, \psi_l\}_{W_2^{(m)}}. \tag{4.18}$$

Here ψ_l is a function uniquely determined from the functional l. In particular, each l in $W_2^{(m)*}$ decomposes into the sum

$$(l, \varphi) = (l_1, \varphi) + (l_2, \varphi), \tag{4.19}$$

where

$$(l_1, \varphi) = (\vec{\varphi} S_r^{-1} S)(\vec{\psi}_l S_r^{-1} S)^*, \tag{4.20}$$

$$(l_2, \varphi) = \int \sum_{|\alpha|=m} \frac{m!}{\alpha!} D^\alpha \varphi D^\alpha \psi_l \, dx. \tag{4.21}$$

Clearly, l_1 coincides with the composition $l\Pi$. Indeed, for every function $\varphi \in W_2^{(m)}$ by (4.18) we have

$$(l, \Pi\varphi) = \{\Pi\varphi, \psi_l\}_{W_2^{(m)}} = (l_1, \varphi).$$

Whence and from (4.19), conclude that $l_2 = l(I - \Pi)$.

Using (4.19)–(4.21), we may derive the following important theorem.

***Theorem* 1.7** (I. Babuška). *Assume that an error $l(x)$ is defined on $L_2^{(m)}$, i.e. $l(x)$ vanishes at every polynomial of degree less than m. If $l(x)$ is $L_2^{(m)}$-optimal, i.e. $l(x)$ has a minimal $L_2^{(m)*}$ norm among all errors with a given system of nodes $x^{(1)}, \dots, x^{(N)}$; there exists an extremal function of $l(x)$ belonging to $L_2^{(m)}$ and vanishing at every node $x^{(k)}$.*

PROOF. Using the given system of nodes, arrange the projection

$$\Pi\varphi = \vec{\varphi} S_r^{-1} x^\alpha$$

and the respective inner product (4.17). Now $W_2^{(m)}$ is a Hilbert space.

As is well known, the projection g_1 of an element g in a Hilbert space $\mathbf{H}$ to a closed subspace $\mathbf{H}_1 \subset \mathbf{H}$ serves as the best approximant to g among the elements of $\mathbf{H}_1$. In this event, the difference $g - g_1$ is orthogonal in $\mathbf{H}$ to every element of $\mathbf{H}_1$.

Take as $\mathbf{H}$ the space $W_2^{(m)}$, and as $\mathbf{H}_1$, the subspace of linear combinations $\sum_{k=1}^N c_k u_k(x)$, with $u_k(x)$ the function in $W_2^{(m)}$ corresponding to the functional $\delta(x - x^{(k)})$ by the Riesz Theorem. For every function $\varphi \in W_2^{(m)}$ we now have

$$\{\varphi, u_k\}_{W_2^{(m)}} = (\delta(x - x^{(k)}), \varphi(x)) = \varphi(x^{(k)}).$$

The subspace $\mathbf{H}_1$ is clearly closed in $\mathbf{H}$.

By the Riesz Theorem, to the indicator $\chi_\Omega(x)$ there corresponds $g \in \mathbf{H}$. Moreover, for each $g_1 \in \mathbf{H}_1$ the difference $g - g_1$ defines the error of some cubature formula. The norm of this error is minimal if the function $\psi_l = g - g_1$ is orthogonal to all functions $u_k(x)$, $k = 1, \dots, N$, i.e., if

$$\{\psi_l, u_k\}_{W_2^{(m)}} = \psi_l(x_k) = 0.$$

In this event, $\psi_l(x)$ is the extremal function for $l(x)$ in $W_2^{(m)}$. It turns out that the same function represents $l(x)$ in $L_2^{(m)}$.

Indeed, by condition, the functional $l(x)$ vanishes at every polynomial of degree less than m. Therefore, the superposition $l_1 = l\Pi$ coincides with the zero operator in $W_2^{(m)}$. Whence and from (4.19)–(4.21) it follows that for every $\varphi \in W_2^{(m)}$ the next equality is valid

$$(l,\varphi) = (l_2,\varphi) = \int \sum_{|\alpha|=m} \frac{m!}{\alpha!} D^\alpha \varphi(x) D^\alpha \psi_l(x)\, dx.$$

The proof of Theorem 1.7 is complete.

It is worth observing that Theorem 1.7 ensues from a more general duality theorem for functionals by S. M. Nikol'skiĭ (see, for instance, [101, p. 26]).

§5. Expansion of the $L_2^{(m)*}$ Norm of an Error with Arbitrary Nodes

As we have already noted, in practice it is useful to know the norm of the error of a cubature formula for every fixed number of nodes N. Then we may find an approximate value of an integral with prescribed accuracy without redundant operations. In the current section we give two expressions for the norm of an error acting on $L_2^{(m)}$. These expressions may in particular be used for computations. By means of these expressions for the norm, we also obtain a system of linear equations for determining the weights of $L_2^{(m)}$-optimal cubature formulas and inspect the simplest properties of the system.

A cubature formula with the error $l(x)$, if considered in $L_2^{(m)}$, may be characterized in two ways.

This cubature formula is determined by the extremal function $\psi_l(x)$ that results as solution to the equation

$$\Delta^m u = (-1)^m l(x). \tag{5.1}$$

The function $\psi_l(x)$ belongs to $L_2^{(m)}$ and may be written down as the convolution

$$\psi_l(x) = G(x) * l(x) + P(x). \tag{5.2}$$

Here $G(x)$ stands for the function $(-1)^m G_{m,n}(x)$, where $G_{m,n}(x)$ is the fundamental solution to the polyharmonic equation; an explicit expression for $G_{m,n}(x)$ is given in § 2. The second summand $P(x)$ in (5.2) is an arbitrary polynomial of degree less than m. The norms of $l(x)$ and $\psi_l(x)$ are related as follows

$$\| l \mid L_2^{(m)*} \|^2 = \int \sum_{|\alpha|=m} \frac{m!}{\alpha!} |D^\alpha \psi_l|^2\, dx. \tag{5.3}$$

The integral on the right side of (5.3) is taken with $\psi_l(x)$ a polyharmonic function at all points of $\mathbb{R}^n$ except for the boundary of Ω and the nodes $x^{(1)}, x^{(2)}, \dots, x^{(N)}$.

If $m = 1$, then it coincides with the conventional *Dirichlet integral.* The difference between the extremal functions of the errors $l_1(x)$ and $l_2(x)$ is a function polyharmonic everywhere in $\mathbb{R}^n$ except for the union of nodes corresponding to $l_1(x)$ and $l_2(x)$.

The cubature formula is determined by its weights c_k, $1 \le k \le N$, connected with the extremal function $\psi_l(x)$ by (5.2). Inserting (5.2) in (5.3), we see that the norm square of the error is a quadratic form in its coefficients

$$\|\psi_l \mid L_2^{(m)*}\|^2 = \sum_{k,k'=1}^{N} G(x^{(k)} - x^{(k')})c_k c_{k'}$$

$$-2\sum_{k=1}^{N} c_k \int_{\Omega} G(x - x^{(k)})\,dx + \int\int \chi_{\Omega}(x)\chi_{\Omega}(y)G(x-y)\,dx dy \equiv \psi(c). \quad (5.4)$$

Recall that the weights c_k in (5.4) obey the linear system (1.7) which is equivalent to the requirement that the cubature formula be exact for all polynomials of degree less than m,

$$(l(x), x^{\alpha}) = 0, \quad |\alpha| < m. \quad (5.5)$$

We now formulate the conditions under which the quadratic form $\psi(c)$ attains a minimum on the set of vectors c subject to the constraints (5.5). To this end, we apply the *Lagrange method of multipliers.* Consider the auxiliary function

$$\psi_1(c, v) = \psi(c) + 2\sum_{j=1}^{M} v_j (l(x), x^{\alpha^{(j)}})$$

$$= \psi(c) - 2\sum_{j=1}^{M}\sum_{k=1}^{N} v_j c_k (x^{(k)})^{\alpha^{(j)}} + 2\sum_{j=1}^{M} v_j \int_{\Omega} x^{\alpha^{(j)}}\,dx.$$

Equating to 0 the partial derivatives of $\psi_1(c, v)$, obtain the system of equations

$$\begin{aligned} &\sum_{k'=1}^{N} G(x^{(k)} - x^{(k')})c_{k'} - \sum_{j=1}^{M} (x^{(k)})^{\alpha^{(j)}} v_j = m_k, \quad 1 \le k \le N; \\ &\sum_{k=1}^{N} (x^{(k)})^{\alpha^{(j)}} c_k = f_j, \quad 1 \le j \le M. \end{aligned} \quad (5.6)$$

Here

$$m_k = \int_{\Omega} G(x - x^{(k)})\,dx, \quad f_j = \int_{\Omega} x^{\alpha^{(j)}}\,dx.$$

A solution to (5.6) which we denote by $c_k^{(0)}$ and $v_j^{(0)}$ is a stationary point of $\psi_1(c, v)$. From the theory of local extrema, we know a sufficient condition under which this solution provides a local minimum of $\psi(c)$ on the manifold (5.5). This condition is positive definiteness of the quadratic form

$$\Phi(c) = \sum_{k,k'=1}^{N} \frac{\partial^2 \psi_1}{\partial c_k \partial c_{k'}} c_k c_{k'} \tag{5.7}$$

on the set of vectors c subject to the constraint

$$Sc = 0. \tag{5.8}$$

Check that this condition is satisfied in the case under study.

Lemma 1.8. *For every nonzero vector $c \in \mathbb{R}^n$ lying in the subspace $Sc = 0$, the function $\Phi(c)$ is strictly positive.*

PROOF. From the definition of $\psi_1(c)$ and (5.7) it follows that

$$\Phi(c) = \sum_{k,k'=1}^{N} G(x^{(k)} - x^{(k')}) c_k c_{k'}. \tag{5.9}$$

Consider the linear combination of delta functions

$$\delta_c(x) = \sum_{k=1}^{N} c_k \delta(x - x^{(k)}). \tag{5.10}$$

By (5.8) this functional belongs to $L_2^{(m)*}$. It thus has the extremal function $u_c(x) \in L_2^{(m)}$ serving as a solution to the equation

$$\Delta^m u_c(x) = (-1)^m \delta_c(x). \tag{5.11}$$

Clearly, we may take as $u_c(x)$ the next linear combination of translates of the fundamental solution

$$u_c(x) = \sum_{k=1}^{N} c_k G(x - x^{(k)}).$$

The square of its norm in $L_2^{(m)}$ coincides with $\Phi(c)$, namely,

$$\left\| u_c \mid L_2^{(m)} \right\|^2 = (\delta_c(x), u_c(x)) = \sum_{k,k'=1}^{N} G(x^{(k)} - x^{(k')}) c_k c_{k'}.$$

This readily implies that, for c nonzero, $\Phi(c)$ is a strictly positive function. The proof of Lemma 1.8 is complete.

If the nodes $x^{(1)}, x^{(2)}, \dots, x^{(N)}$ are chosen so that the matrix S has a right inverse, then (5.6) has a unique solution.

Lemma 1.9. *If the matrix S has a right inverse then the matrix Q of (5.6) is nondegenerate.*

PROOF. Denote the matrix of the quadratic form (5.9) by G, and write the homogeneous system that corresponds to (5.6) as

$$Q\begin{pmatrix} c \\ v \end{pmatrix} = \left(\begin{array}{c|c} G & S^* \\ \hline S & 0 \end{array}\right)\begin{pmatrix} c \\ v \end{pmatrix} = 0. \tag{5.12}$$

Check that the sole solution of (5.12) is identically 0.

So, let c and v make a solution to (5.12). Using (5.10), arrange the generalized function $\delta_c(x)$ that corresponds to the vector c. This function obviously belongs to $L_2^{(m)*}$. As the extremal function for $\delta_c(x)$ take the following natural spline

$$u_c(x) = \sum_{k=1}^{N} G(x - x^{(k)})c_k + \sum_{j=1}^{M} v_j x^{\alpha^{(j)}}.$$

This is correct since $u_c(x)$ belongs to $L_2^{(m)}$ and solves (5.11). The first N equations of (5.12) mean that $u_c(x)$ vanishes at every node $x^{(k)}$. Using this, find the $L_2^{(m)*}$ norm of $\delta_c(x)$, namely,

$$\left\|\delta_c \mid L_2^{(m)*}\right\|^2 = (\delta_c, u_c) = \sum_{k=1}^{N} c_k u(x^{(k)}) = 0,$$

which is possible only when $c = 0$. Considering this, from the first N equations of (5.12) find

$$S^* v = 0. \tag{5.13}$$

By hypothesis, the matrix S has a right inverse, but then S^* has a left inverse. Whence and from (5.13) infer that the solution v is also equal to zero. The proof of Lemma 1.9 is complete.

So, (5.6) has the only solution $c^{(0)}$ and $v^{(0)}$. In this event, $c^{(0)}$ gives a local minimum to the quadratic form $\psi(c)$ on the solution set of (5.5). The entries of this solution are thus coincident with the weights of an $L_2^{(m)}$-optimal cubature formula.

The norm of the optimal error is a function of the number of nodes N and the order m. It is a hard problem to find explicit expressions for this function which are at least asymptotically exact as N and m tend to infinity. While the problem remains unsolved, we have no grounds for deciding on how worse a concrete cubature formula is as compared with an optimal formula. Observe that in the case of the cubature formulas corresponding to the lattice of nodes $hH\gamma$, we may obtain (see Chapter 5) an expansion of the norm of the $L_2^{(m)}$-optimal error which is asymptotically exact as $h \to 0$ at a fixed m. In the case of m tending to infinity, we expose (see Chapter 7) only upper bounds on the norm of the optimal errors and only in the one-dimensional case.

§6. The Weights of Optimal Cubature Formulas on a Given Lattice

We have shown in § 2 how to construct asymptotically optimal cubature formulas on a given lattice of nodes $hH\gamma$. Using our approach to constructing formulas with regular boundary layer, we obtain the number of operations required for finding weights which is $1/h$ times less than the number of operations needed for fulfilling approximate integration.

In some cases, for instance when Ω is a rational polyhedron, this number of operations remains finite even as $h \to 0$. However, this approach does not show the asymptotic behavior of the genuine optimal weights $c_0[\gamma]$ as $h \to 0$.

To clarify the question, consider the linear system for the weights of an $L_2^{(m)}$-optimal cubature formula on a given lattice of nodes. In this case (5.6) takes the shape

$$h^n \sum_{\gamma' \in B} c_0[\gamma] G(hH(\gamma - \gamma')) - \sum_{|\alpha| < m} v_\alpha (hH\gamma)^\alpha = \int_\Omega G(hH\gamma - y)\, dy, \quad \gamma \in B,$$
$$h^n \sum_{\gamma' \in B} c_0[\gamma'](hH\gamma')^\alpha = \int_\Omega y^\alpha\, dy, \quad |\alpha| < m, \tag{6.1}$$

where B is the set of multi-indices γ such that $hH\gamma$ lies in the closure $\overline{\Omega}$.

This is a system with the number of indeterminates which grows like h^{-n} as $h \to 0$. The coefficients of (6.1) do not vanish and straightforward solution of the system becomes complicated; the same applies to other study of the system. We thus propose another, nonalgebraic, approach to solving (6.1).

Extend $c_0[\gamma]$ to a function of a discrete argument ranging over all γ by setting it 0 for γ not in B. Assume further that $G[\gamma] = G(hH\gamma)$. Then (6.1) may be rewritten in convolution form as

$$h^n c_0[\gamma] * G[\gamma] = w[\gamma], \quad \gamma \in B,$$
$$h^n \sum_\gamma c_0[\gamma](hH\gamma)^\alpha = f_\alpha, \quad |\alpha| < m, \tag{6.2}$$

where

$$w[\gamma] = \int_\Omega G(hH\gamma - y)\, dy + \sum_{|\alpha| < m} v_\alpha (hH\gamma)^\alpha, \quad f_\alpha = \int_\Omega y^\alpha\, dy.$$

We find it convenient to consider a more general problem. In addition to $c_0[\gamma]$, introduce one more unknown function $w[\gamma]$, and replace (6.2) by a similar system with an arbitrary right side.

PROBLEM B_1. *Find a solution to the system of equations*

$$\begin{aligned} h^n c[\gamma] * G[\gamma] &= w[\gamma], \quad hH\gamma \in \mathbb{R}^n, \\ c[\gamma] &= 0, \quad hH\gamma \notin \overline{\Omega}, \\ h^n \sum_{\gamma} c[\gamma](hH\gamma)^\alpha &= f_\alpha, \quad |\alpha| < m, \\ w[\gamma] = f[\gamma] + \sum_{|\alpha|<m} v_\alpha (hH\gamma)^\alpha, &\quad hH\gamma \in \overline{\Omega}, \end{aligned} \tag{6.3}$$

with unknowns $c[\gamma]$, $w[\gamma]$, *and* v_α. *In this system* f_α *are given numbers and* $f[\gamma]$ *is known in the closure* $\overline{\Omega}$ *of* Ω.

In $\overline{\Omega}$, the function $c[\gamma]$ and the polynomial

$$P[\gamma] = \sum_{|\alpha|<m} v_\alpha (hH\gamma)^\alpha$$

are unknown. The function $w[\gamma]$ in $\overline{\Omega}$ is expressed through the polynomial $P[\gamma]$ and the function $f[\gamma]$ by means of the last of the relations (6.3).

Problem B_1 admits a continuous analog whose inspection yields a key to solving the system (6.3).

Replace $c[\gamma]$ and $w[\gamma]$ with some functions of a continuous argument $c(x)$ and $w(x)$, and instead of $G[\beta]$ and $f[\beta]$ use $G(x)$ and $f(x)$. Then arrive at the next

PROBLEM A_1. *In a domain* Ω, *find a generalized function* $c(x)$ *and a polynomial* $P(x) = \sum_{|\alpha|<m} v_\alpha x^\alpha$; *in the complement* Ω^* *of* Ω, *find a function* $w(x)$ *satisfying the system of equalities*

$$\begin{aligned} c(x) * G(x) &= w(x), \quad x \in \mathbb{R}^n, \\ c(x) &= 0, \quad x \notin \overline{\Omega}, \\ \int c(x) x^\alpha \, dx &= f_\alpha, \quad |\alpha| < m, \\ w(x) &= f(x) + P(x), \quad x \in \Omega. \end{aligned} \tag{6.4}$$

A solution of Problem A_1 is not unique, nor is a solution to Problem B_1. To obviate this, we impose additional conditions on the functions. These conditions allow us to consider Problem A_1 in a sense as a limit case of Problem B_1. In particular, we may make conclusions about asymptotic behaviour of the weights of optimal formula as $h \to 0$.

We require that the derivatives of order m of $w(x)$ be square integrable over every bounded domain Ω. The second of the equalities (6.4) gives some constraints on the behavior of $w(x)$ in a neighborhood about the point at infinity.

Since $w(x)$ is a polyharmonic function in the exterior of Ω, it follows that $w(x)$ grows at infinity not faster than $|x|^{2m-n}\log|x|$. The second of the equalities (6.4) now yields an explicit expression for a few first terms of the expansion of $w(x)$ in a series in the decreasing powers of $|x|$.

Indeed, $w(x)$ in the exterior of Ω may be written as convolution

$$w(x) = \int G(x-y)c(y)\,dy.$$

Expanding $G(x-y)$ in a power series in y, find that

$$w(x) = \sum_{|\alpha|<m} \int D^\alpha G(x)\frac{(-y)^\alpha}{\alpha!}c(y)\,dy + w_1(x). \tag{6.5}$$

The Taylor series for $G(x-y)$ converges when $y \in \Omega$ and x is sufficiently large. With the aid of (6.4), we may rewrite (6.5) as

$$w(x) = \sum_{|\alpha|<m} (-1)^{|\alpha|} f_\alpha \frac{D^\alpha G(x)}{\alpha!} + w_1(x). \tag{6.6}$$

For n odd or $n > m$, the function $w_1(x)$ in a neighborhood about the point at infinity does not exceed $K|x|^{m-n}$; i.e., for all $x \in \mathbb{R}^n$, the inequality holds

$$|w_1(x)| \leq K(1+|x|^2)^{(m-n)/2}. \tag{6.7}$$

If n is even and $n \leq m$ then the factor $\log|x|$ appears in the right side of (6.7).

Clearly, $w_1(x)$ has the derivatives of order m square integrable in a neighborhood about the point at infinity.

Therefore, the function $w(x)$ is a solution to the next *extension problem*.

PROBLEM $\mathrm{T_c}$. *Find a function $w(x)$ satisfying the conditions:*

(1) *in the exterior of Ω, the function $w(x)$ is polyharmonic,*

$$\Delta^m w(x) = 0, \quad x \notin \overline{\Omega}; \tag{6.8}$$

(2) *in the domain Ω, the equality holds*

$$w(x) = f(x) + \sum_{|\alpha|<m} v_\alpha x^\alpha,$$

where $f(x)$ is a known function;

(3) *in the exterior of* Ω, *the function* $w(x)$ *may be written as the sum* (6.6) *with the summand* $w_1(x)$ *satisfying* (6.7).

According to the routine Embedding Theorem, all derivatives of $w(x)$ up to order $m-1$ remain continuous in crossing the surface $\partial\Omega$. This gives rise to the boundary conditions that must be fulfilled on $\partial\Omega$ by the function $w_1(x)$ polyharmonic in the exterior of $\overline{\Omega}$. It is easy to check that these conditions take the form

$$D^\beta w_1|_{\partial\Omega} = -D^\beta\left\{\sum_{|\alpha|<m}(-1)^{|\alpha|}f_\alpha\frac{D^\alpha G(x)}{\alpha!}\right\}\Bigg|_{\partial\Omega}$$

$$+D^\beta\left\{f(x)+\sum_{|\alpha|<m}v_\alpha x^\alpha\right\}\Bigg|_{\partial\Omega}, \quad |\beta|<m. \tag{6.9}$$

The theory of the polyharmonic equation establishes that a function $w_1(x)$, satisfying (6.8) in the exterior of $\overline{\Omega}$ and (6.9) on the boundary $\partial\Omega$, exists only at a special choice of v_α in which case $w_1(x)$ is unique. The coefficients v_α are determined uniquely. Consequently, Problem T_c is solvable.

We find the function $c(x)$ by applying the operator Δ^m to $w(x)$. The function $c(x)$ is as a rule a generalized function even if $f(x)$ has continuous derivatives of arbitrary order and vanishes on the boundary of the domain Ω.

Indeed, the derivatives of $w(x)$ of order at least m jump in crossing $\partial\Omega$. Therefore, the polyharmonic operator Δ^m acts on $w(x)$ as follows

$$\Delta^m w(x) = \Delta^m f(x) + \sum_{|\alpha|<m} a_\alpha(x) D^\alpha \delta_{\partial\Omega}(x), \tag{6.10}$$

where $\delta_{\partial\Omega}(x)$ is the Dirac function supported by the boundary of Ω, and $\Delta^m f(x)$ is a continuous function in Ω.

Formula (6.10) prompts us the reason for appearance of a boundary layer in the problem of optimal cubature formulas, thus providing a natural explanation for asymptotic optimality of formulas with regular boundary layer. Indeed, the first summand in (6.10) is a smooth function and in our particular case it is close to a constant function. The summands containing the derivatives of $\delta_{\partial\Omega}(x)$ are approximated by finite differences, i.e., oscillating quantities.

We now turn to considering problems with functions of a discrete argument.

The operator of convolution with $G[\beta]$ admits an inverse which is also a convolution but with the function $\sigma[\beta]$ computable by using the Fourier transform. The function $\sigma[\beta]$ depends not only on the discrete argument β but also on the matrix H and the mesh-size h. We list the main properties of $\sigma[\beta]$.

1. As $|\beta|$ tends to infinity, the function $\sigma[\beta]$ decreases exponentially

$$|\sigma[\beta]| \le K e^{-\eta|\beta|}.$$

Here the constants K and η are positive and independent of β.

2. The convolution of $\sigma[\beta]$ with $G[\beta]$ exists and coincides after multiplication by h^{-2m} with the function $\delta[\beta]$, a *discrete analog of the Dirac delta function*,

$$h^{-2m}\sigma[\beta] * G[\beta] = \delta[\beta].$$

3. The convolution of $h^{-2m}\sigma[\beta]$ exists with an arbitrary function $\varphi(hH\beta)$ growing at infinity not faster than every polynomial (i.e. with a *tempered function*), and the limit of this convolution as $h \to 0$ coincides with the value of the polyharmonic operator at $\varphi(x)$, namely,

$$\lim_{h\to 0} h^{-2m}\sigma[\beta] * \varphi(hH\beta) = \Delta^m \varphi(hH\beta).$$

4. The function $\sigma[\beta]$ admits an *expansion of "divergence" type*, namely,

$$\sigma[\beta] = \sum_{|\alpha|=m} L^{(\alpha)}[\beta] * L^{(\alpha)}[-\beta],$$

where the functions $L^{(\alpha)}[\beta]$ decrease exponentially at infinity. The limit of the convolution $L^{(\alpha)}[\beta]$ with $\varphi(hH\beta)$ as $h \to 0$ equals the partial derivative $D^\alpha\varphi(x)$, namely,

$$\lim_{h\to 0} h^{-m} L^{(\alpha)}[\beta] * \varphi(hH\beta) = D^\alpha \varphi(hH\beta).$$

5. The next formula is valid which is analogous to *Green's identity for the polyharmonic operator*

$$\sum_{\beta} \varphi[\beta](\sigma[\beta] * \varphi[\beta]) = \sum_{\beta} \sum_{|\alpha|=m} (L^{(\alpha)}[\beta] * \varphi[\beta], L^{(\alpha)}[\beta] * \varphi[\beta]).$$

The properties of $\sigma[\beta]$ show that the operator of convolution with this function is a discrete analog of the taking of the polyharmonic operator. We have thus indicated a new method for introducing a difference approximation to the polyharmonic operator which is a contrast with those used in theoretic research. It is customary to take as approximation to the polyharmonic operator a difference scheme with finitely many nodes. In other words, there is constructed a compactly-supported function $\sigma^*[\beta]$ of a discrete argument such that the convolution with it transforms into Δ^m as $h \to 0$. Our approach differs in the fact that we begin with constructing not a difference approximation to Δ^m but rather a discretization of the corresponding potential $G(x)$. In this event the convolution of continuous functions with $G(x)$ is replaced by the convolution of discrete functions with $G[\beta]$. Only after that we do invert the discrete potential. This approach leads to not-compactly-supported approximation Δ^m.

We now state the problem that is a discrete analog of the extension Problem $\mathrm{T_c}$.

PROBLEM $\mathrm{T_d}$. *Find a function $w[\gamma]$ satisfying the three conditions:*

(1) *for $\gamma \in B$ the function $w[\gamma]$ assumes values that are given to within a polynomial summand, i.e.,*

$$w[\gamma] = \int_\Omega G(hH\gamma - y)\,dy + \sum_{|\alpha|<m} v_\alpha x^\alpha;$$

(2) *in the complement of B, the function $w[\gamma]$ satisfies the convolution identity*

$$\sigma[\gamma] * w[\gamma] = 0, \quad \gamma \notin B;$$

(3) *at infinity we have the inequality valid*

$$\left| w[\gamma] - \int_\Omega G(hH\gamma - y)\,dy \right| \leq K|hH\gamma|^{m-n}.$$

To Problem $\mathrm{T_d}$ we may translate all research methods that we use in the continuous case. In particular, we may prove existence and uniqueness for a solution to Problem $\mathrm{T_d}$.

Thus, the problem of determining optimal weights of cubature formulas on a fixed lattice of nodes reduces to a discrete analog of one of the classical boundary value problems of the theory of the polyharmonic equation. Inspecting the problem of optimal weights on the basis of the above-suggested approach prompts us their possible asymptotic behavior. Emphasize that we speak of asymptotic behavior as $h \to 0$ at a fixed m. Our hypothesis is as follows: In the interior points of Ω the weights $c_0[\gamma]$ are close to a constant, differing from the latter by a quantity with exponential decay in the domain as the discrete variable γ becomes more distant from the boundary of Ω. Moreover, the exponent of the exponential depends only on h. Therefore, as h decreases, the width of the boundary layer in an $L_2^{(m)}$-optimal cubature formula decreases in all probability in proportion to h. In the case when m and h change agreeably, for instance, when $mh = q < 1$, the hypothesis we stated must be substantially improved.

Chapter 2

Cubature Formulas of Finite Order

In the theory of computations we traditionally consider two problems, interpolation and approximate integration of functions. We have seen that these problems are tightly woven with one another if addressed in a purely algebraic manner. In this case the problem of approximate integration reduces to construction of cubature formulas exact for polynomials of degree m. The cubature formulas arising on this way are often called *formulas of interpolatory type*. We consider them in § 1.

If the integration domain admits a sufficiently simple finite group of transformations to itself, we may pose a problem of seeking for cubature formulas invariant under this transformation group. Calculation of the nodes and weights of these formulas may be very economical which allows us to attain high accuracy at a given cost. Below in § 2 and § 3 we study exact formulas that are invariant under finite groups of space rotations.

§1. Formulas of Interpolatory Type

Assume that $\varphi(x)$ is a function continuous in some domain $\widetilde{\Omega}$ in $\mathbb{R}^n$ and integrable over a bounded subdomain Ω with $\overline{\Omega} \subset \widetilde{\Omega}$. We approximately replace the integral over Ω of $\varphi(x)$ by a linear combination of the values of $\varphi(x)$ at the points of the set $K = \{x^{(1)}, x^{(2)}, \dots, x^{(N)}\} \subset \widetilde{\Omega}$. We do not exclude the case in which some of the nodes of K lie outside Ω.

A *cubature formula* for approximate calculation of the integral

$$I(\varphi) = \int\limits_{\Omega} \varphi(x)\, dx \tag{1.1}$$

is the sum

$$I_N(\varphi) = \sum_{k=1}^{N} c_k \varphi(x^{(k)}), \;\; x^{(k)} \in K, \tag{1.2}$$

that provides the approximate equality

$$I(\varphi) \cong I_N(\varphi).$$

The points of K are the *nodes* of the cubature formula, and the numbers $c_1, \ldots, c_N$ are its *weights*. The *error* of this cubature formula is given by the difference

$$(l_N, \varphi) = I(\varphi) - I_N(\varphi). \tag{1.3}$$

We say that a *cubature formula is exact for* $\varphi(x)$ provided that (1.3) equals 0. Thus, the functions for which the cubature formula (1.2) is exact span the *kernel* of the error l_N.

It seems natural to consider sequences of cubature formulas indexed with the number of nodes N rather than individual formulas. In this case we speak of a *cubature process*. We say that a cubature process with somehow defined nodes and weights *converges*, if for every function φ continuous in $\widetilde{\Omega}$ and integrable over Ω the quantity $I_N(\varphi)$ converges to $I(\varphi)$ as $N \to \infty$.

In an algebraic setting, various approaches are used for determining the nodes and weights of cubature formulas. We describe those that are in most frequent use.

PROBLEM 1. Given a system of nodes K find the weights $c_1, c_2, \ldots, c_N$ of a cubature formula so that it be exact for all polynomials of degree m (the space they span we denote by $\mathbf{P}_m$) with m as large as possible.

PROBLEM 2. Find the weights $c_1, c_2, \ldots, c_N$ and nodes $x^{(1)}, x^{(2)}, \ldots x^{(N)} \in K$ of a cubature formula so that it be exact for all polynomials in $\mathbf{P}_m$ with m as large as possible.

PROBLEM 3. Given the weights $c_k = |\Omega|/N$, $k = 1, \ldots, N$, find the nodes $x^{(1)}, x^{(2)}, \ldots, x^{(N)} \in K$ of a cubature formula so that it be exact for all polynomials in $\mathbf{P}_m$ with m as large as possible.

PROBLEM 4. Given the nodes $x^{(i)} \in K$, $i = 1, 2, \ldots, N_1$, with $1 \leq N_1 \leq N$, find the weights $c_1, c_2, \ldots, c_N$ and the nodes $x^{(N_1+1)}, \ldots, x^{(N)}$ of a cubature formula so that it be exact for all polynomials in $\mathbf{P}_m$ with m as large as possible.

By analogy with the one-dimensional case the cubature formulas, if they exist and solve Problems 1–4, are called the *cubature formulas of Newton–Cotes, Gauss, Chebyshev or Markov type*, respectively.

Stating Problems 1–4, instead of $\mathbf{P}_m$ we may take another finite-dimensional function space, for instance, the space of trigonometric polynomials of a given degree m. In this event the dimension of the space is also uniquely determined from m.

Cubature formulas for approximate calculation of the integral (1.1) are not exhausted by the expressions like (1.2). Also under consideration are the *Hermitian cubature formulas* in which to the sum (1.2) some summands are added that contain the nodal values of the derivatives of $\varphi(x)$.

An integer m is called the *order* of a cubature formula (1.2) if, first, the corresponding error is exact for the polynomials of degree less than m, and, second, there is a polynomial of degree m at which the error (1.3) is other than 0.

In the case when we consider a cubature process, the order m depends on the number of nodes N, i.e., $m = \psi(N)$. As a rule, the function $m = \psi(N)$ increases and tends to infinity as $N \to \infty$.

We now state a criterion for convergence of the errors of cubature formulas on continuous functions.

Theorem 2.1. *A cubature process (1.3) converges for every function φ in $C(\Omega)$ if and only if*

(1) there is a constant $L > 0$ such that

$$\sum_{k=1}^{N} |c_k| \leq L \tag{1.4}$$

uniformly in N,

(2) the sequence $I_N(\varphi)$ converges to $I(\varphi)$ for all φ in some everywhere dense subset of $C(\Omega)$.

The proof is simple, basing on appeal to the classical *Banach–Steinhaus Theorem* [265].

In the case when to the cubature process under study there corresponds the function $m = \psi(N)$ assigning to a given N the corresponding order, as a subset dense in $C(\Omega)$ we may take the space of polynomials. Inequality (1.4) amounts now to estimation of the norms of the sequence of errors l_N in $C(\Omega)^*$.

Theorem 2.2. *For the system of nodes $x^{(k)}$, $k = 1, \dots, N$, to admit of a cubature formula (1.2) exact for all polynomials P in $\mathbf{P}_m$, it is necessary and sufficient that the integrals be 0 of all those polynomials P that vanish at every point $x^{(k)}$.*

PROOF. The necessary and sufficient condition for the system (1.1.7) to be solvable is the orthogonality of b to all solutions of the homogeneous adjoint system

$$aS = 0. \tag{1.5}$$

The set of such solutions a defines the subspace of polynomials P of degree m. By (1.5) each of these polynomials vanishes at all nodes $x^{(k)}$. The orthogonality condition for the vectors a and b in $\mathbb{R}^M$ means the vanishing of the integrals $I(P)$. The proof of Theorem 2.2 is complete.

The rank $r(S)$ of the matrix S may equal M or be less than M. In the first case of $r(S) = M$, the problem of determining the weights c_k of a cubature formula is always solvable. This follows from Theorem 2.2, since the sole solution of (1.5) is zero in the case under consideration. If $r(S) = M$ then the size of K is at least M, i.e., $N \geq M$.

In the second case of $r(S) < M$, it is natural to construct cubature formulas on assuming that $r(S) = N$. Then the solvability condition for (1.1.7) is the equality

$$r(S \times b) = r(S) = N, \tag{1.6}$$

with $S \times b$ the ***augmented matrix***. As is known, the condition (1.6) amounts to the vanishing of $M - N$ determinants of the matrix $S \times b$ and so yields $M - N$ conditions on $b_1, b_2, \dots, b_M$. The reals b_j are the ***system of moments*** of the indicator of the domain Ω. If $n = 1$ and the domain Ω is an interval, then the reals b_j are readily written down as functions of the endpoints of the interval. In this case the set of all b is a 2-dimensional manifold in the M-dimensional space; some of the vectors b may satisfy (1.6).

EXAMPLE 1. Assume that $n = 1$, $\Omega = \{x : -1 \leq x \leq 1\}$, $m = 3$, and $x^{(1)} = -1$, $x^{(2)} = 0$, $x^{(3)} = 1$. The system of equations (1.1.7) takes the form

$$\begin{aligned} c_1 + c_2 + c_3 &= 2, \\ -c_1 \qquad + c_3 &= 0, \\ c_1 \qquad + c_3 &= 2/3, \\ -c_1 \qquad + c_3 &= 0. \end{aligned} \tag{1.7}$$

The rank of S equals 3, the number of nodes of the set K, and is less than $M = 4$. Nevertheless, the system (1.7) is solvable. Moreover, $c_1 = 1/3$, $c_2 = 4/3$, and $c_3 = 1/3$. We thus obtain the customary ***Simpson formula***

$$\int_{-1}^{1} \varphi(x)\,dx \cong \frac{1}{3}\varphi(-1) + \frac{4}{3}\varphi(0) + \frac{1}{3}\varphi(1),$$

which is exact for polynomials of degree 3.

In the case when the dimension of the space is greater than 1, to find the vector b becomes more difficult. This is a problem of many-dimensional moment theory. Nevertheless, for a given domain Ω, especially when Ω possesses some symmetry, it is often possible to find the set of nodes of size $N < M$ such that the system (1.1.7) is automatically solvable. In many dimensions, this problem was addressed by various authors, among which we mention I. Radon [168], [103, Chapter 22, § 4] and I. P. Mysovskikh [133].

Stated as one of its instances, Problems 2–4, the problem of determining the weights and nodes of a cubature formula as solutions to (1.1.7) is harder, since it becomes nonlinear. For instance, Problem 2 has $N(n+1)$ undetermined parameters of a cubature formula (1.2). This size of the set of undetermined parameters enables us to increase essentially the degree of a cubature formula with a given number of nodes.

Systematic study of interpolatory type formulas was accomplished in the articles by I. P. Mysovskikh. The reader may get acquaintance with his results and the relevant references in the monograph [133].

§2. Rotation Invariant Cubature Formulas

Let G be a *rotation group* in $\mathbb{R}^n$ comprising the elements $g_1, g_2, \dots, g_M$, with M the order of G. For every $x^{(k)} \in \mathbb{R}^n$ all points of the shape $g_i x^{(k)}$ (some of them may coincide) make an *G-orbit.*

A cubature formula (1.2) is *invariant under* G or, simply, *G-invariant* if the integration domain Ω is invariant under G and the set of nodes $x^{(k)}$ is the union of G-orbits and to the nodes $x^{(k)}$ of the same orbit there are assigned equal weights c_k.

Let L be an N-dimensional linear subspace in $C(\Omega)$ spanned by linear combinations of continuous functions $\varphi_i(x)$. Assume that $\varphi_i(x)$, $i = 1, 2, \dots, N$, are linearly independent. We have the following

Theorem 2.3. *Let G be a rotation group and let L be some finite-dimensional space invariant under G. A G-invariant cubature formula (1.2) is exact for all functions in L if and only if its error l_N vanishes on the subspace comprising G-invariant functions in L.*

PROOF. Take an arbitrary function $\varphi \in C(\Omega)$ and put

$$\varphi_G(x) = \frac{1}{M} \sum_{j=1}^{M} \varphi(g_j x). \tag{2.1}$$

The set comprising $g_1 x, g_2 x, \dots, g_M x$ goes into itself under multiplication from the left by an arbitrary element g_a of G. Consequently, we have the equality

$$\sum_{j=1}^{M} \varphi(g_j x) = \sum_{j=1}^{M} \varphi(g_a g_j x).$$

Therefore, the *mean function* $\varphi_G(x)$ is G-invariant, i.e.,

$$\varphi_G(gx) = \varphi_G(x), \quad g \in G.$$

Let a cubature formula (1.2) be invariant under G. Then we have the relations

$$I(\varphi) = I(\varphi_G), \quad I_N(\varphi) = I_N(\varphi_G).$$

Hence, the error of the formula at a function φ coincides with that error at the corresponding mean function φ_G, i.e.,

$$(l_N, \varphi) = (l_N, \varphi_G). \tag{2.2}$$

If the error l_N is exact for all G-invariant functions then it also vanishes at φ_G. This, together with (2.2), entails our claim. The proof of Theorem 2.3 is complete.

Corollary 2.1. *If the mean of a continuous function φ, taken over a subgroup G^* of G, equals zero then an arbitrary G-invariant error l_N is exact for φ.*

Indeed, the error l_N, being invariant under G, remains the same under G^*. This means that for φ_{G^*} we have an equality analogous to (2.2), which completes the proof of Corollary 2.1.

It is convenient to view a rotation group as a group of transformations of the *unit sphere* S_n of $\mathbb{R}^n$ into itself. As is well known (see, for instance, [59]), there is an infinite series of *cyclic groups* C_α of orders α which are composed of rotations of S_n. For $n \geq 3$ there are finite rotation groups of regular polyhedra, the *finite polyhedral groups.* We denoted the group that corresponds to an n-dimensional N-hedron by G_n^N.

For $n = 3$ we have the following nonequivalent finite rotation groups of the sphere S_3 which keep invariant some regular polyhedron: the *tetrahedral group* G_3^4, the *octahedral group* G_3^8 and the *icosahedral group* G_3^{20} [59, Chapter III, § 20]. The *cubical group* G_3^6 is equivalent to G_3^8, and the *dodecahedral group* G_3^{12}, is equivalent to G_3^{20}.

For $n = 4$ we have the tetrahedral group G_4^5, the octahedral group G_4^{16} and the cubical group G_4^8, with $G_4^8 = G_4^{16}$. Moreover, we also consider the groups G_4^{24}, G_4^{120}, and G_4^{600}, with $G_4^{120} = G_4^{600}$.

For $n \geq 5$ there are only three groups: the tetrahedral group G_n^{n+1}, the cubical group G_n^{2n}, and the octahedral group $G_n^{2^n}$, with $G_n^{2n} = G_n^{2^n}$.

Denote by $\mathbf{P}_{m_G}$ the space of G-invariant polynomials of degree at most m.

If in Problems 1–4 we additionally require that a formula be invariant under a rotation group G, then the problem of determining the nodes and weights of the formula becomes much simpler. Two reasons are behind this. First, the number of independent parameters of an invariant cubature formula with the number of nodes fixed in advance is less than in the general case. Second, the number of linearly independent polynomials for which our cubature formula must be exact equals M_G, the dimension of $\mathbf{P}_{m_G}$, i.e., it is less than M.

Let us find the dimension M_G of the space $\mathbf{P}_{m_G}$ for several rotation groups G. We find it convenient to use the notion of a solid spherical harmonic in a variable x in $\mathbb{R}^n$. Recall that the homogeneous harmonic polynomial $Z_s(x)$ of degree s is also termed the *solid spherical harmonic* of degree s. In this event, the restriction of

$Z_s(x)$ to the unit sphere is the ***spherical harmonic*** of degree s. To learn more about the properties of spherical harmonics, the reader could consult the book [5].

We have the following

Theorem 2.4. *Every homogeneous polynomial ψ_k of degree k is uniquely expressed as the sum*

$$\psi_k(x) = \sum_{s=0}^{[k/2]} \left(\frac{|x|}{2}\right)^{2s} Z_{k-2s}(x), \tag{2.3}$$

with $Z_{k-2s}(x)$ the solid spherical harmonic of degree $k-2s$, $s = 0, 1, \ldots$.

(The brackets stand for the taking of the ***integral part*** of a real.)

We call (2.3) the ***Gaussian decomposition of a homogeneous polynomial.*** To find a proof of Theorem 2.4, the reader may for instance consult [250].

By (2.3) an arbitrary homogeneous polynomial of degree k may be written on the unit sphere as a linear combination of spherical harmonics of degree k. Knowing the number of linear independent G-invariant spherical harmonics of degree k and using (2.3), we may find the dimension M_G.

We fulfill this plan of actions in the case of 3-dimensional space and the ***cyclic group*** C_α of order α. This group is composed of powers of rotations of the sphere about a fixed axis by the angle $2\pi/\alpha$. In this event, two points of the sphere remain fixed, P_N and P_S. One of them, the point P_N, we treat as the ***north pole***; the other, P_S, the ***south pole***. Introduce the polar coordinates θ and φ on the sphere. Then every transformation g_a, a member of C_α, is determined from one of the equalities

$$g_a^s(\theta, \varphi) = \left(\theta, \varphi + \frac{2\pi s}{\alpha}\right), \quad s = 1, 2, \ldots, \alpha. \tag{2.4}$$

As is known, we may make a basis for the space of spherical harmonics of degree k from the functions

$$P_k^{(|m|)}(\cos\theta)e^{im\varphi}, \quad m = 0, \pm 1, \ldots, \pm k, \tag{2.5}$$

with $P_k^{(|m|)}(\cdot)$ an ***adjoint Legendre function.*** Under every transformation of (2.4), each of the functions (2.5) is multiplied by an exponential of the shape $e^{i2s\pi m/\alpha}$. Consequently, invariant under the transformations of (2.4) are those and only those of the spherical harmonics (2.5) for which α divides into m. This observation enables us to find the ***total number*** $S(k)$ ***of linearly independent*** C_α***-invariant spherical harmonics.*** We have

Theorem 2.5. *The number $S(k)$ of linearly independent spherical harmonics of degree k invariant under the cyclic group C_α of order α is expressed by the formula*

$$S(k) = 2\left[\frac{k}{\alpha}\right] + 1. \tag{2.6}$$

Before calculating $S(k)$ for the case of the rotation groups of regular polyhedra in the 3-dimensional space, we consider some simplest properties of these groups.

Every finite group G of rotations of a regular polyhedron, surely containing the identity, consists of rotations about axes passing through a *vertex of the polyhedron*, rotations about axes passing through the centers of its *faces*, and finally rotations about axes passing through the midpoints of its *edges*. Assume that the number of the vertices of the polyhedron equals t_1; the number of faces is t_2, and the number of edges is t_3, with the rotation angles given by the reals $2\pi k_1/q_1$, $2\pi k_2/q_2$, and $2\pi k_3/q_3$ respectively, where k_j ranges from 0 to $q_j - 1$. We have the following

Lemma 2.1. *The order of G may be found by multiplying the parameters t_j and q_j that correspond to one another*

$$M = t_1q_1 = t_2q_2 = t_3q_3. \tag{2.7}$$

PROOF. We may implement a rotation of G as follows: first, we translate a given vertex (face, or edge) to its terminal position; then, rotate the polyhedron about the axis passing through this vertex (the center of the face, or the midpoint of the edge). The rotation angle coincides with $2\pi k_j/q_j$, where k_j may assume one of the values from 0 to $q_j - 1$. Consequently, all possible rotations in G may be implemented in exactly t_jq_j different ways. The proof of Lemma 2.1 is complete.

In 3-dimensional space, we may check (2.7) by straightforward calculation. We list the values of the relevant parameters:

for the tetrahedral group G_3^4:

$$\begin{aligned} &t_1 = 4, \quad q_1 = 3, \\ &t_2 = 4, \quad q_2 = 3, \quad M = 12, \\ &t_3 = 6, \quad q_3 = 2; \end{aligned}$$

for the octahedral group G_3^8:

$$\begin{aligned} &t_1 = 6, \quad q_1 = 4, \\ &t_2 = 8, \quad q_2 = 3, \quad M = 24, \\ &t_3 = 12, \quad q_3 = 2; \end{aligned}$$

for the icosahedral group G_3^{20}:

$$\begin{aligned} t_1 &= 12, \quad q_1 = 5, \\ t_2 &= 20, \quad q_2 = 3, \quad M = 60. \\ t_3 &= 30, \quad q_3 = 2. \end{aligned}$$

The order of G may be also computed in another way.

Lemma 2.2. *The total number of the vertices, edges and faces of a regular polyhedron is greater than the order of its group G by 2, namely,*

$$t_1 + t_2 + t_3 = M + 2. \tag{2.8}$$

PROOF. The group G consists of the identity transformation and the rotations by nonzero angles about each of the axes passing through the vertices, centers of faces and midpoints of the edges of the polyhedron in question. To every pole there correspond $q_j - 1$ such rotations. The total number of the latter is thus given as $\sum(q_j - 1)/2$, with summation taken over all possible poles (to half is necessary: indeed, every rotation is counted twice for each axis has two poles). Therefore, the order of G is given by the formula

$$M = 1 + \frac{1}{2}\sum_{j=1}^{3} t_j(q_j - 1).$$

Whence and from (2.7) we derive (2.8). The proof of Lemma 2.2 is complete.

The number M always divides by $2q_j$. Indeed, the group G contains the subgroups of order $2q_j$ which are generated by the elements of the cyclic subgroup of G of order q_j and the rotation transposing the north and south poles of the rotation axis. Furthermore, the order of a subgroup of a group is always a divisor of the order of the group; i.e., M is divisible by $2q_j$.

We now turn to deriving explicit formulas for the function $S(k)$.

Theorem 2.6. *The number of linearly independent G-invariant spherical harmonics of degree k equals*

$$S(k) = \left[\frac{k}{q_1}\right] + \left[\frac{k}{q_2}\right] + \left[\frac{k}{q_3}\right] + 1 - k. \tag{2.9}$$

PROOF. To derive (2.9) we use the *representation theory of groups*. An elementary exposition of prerequisites is for instance in [77].

As the *representation space* of G we take the space R_{2k+1} of spherical harmonics of degree k. To each rotation g in G we put in correspondence the linear operator

T_g on R_{2k+1}, making its value at a spherical harmonic $Y(\theta, \varphi)$ of degree k equal to the harmonic $Y(g(\theta, \varphi))$. We thus obtain a representation of G of order $2k + 1$. Choosing a basis for R_{2k+1}, we find the respective matrix form of T_g. Denote this matrix by A_g.

As is well known, every finite-dimensional representation decomposes into *irreducible representations.* In our case, this means that R_{2k+1} splits into the direct sum of subspaces each of which is invariant under all linear operators T_g, $g \in G$, namely,

$$R_{2k+1} = R_1 \oplus R_2 \oplus \cdots \oplus R_I. \tag{2.10}$$

Moreover, it is impossible to find in any of the subspaces R_j, $1 \leq j \leq I$, a nontrivial subspace also invariant under all operators T_g, $g \in G$. Accordingly, the matrix A_g also splits into the direct sum

$$A_g = A_g^{(1)} \oplus A_g^{(2)} \oplus \cdots \oplus A_g^{(I)}. \tag{2.11}$$

Generally speaking, among R_j there are listed several one-dimensional subspaces. The corresponding matrix $A_g^{(j)}$ reduces then to a real, now 1. The number of such summands in (2.11) is exactly $S(k)$. Using this, we now derive the needed formula.

The *trace* of A_g, denoted by $\chi(A_g)$, may be written in view of (2.11) as the sum of the traces $\chi(A_g^{(s)})$, namely,

$$\chi(A_g) = \sum_{s=1}^{I} \chi(A_g^{(s)}). \tag{2.12}$$

Recall that $\chi(A_g^{(j)})$ is called the *character of the representation of G in R_j*, and the theorem is valid asserting the orthogonality of the characters corresponding to distinct representations. We state it more exactly.

Let

$$\chi(A_{g_1}^{(j)}), \chi(A_{g_2}^{(j)}), \ldots, \chi(A_{g_M}^{(j)}), \quad \chi(A_{g_1}^{(k)}), \chi(A_{g_2}^{(k)}), \ldots, \chi(A_{g_M}^{(k)})$$

be the sets of the characters of irreducible representations of G in R_j and R_k respectively. Two representations are called *equivalent*, if all matrices $A_{g_s}^{(j)}$ and $A_{g_s}^{(k)}$, $s = 1, 2, \ldots, M$, have the same order and are *similar* to one another

$$A_{g_s}^{(j)} = BA_{g_s}^{(k)}B^{-1}.$$

The following equalities hold

$$\sum_{g \in G} \chi\left(A_g^{(j)}\right) \overline{\chi}\left(A_g^{(k)}\right) = \begin{cases} M, & \text{if } \{A_g^{(j)}\} \text{ and } \{A_g^{(k)}\} \text{ are equivalent,} \\ 0, & \text{otherwise.} \end{cases} \tag{2.13}$$

Let the space R_j be one-dimensional implying that the character $\chi(A_g^{(j)})$ equals 1. Considering this and summing both sides of (2.12) over all $g \in G$, we come to the equality

$$\sum_{g\in G} \chi(A_g) = \sum_{s=1}^{I} \Big\{ \sum_{g\in G} \chi(A_g^{(s)}) \overline{\chi}(A_g^{(j)}) \Big\}. \tag{2.14}$$

By (2.13), the inner sum on the right side of (2.14) equals M at s such that R_s has dimension 1, while vanishing for all remaining s. The number of one-dimensional subspaces R in the sum (2.10) equals $S(k)$. Consequently, we have the formula

$$S(k) = \frac{1}{M} \sum_{g\in G} \chi(A_g). \tag{2.15}$$

This method for determining $S(k)$ and the formula (2.15) were suggested by D. K. Faddeev at the authors' request.

We now calculate the expression on the right side of (2.15). The idea behind the subsequent reasoning is that, given a specific rotation $g \in G$, we should choose a basis for the representation space R_{2k+1} so that the matrix A_g take the simplest form; then we find its trace explicitly.

Introduce cartesian coordinates (x, y, z) in $\mathbb{R}^3$, traversing the axis z along the rotation axis of g. To the chosen system we assign the spherical coordinates (r, θ, φ) and a basis (2.5) for the space R_{2k+1}. Calculate the matrix A_g in this basis. Assume that g is a rotation by the angle $2\pi s/q_j$, with $s = 0, 1, \dots, q_j - 1$; assume also that $Y_{k,m}(\theta, \varphi)$ is the spherical function (2.5). Then the harmonic $Y_{k,m}(g(\theta, \varphi))$ may be found by the formula

$$Y_{k,m}(g(\theta, \varphi)) = e^{i2\pi ms/q_j} Y_{k,m}(\theta, \varphi).$$

This meant that A_g is the diagonal matrix with entries $e^{i2\pi ms/q_j}$. Its trace is thus equal to the sum of these exponentials over all m from $-k$ to k, namely,

$$\chi(A_g) = \sum_{m=-k}^{k} e^{i2\pi ms/q_j} = \frac{\sin((2k+1)\pi s/q_j)}{\sin(\pi s/q_j)}. \tag{2.16}$$

Using (2.16), transform the right side of (2.15). Summation in the latter is carried out over the matrices A_g corresponding to all possible elements $g \in G$. Recall that G consists of t_1 rotations about axes passing through vertices of an invariant polyhedron, of t_2 rotations about axes passing through the centers of its edges, and, finally, of t_3 rotations passing through the midpoints of edges. About each axis, exactly q_j distinct rotations are possible by the angles $2\pi s/q_j$, with

$s = 0, 1, \ldots, q_j - 1$. We take into account all these rotations and sum the traces of the corresponding matrices A_g. In this event, to every element of the group but the identity there correspond two summands: the first appears in counting up rotations about one pole; the second, in counting up rotations about the other pole. To the identity of the group G in the sum we are counting up there correspond $t_1 + t_2 + t_3$ summands. The trace of the matrix A_0 assigned to the identity transformation equals $2k + 1$ by (2.16). Therefore, (2.15) may equivalently be written as

$$S(k) = \frac{1}{M}\left\{\frac{1}{2}\left[\sum_{j=1}^{3} t_j \left(\sum_{s=0}^{q_j-1} \sum_{m=-k}^{k} e^{i2\pi ms/q_j}\right) - (2k+1)(t_1+t_2+t_3)\right] + 2k+1\right\}. \quad (2.17)$$

The double sum in parentheses is easy to calculate, which yields the value

$$\left(2\left[\frac{k}{q_j}\right] + 1\right) q_j.$$

Inserting this expression in (2.17) and also recalling (2.7) and (2.8), we arrive at (2.9). The proof of Theorem 2.6 is complete.

We infer one more description for $S(k)$. Denote by Q^* the set of those q_j that do not divide k.

Theorem 2.7. *The number $S(k)$ is the integral part of*

$$\frac{1}{M}\left(2k - \sum_{q_j \in Q^*} t_j\right) + 1. \quad (2.18)$$

PROOF. Expressing in (2.9) the integral part $[k/q_j]$ through its *fractional part* $\{k/q_j\}$, find

$$S(k) = \sum_{j=1}^{3}\left(\frac{k}{q_j} - \left\{\frac{k}{q_j}\right\}\right) + 1 - k.$$

In virtue of (2.7) and (2.8) we have

$$\sum_{j=1}^{3} \frac{k}{q_j} = k\left(1 + \frac{2}{M}\right), \quad \sum_{j=1}^{3}\left\{\frac{k}{q_j}\right\} = \sum_{q_j \in Q^*}\left\{\frac{k}{q_j}\right\} = \sum_{q_j \in Q^*}\frac{1}{q_j} + \sum_{q_j \in Q^*}\frac{\eta_j}{q_j}.$$

Here $0 \le \eta_j \le q_j - 2$. Consequently,

$$S(k) = 1 + \frac{1}{M}\left(2k - \sum_{q_j \in Q^*} t_j\right) - \sum_{q_j \in Q^*}\frac{\eta_j}{q_j}.$$

Moreover,

$$0 \leq \sum_{q_j \in Q^*} \frac{\eta_j}{q_j} \leq \sum_{j=1}^{3} \frac{q_j - 2}{q_j} = 1 - \frac{4}{M} < 1.$$

Hence, $S(k)$ is an integer smaller than (2.18), with the difference between (2.18) and $S(k)$ not exceeding unity. Therefore, $S(k)$ is the integral part of (2.18). The proof of Theorem 2.7 is complete.

One more property of $S(k)$ is immediate from the description of $S(k)$ as the integral part of (2.18), namely,

$$S\left(k + \frac{M}{2}\right) = S(k) + 1.$$

Hence, we particularly derive the formula

$$S(k) = \begin{cases} p & \text{for } 2k+1 \leq \sum_{q_j \in Q^*} t_j, \\ p+1 & \text{otherwise.} \end{cases} \tag{2.19}$$

Here p stands for the integral part of the whole quotient $(2k+1)/M$.

With this we finish deriving various presentations of the number $S(k)$ and address the question of how to construct all spherical harmonics of given degree k invariant under a group G in 3-dimensional space.

To this end, we need a special system of harmonic polynomials. Alongside the original coordinates (x, y, z), we choose some cartesian coordinates (x_k, y_k, z_k) in $\mathbb{R}^3$. At every choice of the axes x_k, y_k, z_k and every natural n the functions

$$\zeta_k^n = (x_k + iy_k)^n \tag{2.20}$$

are homogeneous harmonic polynomials of degree n in the initial variables (x, y, z). Their restrictions to the unit sphere define a collection of spherical harmonics of degree n. We have the following

Theorem 2.8. *Any* $2n+1$ *polynomials of the shape* (2.20) *taken arbitrarily are linearly independent provided that the directions of their axes* $z_1, z_2, \ldots, z_{2n+1}$ *are distinct.*

PROOF. Use the trick of complex analysis of presenting the points of the unit sphere S_3 by *stereographic projection.* Issuing half-lines from the south pole of the sphere, map the latter in a one-to-one fashion onto the plane passing through the equator. Introduce the polar coordinates (θ, φ) on S_3. Then, given a point (x, y, z) of the sphere, we have the equalities

$$x = \cos\varphi \sin\theta, \quad y = \sin\varphi \sin\theta, \quad z = \cos\theta. \tag{2.21}$$

Assign to the vector (x, y, z) the complex number w

$$w = \tan\frac{\theta}{2}e^{i\varphi}. \tag{2.22}$$

When the point (x, y, z) ranges over the unit sphere, the number w determined from (2.22) ranges over the whole complex plane. This correspondence is one-to-one.

Every rotation of the sphere determines a *linear-fractional transformation* of the plane which has the form

$$w_1 = \frac{aw - \bar{c}}{cw + \bar{a}}, \tag{2.23}$$

with the numbers a and c meeting the normalization condition

$$a\bar{a} + c\bar{c} = 1. \tag{2.24}$$

We express the function $(x + iy)^n$ via w. From (2.21) and (2.22) it follows that

$$(x + iy)^n = \frac{2^n w^n}{(1 + w\bar{w})^n}. \tag{2.25}$$

Let the coordinate system (x_1, y_1, z_1) result from rotating the system (x, y, z) by some angle. To this rotation there corresponds some linear-fractional transformation of the complex plane which meets (2.23) and (2.24). Using this, we express the function $\zeta_1^n = (x_1 + iy_1)^n$ through w. From (2.25) and (2.23) we have

$$x_1 + iy_1 = \frac{2w_1}{1 + w_1\bar{w}_1} = 2a\bar{c}\frac{(w\bar{w} - 1) + w/w^{(0)} - \bar{w}w^{(0)}}{1 + w\bar{w}}, \tag{2.26}$$

with $w^{(0)}$ the ratio $\bar{c}/a$. Raising both sides of (2.26) to the nth power, find

$$\zeta_1^n = \frac{2^n}{(1 + w\bar{w})^n}(a\bar{c})^n \sum_{k+l+m=n} \frac{n!}{k!\, l!\, m!}(w\bar{w} - 1)^k w^l (-\bar{w})^m (w^{(0)})^{m-l}. \tag{2.27}$$

Arranging the summands on the right side of (2.27) in increasing order of the powers of $w^{(0)}$, after easy calculations come to the equality

$$\zeta_1^n = \frac{(2a\bar{c})^n}{(1 + w\bar{w})^n} \sum_{m=-n}^{n} (w^{(0)})^m \left(-\frac{\bar{w}}{w}\right)^{m/2} Q_n^{(|m|)}(w\bar{w}), \tag{2.28}$$

with the function $Q_n^{(s)}(w\bar{w})$ standing for the sum

$$\sum_{\substack{|s| \le j \le n \\ j \equiv s (\mathrm{mod}\, 2)}} \frac{n!}{(n - j)!\, ((s + j)/2)!\, ((j - s)/2)!}(w\bar{w} - 1)^{n-j}(-w\bar{w})^{j/2}.$$

Observe that, for all $s = -n, -n+1, \dots, n$, the function

$$R_n^{(s)}(w\overline{w}) = 2^n \left(-\frac{\overline{w}}{w}\right)^{s/2} \frac{Q_n^{(|s|)}(w\overline{w})}{(1+w\overline{w})^n}, \tag{2.29}$$

when expressed in the variables θ and φ, differs only by a constant factor from the spherical harmonics $e^{-is\varphi}P_n^{(|s|)}(\cos\theta)$. Thus, the system of functions (2.29) generates a basis for the space of all spherical harmonics of degree n.

Let a_k and c_k denote the parameters of the linear-fractional transformation (2.23) corresponding to the coordinate system (x_k, y_k, z_k). Then, using (2.29), we may expand the function ζ_k^n by (2.28) to obtain

$$\zeta_k^n = \sum_{m=-n}^{n} a_k^{n-m}\overline{c}_k^{n+m} R_n^{(|m|)}(w\overline{w}). \tag{2.30}$$

Linear independence of $\zeta_1^n, \dots, \zeta_{2n+1}^n$ takes place if and only if other than 0 is the determinant Δ of the matrix $(a_k^{2n-q}\overline{c}_k^q)$, $k = 1, 2, \dots, 2n+1$, $q = 0, 1, \dots, 2n$, corresponding to the expansion (2.30). We may readily reduce the determinant Δ to the *Vandermonde determinant*, factoring the constant multiplier a_k^{2n} out of every row. We thus readily infer that

$$\Delta = \prod_{i<j}(a_j\overline{c}_i - a_i\overline{c}_j). \tag{2.31}$$

The product in (2.31) is zero only if there are distinct integers k and l such that

$$\frac{a_k}{\overline{c}_k} = \frac{a_l}{\overline{c}_l}.$$

So were the product zero, the same linear-fractional transformation of the shape (2.23) would correspond to two distinct coordinate systems. This is impossible, implying that the determinant (2.31) is other than 0. The proof of Theorem 2.8 is complete.

Using harmonic polynomials (2.20), we construct a basis for the space of G-invariant spherical harmonics of degree n.

Denote the integral part of the ratio $(2n+1)/M$ by p; and the difference $(2n+1) - pM$, by q. Consider an arbitrary set of points $x^{(1)}, \dots, x^{(p)}$ on the unit sphere which are not equivalent under G. Arrange the union of the orbits of these points

$$gx^{(1)}, gx^{(2)}, \dots, gx^{(p)}, \quad g \in G. \tag{2.32}$$

Obviously, the set (2.32) consists of pM distinct vectors. Also enlist the q points

$$x^{(2n+1)}, x^{(2n)}, \dots, x^{(pM+1)}. \tag{2.33}$$

The choice of the latter depends on whether or not the number $2n+1$ is greater than the sum $\sum_{q_j \in Q^*} t_j$. Conditions on the vectors (2.33) are specified in more detail in the sequel.

Given each of the points $x^{(s)}$, $s = 1, 2, \dots, p$, find a rotation $h(s)$ such that the vector $h(s)x^{(s)}$ has the direction of the axis z. Consider the system of harmonic polynomials

$$\zeta(\mathbf{x}) = x + iy, \quad \mathbf{x} = (x, y, z), \quad \zeta^n_{g,s} = \zeta^n(h(s)g^{-1}\mathbf{x}), \quad g \in G. \tag{2.34}$$

The mean function of $\zeta^n_{g,s}(\mathbf{x})$ over G, defined by the formula

$$\frac{1}{M} \sum_{g_1 \in G} \zeta^n_{g,s}(g_1\mathbf{x}),$$

is independent of the element g. This follows from the chain of equalities

$$\frac{1}{M} \sum_{g_1 \in G} \zeta^n_{g,s}(g_1\mathbf{x}) = \frac{1}{M} \sum_{g_1 \in G} \zeta_{g_1^{-1}g,s}(\mathbf{x}) = \frac{1}{M} \sum_{g_3 \in G} \zeta^n_{g_3,s}(\mathbf{x}),$$

the first of which ensues from the definition (2.34); and the second, from the observation that the element $g_3 = g_1^{-1}g$, with g_1 ranging over the entire group G, also becomes each member of this group. Thus, we independently of $g \in G$ define the collection of spherical harmonics

$$Y_n^{(s)}(\theta, \varphi) = \frac{1}{M} \sum_{g_1 \in G} \zeta^n_{g,s}(g_1\mathbf{x}); \quad s = 1, 2, \dots, p, \tag{2.35}$$

each of which is G-invariant. These functions constitute a sought basis. (Here and in what follows we certainly assume that these and analogous equalities are valid only for the points in the unit sphere S_3.)

Lemma 2.3. *If $S(n) = p$ then every G-invariant spherical harmonic $Y_n^*(\theta, \varphi)$ of degree n may be written as a linear combination of the functions* (2.35).

PROOF. Take as (2.33) the vertices of the polyhedron in question, the centers of its faces and the midpoints of its edges such that $q_j \in Q^*$. This is possible since by hypothesis $S(n) = p$, i.e., as follows from (2.19),

$$2n + 1 \le \sum_{q_j \in Q^*} t_j.$$

To each of the points $x^{(s)}$ for $s = p+1, \ldots, 2n+1$, we put into correspondence the harmonic polynomial

$$\zeta_s^n(\mathbf{x}) = \zeta^n(h(s)\mathbf{x}), \quad s = p+1, \ldots, 2n+1. \tag{2.36}$$

Here the rotation $h(s)$ sends the vector $x^{(s)}$ to the generator of the z axis, and the function $\zeta(\mathbf{x})$ is defined by the relation (2.34).

To each point $x^{(s)}$ for $s \geq p+1$ there corresponds the cyclic subgroup C_q of G comprising the rotations about the axis with the endpoint $x^{(s)}$ by the angles $2\pi k/q$, with $k = 0, 1, \ldots, q-1$. The rotation of the system (x_s, y_s, z_s) about the z_s axis by the angle φ multiplies $\zeta_s^n(\mathbf{x})$ by $e^{in\varphi}$. Calculate the mean value of $\zeta_s^n(\mathbf{x})$ over the cyclic subgroup C_α to obtain

$$\sum_{g \in C_\alpha} \zeta_s^n(g\mathbf{x}) = \sum_{k=0}^{\alpha-1} e^{i2\pi kn/\alpha} \zeta_s^n(\mathbf{x}) = 0. \tag{2.37}$$

Choose the subset K of G which is composed of the elements $g^{(k)}, k = 1, \ldots, M/\alpha$, such that G splits into the disjoint cosets $g^{(k)}C_\alpha, k = 1, \ldots, M/\alpha$, of G by C_α. Then

$$\sum_{g \in G} \zeta_s^n(g\mathbf{x}) = \sum_{g_2 \in K} \sum_{g_1 \in C_\alpha} \zeta_s^n(g_1(g_2\mathbf{x})) = 0.$$

Therefore, the mean over the whole group G of every function of the shape (2.36) vanishes

$$\sum_{g \in G} \zeta_s^n(g\mathbf{x}) = 0. \tag{2.38}$$

To the set (2.36) of q homogeneous harmonic polynomials of degree n, we append the pM functions $\zeta_{g,s}^n(\mathbf{x})$ that are defined by (2.34). We thus obtain $p+q = 2n+1$ harmonic polynomials. Each of them is easily written as (2.20), with the axes z_k of the corresponding coordinates (x_k, y_k, z_k) distinct for all k ranging from 1 to $2n+1$. By Theorem 2.8 the initial harmonics $Y_n^*(\theta, \varphi)$ may be written as a linear combination of the polynomials under consideration

$$Y_n^*(\theta, \varphi) = \sum_{s=1}^{p} \sum_{g \in G} a_{g,s} \zeta_{g,s}^n(\mathbf{x}) + \sum_{k=p+1}^{2n+1} a_k \zeta_k^n(\mathbf{x}). \tag{2.39}$$

By hypothesis the harmonic $Y_n^*(\theta, \varphi)$ is invariant under G. Thus, the mean of Y_n^* over G is this harmonic itself. Considering this, from (2.39) we obtain the equality

$$Y_n^*(\theta, \varphi) = \frac{1}{M} \sum_{s=1}^{p} \sum_{g \in G} a_{g,s} \sum_{g_1 \in G} \zeta_{g,s}^n(g_1\mathbf{x}) + \frac{1}{M} \sum_{k=p+1}^{2n+1} a_k \sum_{g_1 \in G} \zeta_k^n(g_1 x) \tag{2.40}$$

or, involving (2.35) and (2.38), the equality

$$Y_n^*(\theta,\varphi) = \sum_{s=1}^{p}\Big(\sum_{g\in G} a_{g,s}\Big) Y_n^{(s)}(\theta,\varphi).$$

The proof of Lemma 2.3 is complete.

Consider the case in which $S(n) = p+1$. Let $x^{(pM+1)}$ be an arbitrary point of the unit sphere distinct from each of the vectors (2.32). Find a rotation $h(pM+1)$ sending $x^{(pM+1)}$ to the generator of the z axis, and put

$$\zeta_{p+1}^n(\mathbf{x}) = \zeta^n(h(pM+1)\mathbf{x}). \tag{2.41}$$

We denote the mean of this function over G by $Y_n^{(p+1)}(\theta,\varphi)$, i.e.,

$$Y_n^{(p+1)}(\theta,\varphi) = \frac{1}{M}\sum_{g\in G} \zeta_{p+1}^n(g\mathbf{x}). \tag{2.42}$$

Clearly, this is an invariant spherical harmonic of degree n.

Lemma 2.4. *If $S(n) = p+1$ then every G-invariant spherical harmonic $Y_n^*(\theta,\varphi)$ of degree n may be written as a linear combination of $Y_n^{(s)}(\theta,\varphi)$, with $s = 1,2,\dots,p,p+1$.*

PROOF. Take as (2.33) the system of equivalent points

$$x^{(pM+1)}, g_2x^{(pM+1)}, \dots, g_qx^{(pM+1)}. \tag{2.43}$$

Let the elements $g_2, g_3, \dots, g_q$ be chosen arbitrarily in G. Appending to them the identity of G, obtain the set K of q elements. Note that the rotation $h(g) = h(pM+1)g^{-1}$ sends the point $gx^{(pM+1)}$ to the vector generating the z axis. Define the harmonic polynomial $\zeta_g^n(\mathbf{x})$ of degree n as

$$\zeta_g^n(\mathbf{x}) = \zeta^n(h(pM+1)g^{-1}\mathbf{x}), \quad g\in K. \tag{2.44}$$

The mean function of $\zeta_g^n(\mathbf{x})$ over G, defined by the formula

$$\frac{1}{M}\sum_{g_*\in G} \zeta_g^n(g_*\mathbf{x}), \tag{2.45}$$

does not depend on g in K. This ensues from the chain of equalities

$$\sum_{g_*\in G} \zeta_g^n(g_*\mathbf{x}) = \sum_{g_*\in G} \zeta_{g_*^{-1}g}^n(\mathbf{x}) = \sum_{g\in G} \zeta_g^n(\mathbf{x}).$$

The first of them follows from (2.44); and the second, from the observation that the element $g_*^{-1}g$ becomes every element of G when g ranges over G. Therefore, the function (2.45) is indeed independent of $g \in K$.

If g coincides with g_1, the identity of G; then $\zeta_{g_1}^n(\mathbf{x}) = \zeta_{p+1}^n(\mathbf{x})$ and so, for every $g \in G$, we have the equality

$$\frac{1}{M}\sum_{g_* \in G} \zeta_g^n(g_*\mathbf{x}) = \frac{1}{M}\sum_{g_* \in G} \zeta_{p+1}^n(g_*\mathbf{x}) = Y_n^{(p+1)}(\theta,\varphi). \tag{2.46}$$

The G-invariant spherical harmonic $Y_n^*(\theta,\varphi)$ may be written as a formula analogous to (2.39)

$$Y_n^*(\theta,\varphi) = \sum_{s=1}^{p}\sum_{g\in G} a_{g,s}\zeta_{g,s}^n(\mathbf{x}) + \sum_{g\in K} a_g \zeta_g^n(\mathbf{x}).$$

Averaging over G both sides of these equalities, obtain

$$Y_n^*(\theta,\varphi) = \frac{1}{M}\sum_{s=1}^{p}\Big(\sum_{g\in G} a_{g,s}\Big)Y_n^{(s)}(\theta,\varphi) + \frac{1}{M}\sum_{g\in K} a_g \sum_{g_1\in G}\zeta_g^n(g_1\mathbf{x}).$$

Using (2.46), we see that $Y_n^*(\theta,\varphi)$ is actually representable as a linear combination of $Y_n^{(s)}(\theta,\varphi)$, $s = 1,\ldots,p+1$. The proof of Lemma 2.4 is complete.

We have just considered only the groups of proper rotations of a polyhedron excluding reflections. The transformation of *reflection* assigning to a point of the unit sphere with coordinates (θ,φ) the point (θ_1,φ_1), with $\theta_1 = \pi - \theta$ and $\varphi_1 = \pi + \varphi$. A group G with the appended reflection generates the full group G^* which contains twice as many elements as G.

Theorem 2.9. *The set of G^*-invariant spherical functions of even order n coincides with the set of G-invariant spherical functions. The group G^* has no invariant harmonics of odd order n.*

PROOF. The transformation of reflection amounts to the change of variables $x' = -x$, $y' = -y$, and $z' = -z$. A homogeneous polynomial of even degree k remains unchanged under this transformation, whereas a homogeneous polynomial of odd degree changes sign. Therefore, a function of the basis (2.5), for n even, is invariant under reflection and, for n odd, it changes sign. The mean over G^* of a harmonic of even degree, i.e., an invariant harmonic of even degree, coincides with the mean function over G. For every harmonic of odd degree, this mean is zero. The proof of Theorem 2.9 is complete.

G. N. Salikhov conducted analogous research in the spaces of dimension greater than 3 [200]. He obtained formulas for linear representations of groups, for rotation angles and for the dimension $S(k)$ in the case of a regular polyhedron in 4-dimensional and 5-dimensional spaces. For example, for a 600-hedron in 4-dimensional space and k even, the function $S(k)$ may be calculated by the formula

$$S(k) = \frac{1}{240}\left\{120\tau_1(k) + 24\tau_2(k) + 24\tau_3(k) + 40\tau_4(k) + \tau_5(k)\right\},$$

with $\tau_j(k)$ given by the equalities

$$\begin{aligned}
\tau_1(k) &= \left(2\left[\frac{k}{10}\right]+1\right)\left(2\left[\frac{k}{6}\right]+1\right) + \left(2\left[\frac{k}{10}\right]+1\right)\left(2\left[\frac{k}{4}\right]+1\right) \\
&\quad + \left(2\left[\frac{k}{6}\right]+1\right)\left(2\left[\frac{k}{4}\right]+1\right), \\
\tau_2(k) &= 4\left(\sum_{r=0}^{4}\left(\left[\frac{k-2r}{10}\right]+1\right)^2 + \left[\frac{k-8}{10}\right] - \left[\frac{k-2}{10}\right] - \left[\frac{k}{10}\right]\right) - 3, \\
\tau_3(k) &= \left(2(k-4) - 5\left[\frac{k-4}{5}\right]\right)\left(1 + \left[\frac{k-4}{5}\right]\right) + k + 1, \\
\tau_4(k) &= \left(2(k-2) - 3\left[\frac{k-2}{3}\right]\right)\left(1 + \left[\frac{k-2}{3}\right]\right) + k + 1, \\
\tau_5(k) &= 13(k+1)^2 - 4(k+1)\left(59 + 48\left[\frac{k}{10}\right] + 40\left[\frac{k}{6}\right] + 30\left[\frac{k}{4}\right]\right) + 15.
\end{aligned}$$

If k is odd then $S(k)$ equals zero.

The problem of constructing cubature formulas invariant under rotation groups was also considered by V. I. Lebedev [105–112], I. P. Mysovskikh [132–134] and S. I. Konyaev [96–99]. V. I. Lebedev suggested to use the system of G-invariant generators of the algebra of polynomials on inventing several original tricks. First, as a new system of indeterminates he suggested to take the set of values of the generators of the algebra at the nodes of a cubature formula. Second, he specifically selected those among all invariant polynomials which vanish on the axes and symmetry planes of G. Transition from the values of such "symmetric" polynomials at the nodes of a lattice to the values of the *elementary symmetric functions* allowed him to simplify the nonlinear system (1.1.7) significantly and to diminish its size substantially. He also suggested a certain method for calculating the dimension of the space of G-invariant polynomials.

EXAMPLE 2. In the space $\mathbb{R}^n$ consider the *octahedral group with reflection* $(G_n^{2^n})^*$. The generators of the algebra of polynomials invariant under this group are the following elementary symmetric function

$$\sigma_1 = \sum_{i=1}^{n} x_i^2, \quad \sigma_2 = \sum_{i<j} x_i^2 x_j^2, \ \ldots, \quad \sigma_n^2 = x_1^2 x_2^2 \ldots x_n^2.$$

Each polynomial $P(x)$ invariant under $(G_n^{2^n})^*$ is uniquely representable as a polynomial in $\sigma_1, \sigma_2, \ldots, \sigma_n$, namely,

$$P(x) = Q(\sigma_1(x), \sigma_2(x), \ldots, \sigma_n(x)).$$

In this event, the dimension of the space of invariant polynomials of degree k coincides with the number of integer solutions to the following equation

$$2\alpha_1 + 4\alpha_2 + \cdots + 2n\alpha_n = k.$$

Each polynomial $P(x) = Q(\sigma(x))$ on the unit sphere of $\mathbb{R}^n$ is obviously a function of $\sigma_2, \sigma_3, \ldots, \sigma_n$. Therefore, the dimension of the space of invariant spherical harmonics of degree k coincides with the number of solutions to the equation

$$4\alpha_2 + \cdots + 2n\alpha_n = k$$

in nonnegative integers. In this case the system of equations (1.1.7) is equivalent to the next system

$$I_N(Q(\sigma)) = I(Q(\sigma)) \tag{2.47}$$

for the unknowns c_k and $\sigma_k = (\sigma_1(x^{(k)}), \ldots, \sigma_n(x^{(k)}))$, $k = 1, 2, \ldots, N$. A finite set to which a polynomial $Q(\sigma(x))$ may belong should be chosen so that the corresponding set of polynomials $P(x) = Q(\sigma(x))$ be a basis for the space $\mathbf{P}_{m_G}$. The number of equations in (2.47) is much far less than in (1.1.7).

EXAMPLE 3. In the space $\mathbb{R}^3$ consider the *icosahedral group with reflection* $(G_3^{20})^*$. The system of generators of the algebra of invariant polynomials is given by the functions

$$I_2 = \sum_{i=1}^{3} x_i^2;$$

$$I_6 = 5x_1^4x_3^2 + 5x_2^4x_3^2 + x_3^6 + 10x_1^2x_2^2x_3^2 - 5x_1^2x_3^4 - 5x_2^2x_3^4 + 2x_1^5x_3 + 10x_1x_2^4x_3 - 20x_1^3x_2^2x_3;$$

$$I_{10} = (4x_1^2 - 6x_1x_3 + x_3^2)(x_1^4 + 5x_2^4 + x_3^4 + 2x_1^3x_3 - 2x_1x_3^3 - 10x_1^2x_2^2 - x_1^2x_3^2 - 30x_1x_2^2x_3 - 25x_2^2x_3^2)(x_1^4 + 5x_2^4 + x_3^4 - 8x_1^3x_3 + 8x_1x_3^3 - 10x_1^2x_2^2 + 14x_1^2x_3^2 - 10x_2^2x_3^2).$$

The number of linearly independent spherical harmonics of degree k in this case is equal to the number of integer solutions to the equation

$$6\alpha_2 + 10\alpha_3 = k.$$

The system of generators of the algebra of polynomials invariant under the *tetrahedral group* G_n^{n+1} is given in [133].

§3. Rotation Invariant Cubature Formulas on the Sphere in $\mathbb{R}^3$

In this section we expose some simplest cubature formulas on the unit sphere of the 3-dimensional space $\mathbb{R}^3$ which are invariant under the group G^* coincident with G_3^{8*} or G_3^{20*}. Moreover, we derive some asymptotic relations between the number of nodes of a cubature formula and the number of linearly independent spherical harmonics integrated exactly by this formula.

Denote by S the unit sphere of $\mathbb{R}^3$, $S = \{(x,y,z) : x^2 + y^2 + z^2 = 1\}$. The integral $I(f)$ of a continuous function $f(x,y,z)$, calculated over S, is approximated by a linear combination

$$I_N(f) = \sum_{k=1}^{N} c_k f(x^{(k)}),$$

with $x^{(k)}$ a node located on S. We construct a cubature formula

$$I(f) = \int_S f(x,y,z)\,dS \cong I_N(f) \tag{3.1}$$

so that it be G^*-invariant. In other words, the set of nodes $x^{(k)}$, $k = 1,2,\dots,N$, which must be the union of the orbits of G^*, and the weights of the formula which correspond to the points $x^{(k)}$ of the same orbit coincide.

Denote by $\mathfrak{M}$ the regular polyhedron whose rotations generate G^*. If $\mathfrak{M}$ is the octahedron, then we write its vertices in the spherical coordinates (θ,φ) as

$$\theta = 0; \quad \theta = \frac{\pi}{2}, \quad \varphi = 0, \frac{\pi}{2}, \pi, \frac{3\pi}{2}; \quad \theta = \pi. \tag{3.2}$$

If $\mathfrak{M}$ is the icosahedron, then its vertices have the following spherical coordinates

$$\theta = 0; \quad \theta = \tan^{-1}2, \quad \varphi = 0, \frac{2\pi}{5}, \frac{4\pi}{5}, \frac{6\pi}{5}, \frac{8\pi}{5};$$

$$\theta = \pi - \tan^{-1}2, \quad \varphi = \frac{\pi}{5}, \frac{3\pi}{5}, \pi, \frac{7\pi}{5}, \frac{9\pi}{5}; \quad \theta = \pi.$$

Note that not only the set of all vertices of $\mathfrak{M}$ remains invariant under G^* but also the set of all centers of faces of $\mathfrak{M}$ as well as the set of all midpoints of edges of $\mathfrak{M}$. We omit the coordinates of the corresponding points.

Consider two methods for constructing the nodes of (3.1). Note that the surface of the polyhedron $\mathfrak{M}$ consists of the isosceles triangles contiguous to the vertices of $\mathfrak{M}$ (at each vertex there are touching 4 triangles in an octahedron and 5 triangles in an icosahedron). We call every of these triangles *principal*.

METHOD 1. Let k be a natural, $k \geq 2$, and let Δ be a principal triangle of the polyhedron $\mathfrak{M}$. On each of the sides of Δ we allocate $k-1$ equidistant points. Through each of these points lying at one side of Δ, we draw two inner straight line segments in parallel to the other two sides of Δ. As a result, we obtain a partition R of the triangle Δ into tiny parts. The vertices of these tiny triangles constitute the *lattice* $R(k)$. The form of this lattice depends on the magnitude of the remainder of division k by 6. The center of the principal triangle and the midpoints of its sides can be included or not included in $R(k)$.

Repeat the same procedure for each of the principal triangles covering the boundary of $\mathfrak{M}$ to obtain some set $R(k)$ of points lying on the surface of the polyhedron $\mathfrak{M}$. Clearly, this set is invariant under G^*. We draw from the center of the sphere S half-lines passing through the points of $R(k)$, and take as the set of the nodes $x^{(k)}$ of (3.1) the intersection points of these half-lines with the sphere S. We say that the resultant cubature formula (3.1) *has lattice of type* I.

METHOD 2 for prescribing the nodes $x^{(k)}$, $k = 1, \ldots, N$, differs little from the first. Given a natural $k \geq 2$, a principal triangle Δ and the partition of every side of Δ into $k-1$ equal parts, from every point of this partition we draw two straight line segments in parallel to the height of the principal triangle rather than to its sides as in Method 1. The form of the resultant partition R depends on evenness of k.

Projecting by half-lines the lattice $R(k)$ which we have on the surface of the polyhedron $\mathfrak{M}$ to points of the sphere S, we obtain a cubature formula with *lattice of type* II. Clearly, as k increases, the sets of nodes of type I as well as those of type II become thicker.

Let us calculate the number of nodes $N = N(k)$ of (3.1), and the number $L = L(k)$ of distinct G^*-orbits which constitute the set of these nodes. For a G^*-invariant cubature formula the number $L(k)$ coincides obviously with the number of independent weights c_k.

For a lattice of type I, there are S_{I} points in a principal triangle, the vertices of the latter inclusively, with

$$S_{\mathrm{I}} = (k+1) + k + (k-1) + \cdots + 1 = \frac{(k+1)(k+2)}{2}.$$

For the group G^* elementary calculations give the relations

$$N_{\mathrm{I}} = t_2 S_{\mathrm{I}}(k) - t_3(k-1) - t_1(q_1 - 1) = \frac{M}{6}k^2 + 2,$$
$$L_{\mathrm{I}} = \begin{cases} (s+1)(k-3s)+1, & \text{if } k = 6s, \\ (s+1)(k-3s), & \text{if } k \neq 6s. \end{cases} \tag{3.3}$$

Similar formulas are true for a lattice of type II

$$S_{\mathrm{II}} = \frac{3k^2 + 3k + 2}{2}.$$

The same calculations as above lead to the formulas

$$N_{\mathrm{II}} = t_2 S_{\mathrm{II}}(k) - t_3(k-1) - t_1(q_1 - 1) = \frac{M}{2}k^2 + 2,$$
$$L_{\mathrm{II}} = \begin{cases} (s+1)^2, & \text{if } k = 2s, \\ (s+1)(s+2), & \text{if } k = 2s+1. \end{cases} \tag{3.4}$$

Eliminating L (with a subscript) from (3.3) and (3.4), we express N (with the same subscript) as a function of L, namely,

$$N_{\mathrm{I}}(L) = 2M(L - \sqrt{3}\,L^{1/2} + \ldots), \quad N_{\mathrm{II}}(L) = 2M(L - 2L^{1/2} + \ldots). \tag{3.5}$$

These formulas are asymptotically accurate as $L \to \infty$ or, which is the same, as $k \to \infty$.

Let the set of nodes of (3.1) be constructed by one of the above methods. Require that (3.1) be exact for all polynomials of degree n or, which is the same by Theorem 2.4, at all spherical harmonics of degree at most n. Among these harmonics there are $(n+1)^2$ linearly independent. Choosing a respective basis, write the requirement that the formula under study is exact for the members of the basis as simultaneous linear equations for its coefficients. This system consists of $(n+1)^2$ equations. Solving it, we may obtain a final form of (3.1).

Note, however, that, pursuing the above approach, we in no way use the invariance of a cubature formula under G^*. Incidentally, as we now show, this condition enables us to diminish considerably the size of the linear system to be solved.

According to Theorem 2.3 a cubature formula (3.1) exact for G^*-invariant spherical harmonics of degree at most n is also exact for all spherical harmonics of degree at most n. The total number $\sigma^*(n)$ of linearly independent G^*-invariant harmonics of degree at most n is considerably less than $(n+1)^2$. This quantity may be calculated by the formula

$$\sigma^*(n) = \sum_{k=0}^{n} S^*(k),$$

with $S^*(k)$ the total number of G^*-invariant spherical harmonics of degree k. By Theorem 2.9 $S^*(k)$ coincides with $S(k)$ for k even and is 0 otherwise. Consequently, to count $\sigma^*(n)$ up, we may use all formulas for $S(k)$ which were derived in the previous section. In particular, from the equality

$$S\left(k+\frac{M}{2}\right)=S(k)+1$$

we readily infer the next relation

$$\sigma^*(n)=(k-1)\sigma^*\left(\frac{M}{2}-1\right)+\sigma^*\left(\frac{M}{2}-2j+1\right)+\frac{k(k-1)}{2}\frac{M}{4}-k(j-1) \quad (3.6)$$

which is valid for $n=kM/2-2j+1$ and $j=1,2,\dots,[M/4]$.

Taking now some system of $\sigma^*(n)$ linearly independent G^*-invariant harmonics of degree at most n, write the condition that (3.1) is exact for these functions. As a result, obtain a collection of $\sigma^*(n)$ equations. The G^*-invariance of (3.1) means that it has not more than L distinct coefficients. Therefore, the matrix of the resultant linear system has size $\sigma^*(n)\times L$. Assume that it is a square matrix, i.e., $\sigma^*(n)=L$. Denote by $n^*(L)$ the greatest number n that satisfies the preceding equality.

As an instance of perfect implementation of the suggested algorithm, we expose an invariant formula under the icosahedral group G_3^{20*}. Its nodes are located at the vertices of the icosahedron and at the centers of its faces. Moreover, $L=2$, and if the coefficient c_1 of every "vertex" node equals $5\pi/42$, and the coefficient c_2 of the other nodes equals $9\pi/70$, then the corresponding formula (3.1) is exact for polynomials of degree 9. (In this case the solution n^* to the equation $\sigma^*(n)=2$ is equal to 9.) The cubature formula we discuss was proposed by V. A. Ditkin and L. A. Lyusternik in the article [55].

The nodes of the second cubature formula are the vertices of the icosahedron, the projections to S of the midpoints of edges, the centers of its faces and the points dividing the median of a principal triangle in proportion of 1 to 2. Moreover, $L=4$ and the formula is exact for polynomials of degree 15 if the weights of the formula are given by the equalities

$$c_1\cong 0.0154324\pi,\ c_2\cong 0.0364968\pi,\ c_3\cong 0.0381004\pi,\ c_4\cong 0.0326315\pi.$$

The further properties of this cubature formula and some of its natural generalizations are exposed in [200, 307].

We now suggest some quantitative criterion for assessing the properties of (3.1). Complete use of all of its free parameters that include two coordinates of each of the

TABLE 1

Parameters of a Cubature Formula Invariant Under G_3^{8*}

I				II			
L	$N(L)$	$n^*(L)$	$(n^*+1)^2$	L	$N(L)$	$n^*(L)$	$(n^*+1)^2$
1	6	3	16	2	14	5	36
2	18	5	36	4	50	9	100
3	38	7	64	6	110	11	144
4	66	9	100	9	194	15	256
5	102	11	144	12	302	19	400
7	146	13	196	16	434	23	576
8	198	15	256	20	590	25	676
10	258	17	324	25	770	29	900
12	326	19	400	30	974	33	1156
14	402	21	484	36	1202	35	1296
16	486	23	576	42	1454	39	1600
19	578	25	676	49	1730	43	1936
21	678	27	784	56	2030	47	2304
24	786	29	900	64	2354	49	2500
27	902	31	1024	72	2702	53	2916
30	1026	32	1089	81	3074	57	3364
33	1158	35	1296				

nodes $x^{(k)}$ and the weights c_k allows us to achieve exactness of a cubature formula for all polynomials of degree n, with n satisfying the equality

$$(n+1)^2 = 3N.$$

In this connection, given a cubature formula having N nodes and exact for polynomials of degree n_*, we introduce the ratio

$$\eta = \frac{(n_* + 1)^2}{3N}$$

which is further referred to as *efficiency*.

We may evaluate the efficiency of the G^*-invariant formulas corresponding to small L on appealing to Tables 1 and 2. If L is large then it stands to reason to use the asymptotic expansion of the coefficient η which we are about to derive.

TABLE 2

Parameters of a Cubature Formula Invariant Under G_3^{20*}

I				II			
L	$N(L)$	$n^*(L)$	$(n^*+1)^2$	L	$N(L)$	$n^*(L)$	$(n^*+1)^2$
1	12	5	36	2	32	9	100
2	42	9	100	4	122	15	256
3	92	11	144	6	272	19	400
4	162	15	258	9	482	25	676
5	252	17	324	12	762	29	900
7	362	21	484	16	1082	35	1296
8	492	23	576	20	1472	41	1764
10	642	27	784	25	1922	47	2304
12	812	29	900	30	2432	51	2704
14	1002	33	1156	36	3002	57	3364
16	1212	35	1296				
19	1442	39	1600				
21	1692	41	1764				
24	1962	45	2116				
27	2252	49	2500				
30	2562	51	2704				
33	2892	55	3136				
37	3242	59	3600				

Let $n = kM/2 - 1$. Then (3.6) implies that

$$\sigma^*(n) = k\sigma^*\left(\frac{M}{2} - 1\right) + \frac{k(k-1)}{2}\frac{M}{4}.$$

In this event, we have the equalities

$$(n+1)^2 = \frac{k^2M^2}{4}, \quad 2M\sigma^*(n) = \frac{k^2M^2}{4} + \left(2M\sigma^*\left(\frac{M}{2} - 1\right) - \frac{M^2}{4}\right)k.$$

Eliminating the parameter k in these equalities, we arrive at the asymptotic formula

$$(n+1)^2 = 2M\sigma^*(n) + \left(\frac{M}{2} - 4\sigma^*\left(\frac{M}{2} - 1\right)\right)\left(2M\sigma^*(n)\right)^{1/2} + O(1).$$

For $n = n_*(L)$ the quantity $\sigma^*(n)$ "almost" coincides with L and so it is large at large L. In the case of the octahedron when $M = 24$, we have

$$(n_* + 1)^2 = 2M(L - \sqrt{4/3}\,L^{1/2} + \dots).$$

In the case of the icosahedron when $M = 60$, the equality holds

$$(n_* + 1)^2 = 2M(L - \sqrt{49/30}\,L^{1/2} + \dots).$$

Whence and from (3.5) it is easy to derive an asymptotic presentation of the efficiency for formulas with lattice of type I and of type II.

We describe a scheme for constructing cubature formulas that possess the highest degree and are invariant under G_3^{8*}. Assume that the vertices of the octahedron $\mathfrak{M}$ lie on the coordinate axes; i.e., they are given by (3.2). A sought cubature formula takes the form

$$I(f) \cong I_N(f) = A_1 \sum_{i=1}^{6} f(a_i^{(1)}) + A_2 \sum_{i=1}^{12} f(a_i^{(2)}) + A_3 \sum_{i=1}^{8} f(a_i^{(3)})$$
$$+ \sum_{k=1}^{N_1} B_k \sum_{i=1}^{24} f(b_i^{(k)}) + \sum_{k=1}^{N_2} C_k \sum_{i=1}^{24} f(c_i^{(k)}) + \sum_{k=1}^{N_3} D_k \sum_{i=1}^{36} f(d_i^{(k)}). \qquad (3.7)$$

The nodes $a_i^{(1)}$ are at the vertices of $\mathfrak{M}$, the points $a_i^{(2)}$ and $a_i^{(3)}$ result from projecting to S the midpoints of edges and the centers of faces. The inverse images of the points $b_i^{(k)}$ and $c_i^{(k)}$ belong to the bisectrices of the angles of principal triangles and to the edges of $\mathfrak{M}$, respectively. The points $d_i^{(k)}$ are nodes of general position.

For a given k, each of the sets $\{a_i^{(k)}\}$, $\{b_i^{(k)}\}$, $\{c_i^{(k)}\}$, and $\{d_i^{(k)}\}$ is invariant under G_3^{8*}. This means that

$$b_i^{(k)} \in \{(\pm l_k, \pm l_k, \pm m_k), (\pm l_k, \pm m_k, \pm l_k), (\pm m_k, \pm l_k, \pm l_k)\},$$

with $2l_k^2 + m_k^2 = 1$. Furthermore,

$$c_i^{(k)} \in \{(\pm p_k, \pm q_k, 0), (\pm p_k, 0, \pm q_k), (0, \pm p_k, \pm q_k), (\pm q_k, \pm p_k, 0),$$
$$(\pm q_k, 0, \pm p_k), (0, \pm q_k, \pm p_k)\},$$

where $p_k^2 + q_k^2 = 1$. Moreover,

$$d_i^{(k)} = \{(\pm r_k, \pm u_k, \pm w_k), (\pm r_k, \pm w_k, \pm u_k), (\pm u_k, \pm r_k, \pm w_k),$$
$$(\pm u_k, \pm w_k, \pm r_k), (\pm w_k, \pm u_k, \pm r_k), (\pm w_k, \pm r_k, \pm u_k)\}.$$

Thus, the set of the sought parameters of a cubature formula (3.7) is completely determined. We should find them from the system of nonlinear equations which is equivalent to the condition that (3.7) is exact for the chosen basis functions.

In the case under study we may considerably specify the form of basis elements. In doing so, it is convenient to use the fact that the restriction to the sphere S of each G_3^{8*}-invariant polynomial is a polynomial in the variables σ_2 and σ_3, with

$$\sigma_2 = x^2y^2 + x^2z^2 + y^2z^2, \quad \sigma_3 = x^2y^2z^2.$$

We mention in passing that the general theorem on representation of invariant polynomials via basis polynomials was established by C. Chevalley [45].

We define a basis for the space $\mathbf{P}_{n_{G^*}}$ when $n \geq 6$ as the union of the four sets A, B, C, and D. The set A consists of the functions

$$a_1 = 9\sigma_3 - 4\sigma_2 + 1, \quad a_2 = \sigma_2 - 9\sigma_3, \quad a_3 = \frac{4}{3}\left(\sigma_3 + \frac{\sigma_2}{3}(1 - 4\sigma_2)\right). \tag{3.8}$$

The set B is composed of the elements

$$b_{4i+6} = 4\sigma_3 b_4^i, \quad b_{4j+12} = 4\sigma_3 b_6 b_4^j, \tag{3.9}$$

with

$$b_4 = \frac{4}{3}\left(\frac{1}{3} - \sigma_2\right), \quad b_6 = \frac{4}{9}\left(9\sigma_3 - 3\sigma_2 + \frac{2}{3}\right),$$

and the indices i and j ranging from 1 to $(n-6)/4$ and from 1 to $(n-12)/4$, respectively. The set C comprises the elements

$$c_{4i} = \sigma_2^i c_{12}, \tag{3.10}$$

with $c_{12} = \sigma_2^2(1 - 4\sigma_2) - \sigma_3(4 + 27\sigma_3 - 18\sigma_2)$, and the index i ranges from 0 to $(n-12)/4$. The members of the set D are the shape

$$d_{ij} = \sigma_2^i \sigma_3^i c_{12}, \quad j \geq 1, \ 6 \leq 4i + 6j \leq n - 12. \tag{3.11}$$

We obtain the sought system of nonlinear equations on substituting the functions (3.8)–(3.11) for f in the equality $I(f) = I_N(f)$. Because of the particular choice of basis elements, the system splits into triangular subsystems.

Let N stand for the total number of the nodes of (3.7). The values of N, N_1, N_2, and N_3 should be chosen so that the total number of the sought quantities be

TABLE 3

Parameters of a Cubature Formula of Highest Degree n
Invariant Under G_3^{8*}

n	N_1	N_2	N_3	N	η
9	0	1	0	38	0.877
11	1	0	0	50	0.960
13	1	1	0	75	0.928
15	2	1	0	86	0.992
19	3	0	1	146	0.913
21	3	1	1	170	0.949
23	4	1	1	194	0.990
25	5	2	1	230	0.980
27	5	1	2	266	0.982
29	6	2	2	302	0.993
31	6	2	3	350	0.975
33	6	3	3	386	0.998
35	7	2	4	434	0.995
37	7	4	4	482	0.999
39	8	3	5	530	1.006
41	9	3	6	590	0.997
43	9	3	7	650	0.998
45	9	5	7	698	1.011
47	10	3	9	770	0.997
49	11	5	9	830	1.004
51	11	5	10	840	1.013
53	12	4	12	974	0.998
59	13	4	16	1202	0.998

equal to the number of equations in the system. To this end, in some cases we are to set the coefficient A_2 equal 0.

In Table 3 we gather the values of the efficiency of a cubature formula (3.7).

We may conveniently give the degree n of a cubature formula (3.7) as expanded in the sum

$$n = 12m + 2l + 1, \quad l = 0, 1, \ldots, 5. \tag{3.12}$$

In the cases of $l = 2$ and $l = 5$, we obtain especially simple expressions for the parameters N and N_j. Namely, if $l = 2$ then

$$N_1 = 3m, \quad N_2 = m, \quad N_3 = m(m-1), \quad N = \frac{n^2+2n+7}{3}, \quad A_2 = 0.$$

Here m is the parameter of the expansion (3.12). If $l = 5$ then

$$N_1 = 3m+1, \quad N_2 = m, \quad N_3 = (n^2+2n+7)/3.$$

It is worth noting that at these values of l the set of nodes of a cubature formula (3.7) coincides with a lattice of type II described at the beginning of the current section to within a G_3^{8*}-invariant continuous transformation.

Recently, V. I. Lebedev and A. L. Skorokhodov have found G_3^{8*}-invariant cubature formulas of degree 41, 47, 53 and 59 [112].

A. D. McLaren obtained a G_3^{20}-invariant cubature formula of Gauss type with 72 nodes of degree 14. Formulas invariant under the icosahedral group were also dealt with by S. I. Konyaev. He found formulas of degree 29.

G. N. Salikhov [192–200] obtain cubature formulas for the 4-dimensional and 5-dimensional spheres which are invariant under finite rotation groups. In particular, for the 4-dimensional sphere S he derived the formula

$$\int_S f(x_1, x_2, x_3, x_4)\, dS = \frac{\pi^2}{60} \sum_{j=1}^{120} f(x^{(j)}), \tag{3.13}$$

invariant under the rotation group G_4^{600} of the 600-hedron. This formula exactly integrates all spherical harmonics of degree at most 11, i.e., 650 harmonics. The set of nodes $x^{(j)}$ at that comprises the vertices of the 600-hedron and 24 points that result from all possible permutations of signs in the vector

$$(\pm 1/2, \pm 1/2, \pm 1/2, \pm 1/2), \quad (\pm 1, 0, 0, 0), \quad (0, \pm 1, 0, 0), \quad (0, 0, \pm 1, 0), \quad (0, 0, 0, \pm 1)$$

and also 96 points that results from even permutations of the entries of the vector

$$\left(\pm\frac{t}{2}, \pm\frac{1}{2}, \pm\frac{t^{-1}}{2}, 0\right),$$

with t standing for some root of the equation $t^2 + t^{-2} = 3$.

The cubature formulas constructed by G. N. Salikhov have high efficiency and integrate all harmonics of degree at most 19 exactly. The efficiency η of a formula of degree 19 equals 0.9965.

I. P. Mysovskikh found cubature formulas invariant under the group G_n^{n+1} of the n-dimensional simplex [133]. The tables of parameters of various invariant cubature formulas are exposed in the monographs by I. P. Mysovskikh [133] and A. N. Stroud [269].

Chapter 3

Formulas with Regular Boundary Layer for Rational Polyhedra

Construction of asymptotically optimal formulas is one of the main achievements of the theory we develop.

As was mentioned above and is rigorously proven in Chapter 5, asymptotically $L_2^{(m)}$-optimal are the errors with regular boundary layer. The definition of these functionals in Chapter 1 is constructive and proposes a method for deriving corresponding formulas. To this end, it is necessary to find a finite set of elementary errors $l_\gamma(y)$ whose size is in general the same as the number of nodes in the boundary layer. For piecewise-smooth boundaries, this quantity is $O(h^{1-n})$. So is the number of distinct weights of a cubature formula.

To calculate an integral, we need to know the values of the integrand $\varphi(x)$ at all nodes of a lattice. The number of nodes is $O(h^{-n})$ which is higher than $O(h^{-n+1})$. Therefore, the determination of elementary errors and weights of a cubature formula in a boundary layer requires operations whose cost is negligible at small h as compared with the total amount of operations used in integration. The determination of all weights still remains a rather bulky problem.

There are cases in which the construction of cubature formulas may be simplified considerably, since it turns out possible to restrict examination to seeking only a finite set of distinct functionals $l_\gamma(y)$ whose size is in general independent of the mesh-size h. The number of distinct weights of such formulas and the cost of their determination are also independent of h. This happens for certain special polyhedra and is the topic of the current chapter. For these polyhedra we may also find the weights of cubature formulas straightforward, skipping the intermediate stage of determining elementary errors.

§1. Rational Polyhedra

In what follows we deal only with convex polyhedra. However, most of the results below may be translated to the case of nonconvex polyhedra.

A *convex polyhedron* $\overline{\mathfrak{M}}$ is the set of points satisfying compatible simultaneous linear inequalities

$$\Lambda_k \equiv a_k x - b_k \geq 0, \quad k = 1, 2, \ldots, M, \tag{1.1}$$

with a_k a row-vector of dimension n and x a usual column-vector. We call (1.1) an *admissible system* if there is at least one point at which all inequalities become strict. The set of points $\mathfrak{M}$ at which all inequalities of (1.1) are strict presents the *interior* of the polyhedron.

If the set of vectors x satisfying (1.1) is bounded then we call the polyhedron a *bounded* or *compact* polyhedron.

From a given system of inequalities (1.1) we may always select in a unique fashion some *independent subsystem*, a set of inequalities none of which can be eliminated without changing the initial polyhedron.

In the theory of linear inequalities there is an algorithm reducing (1.1) to an independent subsystem. This algorithm is essentially as follows: The left side of every inequality of (1.1) is considered in turn as some function given in the domain determined by the other inequalities. We seek for the maximum and the minimum of this function. If the minimum is nonnegative then the corresponding inequality may be eliminated. If the maximum is nonpositive, then the system is inadmissible. In the remaining cases, the inequality under study must be preserved. Eliminating redundant inequalities one by one, we arrive at an independent subsystem.

In the sequel, we presume that (1.1) is an independent system.

The set of points x at which several of the functions Λ_k vanish and the others are nonnegative is a *face of the polyhedron* $\overline{\mathfrak{M}}$. The *dimension of a face* is the number $s = n - r$, with r the rank of the matrix comprising the weights of those functions Λ_k that vanish at every point of the face. In other words, s indicates the number of independent variables on which the points of the face depend.

Clearly, all faces of each dimension of a convex polyhedron $\overline{\mathfrak{M}}$ are also convex polyhedra in subspaces of lesser dimension. We treat a polyhedron itself as face of dimension n.

Denote by $\mu(s)$ the number of faces of dimension s. Enumerate these faces, denoting each by $\overline{\mathfrak{M}}_{j,s}$, $s = 0, 1, \ldots, n$, $j = 1, 2, \ldots, \mu(s)$. Denote by $K_{j,s}$ the set of the indices of those functions Λ_k in (1.1) that vanish on the whole face $\overline{\mathfrak{M}}_{j,s}$, and let $L_{j,s}$ stand for the set of all remaining indices

$$L_{j,s} = \{k : 1 \leq k \leq M\} \setminus K_{j,s}.$$

Then the face $\overline{\mathfrak{M}}_{j,s}$ is determined by the system of relations

$$\Lambda_k = 0, \quad k \in K_{j,s}, \tag{1.2}$$
$$\Lambda_k \geq 0, \quad k \in L_{j,s}. \tag{1.3}$$

In case $s = n-1$, each function Λ_k, $k \in K_{j,s}$, corresponds to a face of dimension $n-1$, and so $K_{j,n-1}$ is a singleton for every j. For other s the size of $K_{j,s}$ may vary but it is at least $n - s$. The conditions of (1.2) may fail to be independent in this event. For instance, 0-dimensional faces, i.e. *vertices*, of a regular octahedron in $\mathbb{R}^3$ are determined each by 4 equations of the faces touching at this vertex, any three of the equations sufficient for finding the vertex. The affine variety (1.2) is called the *plane* $R_{j,s}$ *of the face* $\overline{\mathfrak{M}}_{j,s}$. In the plane $R_{j,s}$, the inequalities of (1.3) define a convex polyhedron which is $\overline{\mathfrak{M}}_{j,s}$. Obviously, the system of inequalities (1.3) is not always independent and part of the inequalities may be eliminated.

In practical search for an independent system of inequalities for a face, we may use the above-mentioned trick for finding an independent system of inequalities describing an n-dimensional polyhedron $\overline{\mathfrak{M}}$.

The inequalities $\Lambda_k > 0$, $k \in L_{j,s}$, determine the set of interior points of the face $\overline{\mathfrak{M}}_{j,s}$, i.e. the *interior* of the s-dimensional polyhedron $\overline{\mathfrak{M}}_{j,s}$.

Consider one more system resulting from (1.2) by making each equality the inequality

$$\Lambda_k \geq 0, \quad k \in K_{j,s}. \tag{1.4}$$

The system of strict inequalities corresponding to (1.4) takes the form

$$\Lambda_k > 0, \quad k \in K_{j,s}. \tag{1.5}$$

The relations (1.4) determine the set which we call the *closed solid angle* $\overline{\Omega}_{j,s}$ of the polyhedron $\overline{\mathfrak{M}}$. The inequalities (1.5) determine the set $\Omega_{j,s}$ of the interior points of $\overline{\Omega}_{j,s}$ which we call a *solid angle* of $\overline{\mathfrak{M}}$. The plane $R_{j,s}$ is the *blade* of the solid angle. The *order* of the solid angle $\overline{\Omega}_{j,s}$ or $\Omega_{j,s}$ is the number $n - s$.

As an example, we consider the plane triangle

$$\overline{\mathfrak{M}} = \{x : x = (x_1, x_2),\ \Lambda_1 \equiv 3x_1 - 2x_2 \geq 0,\ \Lambda_2 \equiv x_2 \geq 0,\ \Lambda_3 \equiv -3x_1 - x_2 + 3 \geq 0\}.$$

The three vertices O, A, and B of $\overline{\mathfrak{M}}$ are the 0-dimensional faces $\{\overline{\mathfrak{M}}_{j,0}\}_{j=1}^3$; the three sides of $\overline{\mathfrak{M}}$ are the one-dimensional faces $\{\overline{\mathfrak{M}}_{j,1}\}_{j=1}^3$; whereas the triangle $\overline{\mathfrak{M}}$ itself is identical with its 2-dimensional face $\overline{\mathfrak{M}}_{1,2}$ (see Table 4).

The *alternating sum* of the numbers $\mu(s)$ plays an important role in characterization of the properties of a polyhedron. Put $\mu(n) = 1$ and arrange the sum

$$\mu(0) - \mu(1) + \cdots + (-1)^n \mu(n) = \sum_{s=0}^{n} (-1)^s \mu(s). \tag{1.6}$$

TABLE 4

s	j		
	1	2	3
0	$\overline{\mathfrak{M}}_{1,0} = \{O\}$ $\Omega_{1,0} = \angle O$ $R_{1,0} = \{O\}$	$\overline{\mathfrak{M}}_{2,0} = \{A\}$ $\Omega_{2,0} = \angle A$ $R_{2,0} = \{A\}$	$\overline{\mathfrak{M}}_{3,0} = \{B\}$ $\Omega_{3,0} = \angle B$ $R_{3,0} = \{B\}$
1	$\overline{\mathfrak{M}}_{1,1} = [AB]$ $\Omega_{1,1}$, the half-plane $AB(O)$ $R_{1,1}$, the line AB	$\overline{\mathfrak{M}}_{2,1} = [OB]$ $\Omega_{2,1}$, the half-plane $OB(A)$ $R_{2,1}$, the line OB	$\overline{\mathfrak{M}}_{3,1} = [OA]$ $\Omega_{3,1}$, the half-plane $OA(B)$ $R_{3,1}$, the line OA
2	$\overline{\mathfrak{M}}_{1,2} = \Delta OAB$ $\Omega_{1,2} = \mathbb{R}^2$ $R_{1,2} = \varnothing$		

Theorem 3.1 (Euler). *For every compact convex polyhedron the alternating sum* (1.6) *equals unity, i.e.,*

$$\sum_{s=0}^{n}(-1)^s \mu(s) = 1. \tag{1.7}$$

The proof of the theorem may be found, for instance, in [188].

Let a polyhedron $\overline{\mathfrak{M}}$ be somehow partitioned into finitely many smaller convex polyhedra. The boundaries between these parts consist of polyhedra of dimension less than n. By a *complex* we mean the set of all polyhedra and all faces of the polyhedra that participate in such partition of $\overline{\mathfrak{M}}$. A complex contains all polyhedra that coincide with the boundaries between smaller polyhedra of dimension n newly born in partitioning $\overline{\mathfrak{M}}$ and also the old faces of $\overline{\mathfrak{M}}$ or the parts into which they are partitioned.

Given such complex, let us calculate the number of elements (faces) of each dimension from 0 to n. Denote the number of elements of dimension s by $\mu(s)$ in much the same way as in the case of a sole polyhedron. Then we have

Theorem 3.2. *The following equality holds*

$$\sum_{s=0}^{n}(-1)^s\mu(s) = 1. \tag{1.8}$$

Observe that (1.7) is a particular instance of (1.8).

We also state one more theorem which relates to noncompact polyhedra.

Theorem 3.3. *Assume that a noncompact convex polyhedron contains no infinite straight line. Further assume that it is partitioned into finitely many convex polyhedra as in Theorem 3.2. The alternating sum (1.6) for this polyhedron is then zero, in symbols,*

$$\sum_{s=0}^{n}(-1)^s\mu(s) = 0. \tag{1.9}$$

Denote by $\chi_{\overline{j,s}}(x)$ the indicator of the closed solid angle $\overline{\Omega}_{j,s}$ containing the points of its faces of every dimension

$$\chi_{\overline{j,s}}(x) = \begin{cases} 1, & x \in \overline{\Omega}_{j,s}, \\ 0, & x \notin \overline{\Omega}_{j,s}. \end{cases}$$

The Euler Theorem and its generalizations entail the following theorem which we find beneficial.

Theorem 3.4. *The indicator $\chi_{\overline{\mathfrak{M}}}(x)$ of a compact polyhedron $\overline{\mathfrak{M}}$ decomposes as follows*

$$\chi_{\overline{\mathfrak{M}}}(x) = \sum_{s=0}^{n}(-1)^s\sum_{j=1}^{\mu(s)}\chi_{\overline{j,s}}(x). \tag{1.10}$$

In other words, the indicator of a polyhedron equals the alternating sum of the indicators of all its solid angles of various order.

PROOF. Let a point x belong to the polyhedron $\overline{\mathfrak{M}}$. Insert x into the right side of (1.10). Then every indicator $\chi_{\overline{j,s}}(x)$ becomes equal to 1. The number of the functions $\chi_{\overline{j,s}}(x)$ pertinent to the faces of dimension s equals $\mu(s)$. Therefore, the right side of (1.10) equals the sum $\sum_{s=0}^{n}(-1)^s\mu(s)$ and by (1.7) it is 1.

Consider now the case in which a point x lies outside the set $\overline{\mathfrak{M}}$. Inserting x into the right side of (1.10), observe that some summands equal 1 and the others vanish. Denote the number of the latter by $\mu_1(s)$. Rewrite the sum of (1.10) as

$$\sum_{s=0}^{n}(-1)^s[\mu(s)-\mu_1(s)] = \sum_{s=0}^{n}(-1)^s\mu(s) - \sum_{s=0}^{n-1}(-1)^s\mu_1(s).$$

Show that the right side of this equality is equal to 0.

Indeed, by the Euler Theorem

$$\sum_{s=0}^{n}(-1)^s\mu(s) - \sum_{s=0}^{n-1}(-1)^s\mu_1(s) = 1 - \sum_{s=0}^{n-1}(-1)^s\mu_1(s).$$

Consider the alternating sum on the right side of this equality. Clearly, the indicator $\chi_{\overline{j,s}}(x)$ of a solid angle vanishes if and only if we may directly see its blade from the point x, i.e., the line joining each point of the blade with x lies outside $\overline{\Omega}_{j,s}$. Draw the pencil of all lines passing through x and meeting the polyhedron $\overline{\mathfrak{M}}$. Intersect this pencil with some $(n-1)$-dimensional plane. The pencil projects the whole polyhedron $\overline{\mathfrak{M}}$ to this plane, giving in result a new $(n-1)$-dimensional polyhedron $\overline{\mathfrak{M}}_1$. This projection transforms those faces that were seen from the point x to some partition of $\overline{\mathfrak{M}}_1$ into finitely many polyhedra with preservation of dimension. Consequently, $\mu_1(s)$ coincides with the number of faces of dimension s in some complex constructed for a polyhedron of dimension $n-1$. In virtue of Theorem 3.2, we have

$$\sum_{s=0}^{n-1}(-1)^s\mu_1(s) = 1.$$

Whence (1.10) follows. The proof of Theorem 3.4 is complete.

Corollary 1.1. *The indicator $\chi_{\mathfrak{M}}(x)$ of the interior of a polyhedron $\overline{\mathfrak{M}}$ equals the alternating sum of the indicators $\chi_{j,s}(x)$ of the interiors of its solid angles,*

$$\chi_{\mathfrak{M}}(x) = \sum_{s=0}^{n}(-1)^s\sum_{j=1}^{\mu(s)}\chi_{j,s}(x). \tag{1.11}$$

Indeed, every interior point of a polyhedron is an interior point of all its solid angles. At these points, the formula is checked in the same way as (1.10). At the boundary points, the formula is checked in the same way as (1.10) in the case of an exterior point x.

It is worth observing that in the alternating sum of the indicators of closed solid angles we may count up the elements of the boundary which belong to some closed face $\overline{\mathfrak{M}}_{j,s}$ by using the Euler formula. Subtracting (1.11) from (1.10), we find an expression for the indicator of the boundary of $\overline{\mathfrak{M}}$. Moreover, for every face $\mathfrak{M}_{j,s}$ the sum of the indicators of solid angles in the plane of this face yields the indicator of the face. In the other points the sum vanishes, which the reader is welcome to check.

Now we return to studying cubature formulas. Let $l(x)$ stand for the error of such formula. Implement an affine change of independent variables by letting $x = By$, with the determinant of B equal to 1. Then the error transforms by the rule

$$l(x) = l(By) = \bar{l}(y),$$

with $\bar{l}(y)$ again the error of a cubature formula. In this event, the lattice of nodes may change in general. We have

***Theorem* 3.5.** *An affine transformation makes a formula with regular boundary layer of order m into a formula with regular boundary layer of the same order but possibly with other constants L and A.*

PROOF. Under the affine transformation $x = By$ the error

$$l(x) = \sum_{\gamma \in B_L} l_\gamma \left(\frac{x}{h} - H\gamma\right) = \chi_\Omega(x) - \sum_{hH\gamma \in \Omega} h^n c[\gamma]\delta(x - hH\gamma),$$

with $l_\gamma(t) = \chi_{\Omega_\gamma}(t) - \sum_{\gamma'} c^\gamma[\gamma']\delta(t - H\gamma')$, becomes

$$\bar{l}(y) = l(By) = \sum_\gamma l_\gamma \left(B\frac{y}{h} - H\gamma\right).$$

Let $\bar{l}_\gamma(z) = l_\gamma(Bz)$ and $\overline{H} = B^{-1}H$. Then

$$l_\gamma\left(\frac{x}{h} - H\gamma\right) = \bar{l}_\gamma\left(\frac{y}{h} - \overline{H}\gamma\right).$$

Since $\chi_{\Omega_\gamma}(t)$ transforms to $\chi_{\overline{\Omega}_\gamma}(z)$, and $\delta(x)$ is a homogeneous generalized function of order $-n$, the error $\bar{l}_\gamma(z)$ is again written as

$$\bar{l}_\gamma(z) = \chi_{\overline{\Omega}_\gamma}(z) - \sum_{\gamma'} c^\gamma[\gamma']\delta(z - \overline{H}\gamma').$$

Thus, the elementary error l_γ with matrix H transforms to the elementary error $\bar{l}_\gamma$ with matrix $\overline{H}$. Clearly,

$$\|\bar{l}_\gamma(z) \mid C^*(\mathbb{R}^n)\| = \|l_\gamma(t) \mid C^*(\mathbb{R}^n)\| \leq A,$$

and the support of $\bar{l}_\gamma$ lies in the domain $|Bz| < L$. We also check that $\bar{l}_\gamma$ is orthogonal to all polynomials of degree less than m. We have

$$(\bar{l}_\gamma(z), P_{m-1}(z)) = (l_\gamma(t), P_{m-1}(B^{-1}t)) = 0,$$

since $P_{m-1}(B^{-1}t)$ is a polynomial of degree less than m. The proof of Theorem 3.5 is complete.

The property that a boundary layer remains regular under an affine transformation allows us to reduce the construction of a cubature formula with a lattice matrix H to the construction of a cubature formula with a cubic lattice. Therefore, throughout this chapter we always consider formulas with a cubic lattice unless it is otherwise stated.

Our nearest aim is to make calculation of the weights of cubature formulas as standard and routine as possible. To this end, we should chose possibly *concomitant* elementary errors that compose $l(x)$. Concomitance for these functionals means that they result from one of them by application of a member of the group G_0 comprising the translations of $\mathbb{R}^n$ which preserve the lattice under study. In the sequel we consider the subgroups $G_{j,s}$ of G_0 the action of each of which keeps invariant the plane $R_{j,s}$ of some face. If such subgroup depends on s parameters then the elementary errors supported closely to the face may be obtained from a few of them by applying the members of $G_{j,s}$. We now begin implementing this idea.

Consider an arbitrary system of integer column-vectors $U_1, U_2, \ldots, U_s$ comprising the matrix U of size $n \times s$. Let the rank of U equal s. The vectors U_j belong to the hyperplane

$$x = U\lambda, \tag{1.12}$$

with λ an arbitrary s-dimensional vector. In this hyperplane there are infinitely many integer points that obviously generate the sublattice Γ_U of the main cubic lattice Γ. For instance, Γ_U contains all vectors that result from applying (1.12) with integer λ. It may happen that the hyperplane (1.12) has no other points of the main cubic lattice. Then the vectors U_j, $j = 1, 2, \ldots, s$, constitute a basis for the sublattice Γ_U. However, this may fail in general. Nevertheless, using a simple process, we may always construct a basis system $U^{(1)}$ for Γ_U which contains s vectors.

The group $G_{U^{(1)}}$ of translations, given by the formula

$$y = x + U^{(1)}\beta, \quad \beta \in \mathbb{Z}^n, \tag{1.13}$$

is a subgroup of the group G_0 of all translations keeping the lattice Γ invariant. An arbitrary transformation belonging to G_0 has the form

$$y = x + \beta, \quad \beta \in \mathbb{Z}^n.$$

We are interested in the polyhedra in which the s-dimensional plane $R_{j,s}$ of a face $\mathfrak{M}_{j,s}$ remains invariant for $s = 1, \ldots, n-1$ under the action of some s-parameter subgroup $G_{j,s}$ of the translation group G_0. Clearly, the plane $R_{j,s}$ of a face in such polyhedron may be given by the relations

$$x = U_{j,s}\lambda + f,$$

with the matrix $U_{j,s}$ determining the translation group $G_{j,s}$ under which $R_{j,s}$ is invariant. The entries of the matrix $U_{j,s}$ are integers. Hence, all components of the vector a_k in the system of inequalities (1.1) determining the face $R_{j,s}$ may be set integers. Let us demonstrate why this is so.

For definiteness, assume that $s = n-1$. Then the plane $R_{j,n-1}$ of $\mathfrak{M}_{j,n-1}$ is given by the relations

$$x = U_{j,n-1}\lambda + f. \tag{1.14}$$

Whence we obtain a system of relations of the shape (1.2) between the independent variables $x_1, \ldots, x_n$ which is equivalent to (1.14). To this end, write the solvability condition on the simultaneous equations (1.14) for the unknowns λ.

From linear algebra we know that (1.14) is solvable for λ if and only if the value of a special matrix R_l, called a *left annihilator for* $U_{j,n-1}$, at the difference of the vectors x and f is zero,

$$R_l x = R_l f. \tag{1.15}$$

By hypothesis, the rank of $U_{j,n-1}$ equals $n-1$, and so there is a left inverse $U^{-1}_{j,n-1,l}$ of $U_{j,n-1}$. The annihilator R_l now takes the form

$$R_l = I - U_{j,n-1}U^{-1}_{j,n-1,l}.$$

All entries of R_l are rational and, after eliminating the denominators in the equalities of (1.15), we arrive at a system of equations with integer coefficients.

For simplicity, in what follows we consider only those polyhedra that have points of the initial cubic lattice on faces of every dimension. Our supposition implies that the vertices of the polyhedron, i.e. its 0-dimensional faces, are also points of the initial lattice. In this event, each plane $R_{j,s}$ clearly includes the s-dimensional sublattice Γ, and the plane of every face is expressed by equations with integer coefficients. These polyhedra we call *rational.*

§2. Constructing Formulas for Rational Polyhedra

In the current and subsequent sections we give a detailed exposition of a method for constructing cubature formulas with regular boundary layer for rational polyhedra.

To begin with, we show that it is possible to standardize the elementary functionals that compose the errors $l(x)$ of a cubature formula, which makes it possible to use only a few distinct elementary errors $l_\gamma(x)$. Moreover, it turns out that if we obtain new lattices by refining old lattices then the number of different small functionals as well as the number of different weights of the resulting cubature formulas remains the same, i.e., independent of the number of nodes.

The weights of a sought formula are the sum of some functions $c^{j,s}[\beta]$ of a discrete argument which correspond to the faces $\mathfrak{M}_{j,s}$ of the initial polyhedron. The support of $c^{j,s}[\beta]$ is included in the sublattice lying in the plane $R_{j,s}$ and in several sublattices parallel and close to $\Gamma_{j,s}$ and meeting the initial polyhedron $\mathfrak{M}$, i.e., located in a sense on the inner side of the plane $R_{j,s}$. In this event, the function $c^{j,s}[\beta]$ is given only at the points lying in the polyhedron, taking a constant value at each sublattice parallel to $R_{j,s}$.

It is convenient to slightly generalize the concept of a cubature formula with regular boundary layer which was given in Chapter 1.

Once again we consider elementary errors

$$l_\gamma(y) = \chi_\gamma(y) - \sum_{|H\gamma'|<L} c^\gamma[\gamma']\delta(y - H\gamma'),$$

satisfying the conditions

$$\|l_\gamma(y) \mid C^*(\mathbb{R}^n)\| \le A; \quad \operatorname{supp} l_\gamma(y) \subset \{y : |y| < L\}; \quad (l_\gamma(y), y^\alpha) = 0, \quad |\alpha| \le k(\gamma).$$

Assume that the domain Ω has a piecewise-smooth boundary. Assume also that on this boundary there are given finitely many piecewise-smooth manifolds $S_{j,s}$ of various dimension s, with $s = 0, 1, \ldots, n-2$. In our case, as $S_{j,s}$ we may take the faces of the polyhedron $\mathfrak{M}$. Encircle each of these manifolds with the neighborhood $\omega_{j,s,L}$ comprising all points at a distance less than Lh from $S_{j,s}$. Arrange the set $B_{j,s}$ of γ at which $hH\gamma$ lies in $\omega_{j,s,L}$ and does not belong to any neighborhood $\omega_{j,s-1,L}$. The number of these γ is $O(h^{-n+s})$.

Suppose that for all $\gamma \in B_{j,s}$ the order $k(\gamma)$ equals $m - n + s$. This means that for all γ such that the points $hH\gamma$ lie in the neighborhood of $S_{j,n-1}$, the order $k(\gamma)$ equals $m-1$; in the neighborhood of $S_{j,n-2}$ it equals $m-2$, and so on. Recall that in Chapter 1 the order $k(\gamma)$ of the elementary error $l_\gamma(x)$ equals $m - 1$ near the whole boundary of Ω. We assume that this is the only distinction from what was considered before. In particular, for all γ in $B_L^{(i)}$, i.e., for γ such that the point $hH\gamma$ is at distance greater than Lh, from the boundary, the elementary errors result from one of them by the formula

$$l_\gamma\left(\frac{x}{h} - H\gamma\right) = l_0\left(\frac{x}{h} - H\gamma\right).$$

The sum of all elementary errors is the error of some cubature formula in the domain Ω, namely,

$$l(x) = \sum_{\gamma \in B_L} l_\gamma\left(\frac{x}{h} - \gamma\right).$$

As above, we call it an *error with regular boundary layer.*

Thus, we have abstracted the concept of an error with regular boundary layer, allowing that part of elementary errors with support near to the boundary may have less order. The proof of asymptotic optimality of errors with regular boundary layer we give in Chapter 5 makes use of this expanded definition.

Revert to considering rational polyhedra. In the class of errors with regular boundary layer we distinguish subclasses that possess even much more standardized elementary errors.

Let $\Gamma_{j,s}$ be a sublattice of a cubic lattice Γ which comprises integer points in the plane $R_{j,s}$, and let $U_{s+1}, \dots, U_n$ be a system of vectors complementing the system $U_1, \dots, U_s$ to a basis for Γ. Then we may write down Γ as the union of sublattices each of which is a translate of $\Gamma_{j,s}$ by an integer linear combination of the vectors $U_{s+1}, \dots, U_n$ and so lies in a plane parallel to $R_{j,s}$. Obviously, the error $l(x)$ may be constructed so that all elementary errors $l_\gamma(x)$ be concomitant for $\gamma \in B_{j,s}$ at the points belonging to the same sublattice parallel to $\Gamma_{j,s}$. If this condition is fulfilled for all $s = n, n-1, \dots, k$ and $j = 1, \dots, \mu(s)$, then we say that the error $l(x)$ has *regular boundary layer up to dimension k.* The regular boundary layer defined in Chapter 1 was regular only up to dimension n.

Theorem 3.6. *Consider a cubature formula with regular boundary layer up to dimension k for a polyhedron $\mathfrak{M}$. The weights of the formula at the points lying in the neighborhood of an s-dimensional face $\mathfrak{M}_{j,s}$, $s = k, k+1, \dots, n$, comprising the points at a distance greater than Lh from all faces of dimension $k-1$ depend only on a sublattice parallel to the face. At each of these sublattices, the weights are constant.*

PROOF. At each point of the lattice which lies in the neighborhood of $\mathfrak{M}_{j,s}$, $s = k, k+1, \dots, n$, comprising the points at a distance from all faces of lesser dimension, the value of the weights $c[\beta]$ results from summing the corresponding weights of $l_\gamma(x/h - \gamma)$, with $h\gamma$ belonging to the same neighborhood. However, these errors are invariant under the translation of γ by each vector of the sublattice $\Gamma_{j,s}$. Consequently, the weights $c[\beta]$ at the points located on the same plane parallel to $R_{j,s}$ are invariant under such translation, simply presenting the sum of the same quantities.

Obviously, the number of the planes parallel to $R_{j,s}$, lying in the neighborhood of the face $\mathfrak{M}_{j,s}$ and including sublattices of the cubic lattice depends only on L but not on h. Therefore, the number of distinct weights in the neighborhood of the face $\mathfrak{M}_{j,s}$ is independent of h. The proof of Theorem 3.6 is complete.

From this theorem it follows that for $k = 0$ the number of distinct weights of such formula is independent of the parameter h, i.e., of the number of nodes. Indeed, the number of distinct weights near all faces of arbitrary dimension does not depend on h. Since the number of faces is also independent of h, our claim is proven.

Consider all solid angles of various order of a rational polyhedron and construct for each of these angles the error of a cubature formula with regular boundary layer. Assume that the construction is performed in the same manner for each angle. This means that if some s-dimensional plane participates in several solid angles, then we take equal elementary errors supported near to this plane at a distance of Lh from every face of dimension $s-1$.

Theorem 3.7. *The error of a cubature formula with regular boundary layer in a rational polyhedron $\mathfrak{M}$ may be written as*

$$l(x) = \sum_{s=0}^{n}(-1)^s \sum_{j=1}^{\mu(s)} l^{(j,s)}(x), \tag{2.1}$$

with $l^{(j,s)}(x)$ standing for the errors with regular boundary layer constructed in the same manner for all solid angles of various dimension in $\mathfrak{M}$.

PROOF. Denote by M the set of elementary errors composing $l(x)$; and by $M_{j,s}$, the set of elementary errors composing $l^{(j,s)}(x)$. To every integer vector γ there may correspond several distinct elementary errors. For instance, to the point $h\gamma$ lying near to the plane of some $(n-1)$-dimensional face, there corresponds the functional $l_0\,(x/h-\gamma)$ entering the formula for the angle of order zero, that is for the whole space, and the functional $l_\gamma\,(x/h-\gamma)$ participating in the expansion of $l^{(j,n-1)}(x)$.

Label with the index t all elementary errors that correspond to the same γ, denoting them by $l_\gamma^t\,(x/h-\gamma)$. On the set of pairs (γ, t) define the indicators of the polyhedron and its solid angles

$$\chi_{\mathfrak{M}}(\gamma,t) = \begin{cases} 1, & \text{if } l_\gamma^t(x/h-\gamma) \in M, \\ 0, & \text{if } l_\gamma^t(x/h-\gamma) \notin M, \end{cases} \tag{2.2}$$

$$\chi_{j,s}(\gamma,t) = \begin{cases} 1, & \text{if } l_\gamma^t(x/h-\gamma) \in M_{j,s}, \\ 0, & \text{if } l_\gamma^t(x/h-\gamma) \notin M_{j,s}. \end{cases} \tag{2.3}$$

For these indicators we have the identity

$$\chi_{\mathfrak{M}}(\gamma,t) = \sum_{s=0}^{n}\sum_{j=1}^{\mu(s)}(-1)^s \chi_{j,s}(\gamma,t) \tag{2.4}$$

which follows from Theorem 3.4. We clarify the formula (2.4) by example. The simplest case corresponds to an interval $[a, b]$ with equidistant nodes $h\gamma$, $\gamma = 0, \pm 1, \ldots$. Here, possible are three types of elementary errors differing in the manner of their intersecting the elementary meshes $\{x : \gamma < x/h < \gamma + 1\}$ with the interval (a, b). To the meshes lying in (a, b), there corresponds an elementary error l_0. If a mesh contains the point a, then we denote the elementary error by l_1; and if it contains the point b, then we denote the elementary error by l_2. The index t may assume one or two values. At each γ the functional l_0 belongs to the set $\{l_\gamma^t\}$. Given γ, put $l_\gamma^1 = l_0$. For simplicity, set $L = 1$. Then for $t = 2$ we have

$$l_\gamma^2 = \begin{cases} l_1 & \text{at a point } h\gamma \text{ on the right of } a, \\ l_2 & \text{at a point } h\gamma \text{ on the left of } b. \end{cases}$$

In the example under consideration, the polyhedron $\overline{\mathfrak{M}}$ is the interval $[a, b]$, the solid angle $\Omega_{1,0}$ is the half-line $[a, \infty)$, the solid angle $\Omega_{2,0}$ is the half-line $(-\infty, b]$, and $\Omega_{1,1} = \mathbb{R}^1$; the formula (2.4) takes the form

$$\chi_{[a,b]}(\gamma, 1) = \chi_{[a,\infty)}(\gamma, 1) + \chi_{(-\infty,b]}(\gamma, 1) - \chi_{\mathbb{R}^1}(\gamma, 1),$$
$$\chi_{[a,b]}(\gamma, 2) = \chi_{[a,\infty)}(\gamma, 2) + \chi_{(-\infty,b]}(\gamma, 2).$$

Both equalities are easy to check.

Continue proving Theorem 3.7. From (2.4) it follows that

$$l(x) = \sum_{\gamma,t} \chi_{\mathfrak{M}}(\gamma, t) l_\gamma^t \left(\frac{x}{h} - \gamma\right) = \sum_{\gamma,t} \sum_{s=0}^{n} \sum_{j=1}^{\mu(s)} (-1)^s \chi_{j,s}(\gamma, t) l_\gamma^t \left(\frac{x}{h} - \gamma\right)$$
$$= \sum_{s=0}^{n} \sum_{j=1}^{\mu(s)} (-1)^s \sum_{\gamma,t} \chi_{j,s}(\gamma, t) l_\gamma^t \left(\frac{x}{h} - \gamma\right) = \sum_{s=0}^{n} \sum_{j=1}^{\mu(s)} (-1)^s l^{(j,s)}(x).$$

The proof of Theorem 3.7 is complete.

§3. A Formal Boundary Layer

Let $l(x)$ be the error of a cubature formula for a bounded polyhedron $\mathfrak{M}$ with a parallelepipedic lattice, i.e. of the shape

$$\chi_{\mathfrak{M}}(x) - \sum_{\gamma \in B} c[\gamma] h^n \delta(x - hH\gamma). \tag{3.1}$$

Assume additionally that the error is orthogonal to every polynomial of degree at most m,

$$(l(x), x^\alpha) = 0, \quad |\alpha| \le m. \tag{3.2}$$

Clearly, this property is preserved for arbitrary linear combinations of such errors.

For generalized functions like (3.1) we have a certain test for orthogonality to polynomials (3.2) which is based on the Fourier transform. A necessary and sufficient condition for orthogonality of $l(x)$ to a polynomial consists in the requirement that at the point $x = 0$ all derivatives vanish of its Fourier transform up to order m. We call the respective conditions *formal orthogonality* of $l(x)$ *to every polynomial* of degree m.

Using the concept of *formal orthogonality to polynomials*, we derive the conditions under which the error of a cubature formula for Ω presenting a rational polyhedron is an error with regular boundary layer.

Construct the errors

$$l^{(j,s)}(x) = \sum_{hH\gamma\in\Omega_{j,s}} c^{j,s}[\gamma]\delta\left(\frac{x}{h} - H\gamma\right)$$

for all solid angles $\Omega_{j,s}$ of a rational polyhedron $\mathfrak{M}$ so that they possess the following properties:

(a) the weights $c^{j,s}[\gamma]$ at all points that lie near to some face of a solid angle $\Omega_{j,s}$ at a distance at most $2Lh$ from it and lie at a distance at least $2Lh$ from the boundary of this face are constant on every hyperplane parallel to the face;

(b) if two solid angles (possibly, of different orders) have a common face then the weights of the errors of both angles coincide near to this face.

Theorem 3.8. *The alternating sum of the errors with the properties* (a) *and* (b) *for the solid angles of a polyhedron* $\mathfrak{M}$ *is the error of a cubature formula for this polyhedron, i.e.,*

$$\sum_{s=0}^{n}(-1)^s \sum_{j=1}^{\mu(s)} l^{(j,s)}(x) = l(x). \tag{3.3}$$

PROOF. Theorem 3.4 implies that it suffices to establish (3.3) for the discrete components of the errors. An appropriate proof of this fact coincides with the proof of Theorem 3.7, if in all arguments we replace the error $l^t_\gamma(x/h-\gamma)$ with the generalized function $c^t[\gamma]\delta(x/h-\gamma)$. The proof of Theorem 3.8 is complete.

A cubature formula with error $l(x)$ of the shape (3.3) is called a *cubature formula with formal boundary layer* provided that each error $l^{(j,s)}(x)$ is formally orthogonal to every polynomial of degree at most $m-n+s$.

From this definition it follows that every error with regular boundary layer for a rational polyhedron is at the same time an error with formal boundary layer provided that all elementary functionals lying in neighborhoods of its faces coincide. We also prove a converse theorem, assuming (a) and (b) valid.

Theorem 3.9. *Every error of a cubature formula that possesses a formal boundary layer is an error with regular boundary layer.*

The usefulness of this theorem is apparent. It implies that errors with formal boundary layer are asymptotically optimal. On the other hand, to construct them is easier than to construct a formula with regular boundary layer.

The proof of Theorem 3.9 was given by L. V. Voĭtishek. It leans upon a succession of lemmas.

Lemma 3.1. *Assume given a trigonometric polynomial $Q^{(k)}(p')$ in k variables $p' = (p_1, \dots, p_k)$ such that its derivatives up to order r vanish at the coordinate origin. Then the trigonometric polynomial*

$$Q(p) = e^{i2\pi\alpha'' p''} Q^{(k)}(p')$$

in the n variables $p = (p_1, \dots, p_k, p_{k+1}, \dots, p_n) = (p', p'')$, with $p'' = (p_{k+1}, \dots, p_n)$, has the derivatives up to order r vanishing at the coordinate origin.

The validity of Lemma 3.1 is perfectly clear.

Introduce the concept of *zero's error*. By the latter we mean a linear combination of the shape

$$m(x) = \sum_{\gamma \in B} c[\gamma]\delta\left(\frac{x}{h} - H\gamma\right)$$

which vanishes at all polynomials of degree m. We speak of a *zero's error with regular boundary layer* provided that

$$m(x) = \sum_{\gamma \in B_L^{(i)}} m_0\left(\frac{x}{h} - H\gamma\right) + \sum_{\gamma \in B_L^{(l)}} m_\gamma\left(\frac{x}{h} - H\gamma\right). \tag{3.4}$$

Here $m_0(x)$ and $m_\gamma(x)$ are the local zero's errors for the elementary meshes of the domain $\mathfrak{M}$, with $m_\gamma(x), \gamma \in B_{j,s}$, orthogonal to every polynomial of degree $m-n+s$ and $m_0(x)$ orthogonal to every polynomial of degree m. The number s stands for the dimension of the face of $\mathfrak{M}$ near which lies the part of the boundary layer containing $m_\gamma(x)$.

It is not hard to check that in a zero's error with regular boundary layer all weights vanish at the points $hH\gamma$ lying in $\mathfrak{M}$ at distance greater than Lh from the boundary of $\mathfrak{M}$. (These weights are the sum of all weights of $m_0(x)$ orthogonal to constant functions by definition.)

Lemma 3.2. *The sum of every error of a cubature formula with regular boundary layer and a zero's error with regular boundary layer is the error of a formula with regular boundary layer.*

PROOF. The sum

$$l_\gamma\left(\frac{x}{h}-H\gamma\right)+m_\gamma\left(\frac{x}{h}-H\gamma\right)=l_\gamma^{(1)}\left(\frac{x}{h}-H\gamma\right)$$

is again an error of the type under consideration which is orthogonal to every polynomial of degree $m-n+s$. We have the obvious equality

$$m(x)+l(x)=\sum_\gamma l_\gamma^{(1)}\left(\frac{x}{h}-H\gamma\right),$$

entailing the claim of Lemma 3.2.

We speak of $m(x)$ as of a *zero's error with formal boundary layer for a polyhedron* provided that

$$m(x)=\sum_{s=0}^{n}(-1)^s\sum_{j=1}^{\mu(s)}m^{(j,s)}(x),$$

and, moreover, each error

$$m^{(j,s)}(x)=\sum_{hH\gamma\in\Omega_{j,s}}a^{j,s}[\gamma]\delta\left(\frac{x}{h}-H\gamma\right) \tag{3.5}$$

is compactly-supported and formally orthogonal to all polynomials of degree $m-n+s$.

Let $a^{j,s}[\gamma]$ satisfy conditions (a) and (b) of Theorem 3.8, and let $a^{j,n}[\gamma]$ be all zero. We then have

Lemma 3.3. *Each zero's error with formal boundary layer is a zero's error with regular boundary layer.*

PROOF. Theorem 3.8 implies that is suffices to prove the lemma for the solid angles of $\mathfrak{M}$. We accomplish the prove by induction, starting from $s=n$. For $s=n$ the initial error $m^{(n)}(x)$ is zero. Hence, it may be written as (3.4), with $m_\gamma^{(n)}(x)\equiv 0$. Assume now that the errors $m^{(j,s)}(x)$ for $s=n,n-1,\dots,k+1$, $j=1,\dots,\mu(s)$ may be written as (3.4). Consider the error $m^{(l,k)}(x)$ for the angle $\Omega_{l,k}$ with k-dimensional blade. Denote by $m_1^{(l,k)}(x)$ the sum of those elementary errors $m_\gamma^{(j,s)}(x)$, $s=n,\dots,k+1$, whose supports are located in $\Omega_{l,k}$. The difference $\overline{m}^{(l,k)}(x)=m^{(l,k)}(x)-m_1^{(l,k)}(x)$ is again a zero's error of the shape (3.5) for the angle $\Omega_{l,k}$ which is formally orthogonal to every polynomial of degree $m-n+k$. The support of the error $\overline{m}^{(l,k)}(x)$ lies on finitely many planes parallel to the plane $R_{l,k}$ serving as the blade of the solid angle $\Omega_{l,k}$. We are left with proving that $\overline{m}^{(l,k)}(x)$ may be written as sum of elementary errors.

An affine transformation does not violate the regularity property of a boundary layer and sends a formal boundary layer to a formal boundary layer. We may thus conduct the proof only in the case of a cubic lattice when $h = 1$ and the blade $R_{l,k}$ lies in the plane $\{x : x_j = 0,\ j = k+1, \dots, n\}$. In this event the formal boundary layer in the error has weights depending only on $\gamma'' = (\gamma_{k+1}, \dots, \gamma_n)$ and $\overline{m}^{(l,k)}(x)$ may be written as

$$\overline{m}^{(l,k)}(x) \equiv c(x) = \sum_{\gamma'} \sum_{\gamma''} a^{(k)}[\gamma'']\delta(x' - \gamma')\delta(x'' - \gamma'')$$
$$= \Phi_0(x') \sum_{\gamma''} a^{(k)}[\gamma'']\delta(x'' - \gamma'').$$

Here $x = (x', x'')$, $x' = (x_1, \dots, x_k)$, and $\gamma = (\gamma', \gamma'')$, while $\Phi_0(x')$ denotes the generalized function $\Phi_0(x_1) \dots \Phi_0(x_k)$ with

$$\Phi_0(x_i) = \sum_{j=-\infty}^{\infty} \delta(x_i - j).$$

The functional $c(x)$ is by hypothesis formally orthogonal to every polynomial of degree $m-n+k$; i.e., its Fourier transform $\tilde{c}(p)$ vanishes at the coordinate origin together with all derivatives up to order $m-n+k$. Calculate this Fourier transform. By definition

$$\tilde{c}(p) = \Phi_0(p') \sum_{\gamma''} a^{(k)}[\gamma'']e^{i2\pi p''\gamma''} = \Phi_0(p')Q^{(k)}(p'').$$

Since we are to calculate the values of derivatives of $\tilde{c}(p)$ only at the coordinate origin, we may replace $\Phi_0(p')$ with $\delta(p')$. The formal orthogonality condition is written as

$$D^{\alpha}[\delta(p')Q^{(k)}(p'')]|_{p=0} = 0, \quad |\alpha| \le m - n + k. \tag{3.6}$$

Expressing the condition that the derivatives with respect to p'' vanish, obtain

$$D^{\alpha''}Q^{(k)}(p'')|_{p''=0} = 0, \quad |\alpha''| \le m - n + k.$$

The other conditions (3.6) are fulfilled automatically for $p'' \ne 0$, because the derivatives $D^{\alpha'}_{p'} D^{\alpha''}_{p''} Q^{(k)}(p'')$ present derivatives with respect to p' of a function that is actually independent of p' and so these derivatives are identically 0.

It is not hard to see that $c(x)$ may be written as

$$c(x) = \sum_{\gamma'} \left\{ \sum_{\gamma''} a^{(k)}[\gamma'']\delta(x - \gamma) \right\} = \sum_{\gamma'} m_{\gamma'}(x).$$

Here every summand is an elementary error formally orthogonal to all polynomials of degree $m-n+k$. Indeed, the Fourier transform of $m_{\gamma'}(x)$ equals $e^{i2\pi\gamma' p'}Q^{(k)}(p'')$ and by what was proven above vanishes at the coordinate origin together with all derivatives up to order $m-n+k$. On the other hand, $m_{\gamma'}(x)$ is a compactly-supported functional since its support is concentrated in the bounded domain $|x''| < L$, $x' = \gamma'$. Consequently, $m_{\gamma'}(x)$ is orthogonal to the polynomials in question actually rather than just formally. Therefore, $c(x)$ is written as the sum of elementary errors orthogonal to all polynomials of degree $m-n+k$, i.e. it is a zero's error with regular boundary layer. The proof of Lemma 3.3 is complete.

We now proceed to proving Theorem 3.9. Let l_F be the error of a cubature formula with formal boundary layer for a polyhedron $\mathfrak{M}$. Considering an arbitrary error l_R with regular boundary layer for this polyhedron, we have

$$l_F(x) = l_R(x) + (l_F(x) - l_R(x)). \tag{3.7}$$

The difference $l_F(x) - l_R(x)$ is obviously a zero's error with formal boundary layer which by Lemma 3.3 has a regular boundary layer. Further, by Lemma 3.2 the sum in (3.7) is in turn an error with regular boundary layer for $\mathfrak{M}$. The proof of Theorem 3.9 is complete.

For solid angles we describe a method for explicit construction of cubature formulas with formal boundary layer and, consequently, with regular boundary layer. It is convenient to construct an error with formal boundary layer by successively arranging boundary layers of various dimension, starting from $n-1$ and proceeding in decreasing order, i.e., for $n-2, n-3, \ldots, 0$.

Observe that an error for every solid angle may be written as the sum of the indicator of this angle and a linear combination of the products of the functions of one independent variable

$$\delta(x_s - \gamma_s), \quad \Phi_0(x_k) = \sum_{j=-\infty}^{\infty} \delta(x_k - j)$$

and

$$\Phi_1(x_t - \gamma_t) = \frac{1}{2}\delta(x_t - \gamma_t) + \sum_{j=1}^{\infty} \delta(x_t - \gamma_t - j).$$

The construction of a needed error thus reduces to determining the weights of this linear combination.

Now a few remarks are in order on the methods for calculating the Fourier transforms of the errors dealt with.

Consider the Fourier transform of the indicator of a solid angle and of the discrete component of this angle separately.

For a start, we introduce the concept of simplest angle. Each solid angle of order $n - s$ generated precisely by s faces of dimension $n - 1$ we call a ***simplest angle***. In 3-dimensional space, the examples of polyhedra whose every solid angle is simplest are the tetrahedron, the cube, the regular dodecahedron, etc. The Fourier transform of the indicator of a simplest angle is very easy to fulfill. By a linear change of variables we transform our simplest angle into the coordinate angle, with the indicator becoming the product of the functions $1/2 + 1/2 \operatorname{sgn} x_j$ whose Fourier transforms are equal to $1/2\delta(p_j) + 1/(i2\pi p_j)$. Whence we obtain an explicit description of the sought transform as the product of $\delta(p_k)$ and $1/2\delta(p_j) + 1/(i2\pi p_j)$. The Fourier transform of the discrete component of an error for a simplest angle is readily calculated.

If a polyhedron includes not only simplest angles, then we may always split it into parts possessing this property, for instance, into simplices. Then we should separately transform the indicators of these simplices and sum the results. The Fourier transform of the indicator of every convex polyhedron is thus implemented explicitly.

Calculating the Fourier transform of the discrete component of an error for a polyhedron, it is again convenient to split the polyhedron into simplest angles so that proceed further with summing the errors for these angles.

To avoid appearance of boundary layers on the common boundaries of simplest inner angles, while composing the error for each of them, we may use formulas with outer boundary layer spreading beyond the simplest solid angle under consideration. Moreover, we may readily guarantee that undesirable boundary layers cancel out when we sum the corresponding errors. This approach was realized for a solid angle presenting the sum of two simplest angles [35]. The problem of constructing a cubature formula with formal boundary layer for an arbitrary rational polyhedron in 3-dimensional space reduces to the case of the solid angles mentioned above.

The problems arising in actual numerical construction of cubature formulas with formal or regular boundary layer for a polyhedron are dealt within [35]. This article also contains a reduction of the problem of constructing a cubature formula for a rational angle to that of constructing a cubature formulas for a coordinate angle with the lattices of nodes which is a union of several cubic lattices. The latter problem was solved in [292].

The weights of cubature formulas for a cube with centered cubic lattice are calculated in the article [34]. A *centered cubic lattice* is composed of two cubic sublattices shifted with respect to one another by a half-mesh-size in every axis. The found cubature formulas are implemented as computer software [30–31].

Chapter 4

The Rate of Convergence of Cubature Formulas

In the current chapter we address the problem of the rate with which the norm of the error decreases as the number of nodes of a cubature formula increases. For the formulas with a given lattice of nodes $hH\gamma$ we establish estimates exact in order for strong convergence on the Banach space of periodic functions meeting several natural conditions. We prove analogous results for periodic Banach spaces without weight in Bessel scales. We also derive estimates for the norms of errors with equidistributed nodes which act in the $L_2^{(m)}$ space of aperiodic functions.

§1. A Universal Lower Bound on the Rate of Convergence

We inspect the norm of an arbitrary error with a given lattice of nodes $hH\gamma$ which acts in a Banach space $\widetilde{B}$ of periodic functions. This space satisfies the following conditions.

1. We assume that $\widetilde{B}$ consists of periodic functions with orthogonal period matrix H, i.e., of functions $\varphi(x)$ such that for every integer multi-index γ the equality holds

$$\varphi(x + H\gamma) = \varphi(x), \quad x \in \mathbb{R}^n.$$

In particular, every constant function belongs to $\widetilde{B}$.

2. The norm of $\widetilde{B}$ is *translation invariant.* In other words, for all $y \in \mathbb{R}^n$ and $\varphi(x) \in \widetilde{B}$ the function $\varphi(x+y)$ belongs to $\widetilde{B}$ and

$$\|\varphi(x+y) \mid \widetilde{B}\| = \|\varphi(x) \mid \widetilde{B}\|. \tag{1.1}$$

3. The space $\widetilde{B}$ admits a compact embedding into the space of continuous functions $C(\overline{\Omega}_0)$, with Ω_0 the fundamental parallelepiped of the matrix H.

4. For every function $\varphi(x)$ with mean 0, i.e. with the *mean* φ_0 over Ω_0 equal 0, namely,

$$\varphi_0 = \frac{1}{|\Omega_0|} \int_{\Omega_0} \varphi(x)\, dx = 0,$$

and every constant C the inequality holds

$$\|\varphi \mid \widetilde{B}\| \leq \|\varphi + C \mid \widetilde{B}\|. \tag{1.2}$$

The space $\widetilde{B}$ may contain complex-valued members. In this event, we require that, together with a function $\varphi(x)$, the space $\widetilde{B}$ contain the *conjugate function* $\overline{\varphi}(x)$ of $\varphi(x)$. The norms of these functions in $\widetilde{B}$ must coincide, namely,

$$\|\varphi(x) \mid \widetilde{B}\| = \|\overline{\varphi}(x) \mid \widetilde{B}\|. \tag{1.3}$$

In $\widetilde{B}$ we consider a cubature formula with lattice of nodes $hH\gamma$ and domain of integration Ω lying in Ω_0. We assume that, for a given h, among the corresponding errors there is a $\widetilde{B}$-*optimal error*, i.e., an error with a minimal $\widetilde{B}^*$ norm.

Note that the conditions imposed on $\widetilde{B}$ are for instance met by the spaces $\widetilde{W}_2^{(m)}$ defined in § 4 of Chapter 1 and also by their generalization, the space $\widetilde{W}_p^{(m)}$, $1 < p < \infty$, which we discuss in the subsequent section.

Define a positive h so that the ratio $1/h$ be a natural and study the corresponding value of the error

$$l_\infty(x) = \chi_{\Omega_0}(x) - \sum_{hH\gamma \in \Omega_0} h^n \delta(x - hH\gamma) \tag{1.4}$$

at a function $\varphi(x) \in \widetilde{B}$. Under the above hypotheses about the space $\widetilde{B}$, we have

Theorem 4.1. *The error $l_\infty(x)$ is asymptotically optimal in $\widetilde{B}$; i.e., the difference between the $\widetilde{B}^*$ norm of $l_\infty(x)$ and the $\widetilde{B}^*$ norm of an $\widetilde{B}$-optimal error is an infinitesimal as $h \to 0$.*

PROOF. By hypothesis, at a given h there exists a $\widetilde{B}$-optimal error $l_0(x)$, i.e., an error having a minimal norm in $\widetilde{B}^*$. Arrange the arithmetic mean of its *translates*, the functionals $l_0(x - hH\gamma)$, over all $hH\gamma \in \Omega_0$. In other words, put

$$l^0(x) = \frac{1}{N} \sum_{hH\gamma \in \Omega_0} l_0(x - hH\gamma), \tag{1.5}$$

with N the total number of the points $hH\gamma$ in Ω_0, i.e., $Nh^n = |H|$. Owing to the properties of the norm of $\widetilde{B}$, particularly, its invariance under translations, the functional $l^0(x)$ is also $\widetilde{B}$-optimal. Moreover, we may readily verify that $l^0(x)$ has the next shape

$$l^0(x) = \chi_{\Omega_0}(x) - h^n c_0(h) \sum_{hH\gamma\in\Omega_0} \delta(x - hH\gamma) = (1 - c_0)\chi_{\Omega_0}(x) + c_0 l_\infty(x). \quad (1.6)$$

Examine how the function $c_0(h)$ in this equality behaves as $h \to 0$.

The error $l_\infty(x)$ vanishes at the constantly-one function. Using this, from (1.6) obtain

$$|\Omega_0|\,|1 - c_0(h)| = |(l^0(x), 1)| \leq K\|l^0 \mid \widetilde{B}^*\| \leq K\|l_\infty \mid \widetilde{B}^*\|. \quad (1.7)$$

Since the embedding of $\widetilde{B}$ to $C(\overline{\Omega}_0)$ is compact, it follows in much the same way as in Lemma 1.5 that the norm $l_\infty(x)$ in $\widetilde{B}^*$ tends to 0 as the mesh-size h vanishes. Whence and from (1.7) derive that

$$c_0(h) = 1 + O(h). \quad (1.8)$$

Compare the norms of $l_\infty(x)$ and $l^0(x)$. By (1.6), we have

$$\|l^0 \mid \widetilde{B}^*\| = \sup_{\varphi\in\widetilde{B}} \frac{|((1 - c_0)\chi_{\Omega_0}(x) + c_0 l_\infty(x), \varphi(x))|}{\|\varphi \mid \widetilde{B}\|}.$$

Take the supremum on the right side over the subset $\widetilde{B}$ composed of φ with the mean φ_0 over Ω_0 equal to 0;

$$\varphi_0 = \frac{1}{|\Omega_0|} \int_{\Omega_0} \varphi(x)\, dx = 0.$$

We then derive

$$\|l_0 \mid \widetilde{B}^*\| \geq \sup_{\varphi\in\widetilde{B},\, \varphi_0=0} \frac{|(c_0 l_\infty(x), \varphi(x))|}{\|\varphi \mid \widetilde{B}\|}. \quad (1.9)$$

Considering that the cubature formula corresponding to $l_\infty(x)$ is exact for the constantly-one function, rewrite the least upper bound on the right side of (1.9) as follows

$$|c_0| \sup_{\varphi\in B} \frac{|(l_\infty, \varphi(x) - \varphi_0)|}{\|\varphi(x) - \varphi_0 \mid \widetilde{B}\|} = |c_0| \sup_{\varphi\in\widetilde{B}} \frac{|(l_\infty, \varphi)|}{\|\varphi(x) - \varphi_0 \mid \widetilde{B}\|}. \quad (1.10)$$

The mean value of $\varphi(x) - \varphi_0$ over Ω_0 equals 0. Whence and from (1.2) it follows that

$$\|\varphi - \varphi_0 \mid \widetilde{B}\| \leq \|\varphi \mid \widetilde{B}\|.$$

Inserting this inequality in (1.10), from (1.9) obtain

$$\|l_0 \mid \widetilde{B}^*\| \geq |c_0| \sup_{\varphi \in \widetilde{B}} \frac{|(l_\infty, \varphi)|}{\|\varphi \mid \widetilde{B}\|} = |c_0| \, \|l_\infty \mid \widetilde{B}^*\|.$$

Whence, from the $\widetilde{B}$-optimality of l_0 and from (1.8), infer

$$1 \leq \frac{\|l_\infty \mid \widetilde{B}^*\|}{\|l_0 \mid \widetilde{B}^*\|} \leq \frac{1}{|c_0|} = 1 + O(h).$$

Passing to the limit as $h \to 0$ observe that $l_\infty(x)$ is indeed an asymptotically $\widetilde{B}$-optimal error. The proof of Theorem 4.1 is complete.

We are able to prove that, at sufficiently small h, the error $l_\infty(x)$ is optimal but not only asymptotically optimal when the norm of $\widetilde{B}$ satisfied a stronger condition than (1.2). Namely, the claim just formulated is valid if the norm of each function $\varphi \in \widetilde{B}$ maintains the relation

$$\|\varphi \mid \widetilde{B}\| = \max\{\|\varphi - \varphi_0 \mid \widetilde{B}\|, |\varphi_0|\}.$$

The fulfillment of this condition or the inequality (1.2) may be achieved on furnishing $\widetilde{B}$ with some new norm by the formula

$$\|\varphi \mid \widetilde{B}\|_1 = \max\{\|\varphi - \varphi_0 \mid \widetilde{B}\|, |\varphi_0|\}.$$

Consider the action in $\widetilde{B}$ of an arbitrary error $l(x)$ with the lattice of nodes $hH\gamma$ and the integration domain Ω with $\overline{\Omega}$ lying in the interior of Ω_0. Let

$$l(x) = \chi_\Omega(x) - h^n \sum_{hH\gamma \in \Omega} c[\gamma]\delta(x - hH\gamma). \tag{1.11}$$

The concept of extremal function for $l(x)$ is easy to define in $\widetilde{B}$. However, we must in general establish existence of an extremal function. To obviate the difficulty, we introduce the concept of maximizing sequence for $l(x)$ in $\widetilde{B}$.

Say that $\{u_j\}$ is a *maximizing sequence for* $l(x)$ *in* $\widetilde{B}$, if every u_j lies in the unit ball of $\widetilde{B}$ and at least one of the equalities holds

$$\|l \mid \widetilde{B}^*\| = \lim_{j\to\infty} \frac{(l, u_j)}{\|u_j \mid \widetilde{B}\|}, \quad \|l \mid \widetilde{B}^*\| = \lim_{j\to\infty} \frac{|(l, u_j)|}{\|u_j \mid \widetilde{B}\|}. \tag{1.12}$$

Existence of a maximizing sequence is immediate from the definition of the norm of $\widetilde{B}^*$. For $l_\infty(x)$ it is possible to choose such sequence in a special manner. Demonstrate that the following claim is true.

Lemma 4.1. *The error $l_\infty(x)$ has a maximizing sequence $\{v_j\}$ possessing the properties:*

(1) every function $v_j(x)$ has real values and is sent by $l_\infty(x)$ to a nonnegative real, i.e.,

$$(l_\infty, v_j) \geq 0; \tag{1.13}$$

(2) the mean value of every function v_j over Ω_0 is equal to 0 and its norm in $\widetilde{B}$ is equal to 1

$$\frac{1}{|\Omega_0|} \int\limits_{\Omega_0} v_j(x)\, dx = 0, \quad \|v_j \mid \widetilde{B}\| = 1; \tag{1.14}$$

(3) every function $v_j(x)$ is not only periodic with period matrix H but also periodic with period matrix hH, i.e.,

$$v_j(x + hH\gamma) = v_j(x), \quad x \in \mathbb{R}^n, \quad \gamma \in \mathbb{Z}^n; \tag{1.15}$$

(4) the lower limit as $j \to \infty$ of the difference between the function $v_j(x)$ and its value at zero is nonnegative, i.e.,

$$\varliminf_{j\to\infty} (v_j(x) - v_j(0)) \geq 0. \tag{1.16}$$

PROOF. Consider a maximizing sequence $\{u_j\}$ for the functional $l_\infty(x)$ in $\widetilde{B}$ which satisfies the first of the equalities (1.12). Transform it so that all conditions (1.13)–(1.16) be met.

Observe that every function $u_j(x)$ may be assumed real. Otherwise, it suffices to take as $\{u_j\}$ the sequence of the real parts of $u_j(x)$ which is, as we show right away, also a maximizing sequence for $l_\infty(x)$. Indeed, convexity of the unit ball in $\widetilde{B}$ and (1.3) give

$$\|\operatorname{Re} u_j \mid \widetilde{B}\| = \left\| \frac{u_j + \bar{u}_j}{2} \mid \widetilde{B} \right\| \leq \|u_j \mid \widetilde{B}\| \leq 1,$$

implying in particular that

$$\frac{|(l_\infty, \operatorname{Re} u_j)|}{\|\operatorname{Re} u_j \mid \widetilde{B}\|} \geq \frac{|(l_\infty, \operatorname{Re} u_j)|}{\|u_j \mid \widetilde{B}\|} = \left| \operatorname{Re} \frac{(l_\infty, u_j)}{\|u_j \mid \widetilde{B}\|} \right|.$$

Passing here to the limit as $j \to \infty$ and using the definition (1.12), obtain

$$\lim_{j\to\infty} \frac{|(l_\infty, \operatorname{Re} u_j)|}{\|\operatorname{Re} u_j \mid \widetilde{B}\|} \geq \|l_\infty \mid \widetilde{B}^*\|.$$

Hence, the sequence $\{\mathrm{Re}\, u_j\}$ satisfies the second of the equalities (1.12) and so it is a maximizing sequence for $l_\infty(x)$.

Considering that the $u_j(x)$ is a real function, construct three new maximizing sequences by putting

$$z_j^{(1)}(x) = u_j(x)\,\mathrm{sgn}(l_\infty, u_j),$$
$$z_j^{(2)}(x) = z_j^{(1)}(x) - \frac{1}{|\Omega_0|}\int\limits_{\Omega_0} z_j^{(1)}(x)\,dx,$$
$$z_j^{(3)}(x) = h^n \sum_{hH\gamma\in\Omega_0} z_j^{(2)}(x - hH\gamma).$$

It is not hard to check that $z_j^{(3)}(x)$ meets the conditions (1.13) and (1.15) and also the first of the conditions (1.14). Estimate the norm of $z_j^{(3)}(x)$ in $\widetilde{B}$. Convexity of the unit ball of $\widetilde{B}$ implies that

$$\|z_j^{(3)} \mid \widetilde{B}\| \le \|z_j^{(2)} \mid \widetilde{B}\|. \tag{1.17}$$

The condition (1.2) provides the estimate

$$\|z_j^{(2)} \mid \widetilde{B}\| \le \|z_j^{(1)} \mid \widetilde{B}\| \le \|u_j \mid \widetilde{B}\|. \tag{1.18}$$

Combining (1.17) and (1.18), derive a sought estimate for the norm $z_j^{(3)}(x)$.

The error $l_\infty(x)$ is obviously invariant under the translation by the vector $hH\gamma \in \Omega_0$ and vanishes at the constantly-one function. Therefore

$$\left(l_\infty, z_j^{(3)}\right) = \left(l_\infty, z_j^{(2)}\right) = |(l_\infty, u_j)|. \tag{1.19}$$

From (1.17)–(1.19) it follows that $\{z_j^{(3)}(x)\}$ is a maximizing sequence in $\widetilde{B}$ for $l_\infty(x)$. Consequently, so is the sequence

$$v_j(x) = \frac{z_j^{(3)}(x)}{\|z_j^{(3)} \mid \widetilde{B}\|}. \tag{1.20}$$

It is easy to check that $\{v_j(x)\}$ satisfies (1.13)–(1.15). We are left with validating (1.16), thus proving that $\{v_j\}$ is a sought sequence.

Take an arbitrary $x \in \Omega_0$ and consider the function $v_j(y+x) - v_j(x)$ of the argument $y \in \mathbb{R}^n$. By (1.15) this difference vanishes at the nodes $hH\gamma$ and so $l_\infty(y)$ sends it to the integral over Ω_0, i.e.,

$$(l_\infty(y), v_j(y+x) - v_j(x)) = \int_{\Omega_0} (v_j(y+x) - v_j(x))\, dy = \int_{\Omega_0} v_j(y)\, dy - |\Omega_0| v_j(x). \tag{1.21}$$

The equality (1.21) is valid in particular at $x = 0$. Whence it follows that

$$(l_\infty(y), v_j(y)) = (l_\infty(y), v_j(y) - v_j(0)) = \int_{\Omega_0} v_j(y)\, dy - |\Omega_0| v_j(0). \tag{1.22}$$

Expressing via this relation the integral of $v_j(y)$ over Ω_0, inserting the result in (1.21) and considering that the value of $l_\infty(y)$ is equal to 0 at the function $v_j(x)$ constant in y, obtain

$$(l_\infty(y), v_j(y+x)) = (l_\infty(y), v_j(y+x) - v_j(x)) = (l_\infty(y), v_j(y)) - |\Omega_0|(v_j(x) - v_j(0)). \tag{1.23}$$

Estimate from above the left side of (1.23). From (1.1) and (1.14) derive

$$|(l_\infty(y), v_j(y+x))| \le \|l_\infty \mid \widetilde{B}^*\| \, \|v_j(y+x) \mid \widetilde{B}\| \le \|l_\infty \mid \widetilde{B}^*\|.$$

Using this inequality together with the fact that $\{v_j(y)\}$ is a maximizing sequence for $l_\infty(y)$ and passing to the lower limit over j on both sides of (1.23), arrive at the sought relation (1.16). The proof of Lemma 4.1 is complete.

Return to the error $l(x)$ defined by (1.11). Let the boundary of the integration domain Ω be piecewise-smooth. Then we have

Theorem 4.2. *The norm of the error $l(x)$ in $\widetilde{B}^*$ admits the following lower bound*

$$\|l(x) \mid \widetilde{B}^*\| \ge \frac{|\Omega|}{|\Omega_0|} \|l_\infty(x) \mid \widetilde{B}^*\| (1 + O(1)), \tag{1.24}$$

with $O(1)$ vanishing as h decreases.

PROOF. Considering $l_\infty(x)$, find a maximizing sequence $v_j(x)$ possessing the properties that are listed in Lemma 4.1. From (1.22) and (1.14) it follows that

$$|v_j(0)| \le \frac{1}{|\Omega_0|} |(l_\infty, v_j)| \le \frac{1}{|\Omega_0|} \|l_\infty \mid \widetilde{B}^*\|.$$

As mentioned above, the last quantity vanishes with h approaching 0. Therefore,

$$\|v_j(x) - v_j(0) \mid \widetilde{B}\| \leq \|v_j \mid \widetilde{B}\| + |v_j(0)| \, \|1 \mid \widetilde{B}\| = 1 + O(1), \tag{1.25}$$

with $O(1)$ tending to 0 uniformly in j. Considering this, we readily estimate the norm of $l(x)$ in $\widetilde{B}^*$ from below. By the definition of norm and (1.25) we have

$$\|l \mid \widetilde{B}^*\| \geq \overline{\lim_{j\to\infty}} \int_\Omega (v_j(x) - v_j(0))\, dx (1 + O(1)). \tag{1.26}$$

Split the integral in (1.26) into the sum of two integrals over disjoint subsets of Ω. One of them, $\Omega^{(1)}$, is the union of all elementary meshes of the lattice of nodes $hH\gamma$ which lie entirely in the interior of Ω, and the other, $\Omega^{(2)}$, complements $\Omega^{(1)}$ to Ω. By (1.16) and the Fatou Lemma, we have

$$\overline{\lim_{j\to\infty}} \int_{\Omega^{(2)}} (v_j(x) - v_j(0))\, dx \geq \int_{\Omega^{(2)}} \overline{\lim_{j\to\infty}} (v_j(x) - v_j(0))\, dx \geq 0.$$

Consequently, discarding the integral over $\Omega^{(2)}$ may only strengthen (1.26). In turn, the integral over $\Omega^{(1)}$ of the difference $v_j(x) - v_j(0)$ is comparable with the integral of the same function but now over Ω_0. If N_1 is the number of elementary meshes comprising $\Omega^{(1)}$ and N is the total number of such meshes in Ω_0; then, by the periodicity condition (1.15),

$$\int_{\Omega^{(1)}} (v_j(x) - v_j(0))\, dx = \frac{N_1}{N} \int_{\Omega_0} (v_j(x) - v_j(0))\, dx. \tag{1.27}$$

Piecewise-smoothness of the boundary of Ω_0 yields the equality

$$N_1 h^n = |\Omega^{(1)}| = |\Omega|(1 + O(h)).$$

Considering that $Nh^n = |\Omega_0|$ and the integral on the right side of (1.27) coincides with the value of $l_\infty(x)$ at $v_j(x) - v_j(0)$, come to the relation

$$\int_{\Omega^{(1)}} (v_j(x) - v_j(0))\, dx = \frac{|\Omega|}{|\Omega_0|} (l_\infty(x), v_j(x))(1 + O(h)).$$

Passing here to the upper limit as $j \to \infty$ and using (1.26), obtain (1.24). The proof of Theorem 4.2 is complete.

REMARK. The claim of Theorem 4.2 remains valid if we replace the domain Ω with a set of the shape of $\Omega^{(1)}$, i.e., a union of finitely many meshes of the lattice of nodes $hH\gamma$. In other words, (1.24) holds for a sequence of errors like (1.11) for which not only the lattice but also the integration domain may depend on h, splitting into an integer number of elementary meshes at each h.

Theorem 4.2 also remains valid in the case when some nodes of the error in question lie in Ω_0 rather than only in Ω, i.e., when

$$l(x) = \chi_\Omega(x) - h^n \sum_{hH\gamma\in\Omega_0} c[\gamma]\delta(x - hH\gamma).$$

We now define a Bessel scale of Banach spaces and estimate from below the norm of $l_\infty(x)$ in each member $\widetilde{B}$ of such scale.

Let $\widetilde{B}_0$ be a Banach space of functions $\varphi(x)$ with orthogonal period matrix H,

$$\varphi(x + H\gamma) = \varphi(x), \quad x \in \mathbb{R}^n, \quad \gamma \in \mathbb{Z}^n.$$

Suppose that an arbitrary function $\varphi(x) \in \widetilde{C}(\overline{\Omega}_0)$, i.e. a continuous periodic function, belongs to $\widetilde{B}_0$, satisfying the estimate

$$\|\varphi \mid \widetilde{B}_0\| \leq K\|\varphi \mid \widetilde{C}(\overline{\Omega}_0)\|. \tag{1.28}$$

Suppose further that each function $\varphi(x) \in \widetilde{B}_0$ expands in the Fourier series

$$\varphi(x) = \sum_\gamma \varphi[\gamma]e^{i2\pi H^{-1}x\gamma}. \tag{1.29}$$

Now, assume given a function $\mu(\xi)$ of a variable ξ in $\mathbb{R}^n$ which we call *weight.* Say that $\varphi(x)$ *belongs to* $\widetilde{B}_0^\mu$ provided that finite is the following quantity

$$\|\varphi \mid \widetilde{B}_0^\mu\| = \left\|\sum_\gamma \varphi[\gamma]\mu(H^{-1*}\gamma)e^{i2\pi\gamma H^{-1}x} \mid \widetilde{B}_0\right\|. \tag{1.30}$$

If $\mu(\xi) \neq 0$ then (1.30) determines a norm on $\widetilde{B}_0^\mu$. We now say that the collection of spaces $\widetilde{B}_0^\mu$ constitutes the *Bessel scale with index-zero space* $\widetilde{B}_0$ *and weight function* $\mu(\xi)$.

We impose a few conditions on the *weight function* $\mu(\xi)$.

1. Assume that $\mu(\zeta)$ is positive and grows at infinity as a power function. More precisely, there exist positive constants K_1, m_1, and m_2 such that $m_1 < m_2$ and for all $\xi \in \mathbb{R}^n$ the inequality holds

$$K_1^{-1}(1 + |\xi|)^{m_1} \leq \mu(\xi) \leq K_1(1 + |\xi|)^{m_2}. \tag{1.31}$$

2. Assume that there is a positive constant K_2 such that for $t \geq K_2$ and $j = 1$ the function

$$f_j(t) = (1+t)^{-m_j} \min_{|\xi|=t} \mu(\xi) \tag{1.32}$$

increases whereas for $t \geq K_2$ and $j = 2$ it decreases.

3. The function $\mu(\xi)$ agrees with the geometry of the lattice of nodes $H^{-1*}\gamma$. Namely, there are constants $R > 1$ and $K > 0$ and nonzero integer multi-index β_0 such that

$$\mu\left(H^{-1*}\frac{\beta_0}{h}\right) \leq K \min_{1/R \leq |\xi| \leq R} \mu\left(\frac{\xi}{h}\right). \tag{1.33}$$

If we take as $\widetilde{B}_0$ the space $\widetilde{L}_p(\Omega_0)$, $1 < p < \infty$, and take the function $\mu(\xi)$ equal $(1 + |2\pi\xi|^2)^{m/2}$, with $m_1 < m < m_2$, then we obtain the space $\widetilde{W}_p^{(m)}$. If the lattice is *cubic* at that, i.e., H is the identity matrix; then all conditions (1.31)–(1.33) are fulfilled.

Observe that (1.33) is the most restrictive among the conditions on the function $\mu(\xi)$. For instance, (1.33) is not met in the case of $n = 2$ when $\widetilde{B}_0^\mu$ coincides with $\widetilde{W}_2^{1,2}(\Omega_0)$ and the matrix H corresponds to the rotation of the plane by the angle $\pi/3$ with the irrational tangent,

$$H = \begin{pmatrix} 1/2 & \sqrt{3}/2 \\ -\sqrt{3}/2 & 1/2 \end{pmatrix}.$$

Theorem 4.3. *The norm of $l_\infty(x)$ acting in an arbitrary Banach space $\widetilde{B}_0^\mu$ of a Bessel scale admits the following lower bound*

$$\left\| l_\infty(x) \mid \widetilde{B}_0^{\mu*} \right\| \geq K \sup_\xi \psi(\xi, h), \tag{1.34}$$

with

$$\psi(\xi, h) = \frac{1}{\mu(\xi)} \frac{1}{[h(1+|\xi|)]^{-m_1} + [h(1+|\xi|)]^{-m_2}}. \tag{1.35}$$

PROOF. By hypothesis $1/h$ is an integer and every continuous periodic function belongs to the space $\widetilde{B}_0$. A particular member of the latter is the exponential with exponent $i2\pi\beta H^{-1}x/h$. Moreover, by (1.28) we have the relation

$$\left\| e^{i2\pi\beta H^{-1}x/h} \mid \widetilde{B}_0 \right\| \leq K \| e^{i2\pi\beta H^{-1}x/h} \mid \widetilde{C} \| = K, \tag{1.36}$$

with K independent of h. By the definition (1.30) of the norm in $\widetilde{B}_0^\mu$ we have

$$\left\| e^{i2\pi\beta H^{-1}x/h} \mid \widetilde{B}_0^\mu \right\| = \mu(H^{-1*}\beta/h) \| e^{i2\pi\beta H^{-1}x/h} \mid \widetilde{B}_0 \|. \tag{1.37}$$

In the case when β is a nonzero integer multi-index, the value of $l_\infty(x)$ at the corresponding exponential equals -1. This, together with (1.36) and (1.37), implies that the norm of $l_\infty(x)$ admits the following lower bound

$$\|l_\infty \mid \widetilde{B}_0^{\mu*}\| \geq \frac{|(l_\infty(x), e^{i2\pi\beta H^{-1}x/h})|}{\|e^{i2\pi\beta H^{-1}x/h} \mid \widetilde{B}_0^{\mu}\|} \geq \frac{K}{\mu(H^{-1*}\beta/h)}. \tag{1.38}$$

Consider the function $\psi(\xi, h)$ defined by (1.35). Elementary arguments show that we have the estimate

$$\sup_\xi \psi(\xi, h) \leq \sup_{t\geq 0} \frac{1}{h^{-m_1}f_1(t) + h^{-m_2}f_2(t)}, \tag{1.39}$$

with $f_j(t)$ given by (1.32). By hypothesis $f_1(t)$ increases and $f_2(t)$ decreases. Using this, we readily obtain two estimates

$$\begin{aligned} \sup_{t\leq t_0} \frac{1}{h^{-m_1}f_1(t) + h^{-m_2}f_2(t)} &= \sup_{t\leq t_0} \frac{h^{m_2}}{f_2(t)} \frac{1}{1 + (h(t+1))^{m_2-m_1}} \leq \frac{h^{m_2}}{f_2(t_0)}, \\ \sup_{t\geq t_0} \frac{1}{h^{-m_1}f_1(t) + h^{-m_2}f_2(t)} &= \sup_{t\geq t_0} \frac{h^{m_1}}{f_1(t)} \frac{1}{1 + (h(t+1))^{m_1-m_2}} \leq \frac{h^{m_1}}{f_1(t_0)}. \end{aligned} \tag{1.40}$$

Choose a point t_0 so that the values of the functions $h^{-m_1}f_1(t)$ and $h^{-m_2}f_2(t)$ at t_0 be the same. This is possible only if $t_0 = 1/h - 1$. Then from (1.39) and (1.40) it follows that

$$\sup_\xi \psi(\xi, h) \leq \max\left\{\frac{h^{m_1}}{f_1(t_0)}, \frac{h^{m_2}}{f_2(t_0)}\right\} = \frac{1}{\min\limits_{|\xi|=t_0} \mu(\xi)} = \frac{1}{\min\limits_{|\xi|=1-h} \mu(\xi/h)}. \tag{1.41}$$

Using (1.33), at small h we obtain

$$\min_{|\xi|=1-h} \mu\left(\frac{\xi}{h}\right) \geq \min_{1/R\leq|\xi|\leq R} \mu\left(\frac{\xi}{h}\right) \geq \mu\left(H^{-1*}\frac{\beta_0}{h}\right).$$

This together with (1.38) and (1.41) entails (1.34), completing the proof of Theorem 4.3.

§2. The Rate of Convergence of a Homogeneous Error

In $\mathbb{R}^n$, assume given a lattice of nodes $hH\gamma$, a set ω with $\overline{\omega}$ in the interior of the fundamental parallelepiped Ω_0 corresponding to H, and the error $l(x)$ resulting from summation of a sole local error $l_0(y)$ over all translations by $hH\gamma \in \omega$, i.e.,

$$l(x) = \sum_{hH\gamma\in\omega} l_0\left(\frac{x}{h} - H\gamma\right). \tag{2.1}$$

Assume that $l_0(y)$ is written as a linear combination of the indicator $\chi_0(y)$ of the set Ω_0 and delta functions, i.e.,

$$l_0(y) = \chi_0(y) - \sum_{|H\gamma'|\leq L} a[\gamma']\delta(y - H\gamma'). \tag{2.2}$$

The error $l(x)$ defined by (2.1) is called *homogeneous* of degree M, if $l_0(y)$ has order M, i.e., $l_0(y)$ vanishes at every polynomial of degree M, i.e.,

$$(l_0(y), y^\alpha) = 0, \quad |\alpha| \leq M.$$

Observe that L in (2.2) is independent of h, and ϖ lies in the interior of Ω_0. Consequently, the support of a homogeneous error (2.1) also lies in the interior of Ω_0 at sufficiently small h. Observe that a homogeneous error $l(y)$ of degree $M \geq m$ is equidistributed in Ω.

Below in § 4 of the current chapter, we inspect the $L_2^{(M)*}$ norm of a more general error than a homogeneous error (2.1) and, in particular, prove existence of a constant K such that

$$\|l \mid L_2^{(M)*}\| \leq K h^M \tag{2.3}$$

for all sufficiently small h.

In the present section we study the action of a homogeneous error in the spaces of periodic functions $\widetilde{W}_p^{(m)}$ which generalize the Hilbert spaces $\widetilde{W}_2^{(m)}$. Also, using (2.3), we prove validity of analogous inequalities for the norms of $l(x)$ in the dual spaces $\widetilde{W}_p^{(m)*}$.

Let $1 < p < \infty$. Then the set $\widetilde{W}_p^{(m)}$ consists of the functions $\varphi(x)$ with period matrix H, i.e.,

$$\varphi(x + H\gamma) = \varphi(x), \quad x \in \mathbb{R}^n, \quad \gamma \in \mathbb{Z}^n.$$

Moreover, each element $\varphi(x)$ expands in the Fourier series

$$\varphi(x) = \sum_\gamma c_\varphi[\gamma] e^{i2\pi\gamma H^{-1}x}$$

and has the following finite norm

$$\|\varphi \mid \widetilde{W}_p^{(m)}\| = \left\{ \int_{\Omega_0} \left| \sum_\gamma c_\varphi[\gamma](1 + |2\pi H^{-1*}\gamma|^2)^{m/2} e^{i2\pi\gamma H^{-1}x} \right|^p dx \right\}^{1/p}.$$

The so-defined $\widetilde{W}_p^{(m)}$ is a Banach space and enters in a Bessel scale with index-zero space $\widetilde{L}_p(\Omega_0)$.

We are interested in upper bounds on the norm of the error (2.1) in the space $\widetilde{W}_p^{(m)*}$ dual to $\widetilde{W}_p^{(m)}$. To find them, introduce two important functions $G_m(x)$ and $\widetilde{G}_m(x)$, the first serving as an analog of the fundamental solution to the polyharmonic equation in $L_2^{(m)}$ and, the second, the corresponding periodic function. Put $m > n$ and

$$G_m(x) = \int (1 + |2\pi\xi|^2)^{-m/2} e^{i2\pi\xi x}\, d\xi, \quad \widetilde{G}_m(x) = \sum_\gamma G_m(x - H\gamma).$$

The function $G_m(x)$ is infinitely differentiable except the coordinate origin and continuous everywhere for $m > n$. At infinity the function $G_m(x)$ and all its derivatives decrease exponentially [3, 41]. The derivatives of higher order of $G_m(x)$ at the coordinate origin have the same singularities as the fundamental solution to the polyharmonic operator $\Delta^{m/2}$. Namely, if $|\alpha| > m - n$ then for all $x \in \mathbb{R}^n$ the inequality holds

$$|D^\alpha G_m(x)\, | \le K|x|^{m-n-|\alpha|}. \tag{2.4}$$

If m is even, then $G_m(x)$ is the fundamental solution to the differential equation

$$(1 - \Delta)^{m/2} G_m(x) = \delta(x).$$

Under these suppositions, we prove the following

Lemma 4.2. *Let an even m be greater than n and let the degree M of a homogeneous error $l(x)$ be greater than $2m - n + 1$. Then for every p, $1 < p < \infty$, there is a constant K such that for all small h the inequality holds*

$$\|l \mid \widetilde{W}_p^{(m)*}\| \le Kh^m. \tag{2.5}$$

PROOF. Consider the convolution of $\widetilde{G}_m(x)$ and the error $l(x)$, i.e., the function

$$\tilde{u}(x) = \widetilde{G}_m(x) * l(x) = \sum_\gamma (l(y), G_m(x - y - H\gamma)). \tag{2.6}$$

Check that the series over γ converges. Denote the value of $l(y)$ at $G_m(x - y)$ by $u(x)$, i.e.,

$$u(x) = (l(y), G_m(x - y));$$

and estimate the function $u(x)$ at a point x lying outside $\overline{\Omega}_0$.

Take an infinitely differentiable function $\zeta(y)$ equal to 1 in a neighborhood about the support of $l(y)$ and 0 beyond some set $\overline{\omega}_1$ in the interior of Ω_0. We defined a function with analogous properties in § 2 of Chapter 1, using the standard averaging kernel. The following relations are valid

$$\begin{aligned} |(l(y), G_m(x - y))| &= |(l(y), \zeta(y) G_m(x - y))| \\ &\le \|l \mid L_2^{(M)*}\| \, \|\zeta(y) G_m(x - y) \mid L_2^{(M)}(\omega_1)\|. \end{aligned} \tag{2.7}$$

By the definition of the norm of $L_2^{(M)}(\omega_1)$, the second factor on the right side of (2.7) may be written as

$$\left\{ \int_{\omega_1} \sum_{|\alpha|=M} \frac{M!}{\alpha!} \left(D_y^\alpha[\zeta(y) G_m(x-y)]\right)^2 dy \right\}^{1/2}. \tag{2.8}$$

By hypothesis, x belongs to $\mathbb{R}^n \backslash \overline{\Omega}_0$. For all $y \in \omega_1$ we therefore have $|x-y| \geq \delta > 0$, with δ independent of h. Whence and from (2.4) it follows that the integrand in (2.8) does not exceed the ratio $K/(1+|x|)^{2(M-m+n)}$, with K a constant independent of h.

We thus established that, for all x outside $\overline{\Omega}_0$, the estimate holds

$$|u(x)| \leq K \| l \mid L_2^{(M)*} \| / (1+|x|)^{M-m+n}. \tag{2.9}$$

The equality (2.6) amounts to the following

$$\tilde{u}(x) = \sum_\gamma u(x - H\gamma). \tag{2.10}$$

Using (2.9) in estimating $u(x - H\gamma)$ at large γ, obtain

$$|\tilde{u}(x)| \leq \sum_{|\gamma| \leq N} |u(x - H\gamma)| + K \| l \mid L_2^{(M)*} \| \sum_{|\gamma| \geq N} \frac{1}{(1 + |x - H\gamma|)^{M-m+n}}.$$

Observe that, at a given x, the series on the right side is dominated by the sum

$$\sum_{\gamma \neq 0} \frac{1}{\left(\gamma_1^2 + \gamma_2^2 + \cdots + \gamma_n^2\right)^{(M-m+n)/2}}. \tag{2.11}$$

Using the routine inequality between the arithmetic mean and the geometric mean

$$\gamma_1^2 + \gamma_2^2 + \cdots + \gamma_n^2 \geq n \gamma_1^{2/n} \gamma_2^{2/n} \ldots \gamma_n^{2/n},$$

we readily validate convergence of (2.11) and (2.6).

By hypothesis, $1 < p < \infty$. Therefore, the exponent $p' = p/(p-1)$ also lies in the interval $(1, \infty)$. Moreover, uniform convexity of the unit ball of $L_p^{(m)}$ enables us to prove that

$$\| l \mid \widetilde{W}_p^{(m)*} \| = \| \tilde{u}(x) \mid L_{p'}(\Omega_0) \|. \tag{2.12}$$

Dominate the norm $\tilde{u}(x)$ in $L_{p'}(\Omega_0)$ by the sum of the norms of the summands of (2.10), which yields

$$\| \tilde{u} \mid L_{p'}(\Omega_0) \| \leq \sum_{\gamma \neq 0} \| u(x - H\gamma) \mid L_{p'}(\Omega_0) \| + \| u(x) \mid L_{p'}(\Omega_0) \|. \tag{2.13}$$

Estimate each summand on the right side of (2.13) separately. If $\gamma \neq 0$ then by means of (2.9) and (2.3) obtain

$$\sum_{\gamma\neq 0} \|u(x - H\gamma) \mid L_{p'}(\Omega_0)\| \leq Kh^M \sum_{\gamma\neq 0} \frac{1}{(1 + |H\gamma|)^{M-m+n}} = K_1 h^M. \tag{2.14}$$

To estimate the norm of $u(x)$ in $L_{p'}(\Omega_0)$, split the homogeneous error $l(y)$ into two summands

$$l(y) = \sum_{\gamma\in B_L^{(1)}} l_0\left(\frac{y}{h} - H\gamma\right) + \sum_{\gamma\in B_L^{(2)}} l_0\left(\frac{y}{h} - H\gamma\right).$$

The sets $B_L^{(1)}$ and $B_L^{(2)}$ depend on the point x taken in the interior of Ω_0. Namely, $B_L^{(1)}$ consists of γ such that $hH\gamma \in \omega$ and $|x - hH\gamma| \leq 2Lh$. On the other hand, if $hH\gamma \in \omega$ but $|x - hH\gamma| > 2Lh$ then the set of these γ is $B_L^{(2)}$. Recall that constant L in accord with (2.2) characterizes the size of the support of $l_0(y)$. We have

$$|u(x)| = |(l(y), G_m(x-y))| \leq \sum_{\gamma\in B_L^{(1)}} \left|\left(l_0\left(\frac{y}{h} - H\gamma\right), G_m(x-y)\right)\right|$$
$$+ \sum_{\gamma\in B_L^{(2)}} \left|\left(l_0\left(\frac{y}{h} - H\gamma\right), G_m(x-y)\right)\right| \equiv \Sigma_1 + \Sigma_2. \tag{2.15}$$

The sum Σ_1 contains at most $(2L+1)^n$ summands. Choose a maximal summand and pass to the maximum over all $x \in \overline{\Omega}_0$ to obtain

$$\Sigma_1 \leq (2L+1)^n \max_{x\in\overline{\Omega}_0,\gamma} \left|\left(l_0\left(\frac{y}{h}\right), G_m(x - y - hH\gamma)\right)\right|. \tag{2.16}$$

The function $G_m(x)$ possesses the property $G_m(-x) = G_m(x)$. Using the latter, transform the maximand in (2.16) as follows

$$\left(l_0\left(\frac{y}{h}\right), G_m(x - y - hH\gamma)\right) = h^n(l_0(y), G_m(hy - x + hH\gamma)). \tag{2.17}$$

Intending to apply to the expression on the right side of (2.17) the Plancherel identity, calculate the Fourier transform in y of $G_m(hy - z)$. By definition

$$F_y[G_m(hy - z)](\xi) = h^{-n} e^{i2\pi\xi z/h} \int G_m(y) e^{i2\pi y\xi/h}\, dy$$
$$= h^{-n} e^{i2\pi\xi z/h}(1 + |2\pi\xi/h|^2)^{-m/2}. \tag{2.18}$$

Denoting by $\tilde{l}_0(\xi)$ the Fourier transform of $l_0(y)$, from (2.17) and (2.18) obtain

$$\left|\left(l_0\left(\frac{y}{h}\right), G_m(x-y-hH\gamma)\right)\right| = \left|\int \left(1+\left|\frac{2\pi\xi}{h}\right|^2\right)^{-m/2} \tilde{l}_0(\xi) e^{i2\pi\xi(x-hH\gamma)/h}\, d\xi\right|$$
$$\leq h^m \int (h^2+|2\pi\xi|^2)^{-m/2} |\tilde{l}_0(\xi)|\, d\xi. \tag{2.19}$$

The Fourier transform $\tilde{l}_0(\xi)$ of a compactly-supported generalized function $l_0(y)$, equal 0 at polynomials of degree M, has a root of order M at the coordinate origin and is bounded on the whole real space. In other words, we have the inequality

$$|\tilde{l}_0(\xi)| \leq K \min\{1, |\xi|^M\}. \tag{2.20}$$

Split the dominating integral in (2.19) into the sum of two integrals: over the ball $|\xi| \leq 1$ and over the exterior of the ball. Then apply (2.20) to obtain

$$\left|\left(l_0\left(\frac{y}{h}\right), G_m(x-y-hH\gamma)\right)\right|$$
$$\leq Kh^m \left\{ \int\limits_{|\xi|\leq 1} \frac{|\xi|^M}{(h^2+|2\pi\xi|^2)^{m/2}} d\xi + \int\limits_{|\xi|\geq 1} \frac{1}{(h^2+|2\pi\xi|^2)^{m/2}} d\xi \right\}. \tag{2.21}$$

The expression in braces is bounded uniformly in h and $x \in \overline{\Omega}_0$. For the integral over the exterior of the unit ball, this is obvious; whereas the integral over the interior of the ball is bounded by the quantity

$$\int\limits_{|\xi|\leq 1} |\xi|^{M-m}\, d\xi$$

finite by the hypothesis $M > m$. From (2.21) and (2.16) it follows that Σ_1 decreases as $h \to 0$ not slower than h^m, i.e.,

$$\Sigma_1 \leq Kh^m, \tag{2.22}$$

with K a constant independent of h.

Consider the summand Σ_2 in (2.15). Recall that x in this formula is a point belonging to Ω_0. Take an arbitrary $\gamma \in B_L^{(2)}$. Then, for every point y in the support of a local error $l_0(y/h - H\gamma)$, we have the inequality

$$|x-y| \geq |x-hH\gamma| - |y-hH\gamma| \geq L_1 h, \tag{2.23}$$

with L_1 a constant independent of h. The point $hH\gamma$ lies in the support of $l_0(y/h - H\gamma)$. Whence and from (2.23) it follows that $G_m(x-y)$ is infinitely differentiable with respect to y in a neighborhood about $y_0 = hH\gamma$. Consequently, $G_m(x-y)$ expands in the Taylor series with remainder in integral form [42]; i.e.,

$$G_m(x-y) = \sum_{|\alpha|\le M} \left.\frac{D^\alpha G_m(x-y_0)}{\alpha!}\right|_{y_0=hH\gamma} (y_0-y)^\alpha + R_M(x,y), \tag{2.24}$$

with $R_M(x,y)$ written as

$$(M+1)\int_0^1 (1-s)^M \sum_{|\alpha|=M+1} (y_0-y)^\alpha D^\alpha G_m(x-hH\gamma+s(hH\gamma-y))/\alpha!\,ds. \tag{2.25}$$

The equality (2.24) holds uniformly in y belonging to the support of the local error $l_0(y/h - H\gamma)$. Whence and from the vanishing of $l_0(y)$ at all polynomials of degree M, it follows that

$$\begin{aligned}\left(l_0\left(\frac{y}{h}-H\gamma\right), G_m(x-y)\right) &= \left(l_0\left(\frac{y}{h}-H\gamma\right), R_M(x,y)\right)\\ &= h^n(l_0(y), R_M(x, hy+hH\gamma)).\end{aligned}$$

The decomposition (2.2) of $l_0(y)$ enables us to assert that

$$\begin{aligned}&\left|\left(l_0\left(\frac{y}{h}-H\gamma\right), G_m(x-y)\right)\right|\\ &\le h^n\|l_0(y)\mid C^*\| \max_{y\in\operatorname{supp} l_0(y)} |R_M(x,hy+hH\gamma)|.\end{aligned} \tag{2.26}$$

Express the maximand as an integral by (2.25) to obtain

$$\begin{aligned}R_M(x,hy+hH\gamma) &= (M+1)h^{M+1}\\ \times\int_0^1 (1-s)^M &\sum_{|\alpha|=M+1} (-y)^\alpha D^\alpha G_m(x-hH\gamma-sh\gamma)/\alpha!\,ds.\end{aligned} \tag{2.27}$$

By (2.4), for $|\alpha| = M+1$ we have

$$|D^\alpha G_m(x-hH\gamma-shy)| \le K|x-hHy-shy|^{m-n-M-1}.$$

The point y lies in the support of $l_0(y)$. Therefore $|y| \le \sqrt{n}\, L$. Whence and from the containment of γ in $B_L^{(2)}$ it follows that, for a given $x \in \overline{\Omega}_0$, we have inequalities

$$|x - hH\gamma| \ge 2\sqrt{n}\, Lh \ge 2|yh|,$$
$$|x - hH\gamma - shy| \ge |x - hH\gamma| - sh|y| \ge 1/2|x - hH\gamma|.$$

Inserting the second of them in the estimate obtained for the modulus of the derivative of G_m and taking account of the condition $M > m - n - 1$, arrive at the estimate

$$|D^\alpha G_m(x - hH\gamma - shy)| \le K|x - hH\gamma|^{m-n-M-1},$$

from which by (2.27) it follows that

$$\max_{y \in \operatorname{supp} l_0(y)} |R_M(x, hy + hH\gamma)| \le Kh^{M+1}/|x - hH\gamma|^{M-m+n+1}.$$

Using this estimate in (2.26) and summing both sides of the resultant inequality over all $\gamma \in B_L^{(2)}$, find a dominant of the needed shape for Σ_2 as follows

$$\Sigma_2 \le Kh^{n+M+1} \sum_{\gamma \in B_L^{(2)}} 1/|x - hH\gamma|^{M-m+n+1}. \tag{2.28}$$

Observe that for a given $x \in \overline{\Omega}_0$ there is an index N such that, for all γ, $|\gamma| \ge N$, the estimate holds

$$|x - hH\gamma| \ge |hH\gamma|/2.$$

If, on the other hand, $|\gamma| \le N$ and $\gamma \in B_L^{(2)}$ then

$$|x - hH\gamma| \ge 2L\sqrt{n}\, h.$$

Inserting the last two inequalities in (2.28), obtain

$$\Sigma_2 \le Kh^m \sum_{\gamma \ne 0} \frac{1}{|\gamma|^{M+1+n-m}} = K_2 h^m. \tag{2.29}$$

The series in γ in this formula converges by the conditions $M \ge 2m - n + 1$, $m > n$.

The inequalities (2.29), (2.22), and (2.15) allow us to estimate the function $u(x)$ at $x \in \overline{\Omega}_0$ and also its norm in $L_{p'}(\Omega_0)$ as follows

$$\|u \mid L_{p'}(\Omega_0)\| \le Kh^m.$$

This estimate, as well as (2.14) and (2.13), for $M \ge 2m - n$ yields a dominant for the norm in $L_{p'}(\Omega_0)$ of $\tilde{u}(x)$. Namely,

$$\|\tilde{u}(x) \mid L_{p'}(\Omega_0)\| \le Kh^m.$$

This, together with (2.12), entails (2.5). The proof of Lemma 4.2 is complete.

Corollary 2.1. *Under the hypotheses of Lemma 4.2 a homogeneous error* $l(x)$ *has an optimal mode of convergence in* $\widetilde{W}_p^{(m)}$, *which means that there are positive constants* K_1 *and* K_2 *such that, for all* h, *the inequalities hold*

$$K_1 h^m \leq \|l \mid \widetilde{W}_p^{(m)*}\| \leq K_2 h^m. \tag{2.30}$$

Proof. The upper bound of (2.30) coincides with the already-established relation (2.5). Check that the lower bound on the norm is valid.

By Theorem 4.2 there is a constant K such that

$$\|l \mid \widetilde{W}_p^{(m)*}\| \geq K \|l_\infty \mid \widetilde{W}_p^{(m)*}\|, \tag{2.31}$$

with the error l_∞ defined by (1.4). Find an asymptotically exact expansion of the $\widetilde{W}_p^{(m)*}$ norm of l_∞ as $h \to 0$. To this end, write down the function $\tilde{u}(x)$ with period matrix H, which is a solution to the equation $(1-\Delta)^{m/2}\tilde{u}(x) = l_\infty(x)$, as convolution

$$\tilde{u}(x) = \int l_\infty(y)\widetilde{G}_m(x-y)\,dy. \tag{2.32}$$

Expand $\widetilde{G}_m(x)$ in the Fourier series

$$\widetilde{G}_m(x) = \sum_\gamma (1 + |2\pi H^{-1*}\gamma|^2)^{-m/2} e^{i2\pi\gamma H^{-1}x}.$$

Inserting this expansion in (2.32), obtain

$$\tilde{u}(x) = \sum_\gamma L[\gamma](1 + |2\pi H^{-1*}\gamma|^2)^{-m/2} e^{i2\pi\gamma H^{-1}x}. \tag{2.33}$$

Here, $L[\gamma]$ stands for the *Fourier coefficient with index* γ of $l_\infty(y)$, i.e., the value of $l_\infty(y)$ at the exponential with exponent $i2\pi\gamma H^{-1}y$. It is not hard to see that in the case when γh is a nonzero integer multi-index, the coefficient $L[\gamma]$ equals $-1/|H|$, whereas for all other γ the coefficient with index γ vanishes. Thus, from (2.33) it follows that

$$\tilde{u}(x) = -h^m \sum_{\beta \neq 0} (h^2 + |2\pi H^{-1*}\beta|^2)^{-m/2} e^{i2\pi\beta H^{-1}x/h}.$$

Using the last equality, we readily find an asymptotically exact expansion of the $L_{p'}(\Omega_0)$ norm of $\tilde{u}(x)$ as $h \to 0$. Namely,

$$\|\tilde{u}(x) \mid L_{p'}(\Omega_0)\| = Kh^m(1 + O(1)).$$

Considering this expansion, the equality (2.12) taken with $l(x) = l_\infty(x)$, and (2.31), we infer (2.30). The proof of Corollary 2.1 is complete.

We now generalize the result on an optimal mode of convergence of a homogeneous error (2.1) to the case of odd and fractional values of m. We need several new definitions.

Let $\widetilde{B}$ be a space of functions with period matrix H which meets the conditions of § 1 of the current chapter. A bounded linear operator M, acting from $\widetilde{B}$ to $\widetilde{B}$ by the formula

$$M : \sum_\gamma \varphi[\gamma] e^{i2\pi\gamma H^{-1}x} \mapsto \sum_\gamma \psi[\gamma]\varphi[\gamma] e^{i2\pi\gamma H^{-1}x}, \tag{2.34}$$

is called a *Fourier multiplier in* $\widetilde{B}$.

We call a $\psi(\xi)$ a *multiplier in* $\widetilde{B}$, if the operator (2.34) constructed from the sequence $\psi[\gamma] = \psi(H^{-1*}\gamma)$ is bounded in $\widetilde{B}$.

Let a space $\widetilde{B}_0^\mu$ belong to a Bessel scale. Given an $h > 0$, define three auxiliary functions by letting

$$\psi(\xi, h) = \frac{1}{\mu(\xi)} \frac{1}{[h(1+|\xi|)]^{-m_1} + [h(1+|\xi|)]^{-m_2}},$$

$$\psi_j(\xi, h) = \psi(\xi, h) \frac{(1+|\xi|^2)^{m_j/2}}{(1+|\xi|)^{m_j}}, \quad j = 1, 2.$$

Assume that $\psi_1(\xi, h)$ and $\psi_2(\xi, h)$ are multipliers in $\widetilde{B}_0$, and the norms of the corresponding linear operators $M^{(1)}$ and $M^{(2)}$ admit the estimate

$$\|M^{(j)} \mid \widetilde{B}_0^*\| \le K \sup_\xi \psi_j(\xi, h), \tag{2.35}$$

with K a constant independent of h. This property is fulfilled, for instance, in the case when $\widetilde{B}$ coincides with $\widetilde{L}_p(\Omega_0)$ and $\mu(\xi)$ is a rational function subject to (1.31)–(1.33).

Assume given a Bessel scale whose index-zero space B_0 satisfies the conditions of § 1 of the current chapter. We agree to *denote by* $\widetilde{B}_0^m$ *the space* $\widetilde{B}_0^\mu$ *corresponding to the weight function*

$$\mu(\xi) = (1 + |\xi|^2)^{m/2}. \tag{2.36}$$

Assume that the space $\widetilde{B}_0^\mu$ belongs to the scale under consideration, and the action of $l(x) \in \widetilde{B}_0^{\mu *}$ on $\varphi(x) \in \widetilde{B}_0^\mu$ is given by the equality

$$(l(x), \varphi(x)) = \sum_\gamma L[\gamma]\overline{\varphi[\gamma]}, \tag{2.37}$$

where $\varphi[\gamma]$ is the Fourier coefficient with index γ of $\varphi(x)$, and $L[\gamma]$ is the Fourier coefficient with index γ of l, i.e., the value of $l(x)$ at the exponential with exponent $i2\pi\gamma H^{-1}x$. Using (2.37) and convergence of the Fourier series of $\varphi(x)$ in the norm of $\widetilde{B}_0^{\mu}$, we readily see that Fourier series with coefficients $L[\gamma]$ converges to $l(x)$ weakly in $\widetilde{B}_0^{\mu *}$, i.e.,

$$l(x) = \sum_{\gamma} L[\gamma] e^{i2\pi\gamma H^{-1}x}. \tag{2.38}$$

Under the above assumptions, we have the following

Theorem 4.4. *Assume that $l_h(x)$ is a bounded linear functional on $\widetilde{B}_0^{\mu}$, and the weight function $\mu(\xi)$ is subject to the conditions (1.31)–(1.33) with exponents m_1 and m_2 and the condition (2.35). Assume further that the norms of $l_h(x)$ in $(\widetilde{B}_0^*)^{-m_1}$ and $(\widetilde{B}_0^*)^{-m_2}$ satisfy the estimate*

$$\left\| l_h \mid (\widetilde{B}_0^*)^{-m_j} \right\| \leq K_j h^{m_j}, \quad j = 1, 2. \tag{2.39}$$

Then the $(\widetilde{B}_0^*)^{1/\mu}$ norm of $l_h(x)$ decreases as $h \to 0$ not slower than the norm of $l_\infty(x)$ in $(\widetilde{B}_0^{\mu})^*$, namely,

$$\left\| l_h \mid (\widetilde{B}_0^*)^{1/\mu} \right\| \leq K \left\| l_\infty \mid \widetilde{B}_0^{\mu *} \right\|. \tag{2.40}$$

Here K depends neither on h nor on l_h.

PROOF. For $j = 1, 2$ let the operator M_j act from $(\widetilde{B}_0^*)^{-m_j}$ to $\widetilde{B}_0^*$ by the rule

$$M_j : \sum_{\gamma} L[\gamma] e^{i2\pi\gamma H^{-1}x} \mapsto \sum_{\gamma} L[\gamma] \frac{e^{i2\pi\gamma H^{-1}x}}{(1 + |H^{-1}\gamma|^2)^{m_j/2}}.$$

For the functional $l_h(x)$ expanded in the series (2.38) and belonging to $(\widetilde{B}_0^*)^{-m_j}$, we then have the equalities

$$\begin{gathered} \left\| M_j(l_h) \mid \widetilde{B}_0^* \right\| = \left\| l_h \mid (\widetilde{B}_0^*)^{-m_j} \right\|, \\ \left\| l_h \mid (\widetilde{B}_0^*)^{1/\mu} \right\| = \left\| \sum_{\gamma} \frac{1}{\mu(H^{-1*}\gamma)} L[\gamma] e^{i2\pi\gamma H^{-1}x} \mid \widetilde{B}_0^* \right\|. \end{gathered} \tag{2.41}$$

Using the definition of $\psi(\xi, h)$, rewrite the second of them as

$$\left\| l_h \mid (\widetilde{B}_0^*)^{1/\mu} \right\| = \left\| h^{-m_1} M^{(1)} M_1(l_h) + h^{-m_2} M^{(2)} M_2(l_h) \mid \widetilde{B}_0^* \right\|. \tag{2.42}$$

The operator $M^{(j)}$ acts from $\widetilde{B}_0^*$ to $\widetilde{B}_0^*$ in accord with (2.34) in which $\psi[\gamma] = \psi_j(H^{-1*}\gamma, h)$. By (2.35) it is bounded and, moreover,

$$\|M^{(j)}\| \leq K \sup_{\xi} \psi(\xi, h),$$

with K independent of h. Whence and from Theorem 4.3 derive

$$\|M^{(j)}\| \leq K \big\| l_\infty(x) \mid \widetilde{B}_0^{\mu *} \big\|.$$

Now from (2.42), (2.41) and (2.39) it follows that

$$\begin{aligned} &\|l_h \mid (\widetilde{B}_0^*)^{1/\mu}\| \leq h^{-m_1} \|M^{(1)}\| \, \|l_h \mid (\widetilde{B}_0^*)^{-m_1}\| \\ &+ h^{-m_2} \|M^{(2)}\| \, \|l_h \mid (\widetilde{B}_0^*)^{-m_2}\| \leq K \big\| l_\infty \mid \widetilde{B}_0^{\mu *} \big\|. \end{aligned} \tag{2.43}$$

Thus, the estimate (2.40) is obtained, which completes the proof of Theorem 4.4.

Corollary 2.2. *Let $m > 2([n/2]+1)$. Then the norm of a homogeneous error $l(x)$ in $\widetilde{W}_p^{(m)*}$, $1 < p < \infty$, satisfies the inequality*

$$\big\| l \mid \widetilde{W}_p^{(m)*} \big\| \leq K \big\| l_\infty \mid \widetilde{W}_p^{(m)*} \big\|. \tag{2.44}$$

PROOF. Putting $\widetilde{B}_0 = \widetilde{L}_p(\Omega_0)$, take m_1 equal to $2([n/2]+1)$ and m_2 equal to any even number greater than m. It is well known that $\widetilde{B}_0^*$ coincides with $\widetilde{L}_{p'}(\Omega_0)$, $p' = p/(p-1)$. Using this, we readily see that the series (2.37) converges and for every m the equality holds

$$(\widetilde{B}_0^*)^{-m} = \widetilde{B}_0^{m*}.$$

The norm of the homogeneous error $l(x)$ of degree $M > 2m_2 - n + 1$ obeys (2.5) with $m = m_1$ and $m = m_2$. In particular, this functional satisfies the estimate (2.39). Assume that the weight function $\mu(\xi)$ is defined by (2.36). Then, as is not hard to note, all hypotheses of Theorem 4.4 are fulfilled. Consequently, we have the inequality (2.40) now in the shape of (2.44). The proof of Corollary 2.2 is complete.

From (2.44) and Theorem 4.2 we infer that, for $1 < p < \infty$ and $m > 2([n/2]+1)$, a homogeneous error has an optimal mode of convergence in $\widetilde{W}_p^{(m)}$.

Consider the family of errors

$$l(x) = \chi_\Omega(x) - h^n \sum_{hH\gamma \in \omega} c[\gamma] \delta(x - hH\gamma) \tag{2.45}$$

such that the sets $\overline{\Omega}$ and $\overline{\omega}$ are in the interior of the fundamental parallelepiped Ω_0. In particular, every homogeneous error may be written so.

A family (2.45) has an *optimal mode of convergence in* $\widetilde{W}_p^{(m)}(\Omega_0)$ *uniformly in* Ω, if there is a constant K such that, for all $h > 0$ and Ω, the estimate holds

$$h^m/K \le \left\| l \mid \widetilde{W}_p^{(m)*} \right\| \le K h^m. \tag{2.46}$$

Assume that the hypotheses of Theorem 4.4 are fulfilled and, in particular, the function $\mu(\xi)$ obeys (1.31)–(1.33) and (2.35) with exponents m_1 and m_2. Then we have the following

Theorem 4.5. *If an error $l(x)$ has an optimal mode of convergence in $\widetilde{W}_p^{(m)}$, $1 < p < \infty$, uniformly in Ω for $m = m_1, m_2, m_3$, where $m_3 > m_2$, and the space $\widetilde{B}_0$ coincides with $\widetilde{L}_q(\Omega_0)$, $q \ge p$; then the inequality holds*

$$\left\| l \mid \widetilde{B}_0^{\mu*} \right\| \le K |\Omega|^{1/p - 1/q} \left\| l_\infty \mid \widetilde{B}_0^{\mu*} \right\| (1 + O(h^{m_3 - m_2})), \tag{2.47}$$

with K a constant independent of Ω and h.

PROOF. For $\widetilde{B}_0 = \widetilde{L}_q(\Omega_0)$, the space $\widetilde{B}_0^{\mu*}$ coincides with $\left(\widetilde{B}_0^*\right)^{1/\mu}$ and the space $\widetilde{B}_0^m$ coincides with $\widetilde{W}_q^{(m)}$. Applying (2.43) to $l(x)$, arrive at

$$\left\| l \mid \widetilde{B}_0^{\mu*} \right\| \le K \left\| l_\infty \mid \widetilde{B}_0^{\mu*} \right\| \left\{ h^{-m_1} \left\| l \mid \widetilde{W}_p^{(m_1)*} \right\| + h^{-m_2} \left\| l \mid \widetilde{W}_p^{(m_2)*} \right\| \right\} \tag{2.48}$$

Here K is a constant independent of h and l. Dominate the sum in braces.

Suppose that $m = m_1$ or $m = m_2$. Put

$$\tilde{u}_m(x) = (l(y), \widetilde{G}_m(x - y))$$

and express the $\widetilde{W}_p^{(m)*}$ norm of $l(x)$ by the formula (2.12) as

$$\left\| l \mid \widetilde{W}_q^{(m)*} \right\| = \left\{ \int_{\Omega_0} |\tilde{u}_m(x)|^{q'} dx \right\}^{1/q'}, \tag{2.49}$$

with $q' = q/(q-1)$.

Take $\varepsilon > 0$ and denote by Ω_ε the set of points at a distance at most ε from Ω. Suppose that ε is so small that some neighborhood about $\Omega_{2\varepsilon}$ lies in the interior of Ω_0 and the volume of $\Omega_{2\varepsilon}$ does not exceed the double volume of Ω, in symbols,

$$|\Omega_{2\varepsilon}| \le 2|\Omega|. \tag{2.50}$$

Moreover, assume that $\omega \subset \Omega_\varepsilon$. The equality (2.49) implies the estimate

$$\left\| l \mid \widetilde{W}_q^{(m)*} \right\| \le \left\| \tilde{u}_m \mid L_{q'}(\Omega_{2\varepsilon}) \right\| + \left\| \tilde{u}_m \mid L_{q'}(\Omega_0 \setminus \Omega_{2\varepsilon}) \right\|. \tag{2.51}$$

Considering that by hypothesis the error $l(x)$ is bounded in $\widetilde{W}_p^{(m)}$, i.e., $\tilde{u}_m(x) \in L_{p'}(\Omega_0)$, with $p' = p/(p-1)$, estimate the first summand on the right side of (2.51) by the Hölder inequality

$$\|\tilde{u}_m \mid L_{q'}(\Omega_{2\varepsilon})\| \leq |\Omega_{2\varepsilon}|^{1/p-1/q} \|\tilde{u}_m \mid L_{p'}(\Omega_0)\| = |\Omega_{2\varepsilon}|^{1/p-1/q} \big\|l \mid \widetilde{W}_p^{(m)*}\big\|.$$

Taking here $m = m_1$ or $m = m_2$ and using (2.50) and (2.46), come to the relation

$$h^{-m} \|\tilde{u}_m(x) \mid L_{q'}(\Omega_{2\varepsilon})\| \leq K|\Omega|^{1/p-1/q}, \tag{2.52}$$

with K a constant independent of h and Ω.

Consider now the second summand on the right side of (2.51). Let an infinitely differentiable function $\zeta(y)$ equal 1 in Ω_ε and 0 beyond $\Omega_{2\varepsilon}$. Then $\tilde{u}_m(x)$ may be written as

$$\tilde{u}_m(x) = (l(y), \zeta(y)\widetilde{G}_m(x-y)). \tag{2.53}$$

Observe that for all x in $\Omega_0 \setminus \Omega_{2\varepsilon}$ and y in the support of $l(y)$ embedded in Ω_ε, we have the relation $|x-y| \geq \varepsilon$. This, together with (2.53) and the properties of $G_m(x)$, implies that $\tilde{u}_m(x)$ belongs to $W_p^{(M)}(\Omega_0 \setminus \Omega_{2\varepsilon})$ for some $M \geq m_3$. Take M even and estimate the norm of $\tilde{u}_m(x)$ in $L_{q'}(\Omega_0 \setminus \Omega_{2\varepsilon})$ as follows

$$\begin{aligned} \|\tilde{u}_m \mid L_{q'}(\Omega_0 \setminus \Omega_{2\varepsilon})\| &\leq |\Omega_0|^{1/q'} \max_{x\in\Omega_0\setminus\Omega_{2\varepsilon}} |\tilde{u}_m(x)| \\ &\leq |\Omega_0|^{1/q'} \|l(y) \mid W_p^{(M)}(\Omega_\varepsilon)^*\| \\ &\times \max_{x\in\Omega_0\setminus\Omega_{2\varepsilon}} \big\|\zeta(y)\widetilde{G}_m(x-y) \mid W_p^{(M)}(\Omega_\varepsilon)\big\|. \end{aligned} \tag{2.54}$$

The support of $l(y)$ is embedded in Ω_ε. Consequently,

$$\big\|l(y) \mid W_p^{(M)}(\Omega_\varepsilon)^*\big\| = \big\|l(y) \mid \widetilde{W}_p^{(M)}(\Omega_0)^*\big\|.$$

Since $M \geq m_3$, the norm on the right side is less than the norm of $l(x)$ in $\widetilde{W}_p^{(m_3)*}$, i.e.,

$$\big\|l \mid W_p^{(M)}(\Omega_\varepsilon)^*\big\| \leq \big\|l \mid \widetilde{W}_p^{(m)*}\big\| \leq Kh^{m_3}, \tag{2.55}$$

with K independent of h and Ω. The last inequality is valid by hypothesis.

Let $x \in \Omega_0 \setminus \Omega_{2\varepsilon}$. Then

$$\big\|\zeta(y)\widetilde{G}_m(x-y) \mid W_p^{(M)}(\Omega_\varepsilon)\big\| \leq K\Big\{\int_{\Omega_\varepsilon} \sum_{|\alpha|\leq M} |D_y^\alpha[\zeta(y)\widetilde{G}_m(x-y)]|^p \, dy\Big\}^{1/p}. \tag{2.56}$$

For every $y \in \Omega_\varepsilon$ the point $x - y$ lies in a bounded set with the deleted ball of radius ε centered at the coordinate origin. Put

$$c(\varepsilon) = \max_{x \in \overline{\Omega}_0 \setminus \Omega_{2\varepsilon},\, y \in \Omega_\varepsilon} \{|D^\alpha[\zeta(y)\widetilde{G}_m(x-y)]|, |\alpha| \leq M\}.$$

From the properties of $G_m(x)$ it follows that $c(\varepsilon)$ is finite. Continuing the estimate (2.56) and using (2.50), obtain

$$\max_{x \in \Omega_0 \setminus \Omega_{2\varepsilon}} \|\zeta(y)\widetilde{G}_m(x-y) \mid W_p^{(M)}(\Omega_\varepsilon)\| \leq Kc(\varepsilon)|\Omega|^{1/p}. \tag{2.57}$$

Since $|\Omega| \leq |\Omega_0|$, we have the relation

$$|\Omega|^{1/p} \leq |\Omega_0|^{1/q}|\Omega|^{1/p-1/q}.$$

Inserting it in (2.57), arrive at the inequality

$$\max_{x \in \Omega_0 \setminus \Omega_{2\varepsilon}} \|\zeta(y)\widetilde{G}_m(x-y) \mid W_p^{(M)}(\Omega_\varepsilon)\| \leq K_1(\varepsilon)|\Omega|^{1/p-1/q},$$

with $K_1(\varepsilon)$ a constant independent of Ω. Whence and from (2.54) and (2.55) derive

$$h^{-m}\|\tilde{u}_m \mid L_{q'}(\Omega_0 \setminus \Omega_{2\varepsilon})\| \leq K(\varepsilon)h^{m_3-m}|\Omega|^{1/p-1/q}.$$

Now, from (2.51) and (2.52) infer that

$$h^{-m}\|l \mid \widetilde{W}_q^{(m)*}\| \leq K|\Omega|^{1/p-1/q}(1 + O(h^{m_3-m})).$$

Putting here $m = m_1$ and $m = m_2$ and inserting the result in (2.48), arrive at (2.47). The proof of Theorem 4.5 is complete.

§3. The Bakhvalov Theorem

In a bounded domain Ω with piecewise-smooth boundary $\partial\Omega$, assume given a sequence of cubature formulas with the error

$$(l_N(x), \varphi(x)) = \int \Big(\chi_\Omega(x) - \sum_{k=1}^{N} c_k^{(N)}\delta(x - x^{(k,N)})\Big)\varphi(x)\,dx \tag{3.1}$$

and the variable number of nodes N. Denote the volume of Ω by $|\Omega|$ and let

$$h = \left(\frac{|\Omega|}{N}\right)^{1/n}. \tag{3.2}$$

We are interested in the behavior of the $L_2^{(m)*}$ norm of $l_N(x)$ as the number of nodes N tends to infinity. The main theorem of the current section deals with estimating the norm of $l_N(x)$ from below and belongs to N. S. Bakhvalov. Our proofs are practically indistinguishable from those by Bakhvalov.

Before stating the theorem, we give a few definitions and prove an important lemma.

Let k be a constant. Construct the domain Ω_k'' that comprises all points of $\mathbb{R}^n$ at a distance at most k from Ω and the domain Ω_k' that comprises all points of Ω at a distance greater than k from the complement to Ω. Denote $\Omega_k = \Omega_k'' \setminus \Omega_k'$. If there is a constant K such that for all $k \le k_0$ the inequality holds

$$|\Omega_k| = |\Omega_k'' \setminus \Omega_k'| \le Kk|\partial\Omega|,$$

then we say that Ω *has a regular boundary*. It is easy to check that if the boundary of Ω is Lipschitz then this domain has a regular boundary.

Let the number N range over an increasing sequence of integers $N^{(1)}, N^{(2)}, \dots$. Given N, in the domain Ω consider a system of cubes $\Omega_{i,N}$ with vertices the points $a^{(i,N)}$ and edge k_N, namely,

$$\left|x_j - a_j^{(i,N)}\right| < k_N/2, \quad j = 1, 2, \dots, n; \quad i = 1, 2, \dots, N_1(N).$$

We call this system a *system with insufficient data for* $l_N(x)$, if the following conditions are fulfilled:

(1) the value of $l_N(x)$ at the indicator of each of the cubes $\Omega_{i,N}$ is strictly positive, namely,

$$\int \chi_{\Omega_{i,N}}(x) l_N(x)\, dx > 0;$$

(2) all nodes $x^{(k,N)}$ of a cubature formula with the error $l_N(x)$ lie either in the interior of the cube $\Omega_{i,N}$ or in the exterior of the cube but not on the boundary.

Say that a series of such systems *has insufficient data for* $l_N(x)$ *uniformly in* N *entries of the sequence* $\{N^{(j)}\}_{j=1}^{\infty}$ if there are two positive constants η_0 and η_1 less than 1 and such that the following conditions are fulfilled:

(1) the error in integrating unity over each cube of every system of the series under study is a finite fraction of the volume of the cube, i.e.,

$$\int \chi_{\Omega_{i,N}}(x) l_N(x)\, dx \ge \eta_0 k_N^n; \tag{3.3}$$

(2) all nodes of cubature formula belonging to $\Omega_{i,N}$ lie within the inner cube defined by the inequalities

$$\left|x_j - a_j^{(i,N)}\right| \le \eta_1 k_N/2, \quad j = 1, 2, \dots, n. \tag{3.4}$$

We have the following

Lemma 4.3. *Assume that, for every N in the sequence $N^{(1)}, N^{(2)}, \ldots$, the sum of the volumes of nonoverlapping cubes of the corresponding system belonging to a series with uniformly insufficient data for $l_N(x)$ is greater than some positive constant T independent of N, i.e.,*

$$N_1 k_N^n > T, \tag{3.5}$$

with N_1 the number of nonoverlapping cubes. Then we may find a positive constant K depending on η_0, η_1, n, and m, but not on N and such that

$$\left\| l_N \mid L_2^{(m)*} \right\| \geq K k_N^m \sqrt{T}.$$

PROOF. To begin with, find a positive constant $\eta < 1$ such that

$$(1-\eta) \geq \eta_1, \quad (1-\eta)^n - (1-\eta_0) = \varepsilon_0 > 0.$$

Here η_0 and η_1 are the constants of (3.3) and (3.4). With this η available, construct an infinitely differentiable function $\lambda(\xi)$ of a one-dimensional argument ξ with the following properties

$$\begin{aligned} \lambda(\xi) &= 1 \quad \text{for } |\xi| \leq (1-\eta)/2, \\ \lambda(\xi) &= 0 \quad \text{for } |\xi| \geq 1/2, \\ 0 \leq \lambda(\xi) &\leq 1 \quad \text{for other } \xi. \end{aligned} \tag{3.6}$$

Such function obviously exists and may be constructed by the same scheme as the standard averaging kernel we first mentioned in Chapter 1.

Construct the set of the functions $\omega_{i,N}(x)$ by letting

$$\omega_{i,N}(x) = \prod_{j=1}^{n} \lambda\left(\frac{x_j - a_j^{(i,N)}}{k_N}\right).$$

It is clear that the function $\omega_{i,N}(x)$ vanishes outside the cube $\Omega_{i,N}$. Calculate the value of $l_N(x)$ at $\omega_{i,N}(x)$.

From the inequality (3.4) and the properties (3.6) of $\lambda(\xi)$ it follows that, for every node $x^{(k,N)}$ in $\Omega_{i,N}$, the quantity $\omega_{i,N}(x^{(k,N)})$ equals 1. Consequently,

$$(l_N, \omega_{i,N}) = \int\limits_{\Omega_{i,N}} \omega_{i,N}(x)\, dx - \sum_{x^{(k,N)} \in \Omega_{i,N}} c_k^{(N)}. \tag{3.7}$$

Using the first of the properties (3.6), obtain

$$\int_{\Omega_{i,N}} \omega_{i,N}(x)\,dx > (1-\eta)^n k_N^n. \tag{3.8}$$

Since the error $l_N(x)$ is given in $\Omega_{i,N}$ with insufficient data, from (3.3) it follows that

$$\sum_{x^{(k,N)}\in\Omega_{i,N}} c_k^{(N)} \le k_N^n(1-\eta_0).$$

Whence and from (3.7) and (3.8) derive

$$(l_N, \omega_{i,N}) > k_N^n[(1-\eta)^n - (1-\eta_0)] = \varepsilon_0 k_N^n. \tag{3.9}$$

The norm of $\omega_{i,N}(x)$ in $L_2^{(m)}$ is easy to estimate from above, using the definition

$$\left\|\omega_{i,N}(x) \mid L_2^{(m)}\right\| \le K k_N^{n/2-m}.$$

Denoting by N_1 the total number of nonoverlapping cubes $\Omega_{i,N}$ at a given N, estimate the value of $l_N(x)$ at the sum

$$\omega_N(x) = \sum_{i=1}^{N_1} \omega_{i,N}(x).$$

From (3.9) obtain

$$(l_N, \omega_N) = \sum_{i=1}^{N_1} (l_N, \omega_{i,N}) \ge N_1 \varepsilon_0 k_N^n.$$

The cubes $\Omega_{i,N}$ do not overlap. Consequently,

$$\left\|\omega_N(x) \mid L_2^{(m)}\right\|^2 = \sum_{i=1}^{N_1} \left\|\omega_{i,N}(x) \mid L_2^{(m)}\right\|^2 \le K N_1 k_N^{n-2m}.$$

Using the hypotheses (3.5) of the lemma, arrive at the final formula

$$\frac{|(l_N, \omega_N)|}{\left\|\omega_N \mid L_2^{(m)}\right\|} \ge K\sqrt{N_1}\, k_N^{n/2+m} \ge K\sqrt{T}\, k_N^m. \tag{3.10}$$

The proof of Lemma 4.3 is complete.

Theorem 4.6. *For every sequence of errors $l_N(x)$ like (3.1) there is a constant $K > 0$ such that*

$$\left\| l_N \mid L_2^{(m)*} \right\| \geq K h^m. \tag{3.11}$$

PROOF. Cover Ω with a system of cubes $\Omega_{i,h/2}$ generating a cubic lattice with side $h/2$. Calculate the number of cubes lying entirely in the interior of Ω.

Let B_2 be the set of i such that $\Omega_{i,h/2}$ has at least one common point with Ω, and let B_3 be the set of i such that $\Omega_{i,h/2}$ has at least one common point with the boundary $\partial\Omega$. Denote the size of B_2 (B_3) by N_2 (N_3). Let B_1 be the set of points B_2 not belonging to B_3; i.e., $B_1 = B_2 \setminus B_3$. It is clear that B_1 is the set comprising the indices of the cubes lying entirely in the interior of Ω. The total number of these cubes is $N_1 = N_2 - N_3$.

The volume of the union of the cubes $\Omega_{i,h/2}$ over all $i \in B_2$ is at least the volume of Ω, namely,

$$\left| \bigcup_{i \in B_2} \Omega_{i,h/2} \right| \geq |\Omega|.$$

The domain Ω has a regular boundary. Consequently, the inequality holds

$$\left| \bigcup_{i \in B_3} \Omega_{i,h/2} \right| \leq \left|\Omega''_{\sqrt{n}\,h/2} \setminus \Omega'_{\sqrt{n}\,h/2}\right| = |\Omega_{\sqrt{n}\,h/2}| \leq Kh.$$

Since the volume of each elementary cube equals $(h/2)^n$, obtain

$$N_2 \geq |\Omega| 2^n / h^n, \quad N_3 \leq K / h^{n-1}.$$

These relations allow us to estimate from below the number N_1. Obviously,

$$N_1 \geq 2^n \frac{|\Omega|}{h^n} - \frac{K}{h^{n-1}} = N 2^n \left[1 + O\left(\frac{1}{N^{1/n}} \right) \right]. \tag{3.12}$$

Denote by N_0 the number of cubes lying in the interior of Ω and having no nodes of (3.1). Estimate N_0 from below. Let $0 < \varepsilon < 1$. Then from (3.12) it follows that

$$N_0 \geq N_1 - N \geq (2^n - 1 - \varepsilon) N \tag{3.13}$$

for all sufficiently large N. In each of these N_0 cubes, choose a smaller inner cube with side $h/3$. Thus, given N we constructed the system of cubes

$$\Omega_{j,h/3}, \quad j = 1, 2, \ldots, N_0,$$

so that the series of these systems has uniformly insufficient data for $l_N(x)$. Further, all cubes $\Omega_{j,h/3}$ do not overlap. Therefore, their total volume $|\Omega^*|$ is bounded from below by some fraction of the volume $|\Omega|$ independently of N i.e.,

$$|\Omega^*| \geq N_0(h/3)^n \geq \varepsilon_0|\Omega|.$$

The last inequality holds by (3.13) and the definition (3.2) with the constant $\varepsilon_0 = (2^n - 1 - \varepsilon)/3^n > 0$. Therefore, the hypotheses of Lemma 4.3 are fulfilled. Consequently, the norm of the error admits the estimate

$$\| l_N \mid L_2^{(m)*} \| \geq K(h/3)^m = K_1 h^m.$$

The proof of Theorem 4.6 is complete.

***Corollary* 3.1.** *If in a domain Ω there is given a system of cubes $\Omega_{i,N}$ with uniformly insufficient data for the error $l_N(x)$, with the sequence $\{k_N\}$ such that $k_N/h \to \infty$ as $h \to 0$ and the sum of the volumes of $\Omega_{i,N}$ is greater than $\Theta|\Omega|$; then the error of cubature formulas decreases slower than the power function h^m.*

By now we have considered only errors with insufficient data on a system of cubes. By analogy, we may also consider a system of cubes $\Omega'_{i,N}$ with excessive data for the error $l_N(x)$. Say that *data for $l_N(x)$ are excessive on a system of $\Omega'_{i,N}$ by η_0* if

$$\int \chi_{\Omega'_{i,N}}(x) l_N(x)\, dx < -\eta_0 k_N^n.$$

Repeating word by word the proofs of Lemma 4.3 and Theorem 4.6, obtain

$$(l_N(x), \omega_{i,N}(x)) < -\varepsilon_0 k_N^n.$$

Continuing estimation in much the same way as before, show that (3.10) remains valid also and in the case when instead of a system of cubes with insufficient data for $l_N(x)$ there is a system of cubes with excessive data and total volume at least a fixed fraction of the volume of $|\Omega|$.

Given an arbitrary system of N points, we may construct a system of cubes with excessive data for $l_N(x)$ so that hypotheses similar to those of Lemma 4.3 be fulfilled.

We show now that there are formulas such that the lower bound of the rate of convergence we indicated is practically achieved.

§4. The Rate of Convergence of an Equidistributed Error

Let the error $l(x)$ belong to a sequence of equidistributed errors over a domain Ω. In the present section we demonstrate that the norm of $l(x)$ satisfies the inequality reverse to (3.11); i.e., $l(x)$ has an *optimal mode of convergence* in $L_2^{(m)*}$.

By the definition of § 2 of Chapter 1, an equidistributed error over Ω may be written as the sum of local errors

$$l(x) = \sum_{\gamma \in B_L} l_\gamma \left(\frac{x}{h} - H\gamma \right). \tag{4.1}$$

Moreover, we assume that every $l_\gamma(y)$ has order $m+1$, its support lies in the ball of radius L centered at the coordinate origin, and the norm in C^* is bounded by a constant A the same for all γ

$$(l_\gamma(y), y^\alpha) = 0, \ |\alpha| \leq m; \quad \operatorname{supp} l_\gamma(y) \subset \{y : |y| \leq L\}, \quad \|l_\gamma(y) \mid C^*\| \leq A. \tag{4.2}$$

The set B_L consists of γ such that the distance from $hH\gamma$ to Ω is at most Lh.

Before estimating the norm of (4.1), we make a simple remark on how the $L_2^{(m)*}$ norm of an arbitrary error changes under *scaling* of an independent variable. Let $l_0(x) \in L_2^{(m)*}$. Then we have the formula

$$\left\| l_0 \left(\frac{x}{h} \right) \mid L_2^{(m)*} \right\| = h^{n/2+m} \| l_0(x) \mid L_2^{(m)*} \|. \tag{4.3}$$

Prove it.

Take $\varphi(x) \in L_2^{(m)}$. By definition, the following equalities are valid

$$\left(l_0 \left(\frac{x}{h} \right), \varphi(x) \right) = h^n (l_0(y), \varphi(hy)), \quad (l_0(y), \varphi(y)) = h^{-n} \left(l_0 \left(\frac{x}{h} \right), \varphi \left(\frac{x}{h} \right) \right). \tag{4.4}$$

Moreover, straightforward calculation readily yields the relations

$$\begin{aligned} \|\varphi(x) \mid L_2^{(m)}\| &= h^{n/2-m} \|\varphi(hy) \mid L_2^{(m)}\|, \\ \|\varphi(y) \mid L_2^{(m)}\| &= h^{m-n/2} \left\| \varphi \left(\frac{y}{h} \right) \mid L_2^{(m)} \right\|. \end{aligned} \tag{4.5}$$

Dividing both sides of the first of the equalities (4.4) by the corresponding sides of the first of the relations (4.5), obtain

$$\frac{(l_0(x/h), \varphi(x))}{\|\varphi(x) \mid L_2^{(m)}\|} = h^{m+n/2} \frac{(l_0(y), \varphi(hy))}{\|\varphi(hy) \mid L_2^{(m)}\|} \leq h^{m+n/2} \|l_0(y) \mid L_2^{(m)*}\|.$$

Since $\varphi(x) \in L_2^{(m)}$ is taken arbitrarily, it follows that

$$\left\| l_0 \left(\frac{x}{h} \right) \mid L_2^{(m)*} \right\| \leq h^{m+n/2} \|l_0(y) \mid L_2^{(m)*}\|. \tag{4.6}$$

Further, dividing both sides of the second of the equalities (4.4) by the corresponding sides of the second of the relations (4.5), obtain

$$\frac{(l_0(y),\varphi(y))}{\|\varphi(y) \mid L_2^{(m)}\|} = h^{-m-n/2}\frac{(l_0(x/h),\varphi(x/h))}{\left\|\varphi(x/h) \mid L_2^{(m)}\right\|} \le h^{-m-n/2}\left\|l_0\left(\frac{x}{h}\right) \mid L_2^{(m)*}\right\|.$$

Whence follows the inequality reverse to (4.6). This means that (4.3) is indeed valid.

Consider an arbitrary summand $l_\gamma(y)$ in the decomposition (4.1). Since $m > n/2$, the $L_2^{(m)}$ space is embedded in $C(\Omega_*)$ for every compact set Ω_*. In this event the embedding operator is compact. Whence and from (4.2) it follows that

$$\|l_\gamma(y) \mid L_2^{(m)*}\| \le K\|l_\gamma(y) \mid C^*\| \le KA,$$

with K a constant independent of γ. Using this inequality and (4.3), obtain

$$\left\|l_\gamma\left(\frac{x}{h} - H\gamma\right) \mid L_2^{(m)*}\right\| \le KAh^{m+n/2}. \tag{4.7}$$

We give a few reasons that allow us to conjecture an upper bound on the norm of an equidistributed error in Ω. Each of the local errors $l_\gamma(x/h - H\gamma)$ is realizable in the Hilbert $L_2^{(m)}$ space as the extremal function $\psi_\gamma(x)$. If $\psi_\gamma(x)$ are mutually orthogonal functions in $L_2^{(m)}$, then

$$\|l(x) \mid L_2^{(m)*}\|^2 = \sum_{\gamma\in B_L}\left\|l_\gamma\left(\frac{x}{h} - H\gamma\right) \mid L_2^{(m)*}\right\|^2. \tag{4.8}$$

The number N_1 of the elements of the set B_L at sufficiently small h satisfies the inequality $N_1 h^n \le 2|\Omega|$. Whence and from (4.7) and (4.8) it follows that

$$\|l(x) \mid L_2^{(m)*}\|^2 \le K^2A^2N_1h^{n+2m} \le 2K^2A^2|\Omega|h^m.$$

The last inequality enables us to conjecture that the norm of an equidistributed error decreases like h^m as the mesh-size decreases. We demonstrate the validity of this conjecture without the assumption that the extremal functions $\psi_\gamma(x)$, $\gamma \in B_L$, are mutually orthogonal. Instead of making the assumption, we give an estimate for the inner product of these extremal functions and check that the product decreases as the supports of the local errors become more distant.

To prove the main theorem we state below, we need inspect the triple convolution of local errors with the function $G(x)$ serving as a solution to equation

$$\Delta^m G(x) = (-1)^m\delta(x).$$

Let $l_1(x)$ and $l_2(x)$ be two nontrivial compactly-supported generalized functions defining bounded linear functionals on the space $C(\mathbb{R}^n)$ with

$$\operatorname{supp} l_j(x) \subset \{x : |x| \leq L_j\}, \quad \|l_j \mid C^*\| \leq A_j, \quad j = 1, 2. \tag{4.9}$$

Also assume that the value of $l_j(x)$ at a polynomial of degree less than s_j equals 0,

$$(l_j(x), x^\alpha) = 0, \quad |\alpha| < s_j, \quad j = 1, 2. \tag{4.10}$$

We now state and prove a succession of the simplest properties of convolution. Put $l_3(x) = l_1(x) * l_2(x)$.

1. *Since both $l_j(x)$ have compact support, so is the functional $l_3(x)$. Moreover, the support of $l_3(x)$ lies in the ball of radius $L_1 + L_2$ centered at the coordinate origin*

$$\operatorname{supp} l_3(x) \subset \{x : |x| \leq L_1 + L_2\}. \tag{4.11}$$

Fix a point x beyond this ball, i.e., such that $|x| > L_1 + L_2$. Then the supports of the generalized functions $l_1(x - y)$ and $l_2(y)$ of a variable y are disjoint by (4.9). This means that

$$\int l_1(x - y) l_2(y)\, dy = 0,$$

i.e., the embedding (4.11) is in effect.

2. *The functional $l_3(x)$ is bounded in $C(\mathbb{R}^n)$, and the norm of it is at most the product of the norms of the factors*, namely,

$$\|l_3(x) \mid C^*\| \leq \|l_1(x) \mid C^*\| \, \|l_2(x) \mid C^*\|. \tag{4.12}$$

Let $\varphi(x)$ be an arbitrary continuous and bounded function on $\mathbb{R}^n$. Then $\varphi(-x)$ also has the same properties. Arrange the two convolutions

$$\varphi_1(x) = \varphi(-x) * l_1(x), \quad \varphi_2(x) = \varphi_1(x) * l_2(x).$$

Since $l_1(x) \in C^*$, the function $\varphi_1(x)$ is continuous and bounded on $\mathbb{R}^n$, with

$$|\varphi_1(x)| \leq \|l_1(y) \mid C^*\| \|\varphi(y - x) \mid C\| = \|l_1 \mid C^*\| \|\varphi \mid C\|.$$

The function $\varphi_2(x)$ also has the same properties,

$$|\varphi_2(x)| \leq \|l_2 \mid C^*\| \, \|\varphi_1 \mid C\| \leq \|l_2 \mid C^*\| \|l_1 \mid C^*\| \|\varphi \mid C\|. \tag{4.13}$$

Since the functionals $l_j(x)$ have compact support, the following equality is valid

$$\varphi_2(x) = (\varphi(-x) * l_1(x)) * l_2(x) = \varphi(-x) * l_3(x). \tag{4.14}$$

Setting here x equal 0, from (4.13) obtain

$$|(l_3(y), \varphi(y))| = |\varphi_2(0)| \leq \|l_1 \mid C^*\| \|l_2 \mid C^*\| \|\varphi \mid C\|,$$

which immediately entails (4.12).

3. *The functional $l_3(x)$ has order $s_1 + s_2$, i.e., its value at every polynomial of degree less than $s_1 + s_2$ is zero*, namely,

$$(l_3(x), x^\alpha) = 0, \quad |\alpha| < s_1 + s_2. \tag{4.15}$$

From the definition of convolution and (4.10), it is easy that the convolution of each polynomial of degree s with $l_1(x)$ is again a polynomial but now of degree $s - s_1$, namely,

$$x^\alpha * l_1(x) = ((x-y)^\alpha, l_1(y)) = \sum_{\beta \leq \alpha} c_{\alpha,\beta} x^{\alpha-\beta} (l_1(y), (-y)^\beta) = \sum_{|\gamma| \leq |\alpha| - s_1} a[\gamma] x^\gamma.$$

An analogous claim is also valid for the functional $l_2(x)$. Taking the monomial x^α with $|\alpha| < s_1 + s_2$ as $\varphi(x)$ in (4.14), we now come to (4.15).

4. *The equalities hold*

$$\left\| l_j \left(\frac{x}{h}\right) \mid C^* \right\| = h^n \|l_j(y) \mid C^*\|, \quad l_3\left(\frac{x}{h}\right) = h^{-n} l_1\left(\frac{x}{h}\right) * l_2\left(\frac{x}{h}\right). \tag{4.16}$$

By a routine change of variable we readily check the validity of (4.16).

Assuming the functionals $l_1(x)$ and $l_2(x)$ to meet the conditions (4.9) and (4.10), estimate the modulus of a solution $\Phi(x \mid l_1, l_2)$ to the equation

$$\Delta^m \Phi = (-1)^m h^n l_3\left(\frac{x}{h}\right). \tag{4.17}$$

From (4.16) it follows that this solution may be written as ***triple convolution***

$$\Phi(x \mid l_1, l_2) = G(x) * l_1\left(\frac{x}{h}\right) * l_2\left(\frac{x}{h}\right). \tag{4.18}$$

Since the functionals $l_j(x)$ are of compact support, this convolution is associative. We have the following

Lemma 4.4. *If n is odd or $s_1 + s_2 > 2m - n$, then the function $\Phi(x \mid l_1, l_2)$ satisfies the inequality*

$$|\Phi(x \mid l_1, l_2)| \leq K \frac{A_1 A_2 h^{2n+s_1+s_2}}{(h^2 + |x|^2)^{-m+(n+s_1+s_2)/2}}, \tag{4.19}$$

with K a constant independent of h, l_1, and l_2.

PROOF. Suppose that $A_1 = 1$ and $A_2 = 1$. If this is not so, we should consider $l_j(x)/A_j$ instead of $l_j(x)$. The right side of (4.17) is a compactly-supported generalized function. Lemma 1.1 gives an estimate (1.2.20) for a solution to a similar equation. Before applying it to our case, some simple reasoning is in order. Writing down the function $\Phi(x \mid l_1, l_2)$ as convolution and carrying out the change of variables $z = y/h$, obtain

$$\Phi(x \mid l_1, l_2) = h^n \int G(x-y) l_3\left(\frac{y}{h}\right) dy = h^{2n} \int G\left(h\left(\frac{x}{h} - z\right)\right) l_3(z)\, dz. \qquad (4.20)$$

Further, by (1.2.18) and (1.2.19) the function $G_{m,n}(w) = (-1)^m G(w)$ for every positive h satisfies the equality

$$G(hw) = h^{2m-n} G(w) + (-1)^m \varkappa_{m,n} (\log h) P_{|2m-n|}(hw). \qquad (4.21)$$

Here $P_{|2m-n|}(w)$ is a polynomial of degree $|2m-n|$. This polynomial is zero if n is odd or $2m < n$ and agrees with $|w|^{2m-n}$ if n is even and $2m \geq n$. Inserting (4.21) in (4.20) and taking account of (4.15), obtain

$$\Phi(x \mid l_1, l_2) = h^{2m+n} \int G\left(\frac{x}{h} - z\right) l_3(z)\, dz, \qquad (4.22)$$

with the convolution integral on the right side standing for the value at the point x/h of a solution $u(x)$ to the equation

$$\Delta^m u(x) = (-1)^m l_3(x).$$

Applying (1.2.20) to $u(x/h)$ and inserting the result in (4.22), come to the sought relation (4.19). The proof of Lemma 4.4 is complete.

REMARK. For $n = 1$ Lemma 4.4 may be strengthen. In this case the function $G(x)$ looks like $|x|^{2m-1}/2(2m-1)!$ and, consequently, $\Phi(x \mid l_1, l_2) = 0$ for $|x| > (L_1 + L_2)h$. If, on the other hand, $|x| \leq (L_1 + L_2)h$ then

$$|\Phi(x \mid l_1, l_2)| \leq h^{2m+1} \frac{(2L_1 + 2L_2)^{2m-1}}{2(2m-1)!} A_1 A_2.$$

We give an explicit estimate for the constant K in (4.19) with $n \geq 2$. In [299] there is proven that, for $2m - n < s_1 + s_2 \leq 4m + 2$, the inequality (4.19) holds with the constant K equal to

$$|\varkappa_{m,n}| \left(\frac{n}{4}\right)^{[(n-3)/2]/2} 2^n \left(\frac{3}{2}\right)^{2m} (1 + (4L)^2)^{(s_1+s_2)/2},$$

where L is the maximum of L_1 and L_2, and the constant $\varkappa_{m,n}$ is defined in § 2 of Chapter 1.

Lemma 4.5. *If n even and $s_3 = s_1 + s_2 \leq 2m - n$, then the estimate holds*

$$|\Phi(x \mid l_1, l_2)| \leq K \begin{cases} h^{2m+n}|\log h|, & |x| \leq 2(L_1 + L_2)h, \\ |x|^{2m-n-s_3}|\log|x||h^{2n+s_3}, & |x| > 2(L_1 + L_2)h. \end{cases}$$

The proof of this lemma is carried out by the same scheme as the proofs of Lemmas 4.4 and 1.1. We thus omit the proof.

Theorem 4.7. *Let a domain Ω have a regular boundary. Assume further that a generalized function $l(x)$ given by (4.1) belongs to a sequence of equidistributed errors in the domain Ω, with the corresponding parameters L and A independent of h. Then the $L_2^{(m)*}$ norm of $l(x)$ satisfies the inequality*

$$\|l \mid L_2^{(m)*}\| \leq K\sqrt{|\Omega|}\, h^m \left(1 + hK_1 \frac{|\partial\Omega|}{|\Omega|}\right), \tag{4.23}$$

with K and K_1 constants depending on A and L but not on h and Ω.

PROOF. The error $l(x)$ on $L_2^{(m)}$ possesses the $L_2^{(m)}$-extremal function

$$\psi_l(x) = (l(y), G(x - y)) = l(x) * G(x).$$

Moreover, the norm square of $l(x)$ in $L_2^{(m)*}$ coincides with the value of $l(x)$ at $\psi_l(x)$, namely,

$$\|l \mid L_2^{(m)*}\|^2 = (l(x), \psi_l(x)) = (l(x), l(x) * G(x)).$$

Inserting here the decomposition (4.1), obtain

$$\|l \mid L_2^{(m)*}\|^2 = \sum_{\gamma,\gamma' \in B_L} \left(l_\gamma\left(\frac{x}{h} - H\gamma\right), l_{\gamma'}\left(\frac{x}{h} - H\gamma'\right) * G(x)\right). \tag{4.24}$$

Write the summand on the right side of (4.24) which corresponds to γ and γ', using the notation of (4.18)

$$\left(l_\gamma\left(\frac{x}{h} - H\gamma\right), l_{\gamma'}\left(\frac{x}{h} - H\gamma'\right) * G(x)\right) = \Phi(hH(\gamma' - \gamma) \mid l_\gamma, \widehat{l}_{\gamma'}). \tag{4.25}$$

Here $\widehat{l}_{\gamma'}(x)$ is defined as $l_{\gamma'}(-x)$.

By the definition of equidistributed error, the corresponding elementary generalized functions $l_\gamma(y)$, $\gamma \in B_L$, vanish at all polynomials of degree less than $m + 1$, are of compact support and have the same A as a bound for their norms in C^*.

Therefore, to estimate the quantities of the shape (4.25), we may apply Lemma 4.4. For every γ, $\gamma' \in B_L$ we then have the relation

$$|\Phi(hH(\gamma'-\gamma) \mid l_\gamma, \hat{l}_{\gamma'})| \leq KA^2 \frac{h^{2n}}{(h^2+|hH(\gamma-\gamma')|^2)^{n/2+1}} h^{2m+2}. \tag{4.26}$$

Here K and A are independent of h, γ, and γ'.

Inserting (4.25) and (4.26) in (4.24), we readily see that the norm under study decreases as $h \to 0$ not slower than the product of $h^{m+n/2}$ by the square root of the number of points $N_L(h)$ in B_L, namely,

$$\|l \mid L_2^{(m)*}\|^2 \leq KA^2h^{2m+n} \sum_{\gamma,\gamma' \in B_L} \frac{1}{(1+|H(\gamma-\gamma')|^2)^{n/2+1}} \leq KA^2h^{2m+n}N_L(h).$$

To specify the dependence of the found dominant on Ω, estimate the right side of (4.26) by some integral. We write down the latter explicitly. Denote by $\Omega_{h,\gamma}$ an elementary mesh of the lattice of nodes $hH\gamma$. It is not hard to see that there is a positive constant M such that, for all γ, $\gamma' \in B_L$ and points $x \in \Omega_{h,\gamma}$, $y \in \Omega_{h,\gamma'}$ the inequality holds

$$|x-y|^2 + h^2 \leq M(|hH(\gamma-\gamma')|^2 + h^2). \tag{4.27}$$

Further, by the Intermediate Value Theorem for every two sets $\Omega_{h,\gamma}$ and $\Omega_{h,\gamma'}$ there are points $x_* \in \Omega_{h,\gamma}$ and $y_* \in \Omega_{h,\gamma'}$ such that

$$\int_{\Omega_{h,\gamma}} \int_{\Omega_{h,\gamma'}} \frac{dx dy}{(|x-y|^2+h^2)^{n/2+1}} = \frac{h^{2n}|\Omega_0|^2}{(|x_*-y_*|^2+h^2)^{n/2+1}}.$$

Applying (4.27) to the ratio on the right side and inserting the result in (4.26), obtain

$$|\Phi(hH(\gamma'-\gamma) \mid l_\gamma, \hat{l}_{\gamma'})| \leq Kh^{2m+2} \int_{\Omega_{h,\gamma}} \int_{\Omega_{h,\gamma'}} \frac{dx dy}{(|x-y|^2+h^2)^{n/2+1}}. \tag{4.28}$$

The union of the sets $\Omega_{h,\gamma}$ over all $\gamma \in B_L$ consists of the points at a distance at most $L_1 h$ from Ω, with $L_1 = (\sqrt{n}+1)h$. In other words, this union lies in the set $\Omega''_{L_1 h}$. Considering this, substitute (4.28) in (4.25). Inserting the result in (4.24), we then arrive at the estimate

$$\|l \mid L_2^{(m)*}\|^2 \leq Kh^{2m+2} \int_{\Omega''_{L_1 h}} \int_{\Omega''_{L_1 h}} \frac{dx dy}{(|x-y|^2+h^2)^{n/2+1}}. \tag{4.29}$$

Here K is a constant independent of h and Ω. Passing to the spherical coordinates ρ and θ with center the point y in the inner integral with respect to dx, obtain

$$\int\limits_{\Omega''_{L_1 h}} \frac{dx}{(|x-y|^2+h^2)^{n/2+1}} \leq \omega_n \int\limits_0^\infty \frac{\rho^{n-1}\,d\rho}{(\rho^2+h^2)^{n/2+1}} = \frac{\omega_n}{h^2}\int\limits_0^\infty \frac{\rho^{n-1}\,d\rho}{(1+\rho^2)^{n/2+1}},$$

with ω_n denoting the area of the unit sphere in $\mathbb{R}^n$. Integrating both sides of the last inequality with respect to $y \in \Omega''_{L_1 h}$ and inserting the result in (4.29), find

$$\left\|l \mid L_2^{(m)*}\right\|^2 \leq K\left|\Omega''_{L_1 h}\right| h^{2m}. \tag{4.30}$$

By hypothesis the set Ω has a regular boundary. Consequently,

$$\left|\Omega''_{L_1 h}\right| \leq |\Omega| + K_1 L_1 h |\partial\Omega|.$$

Whence and from (4.30) we immediately have (4.23). The proof of Theorem 4.7 is complete.

REMARK. In Theorem 4.7 we estimated the norm of an error composed of elementary errors of order $m+1$. If we require the validity of a weaker condition, namely, the vanishing of the local errors at all polynomials of degree $m-1$; then, likewise (4.29), we derive the inequality

$$\left\|l \mid L_2^{(m)*}\right\|^2 \leq Kh^{2m} \int\limits_{\Omega''_{L_1 h}} \int\limits_{\Omega''_{L_1 h}} \frac{dx\,dy}{(|x-y|^2+h^2)^{n/2}}$$

which implies that

$$\left\|l \mid L_2^{(m)*}\right\| \leq Kh^m\sqrt{|\log h||\Omega|}.$$

This inequality is weaker than (4.23).

Let an arbitrary error $l(x)$ in a sequence of errors equidistributed over Ω be given by the formula (1.2.14). Then, arguing as in the proof of Theorem 4.7, arrive at the estimate

$$\left\|l \mid L_2^{(m)*}\right\| \leq Kh^m\sqrt{|\Omega|}\left(1+K_2\sqrt{h|\log h|}\right),$$

with K a constant independent of h and Ω.

Chapter 5

Cubature Formulas with Regular Boundary Layer

In this chapter we show that cubature formulas with regular boundary layer are asymptotically $L_2^{(m)}$-optimal as the lattice mesh-size vanishes.

It is the Bakhvalov Theorem proven in § 3 of Chapter 4, and the theorem on existence of a cubature formula whose error has the $L_2^{(m)*}$ norm decreasing like h^m that prompt us a way to constructing best cubature formulas. Such formulas which are best in some definite sense possess the error distributed through the integration domain in a most consistent and uniform manner.

§1. The Properties of the Extremal Function of an $\widetilde{L}_2^{(m)}$-Optimal Error

To begin with, consider the problem of approximate integration of periodic functions in $\widetilde{L}_2^{(m)}(H)$. Recall that $\varphi(x)$ belongs to $\widetilde{L}_2^{(m)}(H)$, if it is a *periodic function with period matrix* H, i.e.,

$$\varphi(x + H\gamma) = \varphi(x), \quad x \in \mathbb{R}^n, \quad \gamma \in \mathbb{Z}^n,$$

and, moreover, the following quantity is finite

$$\|\varphi \mid \widetilde{L}_2^{(m)}(H)\| = \left\{ \int_{\Omega_0} \sum_{|\alpha|=m} \frac{m!}{\alpha!} |D^\alpha \varphi(x)|^2 \, dx \right\}^{1/2}.$$

Also recall that by Ω_0 we mean the *fundamental parallelepiped* of H, i.e., the image of the unit cube Q under the linear transformation $x = Hy$,

$$Q = \{y \in \mathbb{R}^n : 0 \leq y_j < 1, \, j = 1, 2, \ldots, n\} \subset \mathbb{R}^n.$$

We have proven earlier that, in the space $\widetilde{L}_2^{(m)}$ of periodic functions with translation invariant norm, optimal is the error with equal weights,

$$l_\infty(x) = \chi_{\Omega_0}(x) - h^n \sum_{hH\gamma\in\Omega_0} \delta(x - hH\gamma). \tag{1.1}$$

From now on we consider only h such that $1/h$ is an integer.

An $\widetilde{L}_2^{(m)}$-optimal error is unique. Indeed, assume given two errors each with a minimal norm in $\widetilde{L}_2^{(m)}$. Consequently, l_1 and l_2 have the same norm. By the *Apollonius identity*, also called the *parallelogram law*, we have

$$\left\|\frac{l_1 + l_2}{2}\right\|^2 = \frac{1}{2}\|l_1\|^2 + \frac{1}{2}\|l_2\|^2 - \left\|\frac{l_1 - l_2}{2}\right\|^2.$$

So, were $l_1 \neq l_2$, the error equal to the arithmetic mean of l_1 and l_2 would have the norm less than either of the l_j, namely, $\|(l_1 + l_2)/2\| < \|l_1\| = \|l_2\|$. The latter is impossible, contradicting the optimality of l_j. Consequently, l_1 and l_2 coincide.

We study the properties of the $\widetilde{L}_2^{(m)}$-extremal function of the optimal error (1.1).

In Chapter 1 we showed that the $\widetilde{L}_2^{(m)}$-extremal function $u(x)$ of every bounded error $l(x)$ is a solution in $L_2^{(m)}$ to the *polyharmonic equation*

$$\Delta^m u(x) = (-1)^m l(x). \tag{1.2}$$

A periodic solution to this equation is determined uniquely to within an additive constant. Expanding $l(x) \in \widetilde{L}_2^{(m)*}$ in the Fourier series

$$l(x) = \sum_{\beta\neq 0} L[\beta] e^{i2\pi H^{-1}x\beta},$$

obtain an analogous expansion for the $\widetilde{L}_2^{(m)}$-extremal function

$$u(x) = c_0 + \sum_{\beta\neq 0} \frac{L[\beta]}{|2\pi H^{-1*}\beta|^{2m}} e^{i2\pi H^{-1}x\beta}.$$

It is not hard to calculate that the optimal error $l_\infty(x)$ has the following Fourier coefficients

$$L[\beta] = \begin{cases} 0, & \text{if } \beta = 0 \text{ or } \beta \text{ is not a multiple of } 1/h, \\ -1, & \text{otherwise.} \end{cases}$$

Thus, the $\widetilde{L}_2^{(m)}$-extremal function for $l_\infty(x)$ is periodic with period matrix hH and is given by the equality

$$u_\infty(x) = c - h^{2m} \sum_{\beta \neq 0} \frac{1}{|2\pi H^{-1*}\beta|^{2m}} e^{i2\pi H^{-1}x\beta/h}.$$

Combine the summands of the sum in this equality pairwise, taking with the index β the corresponding index $-\beta$. Then, we may expand $u_\infty(x)$ to obtain

$$u_\infty(x) = c - h^{2m} \sum_{\beta \neq 0} \frac{\cos 2\pi H^{-1}x\beta/h}{|2\pi H^{-1*}\beta|^{2m}}.$$

This function is thus real-valued provided that the constant c is real. Choosing c as follows

$$c = h^{2m} \sum_{\beta \neq 0} \frac{1}{|2\pi H^{-1*}\beta|^{2m}},$$

we readily see that

$$u_\infty(x) = 2h^{2m} \sum_{\beta \neq 0} \frac{1}{|2\pi H^{-1*}\beta|^{2m}} \sin^2 \pi H^{-1}x\beta/h. \tag{1.3}$$

The function $u_\infty(x)$ is greater than 0 everywhere but the nodes $hH\gamma$ at which it vanishes.

The $\widetilde{L}_2^{(m)*}$ norm of $l_\infty(x)$ coincides with the square root of (l_∞, u_∞). Calculating the latter, arrive at the equality

$$\|l_\infty \mid \widetilde{L}_2^{(m)*}\| = \left(\frac{h}{2\pi}\right)^m \sqrt{\sum_{\beta \neq 0} \frac{1}{|H^{-1*}\beta|^{2m}}}, \tag{1.4}$$

presenting the exact value of the minima of the norms in $\widetilde{L}_2^{(m)*}$ of all possible errors with a given lattice of nodes $hH\gamma$.

The $\widetilde{L}_2^{(m)}$-extremal function $u(x)$, a periodic solution to the polyharmonic equation (1.2), is also obtainable by summing the local errors with small support.

A *local error* is a generalized function

$$l^{(0)}(y) = \chi_{\Omega_0}(y) - \sum_{\gamma \in B_0} c[\gamma]\delta(y - H\gamma), \tag{1.5}$$

with B_0 a finite set of integer vectors. We consider only $l^{(0)}(y)$ vanishing at all polynomials of degree $2m$, in symbols,

$$(l^{(0)}(y), y^\alpha) = 0, \quad |\alpha| < 2m + 1. \tag{1.6}$$

Let a local error $l(y)$ of the shape (1.5) satisfy the conditions

$$\operatorname{supp} l(y) \subset \{y : |y| \leq L\}, \tag{1.7}$$

$$\| l(y) \mid C^* \| \leq A, \tag{1.8}$$

$$(l(y), y^\alpha) = 0, \quad |\alpha| < s. \tag{1.9}$$

Denote the set of all these errors by $R(L, A, s)$. A local error (1.5) by definition belongs to the class $R(L, A, 2m+1)$. Since $2m > n$, the functional $l^{(0)}(y)$ is defined at every continuous function and has a finite C^* norm. Observe that, for a given s, the class $R(L, A, s)$ is nonempty only provided that L and A are sufficiently large. The greatest lower bound of the admissible values of L and A depends on s and tends to infinity as s increases indefinitely.

The C^* norm of $l_\gamma(x) = l^{(0)}(x/h - H\gamma)$ is a homogeneous function of h, i.e.,

$$\left\| l^{(0)}\left(\frac{x}{h} - H\gamma\right) \mid C^* \right\| = h^n \| l^{(0)}(y) \mid C^* \|.$$

The support of $l_\gamma(x)$ lies in the ball of radius Lh with center the node $hH\gamma$, in symbols,

$$\operatorname{supp} l_\gamma(x) \subset \{x : |x - hH\gamma| \leq Lh\}.$$

As before, the functional $l_\gamma(x)$ is orthogonal to every polynomial of degree $2m$.

Arrange the sum of local errors

$$\sum_{hH\gamma' \in \Omega_0} l^{(0)}\left(\frac{x}{h} - H\gamma'\right)$$

$$= \sum_{hH\gamma' \in \Omega_0} \left\{ \chi_{\Omega_0}\left(\frac{x}{h} - H\gamma'\right) - \sum_{\gamma \in B_0} c[\gamma] h^n \delta(x - hH\gamma - hH\gamma') \right\}.$$

The indicators of elementary meshes, when summed up, yield the indicator of the entire fundamental parallelepiped

$$\sum_{hH\gamma' \in \Omega_0} \chi_{\Omega_0}\left(\frac{x}{h} - H\gamma'\right) = \chi_{\Omega_0}(x).$$

Consequently, we have the equality

$$\sum_{hH\gamma' \in \Omega_0} l^{(0)}\left(\frac{x}{h} - H\gamma'\right) = \chi_{\Omega_0}(x) - \sum_{hH\gamma' \in \Omega_0} \sum_{\gamma \in B_0} c[\gamma] h^n \delta(x - hH(\gamma + \gamma')).$$

Transposing the order of summation and substituting γ'' for $\gamma+\gamma'$ in the inner sum, obtain

$$\sum_{hH\gamma'\in\Omega_0} l^{(0)}\left(\frac{x}{h}-H\gamma'\right)=\chi_{\Omega_0}(x)-\sum_{\gamma''}\left[h^n\delta(x-hH\gamma'')\sum_{\gamma\in B_0}c[\gamma]\right].$$

It is possible to calculate the value of the error $l(x)$ on the right side of this equality at an arbitrary function $\varphi(x)\in\widetilde{L}_2^{(m)}$ by the formula

$$(l,\varphi)=\int_{\Omega_0}\varphi(x)\,dx-\sum_{hH\gamma'\in\Omega_0}h^n\varphi(hH\gamma)\Bigl(\sum_{\gamma\in B_0}c[\gamma]\Bigr).$$

The inner sum on the right side equals 1, which follows from the condition that the local error $l^{(0)}$ vanishes at constantly-one function. Finally, we have

$$\sum_{hH\gamma'\in\Omega_0} l^{(0)}\left(\frac{x}{h}-H\gamma'\right)=\chi_{\Omega_0}(x)-\sum_{hH\gamma''\in\Omega_0}h^n\delta(x-hH\gamma'')=l_\infty(x).$$

Whence it follows that we may construct the $\widetilde{L}_2^{(m)}$-extremal function $u_\infty(x)$ as the sum of the series

$$u_\infty(x)=\sum_{hH\gamma'\in\Omega_0}u_{\gamma'}(x)+c, \tag{1.10}$$

where $u_{\gamma'}(x)=u_0(x-hH\gamma')$, and the function $u_0(x)$ is a solution to the equation

$$\Delta^m u_0(x)=(-1)^m l^{(0)}\left(\frac{x}{h}\right) \tag{1.11}$$

which vanishes at infinity. More exactly, $u_0(x)$ is expressed as

$$u_0(x)=\left(l^{(0)}\left(\frac{y}{h}\right),G(x-y)\right)=\int l^{(0)}\left(\frac{y}{h}\right)G(x-y)\,dy,$$

where $(-1)^m G(x)$ is the fundamental solution to the polyharmonic operator which is defined in Chapter 1. It is convenient to study the behavior of the convolution of the fundamental solution $G(x)$ with a compactly-supported generalized function $\rho(y)$ subject to the conditions (1.7)–(1.9), i.e., the behavior of

$$u(x)=\int G(x-y)\rho(y)\,dy,$$

on using the Taylor series for the function $G(x-y)$ and its derivatives in the powers of y, namely,

$$D^{\beta}G(x-y) = \sum_{|\alpha|\le s-1} D^{\alpha+\beta}G(x)\frac{(-y)^{\alpha}}{\alpha!} + R_s(y,x). \tag{1.12}$$

For a fixed x and for $|y| \le L$, the remainder $R_s(y,x)$ may be dominated as follows

$$|R_s(y,x)| \le \frac{|y|^s}{s!} \max_{|z|\le L} |T_{s+|\beta|}(G(x-z))|,$$

where

$$|T_s(G)|^2 = \sum_{|\gamma|=s} \frac{s!}{\gamma!}|D^{\gamma}G|^2.$$

Whence we readily obtain an estimate for $R_s(y,x)$ in the domain $|y| \le L$ and $|x| > 2L$ as follows

$$|R_s(y,x)| \le K|x|^{2m-n-|\beta|-s}\big|\log|x|\big|.$$

This estimate may be improved for odd n and in the case when $2m-n-|\beta|-s<0$. In both cases the logarithmic term disappears.

Establish an important property of $u(x)$ which is of often use in the sequel.

Theorem 5.1. *Let $\rho(x)$ be a compactly-supported generalized function that satisfying the inequality*

$$|(\rho(x), f(x))| \le K \max_{x\in \operatorname{supp}\rho} |f(x)| \tag{1.13}$$

for every infinitely differentiable function $f(x)$. Assume further that $\rho(x)$ vanishes at all polynomials of degree $s-1$, in symbols,

$$(\rho(x), x^{\alpha}) = 0, \quad |\alpha| < s. \tag{1.14}$$

Then, at $|x|$ sufficiently large, the derivatives of order l of the convolution

$$u(x) = \int G(x-y)\rho(y)\,dy \tag{1.15}$$

decrease like the function $|x|^{2m-n-l-s}$ if $l+s>2m-n$ or n is odd and decrease like the function $|x|^{2m-n-l-s}\log|x|$ if $l+s\le 2m-n$ and n is even.

PROOF. Differentiate l times both sides of (1.15) and substitute the expansion (1.12) for $D^\beta G(x-y)$, $|\beta| = l$ in the result to obtain

$$D^\beta u(x) = (\rho(y), D^\beta G(x-y))$$
$$= \left(\rho(y), \sum_{|\alpha|\le s-1} D^{\alpha+\beta} G(x) \frac{(-y)^\alpha}{\alpha!}\right) + (\rho(y), R_s(y,x)). \tag{1.16}$$

In the first summand on the right side of (1.16), all integrals vanish by (1.14). Using (1.13) and the estimate for the remainder $R_s(y,x)$ with $|x| > 2A$ and $|y| \le A$, in the case of even n and $l \le 2m - n - s$ we have

$$D^\beta u(x) = O(|x|^{2m-n-s-|\beta|} \log |x|). \tag{1.17}$$

In the case of $l + s > 2m - n$ or n odd, we obtain a sharper estimate

$$D^\beta u(x) = O(|x|^{2m-n-s-|\beta|}). \tag{1.18}$$

The proof of Theorem 5.1 is complete.

For $|x - hH\gamma| \ge 2Lh$ this theorem entails an estimate for the function $u_\gamma(x)$ of (1.10) as follows

$$|u_\gamma(x)| \le \frac{KAh^{n+2m+1}}{(L^2h^2 + |x - hH\gamma|^2)^{(n+1)/2}}. \tag{1.19}$$

The estimate (1.19) shows that the series $\sum_\gamma u_\gamma(x)$ converges for every x, since its general term decreases faster than $|\gamma|^{-(n+1)}$.

It is worth noting that we may obtain the found solution $u_0(x)$ to (1.11) by an immediate change of variable of a solution $u^{(0)}(y)$ to the equation

$$\Delta^m u^{(0)}(y) = (-1)^m l^{(0)}(y).$$

Indeed, $u^{(0)}(x)$ may be written as convolution

$$u^{(0)}(x) = \int l^{(0)}(y) G(x-y)\, dy.$$

On the other hand, the solution $u_0(x)$ to (1.11) may be written as

$$u_0(x) = h^n \int l^{(0)}(y) G\left(\left(\frac{x}{h} - y\right) h\right) dy.$$

The fundamental solution $G(x)$ for n odd is a homogeneous function of order $2m-n$. Consequently, for such n the equality holds

$$u_0(x) = h^{2m}\left(l^{(0)}(y), G\left(\frac{x}{h} - y\right)\right) = h^{2m} u^{(0)}\left(\frac{x}{h}\right). \tag{1.20}$$

For n even, the same result appears, if we use the expansion

$$\begin{aligned} G(x) &= (-1)^m \varkappa_{m,n} \left|\frac{x}{h}\right|^{2m-n} h^{2m-n} \left\{\log\left(\frac{|x|}{h}\right) + \log h\right\} \\ &= h^{2m-n} G\left(\frac{x}{h}\right) + (-1)^m \varkappa_{m,n} |x|^{2m-n} \log h. \end{aligned}$$

Since the function $|x|^{2m-n}$ here is a polynomial of degree $2m-n$, i.e., of degree less than $2m+1$, we have

$$\left(l^{(0)}(y), G\left(\left(\frac{x}{h} - y\right) h\right)\right) = h^{2m-n}\left(l^{(0)}(y), G\left(\frac{x}{h} - y\right)\right),$$

which readily entails (1.20).

Inspect the behavior of the product of the $\widetilde{L}_2^{(m)}$-extremal function $u_\infty(x)$ and the truncator $\psi_h(x)$ introduced in § 2 of Chapter 1. Given a bounded domain Ω with piecewise-smooth boundary, put

$$\psi_h(x) = \varkappa h^{-n} \int\limits_{\Omega'_{2h}} \omega\left(\frac{x - x'}{h}\right) dx'.$$

Here $\varkappa$ is a constant, $\omega(x)$ is the standard averaging kernel, and integration is implemented over the sets of points of Ω at a distance at least $2h$ from the boundary. Recall that $\psi_h(x) = 1$ at all points of Ω at a distance greater than $3h$ from the boundary. The function $\psi_h(x)$ is infinitely differentiable, with derivatives satisfying the inequality

$$|D^\gamma \psi_h(x)| \le K h^{-|\gamma|}. \tag{1.21}$$

Moreover, it vanishes beyond Ω and at the points of Ω at a distance at most h from the boundary. Inspect the product

$$\varphi^{(h)}(x) = \psi_h(x) u_\infty(x).$$

Estimate the norm of $\varphi^{(h)}(x)$ in $L_2^{(m)}(\Omega)$. Differentiating $\varphi^{(h)}(x)$, obtain

$$D^\alpha \varphi^{(h)}(x) = \begin{cases} D^\alpha u_\infty(x), & \text{if } x \in \Omega'_{3h}, \\ J_\alpha(x), & \text{if } x \in \Omega'_h \setminus \Omega'_{3h}, \\ 0, & \text{otherwise.} \end{cases}$$

Here, by $J_\alpha(x)$ we denote the sum

$$J_\alpha(x) = \sum_{0\le\beta\le\alpha} \frac{\alpha!}{\beta!(\alpha-\beta)!} D^\beta \psi_h(x) D^{\alpha-\beta} u_\infty(x).$$

Estimate the $L_2(\Omega_\gamma)$ norm of $J_\alpha(x)$, where Ω_γ is some parallelepiped of the lattice matrix hH having common points with the set $\Omega'_h \setminus \Omega'_{3h}$. Obviously,

$$\|J_\alpha \mid L_2(\Omega_\gamma)\| \le \sum_{0\le\beta\le\alpha} \frac{\alpha!}{\beta!(\alpha-\beta)!} \|D^\beta \psi_h(x) D^{\alpha-\beta} u_\infty(x) \mid L_2(\Omega_\gamma)\|.$$

By (1.3) we have the estimate

$$|D^{\alpha-\beta} u_\infty(x)| \le K h^{2m-|\alpha-\beta|},$$

with K a constant independent of h. Using the last relation and (1.21), for $|\alpha| = m$ obtain

$$\|D^\beta \psi_h D^{\alpha-\beta} u_\infty \mid L_2(\Omega_\gamma)\|^2 \le K h^{n+2m}.$$

Thus, the estimate holds

$$\|J_\alpha \mid L_2(\Omega_\gamma)\|^2 \le K h^{n+2m}. \tag{1.22}$$

Denote by $N^{(b)}$ the number of the meshes Ω_γ meeting $\Omega'_h \setminus \Omega'_{3h}$. In virtue of our supposition that the boundary of Ω is piecewise-smooth, the following inequality is valid $N^{(b)} h^n \le K_1 h$. Whence and from (1.22) we have

$$\left\|J_\alpha \mid L_2(\Omega'_h \setminus \Omega'_{3h})\right\|^2 \le K h^{2m+1},$$

with K a constant independent of h. This enables us to assert that the norm of the *truncated function* $\varphi^{(h)}(x)$ in $L_2^{(m)}(\Omega'_h \setminus \Omega'_{3h})$ decreases not slower than $h^{m+1/2}$, namely,

$$\left\|\varphi^{(h)}(x) \mid L_2^{(m)}(\Omega'_h \setminus \Omega'_{3h})\right\|^2 \le K h^{2m+1}.$$

Estimate the norm of $\varphi^{(h)}(x)$ in $L_2^{(m)}(\Omega'_{3h})$. Denote by $N^{(i)}$ the number of the meshes Ω_γ lying entirely in Ω'_{3h}, and by $N^{(l)}$, the number of meshes of the lattice which have at least one common point with Ω'_h. Obviously, for some γ

$$\begin{aligned} 0 &\le \left\|\varphi^{(h)}(x) \mid L_2^{(m)}(\Omega'_{3h})\right\|^2 - N^{(i)} \left\|u_\infty(x) \mid L_2^{(m)}(\Omega_\gamma)\right\|^2 \\ &\le (N^{(l)} - N^{(i)}) \left\|u_\infty \mid L_2^{(m)}(\Omega_\gamma)\right\|^2. \end{aligned}$$

Carrying out as before a change of variables in the integral presenting the norm of the function $u_\infty(x)$ over the fundamental parallelepiped Ω_γ and using the estimates

$$|D^\alpha u_\infty(x)| \le Kh^m, \quad |\alpha| = m; \quad |\Omega| - N^{(i)}h^n \le Kh, \quad (N^{(l)} - N^{(i)})h^n \le Kh,$$

come to the equality

$$\left\|\varphi^{(h)}(x) \mid L_2^{(m)}(\Omega'_{3h})\right\|^2 = B_{n,m}^2(\Omega)h^{2m} + O(h^{2m+1}).$$

Here $B_{n,m}(\Omega)$ stands for

$$B_{n,m}(\Omega) = (2\pi)^{-m}\left\{|\Omega| \sum_{\beta\neq 0} |H^{-1*}\beta|^{-2m}\right\}^{1/2}. \tag{1.23}$$

We have thus established that the $L_2^{(m)}(\Omega)$ norm of the truncated $\widetilde{L}_2^{(m)}$-extremal function may be written as

$$\left\|\psi_h(x)u_\infty(x) \mid L_2^{(m)}(\Omega)\right\| = B_{n,m}(\Omega)h^m(1 + O(h)). \tag{1.24}$$

Further, the norm of the difference $w^{(h)}(x)$ between the $\widetilde{L}_2^{(m)}$-extremal function $u_\infty(x)$ and the truncated function $\varphi^{(h)}(x)$ satisfies the inequality

$$\left\|w^{(h)} \mid L_2^{(m)}(\Omega)\right\| \le Kh^{m+1/2},$$

with K independent of h. Straightforward calculation also yields the following estimate

$$\left|\int_\Omega \sum_{|\alpha|=m} \frac{m!}{\alpha!} D^\alpha\varphi^{(h)}(x)D^\alpha w^{(h)}(x)\,dx\right| \le Kh^{2m+1}$$

for the inner product in $L_2^{(m)}(\Omega)$ of $\varphi^{(h)}(x)$ and $w^{(h)}(x)$. Consequently, the functions $\varphi^{(h)}(x)$ and $w^{(h)}(x)$ are almost orthogonal in $L_2^{(m)}(\Omega)$ at small h.

Using the properties of the truncated function $\varphi^{(h)}(x)$ of the $\widetilde{L}_2^{(m)}$-extremal function $u_\infty(x)$, we sharpen the lower bound on the norm of an arbitrary error with lattice of nodes $hH\gamma$. Let this functional look like

$$l(x) = \chi_\Omega(x) - h^n \sum_{hH\gamma\in\Omega} c[\gamma]\delta(x - hH\gamma),$$

and let the set $\overline{\Omega}$ lie in the interior of the fundamental parallelepiped Ω_0 of the lattice matrix H. The boundary of Ω is piecewise-smooth. We have the following

Theorem 5.2. *The norms of the error $l(x)$ of a cubature formula with lattice of nodes $hH\gamma$ in a bounded domain Ω with piecewise-smooth boundary satisfy the inequalities*

$$\|l \mid \widetilde{L}_2^{(m)*}\| \geq B_{n,m}(\Omega)h^m(1+O(h)), \quad \|l \mid L_2^{(m)*}\| \geq B_{n,m}(\Omega)h^m(1+O(h)). \quad (1.25)$$

PROOF. The norms of $l(x)$ in $\widetilde{L}_2^{(m)*}$ and $L_2^{(m)*}$ are bounded from below by the ratio

$$\frac{(l, \varphi^{(h)}(x))}{\|\varphi^{(h)} \mid L_2^{(m)}(\Omega)\|}.$$

This follows from the coincidence of the norms of the truncated function $\varphi^{(h)}(x)$ in $L_2^{(m)}$, $\widetilde{L}_2^{(m)}(H)$, and $L_2^{(m)}(\Omega)$. We have already examined the behavior of the denominator and obtained (1.24). We now find $(l, \varphi^{(h)})$.

The function $\varphi^{(h)}(x)$, as well as $u_\infty(x)$, equals 0 at the nodes $hH\gamma$. Consequently,

$$(l, \varphi^{(h)}) = \int\limits_\Omega \varphi^{(h)}(x)\,dx.$$

Further, $u_\infty(x)$ and $\psi_h(x)$ are nonnegative functions and, moreover, $u_\infty(x)$ is a periodic function with period matrix hH. Hence,

$$N^{(i)} \int\limits_{\Omega_\gamma} u_\infty(x)\,dx \leq \int\limits_\Omega \varphi^{(h)}(x)\,dx \leq N^{(l)} \int\limits_{\Omega_\gamma} u_\infty(x)\,dx. \quad (1.26)$$

The integral of $u_\infty(x)$ over the mesh Ω_γ may be calculate exactly

$$\int\limits_{\Omega_\gamma} u_\infty(x)\,dx = h^n \int\limits_{\Omega_0} u_\infty(y)\,dy = h^{2m+n}\left(\frac{1}{2\pi}\right)^{2m} \zeta(H^{-1*} \mid 2m).$$

Inserting this expression in (1.26) and using the asymptotic equalities

$$N^{(i)}h^n = |\Omega| + O(h), \quad N^{(l)}h^n = |\Omega| + O(h),$$

obtain

$$|(l, \varphi^{(h)}) - B_{n,m}^2(\Omega)h^{2m}| \leq Kh^{2m+1}.$$

Whence and from (1.24) it follows that

$$\frac{(l, \varphi^{(h)})}{\|\varphi^{(h)} \mid L_2^{(m)}(\Omega)\|} = B_{n,m}(\Omega)h^m + O(h^{m+1/2}). \quad (1.27)$$

The proof of Theorem 5.2 is complete.

In closing, observe that the truncated function $\varphi^{(h)}(x)$ belongs to the closure in $L_0^{(m)}(\Omega)$ norm of the set comprising all functions compactly-supported in the interior of Ω. We denote this space by $\mathring{L}_2^{(m)}(\Omega)$. From (1.27) we infer that the $\mathring{L}_2^{(m)*}(\Omega)$ norm of $l(x)$ also admits an estimate that is analogous to (1.25)

$$\left\|l(x) \mid \mathring{L}_2^{(m)*}(\Omega)\right\| \geq B_{n,m}(\Omega)h^m(1+O(h)). \tag{1.28}$$

Thus, in each of the four spaces $\widetilde{L}_2^{(m)}$, $L_2^{(m)}$, $L_2^{(m)}(\Omega)$, and $\mathring{L}_2^{(m)}(\Omega)$ we have the same estimates from below for the norm of an error.

§2. Errors in the $\mathring{L}_2^{(m)}(\Omega)$ Space of Compactly-Supported Functions

In this section we obtain an expansion of the $\check{L}_2^{(m)*}(\Omega)$ norm of $l_\infty(x)$ which is asymptotically exact as the lattice mesh-size vanishes. This enables us to give an analogous formula for the greatest lower bound of the norm of the error with a fixed lattice of nodes $hH\gamma$. Before formulating the corresponding theorem, we list conditions on the domain Ω.

Let Ω be bounded and lie in the interior of the fundamental parallelepiped Ω_0 of the matrix H. We denote by Ω''_η the set of all points of $\mathbb{R}^n$ at a distance less than η from Ω. Moreover, let Ω'_η be the set of points of Ω at a distance greater than η from the boundary of Ω. It is clear that Ω'_η lies in Ω which, in turn, lies in Ω''_η. Say that Ω *has order* q, $0 < q \leq 1$, *of the width of the boundary layer*, if the volume of the strip between Ω''_η and Ω'_η decreases not slower than $K\eta^q$ as $\eta \to 0$, namely,

$$\left|\Omega''_\eta \setminus \Omega'_\eta\right| \leq K\eta^q.$$

Domains with piecewise-smooth boundary have order 1 of the width of the boundary layer.

Denote by $S_1(\Omega)$ the set of the meshes Ω_γ of the lattice of nodes $hH\gamma$ which have common points with Ω. Obviously, for a domain Ω having order q of the width of the boundary layer the equality holds

$$|S_1(\Omega)| = |\Omega|(1+O(h^q)). \tag{2.1}$$

Assume in what follows that the domain Ω has order q of the width of the boundary layer.

Theorem 5.3. *At each function $\varphi(x) \in \overset{\circ}{L}{}_2^{(m)}(\Omega)$ the error $l_\infty(x)$ assumes the value that admits the following upper bound*

$$|(l_\infty, \varphi)| \le \left(\frac{h}{2\pi}\right)^m [\zeta(H^{-1*} \mid 2m)]^{1/2} |S_1(\operatorname{supp}\varphi)|^{1/2} \|\varphi(x) \mid \overset{\circ}{L}{}_2^{(m)}(\Omega)\|. \tag{2.2}$$

PROOF. Let a function $\varphi(x)$ belong to $\overset{\circ}{L}{}_2^{(m)}(\Omega)$. Extend it to the entire Ω_0, by setting it 0 beyond Ω. The value of l_∞ at φ is expressed by the integral

$$(l_\infty, \varphi) = \int_\Omega \sum_{|\alpha|=m} \frac{m!}{\alpha!} D^\alpha \varphi(x) D^\alpha u_\infty(x)\, dx. \tag{2.3}$$

The $\widetilde{L}_2^{(m)}$-extremal function $u_\infty(x)$ has all derivatives up to order m each of which is a homogeneous function of h. More exactly, we have the equality

$$D^\alpha u_\infty(x) = h^{2m-|\alpha|} (D^\alpha u^{(\infty)}) \left(\frac{x}{h}\right), \quad |\alpha| \le m, \tag{2.4}$$

where $u^{(\infty)}(x)$ stands for the function $u_\infty(x)$ corresponding to the value $h = 1$. The domain of integration in (2.3) is obviously the support of $\varphi(x)$. The whole space $\mathbb{R}^n$ is partitioned into the elementary meshes Ω_γ of the initial lattice of nodes $hH\gamma$, with

$$\Omega_\gamma = \{x \in \mathbb{R}^n : 0 \le (H^{-1}x)_j - h\gamma_j < h,\ j = 1, \dots, n\}.$$

We denote by $B\varphi$ the set of γ such that Ω_γ meets the support of φ. Considering (2.4), we may continue (2.3) as follows

$$(l_\infty, \varphi) = h^m \sum_{\gamma \in B\varphi} \int_{\Omega_\gamma} \sum_{|\alpha|=m} \frac{m!}{\alpha!} D^\alpha \varphi(x) D^\alpha u^{(\infty)} \left(\frac{x}{h}\right) dx. \tag{2.5}$$

Using the Cauchy–Bunyakovskiĭ–Schwarz inequality, for every $\gamma \in B\varphi$ we have

$$\left| \int_{\Omega_\gamma} \sum_{|\alpha|=m} \frac{m!}{\alpha!} D^\alpha \varphi(x) D^\alpha u^{(\infty)} \left(\frac{x}{h}\right) dx \right|$$
$$\le \left\|\varphi(x) \mid L_2^{(m)}(\Omega_\gamma)\right\| \left\| u^{(\infty)} \left(\frac{x}{h}\right) \mid L_2^{(m)}(\Omega_\gamma)\right\|. \tag{2.6}$$

Performing the change of variables $x = hy$ in the corresponding integral, we readily calculate that

$$\left\| u^{(\infty)} \left(\frac{x}{h}\right) \mid L_2^{(m)}(\Omega_\gamma)\right\|^2$$
$$= h^n \int_{\Omega_0} \sum_{|\alpha|=m} \frac{m!}{\alpha!} |D^\alpha u^{(\infty)}(y)|^2\, dy = h^n (l_\infty(y), u^{(\infty)}(y)). \tag{2.7}$$

The functional $l_\infty(y)$ here corresponds to the value $h = 1$. Considering this and assuming that the integral of $u^{(\infty)}(x)$ over the fundamental parallelepiped Ω_0 equals 0, obtain

$$(l_\infty(y), u^{(\infty)}(y)) = -u^{(\infty)}(0) = \left(\frac{1}{2\pi}\right)^{2m} \zeta(H^{-1*} \mid 2m).$$

Inserting this in (2.7) and putting the result in (2.6), come to the inequality

$$\left| \int_{\Omega_\gamma} \sum_{|\alpha|=m} \frac{m!}{\alpha!} D^\alpha \varphi(x) (D^\alpha u^{(\infty)}) \left(\frac{x}{h}\right) dx \right|$$
$$\leq h^{n/2} \left(\frac{1}{2\pi}\right)^m [\zeta(H^{-1*} \mid 2m)]^{1/2} \left\| \varphi(x) \mid L_2^{(m)}(\Omega_\gamma) \right\|.$$

Summing these inequalities over all $\gamma \in B\varphi$, from (2.5) obtain

$$|(l_\infty(x), \varphi(x))|$$
$$\leq h^{n/2+m} \left(\frac{1}{2\pi}\right)^m [\zeta(H^{-1*} \mid 2m)]^{1/2} \sum_{\gamma \in B\varphi} \left\| \varphi(x) \mid L_2^{(m)}(\Omega_\gamma) \right\|. \tag{2.8}$$

Denote by N_1 the number of the meshes Ω_γ having common points with the support of $\varphi(x)$. By the Cauchy inequality for sums, we have

$$\sum_{\gamma \in B\varphi} \left\| \varphi(x) \mid L_2^{(m)}(\Omega_\gamma) \right\| \leq N_1^{1/2} \left\{ \sum_\gamma \left\| \varphi(x) \mid L_2^{(m)}(\Omega_\gamma) \right\|^2 \right\}^{1/2}.$$

Inserting this estimate in (2.8) and considering that the volume of $S_1(\operatorname{supp}\varphi)$ equals $N_1 h^n$, come to (2.2). The proof of Theorem 5.3 is complete.

Corollary 2.1. *In the* $\mathring{L}_2^{(m)}(\Omega)$ *space of compactly-supported functions the following asymptotically exact expansion of the norm of* $l_\infty(x)$ *is valid*

$$\left\| l_\infty \mid \mathring{L}_2^{(m)*}(\Omega) \right\| = B_{n,m}(\Omega) h^m (1 + O(h^q)). \tag{2.9}$$

PROOF. In the preceding section we checked that the difference of

$$\left\| l_\infty(x) \mid \mathring{L}_2^{(m)*}(\Omega) \right\|$$

and $B_{n,m}(\Omega) h^m$ is nonnegative. On the other hand, in accord with (2.2), it is at most

$$\left(\frac{h}{2\pi}\right)^m [\zeta(H^{-1*} \mid 2m)]^{1/2} (|S_1(\Omega)|^{1/2} - |\Omega|^{1/2}).$$

Whence and from (2.1) obtain (2.9).

We may improve (2.9) in some cases. Let Ω be the *multiple part* of the fundamental parallelepiped Ω_0, i.e.,

$$\Omega = \{x : 0 \le (KH^{-1}x)_j < 1/h,\ j = 1, \dots, n\},$$

where K is the diagonal matrix with naturals k_j on the diagonal. For such domain Ω, every function $\varphi(x) \in \overset{\circ}{L}{}_2^{(m)}(\Omega)$ extends to the whole of $\mathbb{R}^n$ to a periodic function with period matrix $K^{-1}H$. Thereafter, we may treat it as the function $\varphi_K(x)$ in $\widetilde{L}_2^{(m)}(H)$. In this event, the ratio of the squares of its norms in the larger and smaller spaces equals the determinant of K, i.e.,

$$\left\|\varphi_K(x) \mid \widetilde{L}_2^{(m)}(H)\right\| = \sqrt{\det K}\left\|\varphi(x) \mid \overset{\circ}{L}{}_2^{(m)}(\Omega)\right\|.$$

Put

$$l_\infty^\Omega(x) = \chi_\Omega(x) - h^n \sum_{hH\gamma\in\Omega} \delta(x - hH\gamma).$$

Estimate the norm of this error in $\overset{\circ}{L}{}_2^{(m)*}(\Omega)$. Observe that the value of $l_\infty^\Omega(x)$ at every function $\varphi(x) \in \overset{\circ}{L}{}_2^{(m)}(\Omega)$ may be found by the formula

$$(l_\infty^\Omega, \varphi) = (l_\infty, \varphi_K)/\det K.$$

Consequently, the equality holds

$$\frac{(l_\infty^\Omega, \varphi)}{\left\|\varphi \mid \overset{\circ}{L}{}_2^{(m)}(\Omega)\right\|} = \frac{1}{\sqrt{\det K}} \frac{(l_\infty, \varphi_K)}{\left\|\varphi_K \mid \widetilde{L}_2^{(m)}(H)\right\|}.$$

Taking the supremum of both sides over the set of all nonzero $\varphi(x)$ in $\overset{\circ}{L}{}_2^{(m)}(\Omega)$, establish the estimate

$$\left\|l_\infty^\Omega \mid \overset{\circ}{L}{}_2^{(m)*}(\Omega)\right\| \le \frac{1}{|\det K|^{1/2}}\left\|l_\infty \mid \widetilde{L}_2^{(m)*}(H)\right\| = B_{n,m}(\Omega)h^m.$$

Obviously, the norms in $\overset{\circ}{L}{}_2^{(m)*}(\Omega)$ of $l_\infty^\Omega(x)$ and $l_\infty(x)$ coincide. Considering this, arrive at some expansion more exact than (2.9), namely,

$$\left\|l_\infty \mid \overset{\circ}{L}{}_2^{(m)*}(\Omega)\right\| = B_{n,m}(\Omega)h^m. \tag{2.10}$$

Leaning on the two-sided estimates for the norms of errors which were established in this and preceding sections, we pronounce the following

Theorem 5.4. *The greatest lower bound of the* $\overset{\circ}{L}{}_2^{(m)}(\Omega)$ *norms of all errors with lattice of nodes* $hH\gamma$ *may be written as the asymptotically exact formula*

$$\inf\left\| l \mid \overset{\circ}{L}{}_2^{(m)*}(\Omega)\right\| = B_{n,m}(\Omega)h^m(1 + O(h^q)). \tag{2.11}$$

PROOF. The expansion (2.11) is immediate from the lower bound (1.28) valid for every error $l(x)$ with lattice of nodes $hH\gamma$ and from (2.9).

Observe that we obtained asymptotically exact estimates for the rate of convergence to zero of the norm of $l_\infty(x)$ provided that h vanishes and m is constant. These estimates thus characterize the rate of the vanishing of the possibly maximal error of a cubature formula at the functions of the unit ball in $\overset{\circ}{L}{}_2^{(m)}(\Omega)$. In this event, to each h there corresponds its own extremal function providing the maximal error and the set of these functions is noncompact as $h \to 0$. Consequently, for a fixed function $\varphi(x)$ in $\overset{\circ}{L}{}_2^{(m)}(\Omega)$, the rate of vanishing as $h \to 0$ turns out to be faster than this is guaranteed by the estimate

$$|(l_\infty, \varphi)| \leq \left\|l_\infty \mid \overset{\circ}{L}{}_2^{(m)*}(\Omega)\right\| \left\|\varphi \mid \overset{\circ}{L}{}_2^{(m)*}(\Omega)\right\|.$$

Closing this section, we establish one more property of the extremal function $u_\infty(x)$. As is known, every function $\varphi(x)$ in $L_2^{(m)}(\Omega)$ extends to the entire $\mathbb{R}^n$ to a function belonging to $L_2^{(m)}(\mathbb{R}^n)$. Among these extensions there is one with a minimal $L_2^{(m)}$ norm which is denoted by $\overline{\varphi}$. The function $\overline{\varphi}$ is polyharmonic beyond Ω and has mth order derivatives decreasing at infinity not slower than $|x|^{-n}\log|x|$.

We apply the just-described extension operation to the function $\chi_\Omega(x)u_\infty(x)$ belonging obviously to $L_2^{(m)}(\Omega)$. Denote the result of this operation by $\overline{u}_\infty(x)$. It turns out that we have the following asymptotically exact expansion

$$\left\|\overline{u}_\infty \mid L_2^{(m)}\right\| = B_{n,m}(\Omega)h^m(1 + O(h^{q/2})). \tag{2.12}$$

Check the validity of (2.12). To this end, we need some knowledge of the structure of $L_2^{(m)}$. In $L_2^{(m)}$ we choose the proper subspace $\overset{\circ}{L}{}_2^{(m)}(\mathbb{R}^n \mid \Omega)$ that comprises the functions supported in the closure $\overline{\Omega}$. This subspace is closed in $L_2^{(m)}$. Consequently, it has the orthogonal complement which we denote by $H_2^{(m)}(\mathbb{R}^n \mid \partial\Omega)$.

The space $H_2^{(m)}(\mathbb{R}^n \mid \partial\Omega)$ consists of the functions $\varphi(x)$ polyharmonic in Ω and in the exterior of Ω and having derivatives of order m which decrease at infinity not slower than $|x|^{-n}\log|x|$. Observe that a function $\varphi(x)$ in $H_2^{(m)}(\mathbb{R}^n \mid \partial\Omega)$ is not necessarily polyharmonic in the entire $\mathbb{R}^n$, since the derivatives of $\varphi(x)$ of order at least m might be discontinuous on the boundary of Ω.

Thus, the function $\overline{u}_\infty(x)$ in $L_2^{(m)}$ may be written as the sum of its projections

$$\overline{u}_\infty(x) = \overline{u}_\infty^{(0)}(x) + \overline{u}_\infty^{(1)}(x), \tag{2.13}$$

where $\overline{u}_\infty^{(0)} \in \overset{\circ}{L}{}_2^{(m)}(\mathbb{R}^n \mid \Omega)$ and $\overline{u}_\infty^{(1)} \in H_2^{(m)}(\mathbb{R}^n \mid \partial\Omega)$. By mutual orthogonality of the summands in (2.13) in $L_2^{(m)}$, the equality holds

$$\left\|\overline{u}_\infty \mid L_2^{(m)}\right\|^2 = \left\|\overline{u}_\infty^{(0)} \mid L_2^{(m)}\right\|^2 + \left\|\overline{u}_\infty^{(1)} \mid L_2^{(m)}\right\|^2. \tag{2.14}$$

The norms of $\overline{u}_\infty^{(0)}$ in $L_2^{(m)}$ and $L_2^{(m)}(\Omega)$ obviously coincide

$$\left\|\overline{u}_\infty^{(0)} \mid L_2^{(m)}\right\|^2 = \left\|\overline{u}_\infty^{(0)} \mid L_2^{(m)}(\Omega)\right\|^2. \tag{2.15}$$

Further, $L_2^{(m)}(\Omega)$ splits into the direct sum of $\overset{\circ}{L}{}_2^{(m)}(\Omega)$ and its orthogonal complement $H_2^{(m)}(\Omega)$ consisting of the functions polyharmonic in Ω. Observe that, for $m = 1$ the first-order derivatives of each function in $H_2^{(m)}(\Omega)$ belongs to the *Bergman space* $b_2(\Omega)$ whose definition and properties are exposed in the book [5]. Consequently, the function $\overline{u}_\infty(x)$ as a member of $L_2^{(m)}(\Omega)$ may be decomposed in the sum of its projections to $\overset{\circ}{L}{}_2^{(m)}(\Omega)$ and $H_2^{(m)}(\Omega)$.

Obviously, as such appear the restrictions of $\overline{u}_\infty^{(0)}(x)$ and $\overline{u}_\infty^{(1)}(x)$ to Ω. Using their orthogonality in $L_2^{(m)}(\Omega)$, come to the relation

$$\left\|\overline{u}_\infty^{(0)} \mid L_2^{(m)}(\Omega)\right\|^2 = \left\|u_\infty \mid L_2^{(m)}(\Omega)\right\|^2 - \left\|\overline{u}_\infty^{(1)} \mid L_2^{(m)}(\Omega)\right\|^2.$$

Inserting it in (2.15) and putting the result in (2.14), obtain

$$\left\|\overline{u}_\infty \mid L_2^{(m)}\right\|^2 = \left\|u_\infty \mid L_2^{(m)}(\Omega)\right\|^2 + \left\|\overline{u}_\infty^{(1)} \mid L_2^{(m)}\right\|^2 - \left\|\overline{u}_\infty^{(1)} \mid L_2^{(m)}(\Omega)\right\|^2. \tag{2.16}$$

We have already found in § 1 an asymptotically exact expansion of the first norm on the right side of (2.16). Recall that

$$\left\|u_\infty \mid L_2^{(m)}(\Omega)\right\|^2 = B_{n,m}^2(\Omega)h^{2m}(1 + O(h^q)). \tag{2.17}$$

Estimate the two remaining summands in (2.16). To this end, find out $w^{(h)}(x)$ in $L_2^{(m)}$ with the same projection to $H_2^{(m)}(\Omega)$ as $\overline{u}_\infty(x)$. Observe that, for the projections to coincide, it suffices that $w^{(h)}(x)$ and $\overline{u}_\infty(x)$ agree in a small neighborhood about the boundary of Ω.

Construct the function $w^{(h)}(x)$ at the points x of Ω. Put

$$w_i^{(h)}(x) = u_{\infty}(x) - \psi_h(x)u_{\infty}(x), \quad x \in \Omega.$$

Here $\psi_h(x)$ is the truncator constructed in § 1. For a domain Ω having order q of the width of the boundary layer, the estimate holds

$$\left\|w_i^{(h)}(x) \mid L_2^{(m)}(\Omega)\right\| \leq K h^{m+q/2}. \tag{2.18}$$

The method of its proving is analogous to that described already in § 1.

Using a similar scheme, construct a function $w_e^{(h)}(x)$ in $L_2^{(m)}(\mathbb{R}^n \setminus \Omega)$ that agrees with $\overline{u}_\infty(x)$ near the boundary of Ω and satisfies the inequality

$$\left\|w_e^{(h)}(x) \mid L_2^{(m)}(\mathbb{R}^n \setminus \Omega)\right\| \leq K h^{m+q/2}. \tag{2.19}$$

Put

$$w^{(h)}(x) = \begin{cases} w_i^{(h)}(x), & \text{if } x \in \Omega, \\ w_e^{(h)}(x), & \text{if } x \in \mathbb{R}^n \setminus \Omega. \end{cases}$$

It is clear that this function is in $L_2^{(m)}$ and by (2.18) and (2.19) its norm is bounded as follows

$$\left\|w^{(h)}(x) \mid L_2^{(m)}\right\| \leq K h^{m+q/2}.$$

The projection of $w^{(h)}(x)$ to $H_2^{(m)}(\mathbb{R}^n \mid \Omega)$ coincides with $\overline{u}_\infty^{(1)}(x)$, and its norm is at most the norm of the initial function, i.e.,

$$\left\|\overline{u}_\infty^{(1)} \mid L_2^{(m)}\right\| \leq K h^{m+q/2}. \tag{2.20}$$

Whence it follows that

$$\left\|\overline{u}_\infty^{(1)} \mid L_2^{(m)}(\Omega)\right\| \leq K h^{m+q/2}. \tag{2.21}$$

Inserting (2.20), (2.21), and (2.17) in (2.16), come to the sought formula (2.12).

§3. Constructing a Formula with Regular Boundary Layer

As we have already shown, the optimal cubature formulas for periodic functions have equal weights and we may construct the corresponding errors by summing the translates of the same local error. In this event the integration domain coincides with the fundamental parallelepiped of the lattice matrix under consideration. Now we describe the process that enables us to construct similar almost optimal formulas for a given lattice of nodes. Moreover, the integration domain may have an arbitrary shape if only its boundary is sufficiently smooth. The error of these formulas is characterized by the norm of the corresponding functional $l(x)$. For this error we derive the following asymptotic expansion

$$\|l \mid L_2^{(m)*}\| = B_{n,m}(\Omega)h^m(1 + O(h)). \tag{3.1}$$

The meshes Ω_γ of the initial lattice of nodes $hH\gamma$ lie partly in Ω and partly beyond Ω. We denote by $B_L^{(i)}$ the set of γ to which there correspond the nodes $hH\gamma$ in Ω at a distance not less that Lh from the boundary $\partial\Omega$ of Ω. If, on the other hand, the point $hH\gamma$ is in the complement of Ω and lies at a distance at least Lh from the boundary of Ω; then we assume γ to belong to $B_L^{(e)}$. All remaining integer multi-indices γ compose the set $B_L^{(l)}$. It is clear that to these indices there correspond the nodes $hH\gamma$ that lie at a distance at most Lh from Ω.

Consider an error of the shape

$$l(x) = \chi_\Omega(x) - \sum_{\gamma \in B_{2L}^{(i)} \cup B_{2L}^{(l)}} h^n c[\gamma]\delta(x - hH\gamma). \tag{3.2}$$

Reasoning to be presented does not depend on whether or not the support of $l(x)$ is entirely in Ω. However, in integrating the functions that are given in Ω, we usually require from a cubature formula that all its nodes belong to the domain over which we integrate, i.e., Ω. In this case we require that the weights $c[\gamma]$ of (3.2), corresponding to the points $hH\gamma$ outside Ω, be all equal to 0.

We callt the generalized function (3.2) the *error with regular boundary layer, width* $2L$, *order* s, *and estimate* A, if it may be written down as

$$l(x) = \sum_{\gamma \in B_L^{(i)}} l^{(0)}\left(\frac{x}{h} - H\gamma\right) + \sum_{\gamma \in B_L^{(e)}} l^{(\gamma)}\left(\frac{x}{h} - H\gamma\right), \tag{3.3}$$

in which the *elementary*, or *local*, errors $l^{(\gamma)}(y)$ and $l^{(0)}(y)$ belong to the classes $R(L, A, s)$ and $R(L, A, s+1)$, respectively.

Recall that the error $l_0(y)$ belongs to $R(L, A, m)$, if the next three conditions are met:

(1) The support of the generalized function $l_0(y)$ is included in the ball of radius L with center the coordinate origin

$$\operatorname{supp} l_0(y) \subset \{y : |y| \le L\}.$$

(2) The possibly maximal error of the formula at continuous functions in the unit ball of the space C is at most A, in symbols,

$$\|l_0(y) \mid C^*\| \le A.$$

(3) The values of $l_0(y)$ at all polynomials of degree less than m equal 0, namely,

$$(l(y), y^\alpha) = 0, \quad |\alpha| < m.$$

We also call a cubature formula with error (3.3) a *cubature formula with regular boundary layer.*

Below we point out a method for constructing the local errors $l^{(0)}(x/h - H\gamma)$ and $l^{(\gamma)}(x/h - H\gamma)$.

Partition the domain Ω in the parts included in different meshes Ω_γ of the lattice of nodes $hH\gamma$, obtaining

$$\Omega = \bigcup_{\gamma \in B} (\Omega \cap \Omega_\gamma),$$

where B is the set of γ such that $\Omega \cap \Omega_\gamma$ is nonempty. Obviously, B lies in $B_L^{(i)} \cup B_L^{(l)}$.

The functional $l^{(0)}(x/h - H\gamma)$ is constructed in the same way as in § 1. For all $\gamma \in B_L^{(i)}$ the support of the functional is in the interior of Ω. If for some $\gamma \in B_L^{(l)}$ this support also lies in the interior of Ω, then as $l^{(\gamma)}(x/h - H\gamma)$ we again take $l^{(0)}(x/h - H\gamma)$. In the opposite case we proceed differently.

Among all bounded domains we distinguish a class of domains with sharp boundary. Say that a domain Ω ***has a sharp boundary*** if there is a positive number q such that, for every point $x^{(0)}$ in $\overline{\Omega}$ and every number $\rho > 0$, in the ball of radius ρ with center the point $x^{(0)}$ there is a cube with side $q\rho$ lying entirely in Ω. This class obviously contains every domain with piecewise-smooth boundary.

Let the integration domain Ω have a sharp boundary. For $\gamma \in B_L^{(l)}$ the support of $l^{(0)}(x/h - H\gamma)$ meets the complement of $\overline{\Omega}$. Construct the local error $l^{(\gamma)}(x/h - H\gamma)$ corresponding to this γ. Put

$$l^{(\gamma)}(y) = \chi_\gamma(y) - \sum_{\gamma' \in B_\gamma} c^{(\gamma)}[\gamma']\delta(y - H\gamma'), \tag{3.4}$$

where $\chi_\gamma(y)$ is the indicator of the part of Ω_0 that results from the set $\Omega \cap \Omega_\gamma$ by the change of variables $x = hH\gamma + hy$, i.e.,

$$\chi_\gamma(y) = \chi_\Omega(hH\gamma + hy)\chi_{\Omega_0}(y).$$

Choose the set B_γ of the admissible values of γ' in (3.4) to be finite. In this event, obviously, there is a ball of radius L with center the coordinate origin including the entire support of $l^{(\gamma)}(y)$. Further, at all points $\gamma' \in B_\gamma$ such that the node $hH(\gamma' + \gamma)$ does not belong to $\overline{\Omega}$, set the weight $c^{(\gamma)}[\gamma']$ equal 0. Then, for h sufficiently small, the support of $l^{(\gamma)}(x/h - H\gamma)$ obviously lies entirely in $\overline{\Omega}$. We designate the still-undefined weights $c^{(\gamma)}[\gamma']$ so as to meet the condition of the orthogonality of $l^{(\gamma)}(y)$ to every polynomial of degree less than s, namely,

$$(l^{(\gamma)}(y), y^\alpha) = 0, \quad |\alpha| < s. \tag{3.5}$$

Demonstrate that this is possible. As was explicated in Chapter 1, the conditions (3.5) are equivalent to the simultaneous linear equations

$$S_\gamma c^{(\gamma)} = f^{(\gamma)} \tag{3.6}$$

for the vector of unknowns $c^{(\gamma)}$ which is composed of the weights $c^{(\gamma)}[\gamma']$ that remain undetermined. A solution to (3.6) certainly exists if the rank of S_γ is sufficiently large and the matrix has a right inverse. In our case, we may achieve this as follows: given s, take a sufficiently large L. Indeed, by hypothesis the domain Ω has a sharp boundary. Consequently, there is a cube with side qLh embedded in Ω and containing s^n nodes of the lattice $hH\gamma'$. The corresponding matrix S_γ is then the Vandermonde matrix with entries $(hH\gamma^{(j)})^{\alpha^{(i)}}$, where $hH\gamma^{(j)}$ ranges over all points of the lattice which lie in the cube. In particular, the system of points $hH\gamma^{(j)}$ includes a Newtonian subsystem. In this case, as is known, the matrix S_γ has a right inverse.

Find a solution to (3.6) and construct (3.4). Since B_γ is a finite set, there is a constant A such that

$$\|l^{(\gamma)}(y) \mid C^*\| \leq A.$$

It is clear that A is bounded from below by the sum of the moduli of the weights $c^{(\gamma)}[\gamma']$, i.e., it depends on the order s.

Thus, for a domain with sharp boundary we constructed the local errors $l^{(\gamma)}(x/h - H\gamma)$ with needed properties and so, for a domain with sharp boundary, we proved the existence of an error with regular boundary layer.

Decomposition of a given error with regular boundary layer in the sum like (3.3) is not unique. As an inner elementary error $l^{(0)}(x/h - H\gamma)$ we may take every generalized function of the class $R(L, A, N)$, with $N > s$. However, we must simultaneously choose the values of L and A appropriate to N and to s as well. To prove this result, we are to consider the difference of two errors with regular boundary layer which are constructed for the same domain Ω. It turns out that this difference is a linear combination of delta functions supported near to the boundary $\partial\Omega$.

Theorem 5.5. *For a given domain Ω the difference of two errors with regular boundary layer of width L and order s may be written as*

$$l_1(x) - l_2(x) = \sum_{\gamma \in B_L^{(l)}} m^{(\gamma)}\left(\frac{x}{h} - H\gamma\right), \tag{3.7}$$

where

$$m^{(\gamma)}(y) = \sum_{\gamma' \in B_\gamma} c^{(\gamma)}[\gamma']\delta(y - H\gamma').$$

The support of the generalized function $m^{(\gamma)}(y)$ lies in the ball of radius L with center the coordinate origin, and its values at all polynomials of degree less than s equal 0,

$$(m^{(\gamma)}(y), y^\alpha) = 0, \quad |\alpha| < s.$$

PROOF. Suppose that the domain Ω lies in the interior of the fundamental parallelepiped Ω_0 of the corresponding lattice matrix H. In this event none of the nodes $hH\gamma$ belongs to the boundary of Ω.

Compare the actions of $l_1(x)$, $l_2(x)$, and $l_\infty(x)$ in $\mathring{L}_2^{(m)}(\Omega)$. Extend each function $\varphi(x) \in \mathring{L}_2^{(m)}(\Omega)$ by 0 to the points of Ω_0 lying outside Ω. Thus, the value of $l_\infty(x)$ at $\varphi(x)$ is easy to calculate.

Assume that to the functional $l_1(x)$ in the decomposition (3.3) there correspond the inner local error $l_1^{(0)}(x/h - H\gamma)$. Reasoning analogous to that carried out in § 1 when deriving (1.10) allows us to assert that

$$l_\infty(x) = \sum_{hH\gamma \in \Omega_0} l_1^{(0)}\left(\frac{x}{h} - H\gamma\right). \tag{3.8}$$

In particular, this formula is valid for all functions $\varphi(x) \in \mathring{L}_2^{(m)}(\Omega)$. Whence and from (3.3) obtain

$$\begin{aligned}(l_1 - l_\infty, \varphi) = &\sum_{\gamma \in B_L^{(l)}} \left(l_1^{(\gamma)}\left(\frac{x}{h} - H\gamma\right) - l_1^{(0)}\left(\frac{x}{h} - H\gamma\right), \varphi\right) \\ &- \sum_{\gamma \in B_L^{(e)}} \left(l_1^{(0)}\left(\frac{x}{h} - H\gamma\right), \varphi\right).\end{aligned} \tag{3.9}$$

In this equality, the sum over $\gamma \in B_L^{(e)}$ is identically 0, since for these γ the support of $l_1^{(0)}(x/h - H\gamma)$ and the support of $\varphi(x)$ are disjoint.

Similar arguments apply to the error with regular boundary layer $l_2(x)$. We so obtain

$$(l_j(x) - l_\infty(x), \varphi) = \sum_{\gamma \in B_L^{(l)}} \left(l_j^{(\gamma)}\left(\frac{x}{h} - H\gamma\right) - l_j^{(0)}\left(\frac{x}{h} - H\gamma\right), \varphi\right),$$

where $j = 1, 2$. Subtracting the first of these equalities from the second, find

$$(l_1 - l_2, \varphi) = \sum_{\gamma \in B_L^{(l)}} \left(m^{(\gamma)} \left(\frac{x}{h} - H\gamma \right), \varphi \right), \tag{3.10}$$

with $m^{(\gamma)}(y) = l_1^{(\gamma)}(y) - l_1^{(0)}(y) - l_2^{(\gamma)}(y) + l_2^{(0)}(y)$. It is not hard to see that $m^{(\gamma)}(y)$ is a linear combination of delta functions possessing the needed properties.

The support of $l_1 - l_2$ obviously lies in Ω. Consequently, (3.7) and (3.10) are equivalent. The proof of Theorem 5.5 is complete.

Assume that a natural N is greater than s, and the numbers L and A are chosen so that the classes $R(L, A, s)$ and $R(L, A, N)$ are nonempty. Take a local error $l_*^{(0)}(y)$ in $R(L, A, N)$. We have

Lemma 5.1. *Every error with regular boundary layer, width $2L$, order s, and estimate A may be written as*

$$l(x) = \sum_{\gamma \in B_L^{(i)}} l_*^{(0)} \left(\frac{x}{h} - H\gamma \right) + \sum_{\gamma \in B_L^{(l)}} l^{(\gamma)} \left(\frac{x}{h} - H\gamma \right), \tag{3.11}$$

where $l^{(\gamma)}(y) \in R(L, A, s)$.

The only distinction between this formula and (3.3) is as follows: the inner elementary errors $l_*^{(0)}(x/h - H\gamma)$ vanish at all polynomials of degree greater than s.

PROOF. Using the algorithm of the beginning of the current section and given a local error $l_*^{(0)}(y)$, construct an error with regular boundary layer, width $2L$, order $N - 1$, and estimate A as follows

$$l_1(x) = \sum_{\gamma \in B_L^{(i)}} l_*^{(0)} \left(\frac{x}{h} - H\gamma \right) + \sum_{\gamma \in B_L^{(l)}} l_1^{(\gamma)} \left(\frac{x}{h} - H\gamma \right). \tag{3.12}$$

This is obviously a regular boundary layer of order s. In accord with Theorem 5.5, the difference $l(x) - l_1(x)$ may be written as

$$l(x) - l_1(x) = \sum_{\gamma \in B_L^{(l)}} m^{(\gamma)} \left(\frac{x}{h} - H\gamma \right).$$

Consequently, for $l(x)$ we have (3.11) in which the boundary elementary errors are generalized functions

$$l^{(\gamma)} \left(\frac{x}{h} - H\gamma \right) = l_1^{(\gamma)} \left(\frac{x}{h} - H\gamma \right) + m^{(\gamma)} \left(\frac{x}{h} - H\gamma \right).$$

Obviously, $l^{(\gamma)}(y)$ belongs to $R(L, A, s)$. The proof of Lemma 5.1 is complete.

Observe that in an error (3.2) with regular boundary layer of width $2L$ the weight $c[\gamma]$ for which the point $hH\gamma$ lies at a distance more than $2Lh$ from the boundary of Ω equals 1. This is easy from comparing $l(x)$ and $l_\infty(x)$ so as this was carried out in the proof of Theorem 5.5.

We call the ***boundary layer*** of (3.2) the set of the points $hH\gamma$ that lie outside Ω or have the coefficient $c[\gamma]$ other than 1. The boundary layer may be ***inner*** if all its points lie in Ω, ***outer*** if all of them lie outside Ω, and ***two-sided*** if it has points of each of the two types.

Every error (3.3) of order greater than 1 has an inner boundary layer.

In the case of an arbitrary domain Ω we make no recommendations to determine whether or not a given cubature formula has a regular boundary layer. However, for rational polyhedra we suggested some approach to solving this question in Chapter 3 where we found out that a formal boundary layer is equivalent to a regular boundary layer of a special shape.

Note in closing that we may find whether or not a cubature formula possesses a formal boundary layer in a finite number of steps independent of h.

§4. Asymptotic Expansion of the Norm of an Error with Regular Boundary Layer

In this section we obtain expansions of the norms of an arbitrary error with regular boundary layer in $L_2^{(m)}$ and $L_2^{(m)}(\Omega)$. These expansions are asymptotically exact as the lattice mesh-size vanishes. Before proving a relevant theorem, we specify the integration domain Ω to be considered.

As was done earlier in the current chapter, assume that $\overline{\Omega}$ is a bounded domain and lies in the interior of the fundamental parallelepiped Ω_0 of the lattice matrix H. Denote by D the diameter of Ω. Take an arbitrary point y lying in Ω or on the boundary of Ω and a positive number $\eta \leq D$. Denote by $B(y, \tau, \eta)$ the set of points x at a distance at most η from the boundary of Ω and belonging to the ball of radius τ with center y,

$$B(y, \tau, \eta) = \{x : \rho(x, \partial\Omega) \leq \eta,\ \rho(x, y) < \tau\}.$$

The volume of this set is a function of the arguments y, τ, and η; denote it by $\psi(y, \tau, \eta)$. Say that a domain Ω satisfies the ***regularity condition*** if for some q, $0 < q \leq 1$, the estimate holds

$$\psi(y, \tau, \eta) \leq K \frac{\tau^n \eta^q}{(\tau^2 + \eta^{2q})^{1/2}(D^2 + \tau^2)^{n/2}}. \tag{4.1}$$

The regularity condition expresses the fact that the boundary strip $\Omega_\eta'' \setminus \Omega_\eta'$ is equidistributed over the entire boundary and, for η small, the measure of the portion of the strip which lies in the ball of radius τ is $O(\tau^{n-1}\eta^q)$. Fixing $\tau > 0$ and letting η vanish, from (4.1) we readily derive the following relation

$$\left|\Omega''_\eta \setminus \Omega'_\eta\right| \leq K\eta^q.$$

In other words, a domain Ω satisfying the regularity condition (4.1) surely has order q of the width of the boundary layer. In particular, we may apply to Ω the estimates of § 2 of the current chapter.

For Ω a domain with piecewise-smooth boundary, the regularity condition (4.1) is satisfied with $q = 1$.

Assume now that an error $l(x)$ has a regular boundary layer, width $2L$, order m, and estimate A in Ω. In other words, $l(x)$ may be decomposed in the sum of local errors

$$l(x) = \sum_{\gamma \in B_L^{(1)}} l^{(0)}\left(\frac{x}{h} - H\gamma\right) + \sum_{\gamma \in B_L^{(2)}} l^{(\gamma)}\left(\frac{x}{h} - H\gamma\right), \tag{4.2}$$

where $l^{(0)}(y)$ and $l^{(\gamma)}(y)$ with $\gamma \in B_L^{(2)}$ belong to the classes $R(L, A, 2m+1)$ and $R(L, A, m)$, respectively. We have the following

***Theorem* 5.6.** *Let Ω be a domain satisfying the regularity condition. Then the $L_2^{(m)*}$ norm of (4.2) is expressed as*

$$\left\| l \mid L_2^{(m)*} \right\| = B_{n,m}(\Omega)h^m + O(h^{m+q}), \tag{4.3}$$

where $B_{n,m}(\Omega)$ is defined by (1.23).

We let two lemmas precede the proof of Theorem 5.6.

***Lemma* 5.2.** *The $\widetilde{L}_2^{(m)}$-extremal function $u(x)$ of the error $l(x)$ may be written as a linear combination of the $\widetilde{L}_2^{(m)}$-extremal functions for the local errors $l^{(0)}(y)$ and $l^{(\gamma)}(y)$ of the decomposition (4.2)*

$$u(x) = \sum_{\gamma \in B_L^{(1)}} h^{2m} u^{(0)}\left(\frac{x}{h} - H\gamma\right) + \sum_{\gamma \in B_L^{(2)}} h^{2m} u^{(\gamma)}\left(\frac{x}{h} - H\gamma\right). \tag{4.4}$$

PROOF. As we know, the $\widetilde{L}_2^{(m)}$-extremal functions $u(x)$, $u^{(0)}(y)$, and $u^{(\gamma)}(y)$ satisfy the equations

$$\Delta^m u(x) = (-1)^m l(x), \quad \Delta^m u^{(0)}(y) = (-1)^m l^{(0)}(y),$$
$$\Delta^m u^{(\gamma)}(y) = (-1)^m l^{(\gamma)}(y), \quad \gamma \in B_L^{(2)}.$$

Whence and from (4.2) we readily have (4.4). The proof of Lemma 5.2 is complete.

Observe that the decomposition (4.4) also takes place in an unbounded domain Ω with compact boundary. In this event, the set $B_L^{(1)}$ is infinite but the corresponding series converges absolutely and uniformly. This it is not hard to establish on using the inequality $2m > n$, writing down $u^{(0)}(y)$ as convolution

$$u^{(0)}\left(\frac{x}{h} - H\gamma\right) = \int G\left(\frac{x}{h} - H\gamma - y\right) l^{(0)}(y)\, dy,$$

and involving (1.2.20).

The error $l(x)$ may be consider as a member of $\widetilde{L}_2^{(m)}(H)^*$, since the domain $\overline{\Omega}$ lies in the interior of the fundamental parallelepiped Ω_0. With this in mind, decompose the error $l_\infty(x)$ of an $\widetilde{L}_2^{(m)}(H)$-optimal cubature formula with lattice of nodes $hH\gamma$ in the sum of local errors

$$l_\infty(x) = \sum_{hH\gamma\in\Omega_0} l^{(0)}\left(\frac{x}{h} - H\gamma\right).$$

We then subtract (4.2) from the last equality. In result, we obtain a presentation of the difference $l_*(x) = l_\infty(x) - l(x)$ in the shape of a linear combination of local errors

$$l_*(x) = \sum_{\gamma\in B_L^{(3)}} l^{(0)}\left(\frac{x}{h} - H\gamma\right) + \sum_{\gamma\in B_L^{(2)}} l_*^{(\gamma)}\left(\frac{x}{h} - H\gamma\right). \tag{4.5}$$

Here the set $B_L^{(3)} = B_L^{(e)}$ consists of γ such that the point $hH\gamma$ lies in Ω_0 at a distance greater than Lh from Ω. The local error $l_*^{(\gamma)}(y)$ obviously belongs to the class $R(L, 2A, m)$. The following lemma is immediate from the decomposition (4.5).

Lemma 5.3. *The extremal function $u_*(x)$ of $l_*(x)$ may be written down as*

$$u_*(x) = \sum_{\gamma\in B_L^{(2)}} h^{2m} u_*^{(\gamma)}\left(\frac{x}{h} - H\gamma\right) + \sum_{\gamma\in B_L^{(3)}} h^{2m} u^{(0)}\left(\frac{x}{h} - H\gamma\right), \tag{4.6}$$

where $u^{(0)}(y)$ and $u_*^{(\gamma)}(y)$ are the $\widetilde{L}_2^{(m)}$-extremal functions for $l^{(0)}(y)$ and $l_*^{(\gamma)}(y)$.

Proof of Theorem 5.6. Write down the $\widetilde{L}_2^{(m)}$-extremal function $u(x)$ of $l(x)$ as the difference

$$u(x) = u_\infty(x) - u_*(x) = h^{2m} u^{(\infty)}\left(\frac{x}{h}\right) - u_*(x).$$

Recall that the function $u_\infty(x)$ was defined by (1.3), and the function $u^{(\infty)}(x)$ coincides with $u_\infty(x)$ for $h = 1$. The norm square of the error $l(x)$ may be obtained from the formula

$$\left\| l \mid L_2^{(m)*} \right\|^2 = (l, u) = (l(x), u_\infty(x)) - (l, u_*). \tag{4.7}$$

Consider in turn both quantities on the right side of this equality. Calculating the first of them, use the nonnegativity of $u_\infty(x)$ and also the fact that $u_\infty(x)$ vanishes at the nodes $hH\gamma$. Consequently

$$(l, u_\infty) = \int\limits_\Omega u_\infty(x)\, dx = h^{2m} \int\limits_\Omega u^{(\infty)}\left(\frac{x}{h}\right)\, dx. \tag{4.8}$$

The integral on the right side of (4.8) splits into two: one over the set Ω_1 presenting the union of the elementary meshes Ω_γ of the lattice of nodes $hH\gamma$ which lie entirely in Ω and the other over the set Ω_2 complementing Ω_1 to Ω, namely,

$$\int\limits_\Omega u_\infty(x)\, dx = \int\limits_{\Omega_1} u_\infty(x)\, dx + \int\limits_{\Omega_2} u_\infty(x)\, dx. \tag{4.9}$$

Let Ω_γ be subset of Ω_1. Then from (1.3) we have the equality

$$h^{2m} \int\limits_{\Omega_\gamma} u^{(\infty)}\left(\frac{x}{h}\right)\, dx = h^{2m+n} \int\limits_{\Omega_0} u^{(\infty)}(y)\, dy = h^{2m+n} B_{n,m}^2. \tag{4.10}$$

Let Ω_1 consist of exactly N_1 distinct meshes Ω_γ. Then from (4.10) obtain

$$\int\limits_{\Omega_1} u_\infty(x)\, dx = B_{n,m}^2 N_1 h^{n+2m}. \tag{4.11}$$

Further, replace the integral of $u_\infty(x)$ over Ω_2 with the integral over a larger set, the union of all meshes Ω_γ meeting the boundary of Ω. If the total number of these meshes equals N_2, then the inequality holds

$$\left| \int\limits_{\Omega_2} u_\infty(x)\, dx \right| \le B_{n,m}^2 N_2 h^{n+2m}. \tag{4.12}$$

Recall now that by hypothesis the domain Ω has order q of the width of the boundary layer. Consequently, the next relations are valid

$$N_1 h^n = |\Omega|(1 + O(h^q)), \quad N_2 h^n = O(h^q).$$

Inserting them in (4.11) and (4.12), obtain from (4.8) and (4.9) the asymptotically exact equality

$$(l, u_{nn}) = B^2_{n,m}(\Omega) h^{2m} + O(h^{2m+q}) \tag{4.13}$$

Now estimate the value of $l(x)$ at the function $u_*(x)$. To this end, use the decompositions (4.2) and (4.6) and also the notation $\Phi(x \mid l_1, l_2)$ of Chapter 4 together with (4.4.25). Obtain

$$\begin{aligned}(l, u_*) = &\sum_{\gamma \in B_L^{(1)}} \sum_{\gamma' \in B_L^{(3)}} \Phi\left(x_{\gamma,\gamma'} \mid l^{(0)}, \bar{l}^{(0)}\right) + \sum_{\gamma \in B_L^{(2)}} \sum_{\gamma' \in B_L^{(3)}} \Phi\left(x_{\gamma,\gamma'} \mid l^{(\gamma)}, \bar{l}^{(0)}\right) \\ &+ \sum_{\gamma \in B_L^{(1)}} \sum_{\gamma' \in B_L^{(2)}} \Phi\left(x_{\gamma,\gamma'} \mid l^{(0)}, \bar{l}^{(\gamma')}\right) + \sum_{\gamma \in B_L^{(2)}} \sum_{\gamma' \in B_L^{(2)}} \Phi\left(x_{\gamma,\gamma'} \mid l^{(\gamma)}, \bar{l}^{(\gamma')}\right).\end{aligned}$$

Here, $x_{\gamma,\gamma'}$ stands for the difference $hH\gamma' - hH\gamma$. Applying Lemma 4.4 on estimation of triple convolution, estimate the error (l, u_*) as follows

$$|(l, u_*)| \le K(U_{1,3} + U_{1,2} + U_{2,3} + U_{2,2}), \tag{4.14}$$

where $U_{k,m}$ stands for the double sum

$$U_{k,m} = \sum_{\gamma \in B_L^{(k)}} \sum_{\gamma' \in B_L^{(m)}} \frac{h^{2n+s_{k,m}}}{(h^2 + |x_{\gamma,\gamma'}|^2)^{(n+p_{k,m})/2}}, \tag{4.15}$$

and the numbers $s_{k,m}$ and $p_{k,m}$ are given by the equalities

$$s_{1,3} = 4m + 2, \quad s_{1,2} = s_{2,3} = 3m + 1, \quad s_{2,2} = 2m, \quad p_{k,m} = s_{k,m} - 2m.$$

Along with the sums $U_{k,m}$, consider the system of integrals

$$J_{k,m} = \int\limits_{\Omega^{(k)}} \int\limits_{\Omega^{(m)}} \frac{h^{s_{k,m}}}{(h^2 + |x - y|^2)^{(n+p_{k,m})/2}} \, dx dy. \tag{4.16}$$

Here, $\Omega^{(1)}$, $\Omega^{(2)}$ and $\Omega^{(3)}$ stand for the domain Ω, the boundary strip $\Omega''_{Lh} \setminus \Omega'_{Lh}$ and the complement $\mathbb{R}^n \setminus \Omega$, respectively. Show that there is a constant K such that for all k and m we have the inequality

$$U_{k,m} \le K J_{k,m}. \tag{4.17}$$

To this end, estimate from below the value of $J_{k,m}$ by the sum of the integrals

$$J_{k,m}^{\gamma,\gamma'} = \int\limits_{\Omega^{(\gamma)}} \int\limits_{\Omega^{(\gamma')}} \frac{h^{s_{k,m}}}{(h^2 + |x - y|^2)^{(n+p_{k,m})/2}} \, dx dy \tag{4.18}$$

taken over the elementary meshes Ω_γ and $\Omega_{\gamma'}$ of the lattice of nodes $hH\gamma$. It is not hard to see that as $h \to 0$ we have the formula

$$J_{k,m} = \Big(\sum_{\gamma \in B_L^{(k)}} \sum_{\gamma' \in B_L^{(m)}} J_{k,m}^{\gamma,\gamma'} \Big)(1 + O(h^q)). \tag{4.19}$$

Since the volume of the direct product of the elementary meshes Ω_γ and $\Omega_{\gamma'}$ equals h^{2n}, by the Intermediate Value Theorem, we may write the integral (4.18) as

$$J_{k,m}^{\gamma,\gamma'} = \frac{h^{2n+s_{k,m}}}{(h^2 + |x_c - y_c|^2)^{(n+p_{k,m})/2}}, \tag{4.20}$$

with $x_c \in \Omega_\gamma$ and $y_c \in \Omega_{\gamma'}$.

Further, for all $x_c \in \Omega_\gamma$ and $y_c \in \Omega_{\gamma'}$ the two-sided estimate is valid

$$M_1(|x_{\gamma,\gamma'}|^2 + h^2) \leq |x_c - y_c|^2 \leq M_2(|x_{\gamma,\gamma'}|^2 + h^2).$$

This estimate, together with (4.20), (4.19), and (4.15), entails (4.17).

Estimate the integrals $J_{k,m}$. Start with $J_{1,3}$. Take $x \in \Omega^{(1)}$ and $y \in \Omega^{(3)}$. Introducing the spherical coordinates with center the point x, obtain

$$\int_{\Omega^{(3)}} \frac{h^{4m+2}}{(h^2 + |x-y|^2)^{n/2+m+1}}\, dy \leq K \int_{\rho(x)}^{\infty} \frac{h^{4m+2} r^{n-1}}{(h^2 + r^2)^{n/2+m+1}}\, dr, \tag{4.21}$$

where $\rho(x)$ is the distance from x to the boundary of the domain Ω. Since, for all positive r and h,

$$r < r + h, \quad r^2 + h^2 \geq (r+h)^2/2;$$

we see that the right side of (4.21) admits a dominant of the shape

$$Kh^{4m+2} \int_{\rho(x)}^{\infty} (r+h)^{-2m-3}\, dr = Kh^{4m+2}(\rho(x) + h)^{-2m-2}.$$

Considering this, obtain

$$J_{1,3} \leq K \int_{\Omega} h^{4m+2}(\rho(x) + h)^{-2m-2}\, dx = Kh^{4m+2} \int_0^{D_1} \frac{d\varphi(\rho)}{(\rho + h)^{2m+2}}. \tag{4.22}$$

Here D_1 is a possibly maximal value of $\rho(x)$ for $x \in \Omega$, and $\varphi(\eta)$ is the volume of $\{x \in \Omega : \rho(x, \partial\Omega) \leq \eta\}$.

Estimate $\varphi(\eta)$ by means of (4.1). Take a point x in Ω to find

$$\varphi(\eta) \leq \psi(x, D, \eta) \leq K\eta^q.$$

Consequently,

$$J_{1,3} \leq Kh^{4m+2}\frac{\varphi(D_1)}{(D_1+h)^{2m+2}} + Kh^{4m+2}\int_0^{D_1}\frac{\rho^q\,d\rho}{(\rho+h)^{2m+3}} \leq Kh^{4m+2}$$

$$+Kh^{4m+2}\int_0^{D_1}(\rho+h)^{q-2m-3}\,d\rho \leq Kh^{4m+2}(1+h^{q-2m-2}) \leq Kh^{q+2m}. \quad (4.23)$$

Estimate the rest of the integrals $J_{k,m}$. Obviously, for $j = 1$ or $j = 3$,

$$\int_{\Omega^{(j)}}\frac{h^{3m+1}}{(h^2+|x-y|^2)^{(m+n+1)/2}}\,dx \leq K\int_0^\infty\frac{h^{3m+1}r^{n-1}\,dr}{(h^2+r^2)^{(m+n+1)/2}}$$

$$\leq Kh^{3m+1}\int_0^\infty(r+h)^{-m-2}\,dr \leq Kh^{2m}. \quad (4.24)$$

Setting here j equal 1, from (4.16) obtain

$$J_{1,2} \leq Kh^{2m}|\Omega^{(2)}| \leq Kh^{2m+q}. \quad (4.25)$$

Setting in (4.24) j equal 3, come to the estimate

$$J_{2,3} \leq Kh^{2m}|\Omega^{(2)}| \leq Kh^{2m+q}. \quad (4.26)$$

Now estimate the last integral $J_{2,2}$. Let a point x belong to the strip $\Omega^{(2)}$, let $\rho(x)$ denote the distance from x to the boundary of the domain Ω, and let D_1 be the maximal value of $\rho(x)$ for $x \in \Omega^{(2)}$. By the same arguments as in estimating the integral $J_{1,3}$, obtain the formula

$$\int_{\Omega^{(2)}}\frac{h^{2m}}{(h^2+|x-y|^2)^{n/2}}\,dy = h^{2m}\int_0^{D_1}\frac{d_\rho\psi(x,\rho,(L+1)h)}{(h^2+\rho^2)^{n/2}}.$$

Integrating by parts and using (4.1), come to the inequality

$$\int\limits_{\Omega^{(2)}} \frac{h^{2m}}{(h^2+|x-y|^2)^{n/2}}\,dy \le Kh^{2m+q}.$$

Integrating both sides with respect to $x \in \Omega^{(2)}$, obtain

$$|J_{2,2}| \le Kh^{2m+2q}.$$

In accord with (4.17) the double sums $U_{k,m}$ do not exceed Kh^{2m+q} either. In accord with (4.14), for the error (l, u_*) we also have an analogous estimate

$$|(l, u_*)| \le Kh^{2m+q}.$$

From this estimate, involving (4.7) and (4.13), we infer that the asymptotic expression (4.3) is indeed valid. The proof of Theorem 5.6 is complete.

From (4.3) it follows, in particular, that cubature formulas with regular boundary layer are asymptotically $L_2^{(m)}$-optimal. It is worthwhile to find out whether this property preserves under passage to $L_2^{(m)}(\Omega)$, where Ω is the integration domain possessing one more property in addition to regularity. In § 2 we introduced a special operator that extends $\varphi(x)$ in $L_2^{(m)}(\Omega)$ to $\overline{\varphi}(x)$ in $L_2^{(m)}$. Suppose that this operator is continuous, i.e., there is a constant K_1 such that, for all $\varphi \in L_2^{(m)}(\Omega)$, the inequality holds

$$\|\overline{\varphi}(x) \mid L_2^{(m)}\| \le K_1 \|\varphi \mid L_2^{(m)}(\Omega)\|.$$

Under these hypotheses, we have

Theorem 5.7 (V. I. Polovinkin). *The $L_2^{(m)}(\Omega)^*$ norm of an error $l(x)$ with regular boundary layer in Ω admits the following asymptotic expansion*

$$\|l \mid L_2^{(m)}(\Omega)^*\| = B_{n,m}(\Omega)h^m + O(h^{m+q}). \tag{4.27}$$

PROOF. Observe first that the $L_2^{(m)}(\Omega)^*$ norm of $l(x)$ satisfies a lower bound analogous to (1.28)

$$\|l \mid L_2^{(m)}(\Omega)^*\| \ge B_{n,m}(\Omega)h^m + O(h^{m+q}). \tag{4.28}$$

Thus, to obtain (4.27) it suffices to prove that the right side bounds from above the norm of $l(x)$ in $L_2^{(m)}(\Omega)^*$. Before doing so, derive several properties of the $L_2^{(m)}$-extremal function $\psi_l(x)$ of $l(x)$. As we know, it is given by the convolution

$$\psi_l(x) = \int G(x-y)l(y)\,dy$$

satisfying in $\mathbb{R}^n$ the equation

$$\Delta^m \psi_l(x) = (-1)^m l(x),$$

and its derivatives of order m decrease not slower than $|x|^{-n} \log |x|$. Whence it follows that the extension $\overline{\psi}_l(x)$ of $\psi_l(x)$ from the domain Ω to the whole of $\mathbb{R}^n$ agrees with $\psi_l(x)$ outside Ω as well.

Further, (4.3) is equivalent to the following

$$\{\psi_l, \psi_l\}_{L_2^{(m)}} = B_{n,m}^2(\Omega)h^{2m} + O(h^{2m+q}). \tag{4.29}$$

Here $\{\cdot, \cdot\}_{L_2^{(m)}}$ denotes the inner product in $L_2^{(m)}$.

Consider the function $u_\infty(x)$ defined by (1.3) and extremal for $l_\infty(x)$ in $\widetilde{L}_2^{(m)}$. We inspected the extension $\overline{u}_\infty(x)$ of it in § 2 of the current chapter. It is convenient, considering (4.13), to write the inner product of $\psi_l(x)$ and $\overline{u}_\infty(x)$ as

$$\{\psi_l, \overline{u}_\infty\}_{L_2^{(m)}} = B_{n,m}^2(\Omega)h^{2m} + O(h^{2m+q}). \tag{4.30}$$

Subtracting (4.29) from (4.30), obtain the asymptotic expansion of one more inner product

$$\{\psi_l, \overline{u}_\infty(x) - \overline{\psi}_l(x)\}_{L_2^{(m)}} = O(h^{2m+q}). \tag{4.31}$$

Squaring both sides of (2.12), come to the relation

$$\left\|\overline{u}_\infty \mid L_2^{(m)}\right\|^2 = B_{n,m}^2(\Omega)h^{2m} + O(h^{2m+q}).$$

Subtracting (4.30), write

$$\{\overline{u}_\infty(x) - \psi_l(x), \overline{u}_\infty(x)\}_{L_2^{(m)}} = O(h^{2m+q}). \tag{4.32}$$

Carrying out subtraction again, from (4.32) and (4.31) obtain

$$\{\overline{u}_\infty - \psi_l, \overline{u}_\infty - \psi_l\}_{L_2^{(m)}} = O(h^{2m+q}). \tag{4.33}$$

Decompose the function $\psi_l(x)$ in the sum of projections to the subspaces $\overset{\circ}{L}{}_2^{(m)}(\mathbb{R}^n \mid \Omega)$ and $H_2^{(m)}(\mathbb{R}^n \mid \partial\Omega)$, which yields

$$\psi_l(x) = \psi_l^{(0)}(x) + \psi_l^{(1)}(x).$$

Examine the behavior in $L_2^{(m)}$ of the norms of the projections $\psi_l^{(0)}$ and $\psi_l^{(1)}$ at small h. Obviously,

$$\|\psi_l^{(1)} \mid L_2^{(m)}\| \leq \|\overline{u}_\infty^{(1)} \mid L_2^{(m)}\| + \|\overline{u}_\infty^{(1)} - \psi_l^{(1)} \mid L_2^{(m)}\|.$$

Applying (2.20) to the first norm on the right side, estimating the second summand by (4.33), obtain

$$\|\psi_l^{(1)} \mid L_2^{(m)}\| \leq K h^{m+q/2}. \tag{4.34}$$

The functions $\psi_l^{(0)}$ and $\psi_l^{(1)}$ are orthogonal in $L_2^{(m)}$. Therefore,

$$\|\psi_l^{(0)} \mid L_2^{(m)}\|^2 = \|\psi_l \mid L_2^{(m)}\|^2 - \|\psi_l^{(1)} \mid L_2^{(m)}\|^2.$$

The first norm on the right side is presented by (4.29); the second satisfies (4.34). Consequently, we have the equality

$$\|\psi_l^{(0)} \mid L_2^{(m)}\| = B_{n,m}(\Omega) h^m + O(h^{m+q}). \tag{4.35}$$

Now we are able to estimate the $L_2^{(m)*}(\Omega)$ norm of $l(x)$. Let a function $\varphi(x)$ belong to the unit ball of $L_2^{(m)}(\Omega)$. Decompose it in the sum

$$\varphi(x) = \varphi^{(0)}(x) + \varphi^{(1)}(x),$$

where $\varphi^{(0)}(x) \in \overset{\circ}{L}{}_2^{(m)}(\Omega)$ and $\varphi^{(1)}(x) \in H_2^{(m)}(\Omega)$. Extending $\varphi(x)$ to the whole of $\mathbb{R}^n$ with a minimal norm, find

$$(l, \varphi) = (l, \overline{\varphi}) = \{\psi_l^{(0)}, \overline{\varphi}^{(0)}\}_{L_2^{(m)}} + \{\psi_l^{(1)}, \overline{\varphi}^{(1)}\}_{L_2^{(m)}}. \tag{4.36}$$

Estimate the inner products on the right side of (4.36). Apply first the Cauchy–Bunyakovskiĭ–Schwarz inequality and next (4.34). Then, using the continuity of the extension operator acting from $L_2^{(m)}(\Omega)$ to $L_2^{(m)}$, infer

$$\left|\{\psi_l^{(1)}, \overline{\varphi}^{(1)}\}_{L_2^{(m)}}\right| \leq K h^{m+q/2} \|\varphi^{(1)} \mid L_2^{(m)}(\Omega)\|. \tag{4.37}$$

Bearing in mind that $\|\overline{\varphi}^{(0)} \mid L_2^{(m)}\| = \|\varphi^{(0)} \mid L_2^{(m)}(\Omega)\|$ and using (4.35), estimate the remaining inner product in (4.36) to obtain

$$\left|\{\psi_l^{(0)}, \overline{\varphi}^{(0)}\}_{L_2^{(m)}}\right| \leq [B_{n,m}(\Omega) h^m + O(h^{m+q})] \|\varphi^{(0)} \mid L_2^{(m)}(\Omega)\|. \tag{4.38}$$

Since $\varphi(x)$ belongs to the unit ball of $L_2^{(m)}(\Omega)$, there is an angle θ such that

$$\left\|\varphi^{(0)} \mid L_2^{(m)}(\Omega)\right\| = \cos\theta, \quad \left\|\varphi^{(1)} \mid L_2^{(m)}(\Omega)\right\| = \sin\theta.$$

Considering this, from (4.36)–(4.38) derive

$$\begin{aligned} |(l,\varphi)| &\leq \max_{\theta} \left[K h^{m+q/2} \sin\theta + (B_{n,m}(\Omega)h^m + O(h^{m+q}))\cos\theta\right] \\ &= B_{n,m}(\Omega)h^m + O(h^{m+q}). \end{aligned}$$

In other words, on the norm of $l(x)$ we have the following upper bound

$$\left\|l \mid L_2^{(m)}(\Omega)^*\right\| \leq B_{n,m}(\Omega)h^m + O(h^{m+q}).$$

This and (4.28) entail (4.27). The proof of Theorem 5.7 is complete.

Closing this section, observe that we may obtain the asymptotically exact expansion of the norms of an error with regular boundary layer in the sense of the definition of Chapter 3 by argument analogous to that of the article [150].

§5. The Properties of the Extremal Function of an Error in $L_2^{(m)}(\Omega)$

To the error $l(x)$ of a cubature formula over a domain Ω we may put into correspondence not only the $L_2^{(m)}$-extremal function

$$\psi_l\,(x) = \int G(x-y)l(y)dy$$

but also the $L_2^{(m)}(\Omega)$-extremal function $u_l(x)$. We now demonstrate that the latter is a weak solution to some boundary value problem for the polyharmonic equation in Ω. More precisely, we show that, in a bounded domain of integration Ω with piecewise-smooth boundary, the function $u_l(x)$ is a solution to the equation

$$\Delta^m u = (-1)^m l(x), \tag{5.1}$$

satisfying weakly some boundary conditions on the boundary $\partial\Omega$ of Ω. We derive (5.1) and the shape of conditions over $\partial\Omega$ on using the so-called Green's identity for the polyharmonic operator in Ω.

We have the following

Theorem 5.8. *Let some functions u and v possess continuous derivatives up to order $2m$ in the closure of Ω. The following integral identity holds*

$$\int_{\Omega} \left\{ (-1)^{m+1} v \Delta^m u + \sum_{|\alpha|=m} \frac{m!}{\alpha!} D^\alpha u D^\alpha v \right\} dx$$

$$= \int_{\partial\Omega} \sum_{k=0}^{m-1} \frac{\partial^k v}{\partial \nu^k} B_k(u) dS, \tag{5.2}$$

with ν the outer normal to $\partial\Omega$ and $B_k(u)$ the value of some differential operator of order $2m-k-1$ at u.

The equality (5.2) coincides with Green's identity for the Laplacian in the case of $m = 1$. In this connection, we call (5.2) *Green's identity for the polyharmonic operator* in Ω in the case of an arbitrary m.

PROOF. Assume that u and v have $2m$ continuous derivatives in $\overline{\Omega}$. Define several auxiliary bilinear forms, letting

$$P_j^{(m,k)}(u,v) = \sum_{|\gamma|=m-k-1} \frac{|\gamma|!}{\gamma!} \left[D_j D^\gamma \Delta^k u \right] D^\gamma v, \tag{5.3}$$

$$j = 1, 2, \dots, n; \quad k = 0, 1, \dots, m-1;$$

$$L^{(m,k)}(u,v) = \sum_{j=1}^{n} D_j \left[P_j^{(m,k)}(u,v) \right]. \tag{5.4}$$

Transform (5.4) as follows.

Observe that we may carry out the differentiation D_j easier on rewriting (5.3) so that summation over the multi-index γ be replaced with taking the sums not accounting for the symmetry of the tensor of derivatives. More exactly, write down (5.3) as

$$P_j^{(m,k)}(u,v) = \sum_{j_1=1}^{n} \cdots \sum_{j_{m-k-1}=1}^{n} \left(D_j D_{j_{m-k-1}} \dots D_{j_1} \Delta^k u \right) \left(D_{j_{m-k-1}} \dots D_{j_1} v \right). \tag{5.5}$$

Insert this expansion in (5.4) and substitute j_{m-k}, for j to obtain

$$L^{(m,k)}(u,v)$$

$$= \sum_{j_1=1}^{n} \cdots \sum_{j_{m-k}=1}^{n} \Big[\left(D_{j_{m-k}} D_{j_{m-k-1}} \dots D_{j_1} \Delta^k u \right) \left(D_{j_{m-k}} D_{j_{m-k-1}} \dots D_{j_1} v \right)$$

$$+ \left(D_{j_{m-k}}^2 D_{j_{m-k-1}} \dots D_{j_1} \Delta^k u \right) \left(D_{j_{m-k-1}} \dots D_{j_1} v \right) \Big].$$

Record the right side, considering symmetry of the tensor of derivatives, to find

$$L^{(m,k)}(u,v) = \sum_{|\beta|=m-k} \frac{|\beta|!}{\beta!}(D^\beta \Delta^k u) D^\beta v + \sum_{|\gamma|=m-k-1} \frac{|\gamma|!}{\gamma!}(D^\gamma \Delta^{k+1} u) D^\gamma v. \quad (5.6)$$

Multiplying both sides of (5.6) by $(-1)^k$, sum the results over k ranging from 1 to $m-1$. Cancelling out all intermediate summands in the right side, obtain

$$\sum_{k=0}^{m-1}(-1)^k L^{(m,k)}(u,v) = \sum_{|\alpha|=m} \frac{m!}{\alpha!} D^\alpha u D^\alpha v + (-1)^{m+1} v \Delta^m u.$$

Integrating both sides of this equality over Ω, invoke the definition (5.4) of the form $L^{(m,k)}(u,v)$. The Divergence Theorem in the shape of the Gauss–Ostrogradskiĭ identity reads

$$\int_\Omega \left\{ \sum_{|\alpha|=m} \frac{m!}{\alpha!} D^\alpha u D^\alpha v + (-1)^{m+1} v \Delta^m u \right\} dx$$
$$= \int_{\partial\Omega} \sum_{k=0}^{m-1} (-1)^k P_\nu^{(m,k)}(u,v) dS. \quad (5.7)$$

Here $\nu = (\nu_1, \nu_2, \dots, \nu_n)$ is the outer normal to $\partial\Omega$ and

$$P_\nu^{(m,k)}(u,v) = \sum_{j=1}^{n} \nu_j P_j^{(m,k)}(u,v).$$

We shall briefly denote the integrand on the right side of (5.7) by

$$F_m(u,v,dx) = \sum_{k=0}^{m-1} (-1)^k P_\nu^{(m,k)}(u,v) dS. \quad (5.8)$$

It is well known that calculus provides surface integrals of two kinds. One involves the surface measure dS of a rectifiable surface. The other involves differential forms over the surface under study. Without further specification, we use integrals of both kinds simultaneously. For example, the left side of (5.8) contains a differential form whereas the right side involves the surface measure.

We have

$$\int_\Omega \left(\sum_{|\alpha|=m} \frac{m!}{\alpha!} D^\alpha u D^\alpha v + (-1)^{m+1} v \Delta^m u \right) dx = \int_{\partial\Omega} F_m(u,v,dx). \quad (5.9)$$

Observe that $F_m(u,v,dx)$ is a bilinear form in the derivatives of u and v. In other words,

$$F_m(u,v,dx) = \sum_{\substack{|\beta|\le m-1\\ |\alpha|+|\beta|=2m-1}} c_{\alpha,\beta}(dx) D^\alpha u D^\beta v.$$

This is immediate from (5.3). The coefficients $c_{\alpha,\beta}(dx)$ are the so-called *differential forms* of degree $n-1$ in $dx_1,\dots,dx_n$ which are independent of u and v. Namely,

$$c_{\alpha,\beta}(dx) = \sum_{j=1}^{n} c^{(j)}_{\alpha,\beta} d\widehat{x}_j, \tag{5.10}$$

where the vector $\widehat{x}_j$ results from x by eliminating its jth coordinate

$$d\widehat{x}_j = dx_1 dx_2 \dots dx_{j-1} dx_{j+1} \dots dx_n.$$

The values of the coefficients $c^{(j)}_{\alpha,\beta}$ in the expansion (5.10) are determined by evenness of the entries of $\alpha-\beta-\delta_j$ (here δ_j has all entries zero but the jth entry which equals 1). More exactly, if at least one of the entries of the vector $\alpha-\beta-\delta_j$ is odd then $c^{(j)}_{\alpha,\beta}=0$; otherwise,

$$c^{(j)}_{\alpha,\beta} = (-1)^{|\varepsilon|} \frac{|\beta|!}{\beta!}\frac{|\varepsilon|!}{\varepsilon!}, \qquad \varepsilon = \frac{1}{2}(\alpha-\beta-\delta_j).$$

It is of service to give the integral of $F_m(u,v,dx)$ another record in which all derivatives of v of order from 0 to $m-1$ are taken only along the normal whereas differentiation along the tangent is applied to u and the coefficients.

Assume that the surface $\partial\Omega$ possesses the following properties:

(a) $\partial\Omega$ has the tangent plane continuous;

(b) there is a constant d such that, in the ball $B(x^{(0)},2d)$ of radius $2d$, centered at a point $x^{(0)}$ of $\partial\Omega$, the straight lines parallel to the normal $\nu(x^{(0)})$ to $\partial\Omega$ at $x^{(0)}$, meet $\partial\Omega$ at most once;

(c) if we introduce a local cartesian coordinates $y_1,\dots,y_n$ in a neighborhood of an arbitrary point $x^{(0)}$ in $\partial\Omega$ so that the axis y_n has the direction of the outer normal then the equation of the surface $\partial\Omega$ takes the shape $y_n=\varphi(\widehat{y}_n)$, in these coordinates in $B(x^{(0)},2d)$, with the function φ possessing continuous derivatives up to order $2m+2$.

The set of the balls $B(x^{(0)},d)$, with $x^{(0)}$ ranging over $\partial\Omega$, is a cover of the surface. Choose finitely many points $x^{(1)},x^{(2)},\dots,x^{(N)}$ of $\partial\Omega$ so that $\partial\Omega$ be covered by the balls $B(x^{(1)}, d), \dots, B(x^{(N)}, d)$. This is possible by the Heine–Borel Lemma. Denote the union of all $B(x^{(j)}, d)$ by E, namely,

$$E = \bigcup_{j=1}^{N} B(x^{(j)}, d).$$

From each ball $B(x^{(j)}, d)$ delete every point whose distance to $x^{(j)}$ is greater than the distance to some of the points $x^{(t)}$ $(t \neq j)$. Denote the resultant set by $E(x^{(j)}, d)$. Let $\chi^j(x)$ stand for the indicator of $E(x^{(j)}, d)$. The sum

$$\Psi_{\partial\Omega} = \sum_{j=1}^{N} \chi^j(x)$$

coincides with $\chi_E(x)$ to within the values at the boundary of $E(x^{(j)}, d)$.

The set E obviously includes some layer of finite width about the surface $\partial\Omega$, i.e., all points at a distance to $\partial\Omega$ not exceeding some number, say, $3h$. We may chose $3h$ small enough.

Given $\chi^j(x)$, arrange the mean $\chi_h^j(x)$ of $\chi^j(x)$ by averaging with the radius h. Clearly,

$$\sum_{j=1}^{N} \chi_h^j(x) = \Psi_{\partial\Omega,h}(x).$$

It is easy to check that the mean $\Psi_{\partial\Omega,h}(x)$ assumes 1 at a point at a distance at most h to $\partial\Omega$.

The restrictions of $\chi_h^j(x)$ to $\partial\Omega$ clearly constitute a family of functions $\chi_h^j(\xi)$, on $\partial\Omega$ which is a partition of unity over $\partial\Omega$. Indeed, the sum of these functions is identically 1; in symbols,

$$\sum_{j=1}^{N} \chi_h^j(\xi) = 1.$$

Each $\chi_h^j(\xi)$ is supported in the ball $B(x^{(j)}, 2d)$, being continuously differentiable $2m+2$ times on $\partial\Omega$. (In space, such function would be infinitely differentiable.)

The differential form $F_m(u, v, dx)$ expands into the sum of differential forms each supported in the respective ball $B(x^{(j)}, 2d)$ as follows

$$F_m(u, v, dx) = \sum_{j=1}^{N} \chi_h^j(x) F_m(u, v, dx) = \sum_{j=1}^{N} F_m^{(j)}(u, v, dx). \tag{5.11}$$

Consequently,

$$\int_{\Omega}\left\{(-1)^{m+1}v\Delta^m u+\sum_{|\alpha|=m}\frac{|\alpha|!}{\alpha!}D^\alpha u D^\alpha v\right\}dx=\sum_{j=1}^{N}\int_{\partial\Omega}F_m^{(j)}(u,v,dx).$$

In each of the domains $B(x^{(j)},2d)$ we change coordinates from y to z by the formulas

$$y_j=z_j-z_n\frac{\partial\varphi(\hat{z}_n)/\partial z_j}{\sqrt{1+\sum\limits_{j=1}^{n-1}\left(\partial\varphi(\hat{z}_n)/\partial z_j\right)^2}},$$

$$y_n=\varphi(\hat{z}_n)+\frac{z_n}{\sqrt{1+\sum\limits_{j=1}^{n-1}\left(\partial\varphi(\hat{z}_n)/\partial z_j\right)^2}}.$$

It is easy to see that in the new coordinates the surface $\partial\Omega$ may be written as $z_n=0$. The normal to $y_n=\varphi(\widehat{y}_n)$ transforms to the line $\widehat{z}_n=\text{const}$. Derivatives with respect to z_n of each order simply become derivatives along the normal

$$\left.\frac{\partial^n f}{\partial\nu^n}\right|_{x\in\partial\Omega}=\left.\frac{\partial^n f(x(z))}{\partial z^n}\right|_{z_n=0}.$$

The mapping $x(z)$ is one-to-one and continuous with derivatives up to order $2m+1$ in the layer of width $3h$ about $\partial\Omega$. Write out the integral

$$\int_{B(x^{(j)},2d)\cap\partial\Omega}F_m^{(j)}(u,v,dx)$$

in the new variables. In the coordinates z such integral takes the shape

$$\int_{z_n=0}\Phi_m^{(j)}(u,v,d\widehat{z}_n),$$

where $\Phi_m^{(j)}(u,v,d\widehat{z}_n)$ is again a differential form with coefficients bilinear forms in the derivatives of u and v with respect to z_j. In symbols,

$$\Phi_m^{(j)}(u,v,d\widehat{z}_n)=\sum_{\substack{|\alpha|+|\beta|\le 2m-1\\ |\beta|\le m-1}}a_{\alpha,\beta}^{(j)}(\widehat{z}_n)D_z^\alpha u D_z^\beta v\,d\widehat{z}_n. \tag{5.12}$$

Integrate by parts the right side of (5.12) so as to translate differentiation with respect to $z_1, z_2, \dots, z_{n-1}$ from v to u and $a^{(j)}_{\alpha,\beta}(\widehat{z}_n)$. Constant terms disappear since the form has compact support $B(x^{(j)}, 2d)$. We thus obtain

$$\int\limits_{z_n=0} \Phi^{(j)}_m(u, v, d\widehat{z}_n) = \int\limits_{z_n=0} \sum_{k=0}^{m-1} \sum_{|\gamma|\le 2m-k-1} c^{(j)}_{k,\gamma}(\widehat{z}_n) D^\gamma u \frac{\partial^k v}{\partial z_n^k}\, d\widehat{z}_n.$$

Reverting to the old variables and writing the integral with the surface measure dS, deduce

$$\int\limits_{z_n=0} \Phi^{(j)}_m(u, v, d\widehat{z}_n) = \int\limits_{\partial\Omega} \sum_{k=0}^{m-1} \frac{\partial^k v}{\partial \nu^k} \sum_{\gamma} c^{(j)}_{k,\gamma}(\widehat{x}_n) D^\gamma u\, d\widehat{x}_j$$

$$= \int\limits_{\partial\Omega} \sum_{k=0}^{m-1} \frac{\partial^k v}{\partial \nu^k} B^j_k(u)\, dS,$$

where $B^j_k(u)$ are some differential operators of order $2m-k-1$ in u. After summing these equalities over j from 1 to N and applying (5.11), insert the result in (5.9). We thus come to the sought formula (5.2).

The proof of Theorem 5.8 is complete.

Let φ be an infinitely differentiable function with compact support in Ω. Applying (5.2) to $v = \varphi$ and $u = u_l$, arrive at the relation

$$(l, \varphi) = \int\limits_{\Omega} \sum_{|\alpha|=m} \frac{m!}{\alpha!} D^\alpha \varphi D^\alpha u_l dx = \int (-1)^m \varphi \Delta^m u_l dx.$$

This means that $u_l(x)$ is a weak solution to (5.1) in a neighborhood of each point in the interior of Ω.

The functionals $B_k(u)$, $k = 0, \dots, m-1$, in (5.2) may be defined for an arbitrary function $u(x)$ polyharmonic in Ω rather than for $u(x)$ continuously differentiable $2m$ times in $\overline{\Omega}$. We now explain how to carry this out.

Take some nested sequence of domains Ω_n with piecewise-smooth boundaries tending to Ω. The function $u(x)$ is infinitely differentiable in the closure $\overline{\Omega}_n$ of Ω_n and so the values $B_k(u)$, $k = 0, 1, \dots, m-1$, are clearly defined on $\partial\Omega_n$. Apply (5.2) in Ω_n with a function v in $L^{(m)}_2(\Omega)$ and use absolute continuity of the integral to observe that the next numerical sequence is a Cauchy sequence:

$$\int\limits_{\partial\Omega_n} \sum_{k=0}^{m-1} \frac{\partial^k v}{\partial \nu^k} B_k(u) dS, \quad n = 1, 2, \dots.$$

The limit of this sequence as $n \to \infty$ should be taken as the value of the integral

$$\int\limits_{\partial\Omega} \sum_{k=0}^{m-1} \frac{\partial^k v}{\partial \nu^k} B_k(u) dS.$$

We now describe the boundary conditions that are to be met on $\partial\Omega$ by the $L_2^{(m)}(\Omega)$-extremal function $u_l(x)$ of the error $l(x)$ under consideration.

Let φ be a member of $L_2^{(m)}(\Omega)$. Denote by $\overline{\varphi}$ an extension of φ with a minimal norm to a member of $L_2^{(m)}$. Multiply $\overline{\varphi}$ by an infinitely differentiable, compactly-supported function equal to 1 in some ball about Ω. The latter exists by obvious reasons: it suffices to take the truncator of a ball which is constructed with the standard averaging kernel. Denote the product of $\overline{\varphi}$ and the chosen truncator by $\overline{\varphi}_*$. Clearly, $\overline{\varphi}_*$ belongs to $L_2^{(m)}$ and has compact support.

By the definition of extremal function, we further obtain

$$(l, \varphi) = \int\limits_{\Omega} \sum_{|\alpha|=m} \frac{m!}{\alpha!} D^\alpha \varphi D^\alpha u_l dx,$$

$$(l, \overline{\varphi}_*) = \int \sum_{|\alpha|=m} \frac{m!}{\alpha!} D^\alpha \overline{\varphi}_* D^\alpha \psi_l dx.$$

Subtracting the second of these equalities from thee first and considering that $(l, \varphi) = (l, \overline{\varphi}_*)$, infer

$$\int\limits_{\Omega} \sum_{|\alpha|=m} \frac{m!}{\alpha!} D^\alpha \varphi D^\alpha (u_l - \psi_l) dx = \int\limits_{\Omega^*} \sum_{|\alpha|=m} \frac{m!}{\alpha!} D^\alpha \overline{\varphi}_* D^\alpha \psi_l dx.$$

Here Ω^* stands for the interior of the complement of Ω to $\mathbb{R}^n$. The functions $u_l - \psi_l$ and ψ_l are polyharmonic in Ω and Ω^* respectively. Considering this and applying Green's identity (5.2) twice, arrive to the integral relation

$$\int\limits_{\partial\Omega} \sum_{k=0}^{m-1} \frac{\partial^k \varphi}{\partial \nu^k} B_k^+(u_l - \psi_l) dS = \int\limits_{\partial\Omega^*} \sum_{k=0}^{m-1} \frac{\partial^k \overline{\varphi}}{\partial \nu_*^k} B_k^-(\psi_l) dS.$$

Here ν_* is the outer normal to $\partial\Omega^* = \partial\Omega$, i.e., $\nu_* = -\nu$. Moreover, $B_k^+(u_l - \psi_l)$ and $B_k^-(\psi_l)$ stand for the values of the functional $B_k(\cdot)$, constructed for Ω and the exterior Ω^* of Ω. Considering further that

$$\frac{\partial^k \overline{\varphi}}{\partial \nu_*^k} = (-1)^k \frac{\partial^k \overline{\varphi}}{\partial \nu^k} = (-1)^k \frac{\partial^k \varphi}{\partial \nu^k}, \quad k = 0, 1, \dots, m-1,$$

arrive to the sought boundary conditions of $u_l(*)$ in the shape

$$B_k^+(u_l - \psi_l) = (-1)^k B_k^-(\psi_l), \quad k = 0, 1, \ldots, m-1 \tag{5.13}$$

Observe that these boundary conditions should be understood in a generalized sense, that is weakly, on requiring that, given φ in $L_2^{(m)}(\Omega)$, we have the integral relation

$$\int_{\partial\Omega} \sum_{k=0}^{m-1} \frac{\partial^k \varphi}{\partial \nu^k} \left\{ B_k^+(u_l - \psi_l) + (-1)^{k+1} B_k^-(\psi_l) \right\} dS = 0.$$

For $m = 1$, equation (5.1) with the boundary conditions (5.13) is the Neumann problem for the Laplacian in Ω.

Suppose that a point x lies in Ω and $\Omega(\varepsilon)$ results from Ω by deleting the ball with center x and radius ε. Apply (5.2) first to the functions $(u_l - \psi_l)(y)$ and $(-1)^m G(x-y)$, polyharmonic in $\Omega(\varepsilon)$ and next to the same functions taken in the opposite order. Subtracting the resulting equalities from one another and passing to the limit as $\varepsilon \to 0$, arrive at the following decomposition of the $L_2^{(m)}(\Omega)$-extremal function $u_l(x)$ of the error $l(x)$ under consideration

$$u_l(x) = \psi_l(x) + (-1)^m \int_{\partial\Omega} \sum_{k=0}^{m-1} \left\{ \frac{\partial^k G(x-y)}{\partial \nu^k} B_k((u_l - \psi_l)(y)) \right.$$
$$\left. - \frac{\partial^k (u_l - \psi_l)(y)}{\partial \nu^k} B_k(G(x-y)) \right\} dS_y. \tag{5.14}$$

The first term on the right side is the volume potential with density $(-1)^m l(y)$ and the remaining terms, surface potentials.

The decomposition (5.14) allows us to compare the values of the norms of the error $l(x)$ in the spaces $L_2^{(m)}$ and $L_2^{(m)}(\Omega)$. In particular, if $l(x)$ is an error with regular boundary layer then, as follows from Theorem 5.7, the main contribution to its $L_2^{(m)}(\Omega)$ norm at small h is given by the norm of the first term $\psi_l(x)$ in the appropriate decomposition (5.14).

Chapter 6

Universal Asymptotic Optimality

A cubature formula having an optimal mode of convergence in a given Banach space certainly preserves this property under passage to an equivalent norm. A stronger assertion is often valid: a formula has an optimal mode of convergence in different nonequivalent spaces. For instance, a homogeneous error of degree M has an optimal mode of convergence in all spaces $\widetilde{W}_p^{(m)}$ for $m \in (n, M/2]$ and $p \in (1, \infty)$. A deeper fact was faced: the same cubature formula may be asymptotically optimal simultaneously in many spaces not necessarily equivalent to one another. We saw this by the example of a formula with the error $l_\infty(x)$ acting over the periodic spaces $\widetilde{B}$ defined in § 1 of Chapter 4. Such results drive us to supposing that the concept of asymptotic optimality is stable under the choice of function spaces and to making a conjecture that there exist *universal asymptotically optimal formulas*. In our understanding, these formulas preserve the asymptotic optimality property on a wide class of spaces of integrable functions.

§1. Cubature Formulas with Bounded Boundary Layer in Hilbert Spaces

I. Babuška substantiated the concept of universal asymptotic optimality by the example of the spaces $\widetilde{H}_2^\mu$. S. L. Sobolev established the universal optimality of a cubature formula with error $l_\infty(x)$ in $\widetilde{L}_2^{(m)}$ spaces. Later, many articles appear establishing asymptotic optimality for formulas with regular boundary layer in various specific function spaces.

However, it turns out eventually that the property that a given error is asymptotically optimal depends essentially on the choice of a norm. V. I. Polovinkin constructed the first example that corroborates this. He demonstrated that, in the periodic case, the cubature formula with error $l_\infty(x)$ is not asymptotically optimal in $\widetilde{W}_p^{(m)}$, $p \neq 2$.

We expose the proof of this fact, restricting exposition to the one-dimensional case in which m is even and $p = 2q/(2q-1)$, with $q > 1$ an integer. The norm in $\widetilde{W}_p^{(m)}$ is given by the equality

$$\|f \mid \widetilde{W}_p^{(m)}\| = \left\{ \int_0^1 |f(x) + (-1)^{m/2} D^m f(x)|^p \, dx \right\}^{1/p}.$$

Let $1/h$ be an integer. Then $\widetilde{W}_p^{(m)}$-optimal error has the form

$$l_h^0(x) = 1 - c_0 h \sum_{k=1}^{1/h} \delta(x - hk) = (1 - c_0) + c_0 l_\infty(x) = \sum_s L_0[s] e^{i2\pi xs}.$$

Its *Fourier coefficients* are defined by the equality

$$L_0[s] = \left(l_h^0(x), e^{-i2\pi xs}\right) = \begin{cases} 1 - c_0 & \text{for} \quad s = 0, \\ -c_0 & \text{for} \quad s \text{ a multiple of } 1/h, \\ 0 & \text{for} \quad \text{other } s. \end{cases}$$

Consequently,

$$\|l_h^0 \mid \widetilde{W}_p^{(m)*}\| = \left\| 1 - c_0 - c_0 \sum_{s \neq 0} e^{i2\pi xs/h} \left\{ 1 + \left(\frac{2\pi s}{h} \right)^m \right\}^{-1} \mid \widetilde{L}_{2q} \right\|.$$

We use the convenient notations

$$1 - c_0 = \varepsilon, \quad g_m(x) = \sum_{s \neq 0} e^{i2\pi xs/h} \left\{ 1 + \left(\frac{2\pi s}{h} \right)^m \right\}^{-1}.$$

The function of ε, written as the integral

$$\|l_h^0 \mid \widetilde{W}_p^{(m)*}\|^{2q} = \int_0^1 (\varepsilon - (1-\varepsilon) g_m(x))^{2q} \, dx = \int_0^1 [\varepsilon(1 + g_m(x)) - g_m(x)]^{2q} dx,$$

is a real polynomial of an even degree. A necessary minimality condition for this function may be written at some ε as

$$\frac{d}{d\varepsilon} \int_0^1 [\varepsilon(1+g_m(x)) - g_m(x)]^{2q} \, dx = 2q \int_0^1 [\varepsilon(1+g_m(x)) - g_m(x)]^{2q-1} (1+g_m(x)) \, dx = 0$$

or

$$\sum_{l=0}^{2q-1} \binom{2q-1}{l} \varepsilon^{2q-1-l} \int_0^1 (1+g_m(x))^{2q-l}(-g_m(x))^l \, dx = 0.$$

Observe that $\int_0^1 g_m(x)\,dx = 0$, and for $l > 1$ the asymptotic equality holds

$$\int_0^1 g_m^l(x)\,dx = h^{ml} \sum_{\substack{s_1 \ldots s_l \neq 0 \\ s_1+\cdots+s_l=0}} \prod_{j=1}^{l} [h^m + (2\pi s_j)^m]^{-1} = b_l h^{ml}(1+o(1)),$$

with

$$b_l = \sum_{\substack{s_1 \ldots s_l \neq 0 \\ s_1+\cdots+s_l=0}} \prod_{j=1}^{l} (2\pi s_j)^{-m} > 0.$$

For the unknown $\gamma = \varepsilon h^{-m}$ we thus obtain the following equation

$$\sum_{l=0}^{2q-1} \binom{2q-1}{l} \gamma^{2q-1-l}(-1)^l b_l (1+o(1)) = 0.$$

It is obvious that for $q > 1$ none solution of this equation is $o(1)$ as $h \to 0$ and all feasible solutions γ of this equation are $O(1)$. One of them is $\gamma_0(1+o(1))$ with γ_0 a nonzero constant. To this constant there corresponds the optimal value of c_0 given by the equality

$$c_0(h) = 1 - \gamma_0 h^m (1+o(1)).$$

Consequently, we may write down the estimate

$$\begin{aligned} \left\| l_\infty - l_h^0 \mid \widetilde{W}_p^{(m)*} \right\| &= \left\| (c_0 - 1)(1 - l_\infty) \mid \widetilde{W}_p^{(m)*} \right\| \\ &\geq |1 - c_0| \left(1 - \left\| l_\infty \mid \widetilde{W}_p^{(m)*} \right\|\right) = |\gamma_0| h^m (1+o(1)). \end{aligned}$$

A similar upper bound also holds. In Chapter 4 it is proven that the norm $\| l_\infty \mid \widetilde{W}_p^{(m)*} \|$ is $O(h^m)$ and, moreover,

$$|\gamma_0| h^m (1+o(1)) / \left\| l_\infty \mid \widetilde{W}_p^{(m)*} \right\| = \gamma_1 (1+o(1))$$

with some constant $\gamma_1 > 0$.

Show that l_∞ is not asymptotically $\widetilde{W}_p^{(m)}$-optimal, i.e.,

$$\left\| l_h^0 \mid \widetilde{W}_p^{(m)*} \right\| / \left\| l_\infty \mid \widetilde{W}_p^{(m)*} \right\|$$

does not tend to 1 as $h \to 0$. Assume the contrary and, denoting

$$\|e\| = \|e \mid \widetilde{L}_{2q}\|, \quad \widetilde{G}_m(x) = \sum_k e^{i2\pi ks}/[1+(2\pi k)^m],$$

$$e^0(x) = \frac{(l_h^0(y), \widetilde{G}_m(x-y))}{\|l_\infty \mid \widetilde{W}_p^{(m)*}\|}, \quad e^1(x) = \frac{(l_\infty(y), \widetilde{G}_m(x-y))}{\|l_\infty \mid \widetilde{W}_p^{(m)*}\|},$$

obtain $\|e^0\| = 1 + o(1)$ and $\|e^1\| = 1$. It was proven earlier that

$$\|e^0 - e^1\| = |1 - c_0| / \|l_\infty \mid \widetilde{W}_p^{(m)*}\|(1 + o(1)) = \gamma_1 + o(1), \quad \gamma_1 > 0.$$

Use the generalization of the *Clarkson inequality* which is valid for all $\varphi(x)$, $\psi(x)$, and $\rho = 2q > 2$ (see [250, Chapter 3, § 3])

$$\left\|\frac{\varphi+\psi}{2}\right\|^\rho + \left\|\frac{\varphi-\psi}{2}\right\|^\rho \le \left\{\frac{\|\varphi\|^{\rho/(\rho-1)} + \|\psi\|^{\rho/(\rho-1)}}{2}\right\}^{\rho-1}.$$

Taking $\varphi = e^0$ and $\psi = e^1$, we have

$$\left\|\frac{e^0+e^1}{2}\right\|^{2q} + \left(\frac{\gamma_1}{2}\right)^{2q} + o(1) \le 1 + o(1).$$

From optimality of l_h^0 it follows that

$$\left\|\frac{e^0+e^1}{2}\right\| \ge \|e^0\| = 1 + o(1).$$

Then $1 + (\gamma_1/2)^{2q} \le 1 + o(1)$ or $\gamma_1 = 0$, which contradicts the condition $\gamma_1 > 0$. Consequently, $l_\infty(x)$ is not asymptotically $\widetilde{W}_p^{(m)}$-optimal for $p = 2q/(2q-1)$, $q > 1$.

Describe the class of cubature formulas which we deal with in this chapter. Given a matrix H, consider the cubature formula

$$I^*(f) = h^n|H| \sum_{hH\gamma\in\Omega} c_\gamma(h) f(hH\gamma)$$

with $\overline{\Omega}$ lying in the interior of the fundamental parallelepiped Ω_0 of H, and the error l_{hH}^Ω satisfying

$$\exists K \; \forall h \; \forall \gamma \quad |c_\gamma(h)| \le K, \tag{1.1}$$

$$\exists L \; \forall h \quad \rho(hH\gamma, \mathbb{R}^n \setminus \Omega) \ge Lh \Rightarrow c_\gamma(h) = 1. \tag{1.2}$$

We call them *formulas with bounded boundary layer* (BBL-formulas). The concept of BBL-formula appeared as generalization of the concept of formula with regular boundary layer.

We show that there are universal BBL-formulas asymptotically optimal on the spaces $W_p^{(m)}$ and in the Hölder classes $(\widetilde{C}^{\gamma})^m$ for at least one of the equivalent norms in each of these spaces.

In particular, the formula of the above example is asymptotically optimal in $\widetilde{W}_p^{(m)}$ with the norm

$$\|f\| = \left\| f(x) - \int_0^1 f(x)\,dx \mid \widetilde{W}_p^{(m)} \right\| + \left| \int_0^1 f(x)\,dx \right|.$$

We study asymptotically optimal formulas in Bessel scale of the spaces $\widetilde{B}_0^{\mu} = F^{-1}\mu F \widetilde{B}_0$. The definition of Bessel scale is given in § 1 of Chapter 4. The index-zero space $\widetilde{B}_0$ contains $\widetilde{C}$ and is embedded in $\widetilde{L}_1$. The space $\widetilde{B}_0$ is *weightless* and its norm is invariant under translations of the argument of a function. We require the function μ to meet the polynomial growth condition (4.1.31), the monotonicity condition (4.1.32), the lattice agreement condition (4.1.33) and the condition of being a multiplier (4.2.35).

Require in addition that μ be a *hypoelliptic function*, i.e., $\mu(\xi)$ is infinitely differentiable for $|\xi| \geq T$ and has the derivatives dominated as follows

$$\exists \delta > 0 \ \forall \alpha \ \exists K_\alpha : \quad |D^\alpha \mu(\xi)| \leq K_\alpha (1 + |\xi|)^{-\delta|\alpha| + m_2}. \tag{1.3}$$

The number m_2 in (1.3) is the same as in (4.1.31) and (4.1.32).

Observe that the whole collection of these constraints is satisfied, for instance, for the ordinary spaces $\widetilde{W}_p^{(m)}$ and the Hölder class $(\widetilde{C}^{\gamma})^m$.

Some of the constraints are imposed on μ for technical reasons. In our opinion, most essential are the condition that the norm be translation invariant and the conditions of polynomial growth and hypoellipticity on μ.

Given a bounded domain with piecewise-smooth boundary, we point out a universal asymptotically optimal BBL-formula. Not necessarily does such formula possess a regular boundary layer, but every formula with regular boundary layer of order $M > m_2$ is a universal asymptotically optimal formula.

The main result of the present chapter is in our opinion the derivation of simple sufficient conditions for universal asymptotic optimality. Let us clarify their essence.

In the class of BBL-formulas, one of every two of them differs from the other by a functional supported in a narrow boundary strip $\Omega_h = \{x : \rho(x, \partial\Omega) \leq Lh\}$. Whereas in the natural class, for instance, in the spaces $\widetilde{W}_p^{(m)*}$, the principal term of the asymptotic expansion of the norm of an error having an optimal mode of convergence depends continuously on the volume of the support of this functional. In essence, this is a consequence of the Hölder inequality

$$\|f \mid L_{p_1}(\Omega)\| \le |\Omega|^{1/p_1-1/p_2}\|f \mid L_{p_2}(\Omega)\|, \quad p_1 > p_2,$$

translated to the norms of functionals in dual spaces. Observe also that in the subclass of errors having an optimal mode of convergence, the difference $l_h^{\Omega,1} - l_h^{\Omega,2}$ also has at least the same mode of convergence. A small size of the support of the difference of BBL-formulas makes the norm of the difference of their errors a higher order infinitesimal. Thus, in the intersection of the class of BBL-formulas and the class of errors having an optimal mode of convergence, every two sequences of errors differ from one another by summands presenting a higher order infinitesimal. In other words, over the class of BBL-formulas, optimality with respect to the mode of convergence coincides with asymptotic optimality.

Justification of this claim is given under some additional constraints that are mostly of technical character. We now indicated only one of them which we rank as most principal. To establish continuity of the norms of a sequence of errors in the size of the support of an entry in the space B_1 under consideration, we need optimality with respect to the mode of convergence for these errors in the space B_2 embedded compactly in B_1; moreover, the optimality must be uniform in the size of the support of an entry.

We accomplish the prove for a Bessel scale $\widetilde{B}_0^\mu$ of Hilbert spaces with index-zero space $\widetilde{B}_0 = \widetilde{L}_2$. This scale is customarily denoted by $\widetilde{H}_2^\mu$. The space $\widetilde{H}_2^\mu(\Omega)$ consists of the restrictions of f in $\widetilde{H}_2^\mu$ to the domain Ω lying in the interior of the fundamental parallelepiped Ω_0. In this event the norm is given by the equality

$$\|f \mid \widetilde{H}_2^\mu(\Omega)\| = \inf\|g \mid \widetilde{H}_2^\mu\|.$$

The greatest lower bound is taken over all $g \in \widetilde{H}_2^\mu$ satisfying the condition $g|_\Omega = f|_\Omega$. It turns out that the infimum is attained, and beyond Ω the function, providing the infimum, possesses some additional properties we need.

Assume that P_Ω, an operator acting in the dual $\widetilde{H}_2^{1/\mu}$ of $\widetilde{H}_2^\mu$, is the orthoprojection to the subspace of elements supported in Ω, and P_Ω^* is the transpose of P_Ω acting from $\widetilde{H}_2^\mu$ to $\widetilde{H}_2^\mu$.

Lemma 6.1. *The operator P_Ω^* sends $\widetilde{H}_2^\mu$ onto $\widetilde{H}_2^\mu(\Omega)$, and, moreover,*

$$\|P_\Omega^* f \mid \widetilde{H}_2^\mu(\Omega)\| = \inf\{\|g \mid \widetilde{H}_2^\mu\| : g \in \widetilde{H}_2^\mu,\ g|_\Omega = f|_\Omega\}.$$

Let $|\mu(D)|^2$ be the pseudodifferential operator acting from $\widetilde{H}_2^\mu$ to $\widetilde{H}_2^{1/\mu}$ and defined by the equality

$$|\mu(D)|^2 = F^{-1}|\mu(\xi)|^2 F,$$

where F is the Fourier transform acting from $\widetilde{H}_2^\mu$ to $\widetilde{H}_2^{1/\mu}$. Then, for every function $f \in \widetilde{H}_2^\mu(\Omega)$, the equality holds

$$\left(|\mu(D)|^2 f\right)\Big|_{\Omega_0 \setminus \Omega} = 0.$$

PROOF. The operator P_Ω^* is a projection as the dual of a projection. Denote the inner product in $\widetilde{H}_2^\mu$ by $[\cdot, \cdot]_\mu$. Let $\zeta \in \widetilde{H}_2^\mu$ and $\operatorname{supp}\zeta \subset \Omega_0 \setminus \Omega$. By the definition of P_Ω, for every $f \in \widetilde{H}_2^\mu$ we have

$$(P_\Omega \mid \mu(D)|^2 f, \zeta) = 0.$$

Here $(\cdot, \cdot)$ is the inner product in $\widetilde{L}_2$. Continuing this equality

$$(P_\Omega|\mu(D)|^2 f, \zeta) = \left(|\mu(D)|^2 f, P_\Omega^* \zeta\right) = \left[f, P_\Omega^*\zeta\right]_\mu = \left[P_\Omega^* f, \zeta\right]_\mu = 0,$$

obtain

$$\left\|P_\Omega^* f + \zeta \mid \widetilde{H}_2^\mu\right\|^2 = \left\|P_\Omega^* f \mid \widetilde{H}_2^\mu\right\|^2 + \left\|\zeta \mid \widetilde{H}_2^\mu\right\|^2 \geq \|P_\Omega^* f \mid \widetilde{H}_2^\mu\|^2,$$

i.e., $P_\Omega^* f$ belongs to $\widetilde{H}_2^\mu(\Omega)$ and has the minimal norm.

To prove the second claim of the lemma, we use the fact that every function $f \in \widetilde{H}_2^\mu(\Omega)$ may be written as $f = P_\Omega^* f$. If $\zeta \in C_0^{(\infty)}(\Omega_0 \setminus \Omega)$ and $f \in \widetilde{H}_2^\mu(\Omega)$, then

$$\begin{aligned}(|\mu(D)|^2 f, \zeta) &= (|\mu(D)|^2 P_\Omega^* f, \zeta) = \left[P_\Omega^* f, \zeta\right]_\mu = \left[f, P_\Omega^* \zeta\right]_\mu \\ &= \left(|\mu(D)|^2 f, P_\Omega^* \zeta\right) = (P_\Omega|\mu(D)|^2 f, \zeta) = 0.\end{aligned}$$

The proof of Lemma 6.1 is complete.

We need the following well-known lemma whose proof is omitted.

Lemma 6.2. *Assume that a function $\nu(\xi)$ satisfies the hypoellipticity condition and $\nu(D) = F^{-1}\nu(\xi)F$. If $\nu(D)u(x)|_\omega = 0$, for some domain ω in $\mathbb{R}^n$ then u is infinitely differentiable in ω. Moreover, for a compact subset K of the interior of ω and every α, the estimate holds*

$$\left(\int_K |D^\alpha u(x)|^2 \, dx\right)^{1/2} \leq C_{\alpha,K}\left(\int_\omega |u(x)|^2 \, dx\right)^{1/2}.$$

We now state the main result.

Theorem 6.1. *Let a sequence l_{hH}^{Ω} of errors with bounded boundary layer have optimal mode of convergence in the spaces $\widetilde{W}_p^{(m)}$, where $m \in [m_1, m_3]$, the numbers m_1 and m_3 are even, and $p \in (1,2)$. Let the function $\mu(\zeta)$ be a hypoelliptic function satisfying (4.1.31)–(4.1.33) and (4.2.35) while $m_2 \in (m_1, m_3)$. Then the sequence l_{hH}^{Ω} is asymptotically $\widetilde{H}_2^{\mu}(\Omega)$-optimal.*

PROOF. For simplicity, denote

$$\left\| l \mid \widetilde{H}_2^{\mu *} \right\| = \|l\|_{1/\mu}, \quad \|l_\infty\|_{1/\mu} = \psi(h).$$

Let $l_{hH}^{\Omega,0}$ be an $\widetilde{H}_2^{\mu}(\Omega)$-optimal error supported in $\overline{\Omega}$. Then we have the equality

$$\left\| l_{hH}^{\Omega} \right\|_{1/\mu}^2 - \left\| l_{hH}^{\Omega,0} \right\|_{1/\mu}^2 = \left\| l_{hH}^{\Omega} - l_{hH}^{\Omega,0} \right\|_{1/\mu}^2.$$

We now prove the relation

$$\left\| l_{hH}^{\Omega} - l_{hH}^{\Omega,0} \right\|_{1/\mu} = o(\psi(h)) \quad \text{as} \quad h \to 0.$$

In the Hilbert space $\widetilde{H}_2^{1/\mu}$ take P_1, the orthoprojection to the subspace comprising the elements

$$\sum_{hHk \in \Omega} a_k \delta(x - hHk),$$

and P_2, the orthoprojection to the subspace comprising

$$\sum_{hHk \in \Omega_0} a_k \delta(x - hHk),$$

with h fixed and a_k arbitrary coefficients. We have

$$l_{hH}^{\Omega}(x) - l_{hH}^{\Omega,0}(x) = P_1 \chi_\Omega(x) - h^n |H| \sum_{hHk \in \Omega} c_k(h) \delta(x - hHk) = (P_1 l_{hH}^{\Omega})(x).$$

Take a sufficiently small $\varepsilon > 0$ and put

$$\Gamma_\varepsilon = \{x : x \in \Omega_0 \setminus \Omega,\ \rho(x, \Omega) \le \varepsilon\}, \quad \omega_\varepsilon = \Omega_0 \setminus (\Omega \cup \Gamma_\varepsilon).$$

These sets are nonempty. Consider

$$l_{hH}^{\omega_\varepsilon, h}(x) = \sum_{hHk \in \omega_\varepsilon} l_0 \left(\frac{x}{h} - Hk \right),$$

a homogeneous error of degree $M > 2m_3$, and arrange the difference

$$l_{hH}^{\Gamma_\varepsilon}(x) = l_\infty(x) - l_{hH}^{\Omega}(x) - l_{hH}^{\omega_{\varepsilon,h}}(x).$$

We know that for all $m \in [m_1, m_3]$ the following equalities hold

$$\left\| l_\infty \mid \widetilde{W}_p^{(m)*} \right\| = O(h^m), \qquad \left\| l_h^{\omega_{\varepsilon,h}} \mid \widetilde{W}_p^{(m)*} \right\| = O(h^m).$$

Consequently,

$$\left\| l_{hH}^{\Gamma_\varepsilon} \mid \widetilde{W}_p^{(m)*} \right\| = O(h^m), \quad m \in [m_1, m_3].$$

Applying Theorem 4.5 with $\widetilde{B}_0 = \widetilde{L}_2$, infer

$$\left\| l_{hH}^{\Gamma_\varepsilon} \mid \widetilde{H}_2^{\mu*} \right\| \leq K|\Gamma_\varepsilon|^{1/p-1/2}\psi(h)\{1 + K(\varepsilon)h^{m_3-m_2}\}.$$

Revert to the main estimate. It is clear that

$$\left\| P_1 l_{hH}^{\Omega} \right\|_{1/\mu} \leq \| P_1 l_\infty \|_{1/\mu} + \left\| P_1 l_{hH}^{\Gamma_\varepsilon} \right\|_{1/\mu} + \left\| P_1 l_{hH}^{\omega_{\varepsilon,h}} \right\|_{1/\mu} = \mathrm{I} + \mathrm{II} + \mathrm{III}.$$

To estimate summand I use the Three Perpendicular Theorem in Hilbert Space which yields $P_1 = P_1 P_2$. Further,

$$\mathrm{I} = \left\| P_1 P_2 l_\infty \right\|_{1/\mu} \leq \left\| P_2 l_\infty \right\|_{1/\mu}.$$

By Theorem 4.2 the error l_∞ is asymptotically optimal in $\widetilde{H}_2^{\mu}$. Whence, for $\widetilde{H}_2^{\mu}$-optimal error l_{hH}^0 the asymptotic equality holds

$$\left\| l_\infty - l_{hH}^0 \right\|_{1/\mu} = \left\{ \| l_\infty \|_{1/\mu}^2 - \left\| l_{hH}^0 \right\|_{1/\mu}^2 \right\}^{1/2} = o(\psi(h)).$$

Noting that $P_2 l_\infty = P_2\left(l_\infty - l_{hH}^0\right)$, come to the estimate

$$\mathrm{I} \leq \left\| P_2\left(l_\infty - l_{hH}^0\right) \right\|_{1/\mu} \leq \left\| l_\infty - l_{hH}^0 \right\|_{1/\mu} = o(\psi(h)).$$

To estimate summand II, use the fact that the norm of an orthoprojection is at most 1 and $|\Gamma_\varepsilon| \leq K\varepsilon$. We thus have

$$\mathrm{II} \leq \left\| l_{hH}^{\Gamma_\varepsilon} \right\|_{1/\mu} \leq K\varepsilon^{1/p-1/2}\psi(h)\left\{1 + K(\varepsilon)h^{m_3-m_2}\right\}.$$

Since P_Ω is the orthoprojection in $\widetilde{H}_2^{\mu *}$ to the subspace comprising the elements supported in Ω, from the Three Perpendicular Theorem infer that $P_1 = P_1 P_\Omega$. Consequently,

$$\mathrm{III} = \left\|P_1 P_\Omega l_{hH}^{\omega_\varepsilon,h}\right\|_{1/\mu} \leq \left\|P_\Omega l_{hH}^{\omega_\varepsilon,h}\right\|_{1/\mu} = \sup_f \frac{\left|(P_\Omega l_{hH}^{\omega_\varepsilon,h}, f)\right|}{\|f\|_\mu}$$

$$= \sup_f \frac{\left|(l_{hH}^{\omega_\varepsilon,h}, P_\Omega^* f)\right|}{\|f\|_\mu}. \tag{1.4}$$

By Lemmas 6.1 and 6.2, the function $u = P_\Omega^* f$ obeys the equation

$$\left(|\mu(D)|^2 u\right)\Big|_{\Omega_0\setminus\Omega} = 0$$

in the domain $\Omega_0 \setminus \Omega$. Moreover, for every function $\zeta \in \widetilde{C}_0^{(\infty)}$ with support at a positive distance δ from Ω and every integer m the next inequality is valid

$$\sum_{|\alpha|\leq m} \|D^\alpha(\zeta u) \mid \widetilde{L}_2\| \leq K(m,\delta,\Omega)\|u \mid \widetilde{L}_2\|. \tag{1.5}$$

Continuing (1.4), use (1.5) at $m = m_3$ and $\delta = \varepsilon/4$ with some function ζ equal 1 on the support of $l_{hH}^{\omega_\varepsilon,h}$. It is obvious that

$$\mathrm{III} \leq \sup_f \left\{ \frac{\left|(l_{hH}^{\omega_\varepsilon,h}, \zeta P_\Omega^* f)\right|}{\|\zeta P_\Omega^* f \mid \widetilde{W}_2^{(m_3)}\|} \frac{\|\zeta P_\Omega^* f \mid \widetilde{W}_2^{(m_3)}\|}{\|P_\Omega^* f \mid \widetilde{L}_2\|} \frac{\|P_\Omega^* f \mid \widetilde{L}_2\|}{\left\|P_\Omega^* f\right\|_\mu} \frac{\left\|P_\Omega^* f\right\|_\mu}{\|f\|_\mu} \right\}.$$

By Lemma 4.2 the first factor in braces is $O(h^{m_3})$. Whence and from Theorem 4.3 the first factor is $o(\psi(h))$. The second may be estimated by the constant $K(m_3, \varepsilon/4, \Omega)$ in virtue of (1.5). The third factor is also bounded since $\widetilde{H}_2^\mu$ is embedded in $\widetilde{L}_2$. The fourth is at most the norm of P_Ω^* which is 1. Therefore, summand III is $o(\psi(h))$.

Combining the above estimates, find

$$\left\|l_{hH}^\Omega - l_{hH}^{\Omega,0}\right\| \leq \mathrm{I} + \mathrm{II} + \mathrm{III} \leq K\varepsilon^{1/p-1/2}\psi(h) + K(\varepsilon)o(\psi(h))$$

as $h \to 0$. By construction, ε is an arbitrary positive real and $p \in (1,2)$. Therefore, we have the following asymptotic expansion

$$\left\|l_{hH}^\Omega - l_{hH}^{\Omega,0}\right\|_{1/\mu} = o(\psi(h)).$$

Moreover, by Theorem 4.2 the ratio of $\psi(h)$ to the norm $\|l_{hH}^{\Omega,0}\|_{1/\mu}$ is bounded above as $h \to 0$. Consequently, the difference between l_{hH}^{Ω} and $l_{hH}^{\Omega,0}$ in $\widetilde{H}_2^{\mu *}$ is $o(\|l_{hH}^{\Omega,0}\|_{1/\mu})$. The proof of Theorem 6.1 is complete.

Calculate the principal term in the expansion of the norm of an $\widetilde{H}_2^{\mu}(\Omega)$-optimal error $l_{hH}^{\Omega,0}$ as $h \to 0$. To this end, it suffices to consider asymptotically optimal formulas. First, establish the monotonicity of the norm of the error with respect to order by inclusion of domains over some class of domains of a simple shape. Next, obtain the exact estimates of $\|l_{hH}^{\Omega,0}\|_{1/\mu}$, approximating Ω from inside and from outside by domains of a simple shape.

***Lemma* 6.3.** *Let ω_1 and ω_2 be subdomains of Q with piecewise-smooth boundary such that $\overline{\omega}_1 \subset \omega_2$. Assume that* BBL*-errors $l_{hH}^{\omega_1}$ and $l_{hH}^{\omega_2}$ have optimal modes of convergence in $\widetilde{W}_p^{(m)}$ for some $p \in (1,2)$ and $m \in [m_1, m_3]$. Then*

$$\|l_{hH}^{\omega_1}\|_{1/\mu} \leq \|l_{hH}^{\omega_2}\|_{1/\mu}(1 + o(1)).$$

PROOF. For simplicity, we let H be the identity matrix I. Take a local error l_0 of degree $M > 2m_3 - n + 1$ and arrange the homogeneous error of the domain ω_2, i.e., put

$$l_h^{\omega_2,h}(x) = \sum_{hk \in \omega_2} l_0\left(\frac{x}{h} - k\right).$$

Take $\varepsilon > 0$; arrange the set ω_3 of the points of ω_2 at a distance greater than ε from ω_1. Put

$$l_h^{\omega_3,h}(x) = \sum_{hk \in \omega_3} l_0\left(\frac{x}{h} - k\right).$$

Arrange two more functionals

$$l_h^{\gamma_1} = l_h^{\omega_2,h} - l_h^{\omega_3,h} - l_h^{\omega_1}, \quad l_h^{\gamma_2} = l_h^{\omega_2} - l_h^{\omega_2,h}.$$

It is clear that $l_h^{\gamma_1}$ is supported near the boundary γ_1 of ω_1, and $l_h^{\gamma_2}$ is supported near the boundary γ_2 of the domain ω_2. Moreover, the equality holds

$$l_h^{\omega_2} = l_h^{\omega_1} + l_h^{\gamma_1} + l_h^{\omega_3,h} + l_h^{\gamma_2}.$$

For every functional $l \in \widetilde{H}_2^{\mu *}$ and every $\overline{\omega}$ in the interior of the unit cube we have

$$\|l \mid \widetilde{H}_2^{\mu *}\| \geq \|l \mid H_2^{\mu *}(\omega)\|.$$

Consequently, the following inequalities hold

$$\begin{aligned}
\|l_h^{\omega_2} \mid \widetilde{H}_2^{\mu *}\| &\geq \|l_h^{\omega_2} \mid \widetilde{H}_2^{\mu *}(\omega_1)\| \geq \|l_h^{\omega_1} \mid \widetilde{H}_2^{\mu *}(\omega_1)\| \\
&- \left\{\|l_h^{\gamma_1} \mid \widetilde{H}_2^{\mu *}(\omega_1)\| + \|l_h^{\omega_3,h} \mid \widetilde{H}_2^{\mu *}(\omega_1)\| + \|l_h^{\gamma_2} \mid \widetilde{H}_2^{\mu *}(\omega_1)\|\right\} \\
&\geq \|l_h^{\omega_1} \mid \widetilde{H}_2^{\mu *}\| - \left\{\|l_h^{\gamma_1} \mid \widetilde{H}_2^{\mu *}\| + \|l_h^{\omega_3,h} \mid \widetilde{H}_2^{\mu *}(\omega_1)\| + \|l_h^{\gamma_2} \mid \widetilde{H}_2^{\mu *}\|\right\}.
\end{aligned}$$

The norms of errors $l_h^{\gamma_1}$ and $l_h^{\gamma_2}$ in $W_p^{(m)}$ for $m \in [m_1, m_3]$ are $O(h^m)$. It follows from Lemma 4.2. Whence and from Theorems 4.2 and 4.5 it follows that

$$\|l_h^{\gamma_j} \mid H_2^{\mu*}\| \le K\big|\operatorname{supp} l_h^{\gamma_j}\big|^{1/p-1/2}\|l_h^{\omega_1} \mid \widetilde{H}_2^{\mu*}\|(1+o(1)), \quad j = 1, 2.$$

Moreover, $|\operatorname{supp} l_h^{\gamma_1}| \le K\varepsilon$, and $|\operatorname{supp} l_h^{\gamma_2}| = o(1)$ as $h \to 0$.

Let P_{ω_1} be the orthoprojection of $\widetilde{H}_2^{\mu*}$ to the subspace comprising the functionals with support in ω_1. By Lemma 6.1 $P_{\omega_1}^*$ projects $\widetilde{H}_2^{\mu}$ onto $\widetilde{H}_2^{\mu}(\omega_1)$. Take a compactly-supported infinitely differentiable function ζ equal 1 on the support of $l_h^{\omega_3,h}$ and equal 0 for $\rho(x, \omega_1) < \varepsilon/4$. We then have

$$\|l_h^{\omega_3,h} \mid \widetilde{H}_2^{\mu*}(\omega_1)\| = \sup_{f \in \widetilde{H}_2^{\mu}(\omega_1)} \frac{|(l_h^{\omega_3,h}, f)|}{\|f \mid \widetilde{H}_2^{\mu}\|} = \sup_{f \in \widetilde{H}_2^{\mu}} \frac{|(l_h^{\omega_3,h}, P_{\omega_1}^* f)|}{\|P_{\omega_1}^* f \mid \widetilde{H}_2^{\mu}\|}$$

$$= \sup_{f \in \widetilde{H}_2^{\mu}} \left\{ \frac{|(l_h^{\omega_3,h}, \zeta P_{\omega_1}^* f)|}{\|\zeta P_{\omega_1}^* f \mid \widetilde{W}_2^{(m_3)}\|} \, \frac{\|\zeta P_{\omega_1}^* f \mid \widetilde{W}_2^{(m_3)}\|}{\|P_{\omega_1}^* f \mid \widetilde{H}_2^{\mu}\|} \right\}.$$

By Lemmas 6.1 and 6.2, $\zeta P_{\omega_1}^* f \in \widetilde{W}_2^{(m_3)}$ and

$$\|\zeta P_{\omega_1}^* f \mid \widetilde{W}_2^{(m_3)}\| \le K\|P_{\omega_1}^* f \mid \widetilde{H}_2^{\mu}\|.$$

Whence and from Lemma 4.2 the first factor in braces of the last equality is at most Kh^{m_3}, and the second is bounded uniformly in f. Thus, given $m_3 > m_2$, by Theorem 4.3 we have

$$\|l_h^{\omega_2} \mid \widetilde{H}_2^{\mu*}\| \ge \|l_h^{\omega_1} \mid \widetilde{H}_2^{\mu*}\|(1 - K\varepsilon^{1/p-1/2} + o(1)).$$

Whence Lemma 6.3 ensues by arbitrariness of ε.

Theorem 6.2. *Let Ω be a bounded domain with piecewise-smooth boundary Γ, let H stand for the identity matrix, and let $l_h^{\Omega,0}$ be the corresponding $\widetilde{H}_2^{\mu}(\Omega)$-optimal error. Then*

$$\|l_h^{\Omega,0} \mid \widetilde{H}_2^{\mu*}\| = |\Omega|^{1/2}\|l_\infty \mid \widetilde{H}_2^{\mu*}\|(1+o(1)). \tag{1.6}$$

PROOF. To calculate the principal term in the expansion of the norm in $\widetilde{H}_2^{\mu*}$, replace $l_h^{\Omega,0}$ by a BBL-error $l_h^{\Omega,a}$. Let the norms of $l_h^{\Omega,a}$ in $\widetilde{W}_p^{(m)*}$ for some $p < 2$ and $m \in [m_1, m_3]$ be $O(h^m)$. Whence and from Theorem 6.1 the error $l_h^{\Omega,a}$ is asymptotically optimal in $\widetilde{H}_2^{\mu*}$. We prove the existence of such functional in Theorem 6.6.

Apply Lemma 6.3, specifying the domains ω_1 and ω_2. Take $\varepsilon > 0$ and, given $h \le \varepsilon$, put $h_0 = [\varepsilon/h]h$. Let $h_0(Q+k)$ stand for the cube of

$$\{x : (\exists y \in Q)\; x = h_0(y+k)\}.$$

Assume that the intersection of the interior of the cube $h_0(Q+k)$ with Γ is nonempty. Uniting all these cubes, put

$$\Gamma_{h_0} = \bigcup_{h_0(Q+k)\cap\Gamma\neq\varnothing} h_0(Q+k), \quad \omega_1 = \Omega\setminus\Gamma_{h_0}, \quad \omega_2 = \Omega\cup\Gamma_{h_0}.$$

Let $\psi(h) = \left\|l_\infty \mid \widetilde{H}_2^{\mu*}\right\|$ and consider the homogeneous errors $l_h^{\omega_1}(x)$ and $l_h^{\omega_2}(x)$ given by the equalities

$$l_h^{\omega_j}(x) = \sum_{hk\in\omega_j} l_0\left(\frac{x}{h} - k\right), \quad j = 1, 2.$$

Here $l_0(y)$ is an elementary error of the shape (4.2.2) and order M. The support of $l_0(y)$ lies clearly in the cube with center the origin and side $2L$. Whence and from Lemma 6.3 the following relations are valid

$$\varliminf_{h\to 0} \frac{\left\|l_h^{\omega_1} \mid \widetilde{H}_2^{\mu*}\right\|}{\psi(h)} \le \varliminf_{h\to 0} \frac{\left\|l_h^{\Omega,a} \mid \widetilde{H}_2^{\mu*}\right\|}{\psi(h)} \le \varlimsup_{h\to 0} \frac{\left\|l_h^{\Omega,a} \mid \widetilde{H}_2^{\mu*}\right\|}{\psi(h)} \le \varlimsup_{h\to 0} \frac{\left\|l_h^{\omega_2} \mid \widetilde{H}_2^{\mu*}\right\|}{\psi(h)}.$$

Prove that, as $\varepsilon \to 0$ (i.e., when $h_0 \to 0$), the extreme terms of this chain of inequalities have the same limit equal to $|\Omega|^{1/2}$. We will so arrive at (1.6).

To this end, we calculate the norms of $l_h^{\omega_1}(x)$ and $l_h^{\omega_2}(x)$ in $\widetilde{H}_2^{\mu*}$. The advantage of these errors is in the fact that ω_1 and ω_2 are partitioned into an integral number of cubes $h_0(Q+k)$ (and $h(Q+k)$) for all h.

Let $l_h^{\omega_j}(x)$ be the sum of elementary errors of degree $M \ge 2m_3$. By choice of ω_1 and ω_2, for $\omega = \omega_j$ we have

$$l_h^{h_0}(x) = \sum_{hk\in h_0 Q} l_0\left(\frac{x}{h} - k\right), \quad l_h^{\omega}(x) = \sum_{h_0 s\in\omega} l_h^{h_0}(x - h_0 s).$$

If $\omega = Q$ then $l_h^{\omega}(x)$ coincides with $l_\infty(x)$.

Calculate the norm of l_h^{ω} in $\widetilde{H}_2^{\mu*}$. To this end, denote by $[\cdot,\cdot]_{1/\mu}$ the inner product in the Hilbert space $\widetilde{H}_2^{\mu*} = \widetilde{H}_2^{1/\mu}$ and put

$$\mathrm{I} = \left[l_h^{h_0}(x - h_0 s), l_h^{h_0}(x - h_0 r)\right]_{1/\mu} = \left(l_h^{h_0}(x - h_0 s), \mu^{-2}(D) l_h^{h_0}(x - h_0 r)\right).$$

Here the function $u(x) = \mu^{-2}(D)l_h^{h_0}(x - h_0 r)$ stands for a solution in $\widetilde{H}_2^{\mu}$ to the pseudodifferential equation

$$|\mu(D)|^2 u = l_h^{h_0}(x - h_0 r).$$

The definition of the norm in $\widetilde{H}_2^{\mu}$ implies that

$$\|u \mid \widetilde{H}_2^{\mu}\| = \|l_h^{h_0}(x - h_0 r) \mid \widetilde{H}_2^{\mu *}\| = \|l_h^{h_0}(x) \mid \widetilde{H}_2^{\mu *}\|.$$

Show that if $s \neq r$ then the inner product I is $o(\psi^2(h))$ as $h \to 0$.

Given L and ε, observe that for all $h \leq \varepsilon/(16L+1)$ the interval $(8Lh, h_0/2)$ is surely nonempty. Taking δ in this interval and letting

$$\gamma_\delta = \{x : x \in h_0 Q,\ \rho(x, \mathbb{R}^n \setminus h_0 Q) < \delta\}, \quad l_h^{\gamma_\delta}(x) = \sum_{hk \in \gamma_\delta} l_0\left(\frac{x}{h} - k\right),$$

obtain

$$l_h^{h_0}(x - h_0 s) = l_h^{\gamma_\delta}(x - h_0 s) + \left\{l_h^{h_0}(x - h_0 s) - l_h^{\gamma_\delta}(x - h_0 s)\right\}.$$

Thus, the inner product I admits the estimate

$$|\mathrm{I}| \leq \left|\left(l_h^{\gamma_\delta}(x - h_0 s), \mu^{-2}(D) l_h^{h_0}(x - h_0 r)\right)\right|$$
$$+\left|\left(l_h^{h_0}(x - h_0 s) - l_h^{\gamma_\delta}(x - h_0 s), \mu^{-2}(D) l_h^{h_0}(x - h_0 r)\right)\right|.$$

By Lemma 4.2 and Theorem 4.5 the norm of $l_h^{\gamma_\delta}(x)$ may be estimated as follows

$$\|l_h^{\gamma_\delta} \mid \widetilde{H}_2^{\mu *}\| \leq K \delta^{1/p - 1/2} \psi(h)(1 + o(1)).$$

By Lemma 4.2 and Theorem 4.4 the norm of $l_h^{h_0}(x)$ in $\widetilde{H}_2^{\mu *}$ is $O(\psi(h))$ as $h \to 0$. Considering this, it is easy to estimate the first summand in the inequality for $|\mathrm{I}|$. Estimate the second summand in the inequality for $|\mathrm{I}|$ through the norm of the difference $l_h^{h_0}(x - h_0 s)$ and $l_h^{\gamma_\delta}(x - h_0 s)$ in $\widetilde{W}_2^{(m_3)*}$. By choice of δ, the function $u(x) = \mu^{-2}(D) l_h^{h_0}(x - h_0 r)$ for $r \neq s$ is a solution to the homogeneous pseudodifferential hypoelliptic equation

$$|\mu(D)|^2 u(x) = 0$$

in the $\delta/2$-neighborhood about the support g of the difference. Consequently, by Lemma 6.2 $u(x)$ is an infinitely differentiable function in the $\delta/2$-neighborhood $g''_{\delta/2}$ of g and, moreover,

$$\|u \mid W_2^{(m_3)}(g''_{\delta/2})\| \leq C_{m_3,\delta} \|u \mid L_2(G)\|.$$

Here G is a set whose interior includes the closure of the neighborhood $g''_{\delta/2}$ and whose distance to the support of $l_h^{h_0}(x - h_0 r)$ is at least $\delta/8$. For $r \neq s$ such G ready to exist by choice of δ. The inequality for the norm of $u(x)$ in $W_2^{(m_3)}(g''_{\delta/2})$ is easy to continue by involving (4.1.31)

$$\|u \mid W_2^{(m_3)}(g''_{\delta/2})\| \leq C_{m_3,\delta}\|u \mid \widetilde{L}_2\| \leq C(\delta)\|u \mid \widetilde{H}_2^{\mu}\| = C(\delta)\|l_h^{h_0}(x) \mid \widetilde{H}_2^{\mu *}\|.$$

Therefore, we arrive at the estimate

$$\begin{aligned}
&\left|(l_h^{h_0}(x - h_0 s) - l_h^{\gamma_\delta}(x - h_0 s), u(x))\right| \\
&\leq C\|l_h^{h_0} - l_h^{\gamma_\delta} \mid W_2^{(m_3)*}(g''_{\delta/2})\| \, \|u \mid W_2^{(m_3)*}(g''_{\delta/2})\| \\
&\leq C(\delta)\|l_h^{h_0} - l_h^{\gamma_\delta} \mid \widetilde{W}_2^{(m_3)*}\|\psi(h).
\end{aligned}$$

By Lemma 4.2 the norm of the difference between $l_h^{h_0}$ and $l_h^{\gamma_\delta}$ in $W_2^{(m_3)*}$ is $O(h^{m_3})$. Considering this and Theorem 4.3, estimate the inner product I as follows

$$|\mathrm{I}| \leq K\delta^{1/p-1/2}\psi^2(h) + K(\delta)h^{m_3}\psi(h) \leq K\big(\delta^{1/p-1/2} + C(\delta)h^{m_3-m_2}\big)\psi^2(h).$$

Here K is independent of h and δ. Dividing the above inequality by $\psi^2(h)$, pass to the limit as $h \to 0$, fixing δ. This is possible since $0 < 8Lh < \delta < h_0/2$ by hypothesis, and the parameter h_0 belongs to the interval $[\varepsilon/2, \varepsilon]$ for the chosen positive ε. Also considering that $m_3 > m_2$, arrive at the final estimate for the inner product I in the shape

$$|\mathrm{I}| \leq K\delta^{1/p-1/2}\psi^2(h).$$

Whence, in order notation, the inner product I is $o(\psi^2(h))$ as $h \to 0$.

Show now that, given $\varepsilon > 0$, we have the asymptotic equalities

$$\begin{aligned}
\|l_h^{\omega} \mid \widetilde{H}_2^{\mu *}\|^2 &= \sum_{h_0 s \in \omega} \|l_h^{h_0}(x - h_0 s) \mid \widetilde{H}_2^{\mu *}\|^2 + o(\psi^2(h)), \\
\|l_\infty \mid \widetilde{H}_2^{\mu *}\|^2 &= \sum_{h_0 s \in Q} \|l_h^{h_0}(x - h_0 s) \mid \widetilde{H}_2^{\mu *}\|^2 + o(\psi^2(h)).
\end{aligned}$$

Indeed, if we express the left sides of these equalities through the inner product in $\widetilde{H}_2^{\mu *}$ and substitute for l_h^{ω} and l_∞ the sums of $l_h^{h_0}(x - h_0 s)$, then it suffices to use the just-established estimates for the inner products of various functionals $l_h^{h_0}(x - h_0 s)$, arriving at the sought asymptotic expansions for the norms. Observe now that the norms of $l_h^{h_0}(x - h_0 s)$ do not depend on s. Consequently,

$$\begin{aligned}
\|l_h^{\omega} \mid \widetilde{H}_2^{\mu *}\|^2 &= |\omega| h_0^{-n} \|l_h^{h_0} \mid \widetilde{H}_2^{\mu *}\|^2 + o(\psi^2(h)), \\
\|l_\infty \mid \widetilde{H}_2^{\mu *}\|^2 &= h_0^{-n} \|l_h^{h_0} \mid \widetilde{H}_2^{\mu *}\|^2 + o(\psi^2(h)).
\end{aligned}$$

Therefore, given $\varepsilon > 0$, we have

$$\|l_h^\omega \mid \widetilde{H}_2^{\mu *}\| = |\omega|^{1/2}\|l_\infty \mid \widetilde{H}_2^{\mu *}\|(1+o(1)).$$

Letting $\omega = \omega_1$ and $\omega = \omega_2$, pass to the limit, first, as $h \to 0$, and, second, as $\varepsilon \to 0$. We then come to the equalities

$$\lim_{\varepsilon\to 0}\lim_{h\to 0}\frac{\|l_h^{\omega_j} \mid \widetilde{H}_2^{\mu *}\|}{\psi(h)} = |\Omega|^{1/2}, \quad j = 1, 2.$$

Thus, the proof of Theorem 6.2 is complete.

REMARKS. 1. For an arbitrary lattice matrix H, the equality holds

$$\|l_{hH}^{\Omega,0} \mid \widetilde{H}_2^{\mu *}\| = \left(\frac{|\Omega|}{|H|}\right)^{1/2}\|l_\infty \mid \widetilde{H}_2^{\mu *}\|(1+o(1)).$$

2. For the space $\widetilde{W}_2^{(m)}$ with m an integer, (1.6) looks like

$$\|l_{hH}^{\Omega,0} \mid \widetilde{W}_2^{(m)*}\| = |\Omega|^{1/2}\left(\sum_{k\neq 0}\frac{1}{|2\pi H^{-1*}k|^{2m}}\right)^{1/2} h^m(1+o(1)).$$

This equality is analogous to (5.3.1).

3. A formula (1.6) is also valid for a homogeneous error of degree M in the shape

$$l_{hH}^{\Omega}(x) = \sum_{hHk\in\Omega} l_0\left(\frac{x}{h} - Hk\right).$$

Draw attention to the fact that the rate of convergence of a cubature formula on anisotropic spaces depends essentially on the choice of nodes. We explicate the situation by example. Consider the space of periodic functions in two variables possessing generalized derivatives of order 3 with respect to x_1 and of order 4 with respect to x_2 square summable over the unit cube Q. Introduce a norm in this space, by letting

$$\|f \mid \widetilde{W}_2^{3,4}\| = \left\{\int_Q \left(|f(x)|^2 + \left|D_1^3 f(x)\right|^2 + \left|D_2^4 f(x)\right|^2\right) dx\right\}^{1/2}.$$

The cubature formula with equal weights for a cubic lattice has error l_∞ whose rate of vanishing is defined by the equality

$$\|l_\infty \mid (\widetilde{W}_2^{3,4})^*\| = \left\{\sum_{k\neq 0}\frac{1}{1+(2\pi k_1/h)^6+(2\pi k_2/h)^8}\right\}^{1/2} = O(h^3).$$

This relation remains valid for all lattices with matrices H constant in h.

Consider the lattice with mesh-size-vector $\overline{h} = (h_1, h_2)$ and nodes (k_1h_1, k_2h_2), $k \in \mathbb{Z}^2$. The norm of the error $l_{\overline{h}}$ analogous to l_∞ is given by the equality

$$\left\| l_{\overline{h}} \mid (\widetilde{W}_2^{3,4})^* \right\| = \left\{ \sum_{k \neq 0} \frac{1}{1 + (2\pi k_1/h_1)^6 + (2\pi k_2/h_2)^8} \right\}^{1/2}.$$

The best rate of vanishing for this quantity is achieved at the same rates of the summands $(k_1/h_1)^6$ and $(k_2/h_2)^8$. Put, for instance, $h_1 = h_2^{4/3}$. Then

$$\left\| l_{\overline{h}} \mid (\widetilde{W}_2^{3,4})^* \right\| = O(h_2^4).$$

To have a possibility of comparing the rates of convergence for various lattices, we are to treat as parameter N the number of nodes of a cubature formula rather than the lattice mesh-size-vector. For a scalar mesh-size h, we have $N = h^{-2}$; whereas, for the mesh-size-vector, $N = 1/(h_1h_2)$. The norm of the error with a scalar mesh-size is $O(N^{-3/2})$, and in the case of the error with mesh-size-vector $\overline{h}$ satisfying $h_1 = h_2^{4/3}$ the norm is $O(N^{-12/7})$.

To complete our research, we study the properties of formulas on lattices which give the best rate of convergence on the space $\widetilde{H}_2^\mu$ as the number of nodes increases infinitely. Allowing the arbitrary dependence of the lattice matrix on N, the number of nodes of a cubature formula, we come to the necessity of expanding the definition of asymptotic optimality.

Assume given a space B and a sequence of lattice matrices $H(N)$. *Asymptotically optimal* in B we call a sequence of errors

$$l_h^{\Omega,a}(x) = \chi_\Omega(x) - \sum_{H(N)k \in \Omega} a(k, N)\delta(x - H(N)k),$$

with the property

$$\lim_{N \to \infty} \frac{\left\| l_N^{\Omega,a} \mid B^* \right\|}{\inf\limits_{\{b_k\}} \left\| \chi_\Omega(x) - \sum\limits_{H(N)k \in \Omega} b_k \delta(x - H(N)k) \mid B^* \right\|} = 1. \tag{1.7}$$

Our main problem remains to be that of finding an asymptotically optimal error with bounded boundary layer given Ω, B, and $\{H(N)\}_{N=1}^\infty$. We solve an important particular case of the problem in which the space $\widetilde{H}_2^\mu$ coincides with $\widetilde{W}_2^m$, and H is a diagonal matrix determining the lattice of nodes

$$(h_1k_1, h_2k_2, \ldots, h_nk_n), \quad k_j \in \mathbb{Z}. \tag{1.8}$$

We call the vector $\overline{h} = (h_1, h_2, \ldots, h_n)$ the *mesh-size-vector* of the lattice (1.8) or the *lattice mesh-size-vector.* Minimizing the norm of error over the lattices of such shape, obtain asymptotically optimal lattice formulas for an anisotropic space. They happen to have optimal mode of convergence among all formulas with arbitrarily-placed nodes.

In order that a lattice of nodes agrees with the fundamental parallelepiped of the space of periodic functions under consideration, the numbers $1/h_j$ should be integral, in symbols,

$$N_j = 1/h_j \in \mathbb{Z}. \tag{1.9}$$

Let $\overline{m} = (m_1, \ldots, m_n)$ be a multi-index with integer entries. The space $\widetilde{W}^{\overline{m}_2}$ comprises the functions $f(x)$ in $\widetilde{L}_2$, whose generalized derivatives $D_j^{m_j} f(x)$ of order m_j with respect to x_j, $j = 1, 2, \ldots, n$, are square summable over the unit cube. Introduce a norm into this space by letting

$$\|f \mid \widetilde{W}_2^{\overline{m}}\| = \left\{ \|f \mid \widetilde{L}_2\|^2 + \sum_{j=1}^{n} \|D_j^{m_j} f \mid \widetilde{L}_2\|^2 \right\}^{1/2}.$$

This norm is easily written through the Fourier coefficients $c_f[\beta]$ of f. Using the Parseval identity, obtain

$$\|f \mid \widetilde{W}_2^{\overline{m}}\| = \left\{ \sum_{\beta} |c_f[\beta]|^2 \left(1 + \sum_{j=1}^{n} (2\pi\beta_j)^{2m_j} \right) \right\}^{1/2}.$$

The Banach space $W_2^{\overline{m}}$ is thus coincident with the space $\widetilde{H}_2^{\mu_0}$ of the Bessel scale with zero space $\widetilde{L}_2$ and weight function

$$\mu_0(\varepsilon) = \left\{ 1 + \sum_{j=1}^{n} (2\pi\xi_j)^{2m_j} \right\}^{1/2}.$$

Let $K > 1$ and assume that for all j, $1 \le j \le n$, we have the estimates

$$K^{-1} \left(\frac{1}{h} + (2\pi\xi_j)^{2m_j} \right) \le (1 + \xi_j^2)^{m_j} \le K \left(\frac{1}{h} + (2\pi\xi_j)^{2m_j} \right).$$

Summing the latter over j ranging from 1 to n, find

$$K^{-1} \mu_0^2(\xi) \le \mu^2(\xi) \le K \mu_0^2(\xi).$$

Here, $\mu(\xi)$ stands for the function $\{\sum_{j=1}^{n} (1 + \xi_j^2)^{m_j}\}^{1/2}$. The above inequality allows us to assert that the spaces $\widetilde{H}_2^{\mu_0}$, $\widetilde{H}_2^{\mu}$, and $\widetilde{W}_2^{\overline{m}}$ coincide. Moreover, the norms, generated in these spaces by the weight functions $\mu_0(\xi)$ and $\mu(\xi)$, are equivalent to the initial norm of $\widetilde{W_2^{\overline{m}}}$. Bearing this in mind, we further agree that $\widetilde{W_2^{\overline{m}}}$ coincides with $\widetilde{H}_2^{\mu}$ and has the norm generated by the weight function $\mu(\xi)$. The same agreement is effective in the case when the entries of $\overline{m}$ are positive possibly nonintegral numbers.

Suppose that $p > 1$ and $p \neq 2$. We now define $\widetilde{W_p^{\overline{m}}}$ to be the space $\widetilde{B}_0^{\mu}$ of the Bessel scale with index-zero space $\widetilde{B}_0 = \widetilde{L}_p$ and weight function

$$\mu(\xi) = \left\{ \sum_{j=1}^{n} (1 + \xi_j^2)^{m_j} \right\}^{1/2}.$$

We now describe the class of cubature formulas whose action over $\widetilde{W_2^{\overline{m}}}$ will be under study.

Let $\overline{h} = (h_1, \dots, h_n)$ with h_j meeting the conditions (1.9). Denote by $\mathrm{I}(\overline{h})$ the diagonal matrix with diagonal entries $h_1, \dots, h_n$. Given a domain Ω whose closure is included in the interior of the unit cube, consider a cubature formula with error

$$l_{\overline{h}}^{\Omega}(x) = \chi_{\Omega}(x) - h_1 \dots h_n \sum_{\mathrm{I}(\overline{h})\beta \in \Omega} c_\beta(\overline{h}) \delta(x - I)(\overline{h})\beta).$$

Assume that the weights $c_\beta(\overline{h})$ satisfy the conditions like (1.1)–(1.2), namely,

$$\exists K \; \forall \overline{h} \; \forall \beta \quad |c_\beta(\overline{h})| \leq K,$$
$$\exists L \; \forall \overline{h} \quad \rho(I(\overline{h})\beta, \mathbb{R}^n \setminus \Omega) \geq L \max_j h_j \Rightarrow c_\beta(\overline{h}) = 1.$$

We then call $l_{\overline{h}}^{\Omega}(x)$ an error with bounded boundary layer. For $h_1 = \dots = h_n = h$ this definition amounts to that given above.

We will prove that a sequence of errors with bounded boundary layer is asymptotically optimal in $\widetilde{W_2^{\overline{m}}}$ under the conditions analogous to the hypotheses of Theorem 6.1. We start with a preliminary definition and lemma.

Let $\overline{h} = (h_1, \dots, h_n)$ and assume (1.9) valid. Given x in $\mathbb{R}^n$, denote $(x_1/h_1, \dots, x_n/h_n)$ by $x/\overline{h}$. Take a domain ω with closure included in the interior of the unit cube and consider the error

$$l_{\overline{h}}^{\omega}(x) = \sum_{I(\overline{h})\beta \in \omega} l_0(x/\overline{h} - \beta).$$

Write down $l_0(y)$ in accord with (4.2.2). If the values of $l_0(y)$ at polynomials of degree M are all zero, then we call $l_{\overline{h}}^{\omega}(x)$ a *homogeneous error of degree* M. For $h_1 = \dots = h_n = h$ this definition coincides with that of Section 2 of Chapter 4.

Lemma 6.4. *Let $M_2 > M_1 > n/2$ and let the minimum and the maximum of $m_1, m_2, \dots, m_n$ lie in the interval (M_1, M_2). If for $m = M_1, M_2$ the norm of $l_{\overline{h}}^{\Omega}(x)$ admits the estimate*

$$\|l_{\overline{h}}^{\Omega} \mid \widetilde{W}_p^{(m)*}\| \leq K_1 \max_{1\leq j\leq n} h_j^m, \quad p>1, \tag{1.10}$$

then we have the inequality

$$\|l_{\overline{h}}^{\Omega} \mid \widetilde{W}_p^{\overline{m}*}\| \leq K_2 \max_{1\leq j\leq n} h_j^{m_j}. \tag{1.11}$$

PROOF. Denote $\{\sum_{j=1}^n (1+\xi_j^2)^{m_j}\}^{1/2}$ by $\mu(\xi)$ and let $k/\overline{h}$ stand for the vector $(k_1/h_1, \dots, k_n/h_n)$. Given $s \in \mathbb{Z}^n$, consider the inequality $1/\mu^2(s) \leq KI(s,\overline{h})$ with

$$I(s,\overline{h}) = \sum_{k\neq 0} \frac{1}{\mu^2(k/\overline{h})} \left\{ \frac{1}{\max\limits_j h_j^{M_1}(|s|^2+1)^{M_1/2}} + \frac{1}{\max\limits_j h_j^{M_2}(|s|^2+1)^{M_2/2}} \right\}^2.$$

Observe that this inequality may be checked in the same way as in Theorem 4.3.

Now, consider the Fourier series $\sum_s L[s]e^{i2\pi xs}$ for the functional $l_{\overline{h}}^{\Omega}(x)$ and its $\widetilde{W}_p^{\overline{m}*}$ norm, namely,

$$\left\|\sum_s \frac{L[s]}{\mu(s)} e^{i2\pi xs} \mid \widetilde{L}_{p'}\right\| = \left\|\sum_s \frac{\mu^{-1}(s)}{\sqrt{I(s,\overline{h})}} L[s]\sqrt{I(s,\overline{h})}\, e^{i2\pi xs} \mid \widetilde{L}_{p'}\right\|.$$

As usual, $p' = p/(p-1)$. The linear operator dividing the Fourier coefficients by $\mu(s)\sqrt{I(s,\overline{h})}$ is bounded in $\widetilde{L}_{p'}$ for $p > 1$. Consequently, the $\widetilde{W}_p^{\overline{m}*}$ norm of the error $l_{\overline{h}}^{\Omega}(x)$ is at most

$$K\sqrt{\sum_{k\neq 0} \mu^{-2}(k/\overline{h})} \left\{ (\max_j h_j)^{-M_1} \|l_{\overline{h}}^{\Omega} \mid \widetilde{W}_p^{(M_1)*}\| + (\max_j h_j)^{-M_2} \|l_{\overline{h}}^{\Omega} \mid \widetilde{W}_p^{(M_2)*}\| \right\}.$$

By (1.10) the expression in braces is bounded uniformly in $\overline{h}$, and

$$\sqrt{\sum_{k\neq 0} \mu^{-2}(k/\overline{h})} = O(\max_j h_j^{m_j}).$$

Thus, the proof of Lemma 6.4 is complete.

Corollary 1.1. *If the estimate* (1.10) *is valid for* $m = M_1, M_2$, *then it is also preserved for every* $m \in [M_1, M_2]$.

It suffices to apply Lemma 6.4 for $m_1 = \cdots = m_n = m$ and find what was required.

Lemma 6.5. *Let* M_2, $M_2 > n$, *be even and let* $l^{\omega}_{\overline{h}}(x)$ *be a homogeneous error of degree* $M \geq 2M_2 - n + 1$. *The norms of* $l^{\omega}_{\overline{h}}$ *in* $\widetilde{W}_q^{(m)}$ *for all* $q > 1$ *and even* $m \in (n, M_2]$ *are* $O(\max_j h_j^m)$.

PROOF. Denote the even integer $2([n/2] + 1)$ by M_1. Corollary 1.1 reduces the question to the even $m = M_1, M_2$. Consequently, it suffices to establish the estimates

$$\|l^{\omega}_{\overline{h}} \mid \widetilde{W}_q^{(m)*}\| \leq K \max_j h_j^m, \quad m = M_1, M_2. \tag{1.12}$$

By the definition of norm, we have

$$\|l^{\omega}_{\overline{h}} \mid \widetilde{W}_q^{(m)*}\| = \sup_f \frac{|(l^{\omega}_{\overline{h}}, f)|}{\|f \mid \widetilde{W}_q^{(m)}\|}.$$

In the supremand, perform a change of variables that transforms the lattice (1.8) to a cubic lattice. Let

$$\tau = \max_j h_j; \quad y_j = x_j\tau/h_j; \quad g(y) = f(h_1 y_1/\tau, \dots, h_n y_n/\tau).$$

Denote by ω_τ and $Q\tau/\overline{h}$ the images of ω and Q under the change from x to y. Then

$$\begin{aligned} &|(l^{\omega}_{\overline{h}}, f)| \Big/ \Big\{ \|f \mid \widetilde{L}_q\| + \sum_{|\alpha|=m} \|D^\alpha f \mid \widetilde{L}_q\| \Big\} \\ &= |(l^{\omega_\tau}_{\tau}, g)| \Big/ \Big\{ \|g \mid L_q(Q\tau/\overline{h})\| + \sum_{|\alpha|=m} \tau^m \overline{h}^{-\alpha} \|D^\alpha g \mid L_q(Q\tau/\overline{h})\| \Big\} \\ &\leq K \|l^{\omega_\tau}_{\tau} \mid \widetilde{W}_q^{(m)*}(Q\tau/\overline{h})\|. \end{aligned}$$

The last inequality is valid in virtue of the next two observations. First, the product $\tau^m \overline{h}^{-\alpha}$ is at least 1, for $|\alpha| = m$, which ensues from the definition of τ. Second, the support of $l^{\omega_\tau}_{\tau}(y)$ lies in the interior of $Q\tau/\overline{h}$. This enables us to replace $g(y)$ with an extension of it from the support of $l^{\omega_\tau}_{\tau}(y)$ to a member $\widetilde{g}(y)$ of $\widetilde{W}_q^{(m)}(Q\tau/\overline{h})$.

The mapping $x = I(\overline{h}/\tau)y$ performs a one-to-one correspondence between the points of the set (1.8) in R^n_x and the points τk, $k \in \mathbb{Z}^n$, comprising a cubic lattice in R^n_y. Moreover, to the homogeneous error $l^{\omega}_{\overline{h}}(x)$ with mesh-size-vector $\overline{h}$ there corresponds the error

$$l^{\omega_\tau}_{\tau}(y) = \sum_{\tau\beta\in\omega_\tau} l_0\left(\frac{y}{\tau} - \beta\right)$$

with scalar mesh-size τ. Fixing the entries h_j of the vector $\overline{h}$, consider the homogeneous error of degree M having the shape

$$l^{\omega_\tau}_{h}(y) = \sum_{\tau\beta\in\omega_\tau} l_0\left(\frac{y}{h} - \beta\right), \quad 0 < h < 1.$$

The errors $l^{\omega_\tau}_{h}(y)$ and $l^{\omega_\tau}_{\tau}(y)$ are obviously the same for $h = \tau$. With this in mind, we dominate the norm $l^{\omega_\tau}_{h}(y)$ in $\widetilde{W}^{(m)*}_q(Q_\tau/\overline{h})$, applying the result to $l^{\omega_\tau}_{\tau}(y)$.

Observe that the volume of ω_τ, as a function in $\overline{h}$, may become indefinitely large in a neighborhood about the point $\overline{h} = 0$. Whence we conclude that the sought upper bound on the norm of $l^{\omega_\tau}_{\overline{h}}(y)$ must be uniform in all ω_τ with $0 < \tau < 1$. We demonstrate existence of such a uniform bound by slightly adjusting the proof of Lemma 4.2.

We first agree on the current notation. Let H the diagonal matrix inverse to $I(\overline{h}/\tau)$, with $\Omega_0 = Q\tau/\overline{h}$ the corresponding fundamental parallelepiped. The homogeneous error $l^{\omega_\tau}_{h}(y)$ is clearly bounded in $\widetilde{W}^{(m)}_p(\Omega_0)$, $1 < p < +\infty$. By (4.2.12), the norm of $l^{\omega_\tau}_{h}(y)$ in $\widetilde{W}^{(m)}_p(\Omega_0)^*$ coincides with the $\widetilde{L}_{p'}(\Omega_0)$ norm of the function $\widetilde{u}(x)$ defined by (4.2.6). From (4.2.13) and (4.2.9) it follows that

$$\|\widetilde{u} \mid \widetilde{L}_{p'}\| \leq \|u \mid L_{p'}(\Omega_0)\| + K_1\|l^{\omega_\tau}_{h} \mid L^{(M)*}_2\|,$$

with K_1 a constant independent of h and ω_τ and $u(y)$ a solution to the differential equation

$$(1-\Delta)^{m/2}u(y) = l^{\omega_\tau}_{h}(y).$$

We estimate the $\widetilde{L}_{p'}(\Omega_0)$ norm of $u(y)$ by using (4.2.15), (4.2.22), and (4.2.29). This leads to the inequality

$$\|u \mid L_{p'}(\Omega_0)\| \leq Kh^m,$$

with K a constant independent of h and ω_τ. Arguing further in much the same way as in the proof of Theorem 4.7, we easily find the estimate

$$\|l^{\omega_\tau}_{h} \mid L^{(M)*}_2\| \leq K\left(\frac{|\omega_\tau|}{|Q\tau/\overline{h}|}\right)^{1/2} h^M,$$

with K again a constant independent of h and ω_τ. Whence and from the condition $M > M_2 \geq m$, we conclude that $\widetilde{W}^{(m)}_p(\Omega_0)^*$ norm of $l^{\omega_\tau}_{h}(y)$ is in fact $O(h^m)$ uniformly in ω_τ, $0 < \tau < 1$. Letting $h = \tau$ and $p = q$, we thus come to (1.12).

The proof of Lemma 6.5 is complete.

Take $\overline{h} = (h_1, h_2, \ldots, h_n)$, $h_j > 0$. To this mesh-size-vector we assign the error

$$l_{\overline{h}}(x) = \chi_Q(x) - h_1 \ldots h_n \sum_{I(\overline{h})\beta \in Q} \delta(x - I(\overline{h})\beta). \tag{1.13}$$

In the case of $h_1 = \ldots = h_n = h$ we let $l_{\overline{h}}(x)$ coincide with $l_\infty(x)$.

Lemma 6.6. *Given $\overline{m} = (m_1, \ldots, m_n)$, assume that the minimum of m_j is greater than n. Then the $\widetilde{W_2^{\overline{m}}}{}^*$ norm of $l_{\overline{h}}$ is $O(\max_j h_j^{m_j})$ and $l_{\overline{h}}$ is asymptotically optimal in $\widetilde{W_2^{\overline{m}}}$.*

PROOF. Let the error $l^*_{\overline{h}}(x)$ take the shape

$$l^*_{\overline{h}}(x) = \chi_Q(x) - h_1 \ldots h_n \sum_{I(\overline{h})\beta \in Q} c_\beta(\overline{h}) \delta(x - I(\overline{h})\beta)$$

and belong to $\widetilde{W_2^{\overline{m}}}{}^*$. In particular, the value of $l^*_{\overline{h}}(x)$ at the constantly-one function is then zero. In other words, the weights $c_\beta(\overline{h})$ satisfy the condition

$$h_1 \ldots h_n \sum_{I(\overline{h})\beta \in Q} c_\beta(\overline{h}) = 1. \tag{1.14}$$

The $\widetilde{W_2^{\overline{m}}}{}^*$ norm of $l^*_{\overline{h}}(x)$ is defined as

$$\| l^*_{\overline{h}} \mid \widetilde{W_2^{\overline{m}}}{}^* \| = \left\{ \sum_{\gamma \neq 0} \frac{|L^*_{\overline{h}}[\gamma]|^2}{|\mu(\gamma)|^2} \right\}^{1/2}.$$

Here

$$\mu(\xi) = \left\{ \sum_{j=1}^{n} (1 + \xi_j^2)^{m_j} \right\}^{1/2},$$

and $L^*_{\overline{h}}[\gamma]$ stands for the Fourier coefficient of $l^*_{\overline{h}}(x)$, i.e. the value of the error at the exponential $e^{i2\pi\gamma x}$, $\gamma \in \mathbb{Z}^n$. If the entry γ_j of γ is a multiple of $N_j = 1/h_j$ for all j, then $L^*_{\overline{h}}[\gamma]$ must equal to -1 by (1.14). This enables us to assert that

$$\| l^*_{\overline{h}} \mid \widetilde{W_2^{\overline{m}}}{}^* \| = \left\{ \sum_{\beta \neq 0} \left[\sum_{j=1}^{n} \left[1 + \left(\frac{\beta_j}{h_j} \right)^2 \right]^{m_j} \right]^{-1} + R(\overline{m}, \overline{h}, c_\beta(\overline{h})) \right\}^{1/2}.$$

The summands $R(\overline{m}, \overline{h}, c_\beta(\overline{h}))$ on the right side are always nonnegative. If $c_\beta(\overline{h}) = 1$ for all $\beta \in \mathbb{Z}^n$ satisfying $I(\overline{h})\beta \in Q$ then $l^*_{\overline{h}}(x)$ coincides with $l_{\overline{h}}(x)$. Moreover, the Fourier coefficient $L^*_{\overline{h}}[\gamma]$ vanishes provided that at least for one of the j the entry γ_j is not a multiple of $N_j = 1/h_j$. Consequently, the summand $R(\overline{m}, \overline{h}, c_\beta(\overline{h}))$ vanishes in the case of $l_{\overline{h}}(x)$. Thus, the $\widetilde{W}_2^{\overline{m}*}$ norm of the error $l_{\overline{h}}(x)$ has the shape

$$\|l_{\overline{h}} \mid \widetilde{W}_2^{\overline{m}*}\| = \left\{ \sum_{k \neq 0} \left[\sum_{j=1}^{n} \left[1 + \left(\frac{k_j}{h_j} \right)^2 \right]^{m_j} \right]^{-1} \right\}^{1/2},$$

i.e., this norm is $O(\max_j h_j^{m_j})$. Furthermore, the $\widetilde{W}_2^{\overline{m}*}$ norm of $l_{\overline{h}}(x)$ is not greater than the norm of an arbitrary error $l^*_{\overline{h}}(x)$ with the same mesh-size-vector $\overline{h}$. Consequently, $l_{\overline{h}}(x)$ is asymptotically optimal in $\widetilde{W}_2^{\overline{m}}$ as $|\overline{h}| \to 0$.

The proof of Lemma 6.6 is complete.

Corollary 1.2. *For $m > n$ the error $l_{\overline{h}}(x)$ is asymptotically optimal in $\widetilde{W}_2^{(m)}$ as $|\overline{h}| \to 0$.*

To demonstrate, let $m_1 = \ldots = m_n = m$ under the hypotheses of Lemma 6.6.

Theorem 6.3. *Let Ω be a domain with piecewise-smooth boundary which lies strictly in the interior of Q. Let* BBL*-error $l^\Omega_{\overline{h}}(x)$ have an optimal mode of convergence in the isotropic spaces $\widetilde{W}_p^{(m)}$ for some $p < 2$ and $m = M_1, M_2$, where $M_2 > M_1 > n$, M_1 and M_2 are even, i.e., the estimates hold*

$$\|l^\Omega_{\overline{h}} \mid \widetilde{W}_p^{(m)*}\| \leq K \max_{1 \leq j \leq n} h_j^m, \quad 1 \leq p < 2. \tag{1.15}$$

Then $l^\Omega_{\overline{h}}(x)$ is asymptotically $\widetilde{W}_2^{\overline{m}}$-optimal for $m_j \in (M_1, M_2)$.

PROOF. In the hypotheses of the theorem we may assume that $p > 1$. If $p = 1$ then, in view of the embedding of $\widetilde{W}_p^{(m)}$ in $\widetilde{W}_1^{(m)}$, the estimate (1.12) is valid for all $p > 1$. Without loss of generality, we may thus suppose that $1 < p < 2$. By Corollary 1.1 the estimate (1.12) remains valid for every m in $[M_1, M_2]$, in particular, for m equal to some M_3 between $\max_j m_j$ and M_2.

We proceed with proof along the lines of Theorem 6.1. Let

$$\mu^2(\xi) = \sum_{j=1}^{n} (1 + \xi_j^2)^{m_j}, \quad \|l \mid \widetilde{H}_2^{\mu*}\| = \|l\|_{\overline{m}}, \quad \|l_{\overline{h}}\|_{\overline{m}} = \psi(\overline{h}).$$

Denote by $l^{\Omega,0}_{\overline{h}}(x)$ the error supported in the closure of Ω, whose $\widetilde{W}_2^{\overline{m}*}$ norm is minimal among the $\widetilde{W}_2^{\overline{m}*}$ norms of all errors $l^*_{\overline{h}}(x)$ of the shape

$$l^*_{\overline{h}}(x) = \chi_\Omega(x) - h_1 \ldots h_n \sum_{I(\overline{h})\beta \in \Omega} c_\beta(\overline{h}) \delta(x - I(\overline{h})\beta).$$

If P_1 stands for the projection in $\widetilde{W_2^{\overline{m}*}}$ to the subspace composed of the linear combinations

$$\sum_{I(\overline{h})\beta\in\Omega} a_k\delta(x - I(\overline{h})\beta),$$

then we clearly have $l_{\overline{h}}^{\Omega,0}(x) = \chi_\Omega(x) - P_1\chi_\Omega(x)$. Moreover,

$$\|l_{\overline{h}}^{\Omega}\|_{\overline{m}}^2 - \|l_{\overline{h}}^{\Omega,0}\|_{\overline{m}}^2 = \|l_{\overline{h}}^{\Omega} - l_{\overline{h}}^{\Omega,0}\|_{\overline{m}}^2. \tag{1.16}$$

Show that the right side of the above equality is $o(\psi(\overline{h}))$ as $|\overline{h}| \to 0$.

To this end, take an arbitrary $\varepsilon > 0$ and consider the sets

$$\Gamma_\varepsilon = \{x : x \in \Omega_0 \setminus \Omega,\ \rho(x, \Omega) \leq \varepsilon\}, \quad \omega_\varepsilon = \Omega_0 \setminus (\Omega \cup \Gamma_\varepsilon).$$

To the set ω_ε we assign the homogeneous error

$$l_{\overline{h}}^{\omega_\varepsilon}(x) = \sum_{I(\overline{h})\beta\in\omega_\varepsilon} l_0\left(x/\overline{h} - \beta\right)$$

of degree $M \geq 2M_2 - n + 1$, and consider the difference

$$l_{\overline{h}}^{\Gamma_\varepsilon} \equiv l_{\overline{h}} - l_{\overline{h}}^{\Omega} - l_{\overline{h}}^{\omega_\varepsilon}.$$

By the hypotheses of Theorem 6.3, $l_{\overline{h}}^{\Omega}$ is a BBL-error. Therefore, the volume of the support of $l_{\overline{h}}^{\Gamma_\varepsilon}$ is $O(\varepsilon)$ as $\varepsilon \to 0$. By Lemma 6.5, Corollary 1.2, and the estimates (1.15) the $W_p^{(m)*}$ norm of each of the functionals $l_{\overline{h}}$, $l_{\overline{h}}^{\Omega}$, and $l_{\overline{h}}^{\omega_\varepsilon}$ is $O(\max_j h_j^m)$ for $m = M_1, M_2, M_3$. Consequently, this holds true also for $l_{\overline{h}}^{\Gamma_\varepsilon}(x)$. Whence and from Theorem 4.5, for $m = M_1, M_2$, obtain

$$\|l_{\overline{h}}^{\Gamma_\varepsilon} \mid \widetilde{W_2^{(m)*}}\| \leq K\varepsilon^{1/p-1/2} \max_j h_j^m(1 + o(1)).$$

Applying Lemma 6.4 with $p = 2$ and $|\overline{h}| \to 0$ to $l_{\overline{h}}^{\Gamma_\varepsilon}$, infer the estimate

$$\|l_{\overline{h}}^{\Gamma_\varepsilon} \mid \widetilde{W_2^{\overline{m}*}}\| \leq K\varepsilon^{1/p-1/2} \max_j h_j^{m_j}(1 + o(1)).$$

Revert to estimating the norm in the right side of (1.16). The definition of P_1 implies that

$$\|l_{\overline{h}}^{\Omega} - l_{\overline{h}}^{\Omega,0}\|_{\overline{m}} = \|P_1 l_{\overline{h}}^{\Omega}\|_{\overline{m}}. \tag{1.17}$$

Decompose $l_{\overline{h}}^{\Omega}$ into the sum of $l_{\overline{h}}(x)$, $l_{\overline{h}}^{\omega_\varepsilon}(x)$, and $l_{\overline{h}}^{\Gamma_\varepsilon}$ to obtain

$$\|P_1 l_{\overline{h}}^{\Omega}\|_{\overline{m}} \le \|P_1 l_{\overline{h}}\|_{\overline{m}} + \|P_1 l_{\overline{h}}^{\Gamma_\varepsilon}\|_{\overline{m}} + \|P_1 l_{\overline{h}}^{\omega_\varepsilon}\|_{\overline{m}} = \mathrm{I} + \mathrm{II} + \mathrm{III}. \qquad (1.18)$$

We estimate I and II from above, arguing as in the demonstration of upper bounds on I and II in Theorem 6.1. Doing so, we should bear in mind that the error $l_{\overline{h}}(x)$ is asymptotically optimal in $\widetilde{W}_2^{\overline{m}*}$ by Lemma 6.6. We thus infer

$$\mathrm{I} + \mathrm{II} \le o(\psi(\overline{h})) + \|l_{\overline{h}}^{\Gamma_\varepsilon}\|_{\overline{m}} \le K\varepsilon^{1/p-1/2} \max_j h_j^{m_j}(1 + o(1)).$$

Writing down an expression analogous to (1.4), we conclude that

$$\mathrm{III} \le \|l_{\overline{h}}^{\omega_\varepsilon} \mid W_2^{\overline{m}}(\Omega)^*\|.$$

To estimate the norm in the right side, use an analogous inequality to (1.5), arriving at

$$\|l_{\overline{h}}^{\omega_\varepsilon} \mid W_2^{\overline{m}}(\Omega)^*\| \le \|l_{\overline{h}}^{\omega_\varepsilon} \mid \widetilde{W}_2^{(M_2)*}\|.$$

By the hypotheses of Theorem 6.3 M_2 is even and greater than the maximum of the numbers m_j. Consequently, we may apply Lemma 6.5 to estimating the $\widetilde{W}_2^{(M_2)*}$ norm of $l_{\overline{h}}^{\omega_\varepsilon}$. We hence obtain

$$\mathrm{III} \le K \max_j h_j^{M_2} = o(\max_j h_j^{m_j}).$$

Summing the bounds on I, II, and III in (1.18) and using (1.16)–(1.17), we finally arrive at

$$\begin{aligned} &\left\{\|l_{\overline{h}}^{\Omega}\|_{\overline{m}}^2 - \|l_{\overline{m}}^{\Omega,0}\|_{\overline{m}}^2\right\}^{1/2} \\ &\le \|l_{\overline{h}}^{\Omega} - l_{\overline{h}}^{\Omega,0}\|_{\overline{m}} \le K\varepsilon^{1/p-1/2} \max_j h_j^{m_j} + o(\max_j h_j^{m_j}). \end{aligned}$$

Since ε is arbitrary, it follows that the left side is

$$o(\max_j h_j^{m_j}) = o(\psi(\overline{h})).$$

Whence and from Lemma 6.6 it follows that $l_{\overline{h}}^{\Omega}(x)$ is asymptotically $\widetilde{W}_2^{\overline{m}}$-optimal. The proof of Theorem 6.3 is complete.

For anisotropic spaces, the formula (1.6) for the norm of an asymptotically optimal error remains valid

$$\|l_{\overline{h}}^{\Omega,a} \mid \widetilde{W}_2^{\overline{m}*}\| = |\Omega|^{1/2}\|l_{\overline{h}} \mid \widetilde{W}_2^{\overline{m}*}\|(1 + o(1)). \qquad (1.19)$$

Optimize the right side of this formula over all admissible $\overline{h}$.

Theorem 6.4. *A mesh-size-vector $\overline{h}_0$ optimal with respect to the rate of convergence for the space $\widetilde{W}_2^{\overline{m}}$ is determined from the equalities*

$$h_1^{m_1} = \cdots = h_n^{m_n},$$

provided only that the conditions (1.15) and (1.9) agree. Moreover, we have the equality

$$\| l_{\overline{h}_0}^{\Omega,a} \mid \widetilde{W}_2^{\overline{m}*} \| = |\Omega|^{1/2} N^{-1/\sum_{j=1}^{n} 1/m_j} \left\{ \sum_{k \neq 0} \left[\sum_{j=1}^{n} k_j^{2m_j} \right]^{-1} \right\}^{1/2} (1 + o(1)),$$

with $N = N_1 \dots N_n$ standing for the number of lattice nodes in the cube Q.

PROOF. Optimal among the lattices (1.8) and (1.9) is that to which there corresponds the fastest decrease of $\max_j h_j^{m_j}$ as the number of nodes of a cubature formula increases infinitely.

The number of nodes grows like $|\Omega|/(h_1 \dots h_n) = O(1/(h_1 \dots h_n))$. Consequently, those $h_1, \dots, h_n$ are best that provide the minimum of $\max_j h_j^{m_j}$ for $h_1 \dots h_n = 1/N$. Thus, we come to the problem: find $h_{10}, \dots, h_{n0}$ satisfying the condition

$$\max_j h_{j0}^{m_j} = \min_{h_1 \dots h_n = N^{-1}} \max_j h_j^{m_j}.$$

If at least one of the numbers $h_{j0}^{m_j}$ were less than the others then we would diminish all the rest of them by slightly increasing $h_{j0}^{m_j}$. We thus would diminish $\min(\max_j h_j^{m_j})$, which is impossible. Consequently, all $h_{j0}^{m_j}$ are equal. Thus,

$$h_{j0}^{m_1} = \cdots = h_{j0}^{m_n} = \tau, \quad N_j = 1/h_{j0} = \tau^{-1/m_j}, \quad N = N_1 \dots N_n = \tau^{-\sum_{j=1}^{n} 1/m_j}.$$

Therefore,

$$\tau = N^{-1/\left(\sum_{j=1}^{n} 1/m_j\right)}.$$

The proof of Theorem 6.4 is complete.

In the above example, the norm of the error of the cubature formula with equal weights in $\widetilde{W}_2^{3,4}$ is

$$O(N^{-1/(1/3+1/4)}) = O(N^{-12/7})$$

in the case of the best choice of h_1 and h_2.

Observe also that the requirement of agreement between (1.9) and (1.15) in Theorem 6.4 is of technical character. It may be eliminated.

§2. Cubature Formulas with Bounded Boundary Layer in Hölder Spaces

We intend to translate the theory of asymptotic optimality of lattice cubature formula on the classes of functions continuous with derivatives up to some order. We, however, encounter difficulties in exactly defining the normed spaces of these functions. To clarify the matter, we revert to the theory of differential equations.

The solvability theory for elliptic equations take a perfect shape not in the classical spaces of continuously differentiable functions but rather in the Hölder classes $(\widetilde{C}^\gamma)^m$. Here m is a natural, and the exponent $\gamma > 0$ enters in the Hölder condition which must be met by the higher derivatives of order m of a function in $(\widetilde{C}^\gamma)^m$. This relates to the problem of existence of multipliers of the Fourier transform which are available in $(\widetilde{C}^\gamma)^m$ only for $\gamma > 0$.

In our calculations we lean upon Bessel scales of spaces whose definition is tightly woven with the concept of multiplier and, therefore, we work in Hölder classes.

Deeper reasons for absence of asymptotically optimal cubature formulas in the classical spaces $C^{(m)}$ consist probably in the fact that the unit ball of such space is not strictly convex. This complicates the optimization problem, since the extremal function of an error may fail to exist or to be unique.

We deal with one of the possible equivalent norms of a space. We introduce a bulky but technically convenient norm of the index-zero space generating the Bessel scale of the Hölder spaces.

Let $\widetilde{B}_0 = \widetilde{C}^\gamma$ comprise the periodic functions with period matrix H which are the limits of finite Fourier series

$$f(x) = \sum_\beta f[\beta] e^{i2\pi\beta H^{-1}x}$$

in the norm

$$\|f \mid \widetilde{C}^\gamma\| = \max\left(|f[0]|, \max_x \mathscr{H}^\gamma \sum_\beta (1 + |H^{-1*}\beta|^2)^{-\gamma/2} f[\beta] e^{i2\pi\beta H^{-1}x} \right),$$

with $0 < \gamma < 1$ and

$$(\mathscr{H}^\gamma g)(x) = \sup_y \frac{|g(x+y) - g(x)|}{|y|^\gamma}$$

standing for the Hölder constant of order γ at a point x. It is well known that for all $\gamma \in (0,1)$ and $m > 0$ the spaces $\widetilde{B}_0^m \equiv \widetilde{C}^m$ are the Hölder classes of smoothness m. Observe that the class $\widetilde{C}^m$ with m an integer is close to but not coincident with the class $C^{(m)}$ of m times continuously differentiable functions.

We further require, considering $h \to 0$, not only that $1/h$ be an integer number but also existence of a sequence $h_0 \to 0$ with $1/h_0$ and h_0/h both integers and a possibility that h vanishes at h_0 constant. Such sequence is, for instance, $h_j = 2^{-j}$ in which case $h_0 = 2^{-j_0}$.

Arrange the homogeneous error of degree M that corresponds to the unit cube Q shrank in $1/h_0$ times, i.e., put

$$l_h^{h_0}(x) = \sum_{hk \in h_0 Q} l_0 \left(\frac{x}{h} - k \right).$$

Derive the estimate

$$\left\| l_h^{h_0} \mid \widetilde{C}^{m*} \right\| \le h_0^n \left\| l_\infty \mid \widetilde{C}^{m*} \right\| (1 + o(1))$$

as $h \to 0$. The main obstacle to this consists in the fact that the simplest transformation $x = h_0 y$ sending $l_h^{h_0}$ to l_∞ does not preserve the periodic space of test functions. To obviate this difficulty, given $l_h^{h_0}$, in $\widetilde{C}^m(Q)$ define a sequence of asymptotically norming functions $\Phi_j(x, h)$ with period matrix $h_0 I$ which possess the properties

$$\left\| \Phi_j \mid \widetilde{C}^m \right\| = 1, \quad \lim_{h \to 0} \lim_{j \to \infty} \left\| l_h^{h_0} \mid \widetilde{C}^{m*} \right\| / \left| (l_h^{h_0}, \Phi_j) \right| = 1.$$

If the functions $\Phi_j(x, h)$ are available then by their periodicity with matrix I we have

$$\left| (l_h^{h_0}, \Phi_j) \right| = h_0^n \left| (l_\infty, \Phi_j) \right| (1 + o(1)) \le h_0^n \left\| l_\infty \mid \widetilde{C}^{m*} \right\| (1 + o(1)).$$

Whence and from the definition of $\{\Phi_j\}$ we infer a desired estimate for the norm of $l_h^{h_0}$ in $\widetilde{C}^{m*}$.

Before constructing the needed functions $\Phi_j(x, h)$, establish an additional property of a homogeneous error.

Consider an infinitely differentiable function ζ supported in the closure of Q. Let ζ depend on h so that

$$\left| D_x^\alpha \zeta(x, h) \right| \le K_\alpha |\log h|^{|\alpha|}, \quad |\alpha| \ge 0, \quad 0 < h \le 1.$$

Then we call $\zeta(x)$ a *function of logarithmic truncation* or, in short, a *logarithmic truncator.*

At homogeneous errors l_h^ω, the operator $(1 - \Delta)^m$ and the operator of multiplication by a logarithmic truncator commute to within summands presenting higher order infinitesimals as $h \to 0$. We state our claim exactly. Agree to denote by

$$[(1 - \Delta)^m, \zeta] = (1 - \Delta)^m \zeta - \zeta (1 - \Delta)^m$$

the commutator of $(1 - \Delta)^m$ and the operator of multiplication by ζ. We have the following

Lemma 6.7. *Assume that $l_h^\omega(x)$ is a homogeneous error of degree M, and $\zeta(x,h)$ is a logarithmic truncator. Then as $h \to 0$, for all numbers m_1, m_2, and m_3, $m_j > n$, satisfying the condition $n < 2(m_3 - m_1 - m_2) < M - 1$, we have the relation*

$$\begin{aligned}\big\|(1-\Delta)^{m_1}[(1-\Delta)^{m_2},\zeta](1-\Delta)^{-m_3}l_h^\omega \mid \widetilde{L}_2\big\| \\ = o\big(\big\|(1-\Delta)^{m_1+m_2-m_3}l_h^\omega \mid \widetilde{L}_2\big\|\big).\end{aligned} \tag{2.1}$$

PROOF. By Theorems 6.2 and 4.3, given m in the interval (n, M), observe

$$\big\|(1-\Delta)^{-m/2}l_h^\omega \mid \widetilde{L}_2\big\| = \big\|l_h^\omega \mid \widetilde{W}_2^{(m)*}\big\| \geq Ch^m.$$

It thus suffices to prove that the left side of (2.1) is $o\big(h^{2(m_3-m_1-m_2)}\big)$.

If $z_1(x)$ and $z_2(x)$ are periodic functions, $z_1 \in \widetilde{C}^{(\infty)}$ and $z_2 \in \widetilde{W}_2^{(m)}$, then for all $\varepsilon \in [0, m-n/2)$ and $2r \in (-1, m+1]$ the inequality holds

$$\begin{aligned}\big\|[(1-\Delta)^r, z_1]z_2 \mid \widetilde{L}_2\big\| \leq C_r \sum_{1\leq|\alpha|\leq[|2r|]} \max_x |D^\alpha z_1(x)|\, \big\|z_2 \mid \widetilde{W}_2^{(2r-|\alpha|)}\big\| \\ + C_\varepsilon \sum_{|\alpha|=[|2r|]+1} \max_x |D^\alpha z_1(x)|\, \big\|z_2 \mid \widetilde{W}_2^{(m-\varepsilon)}\big\|.\end{aligned} \tag{2.2}$$

For $[|2r|] = 0$, the first sum on the right side of (2.2) is absent.

To justify (2.2), express the result of the action of the commutator at z_2 through the Fourier coefficients. We have

$$(1-\Delta)^r(z_1(x)z_2(x)) = \sum_k (1+|2\pi k|^2)^r \sum_s z_1[k-s]z_2[s]e^{i2\pi kx}.$$

Move the factor $(1+|2\pi k|^2)^r$ under the summation sign over s and expand it in the Taylor formula at the point s with remainder of degree $[|2r|]+1$ in integral form. To the first term of the Taylor series there corresponds $z_1(x)(1-\Delta)^r z_2(x)$, to all others taken together, the result of the action of the commutator at $z_2(x)$. The summands corresponding to the Taylor series may be written as

$$\sum_{1\leq|\alpha|\leq[|2r|]} \frac{D^\alpha z_1(x)}{\alpha!} \sum_s (D_s^\alpha(|2\pi s|^2+1)^r) z_2[s]e^{i2\pi sx}.$$

The respective $\widetilde{L}_2$ norm is easy to estimate as follows

$$C_r \sum_{1\leq|\alpha|\leq[|2r|]} \max_x |D^\alpha z_1(x)|\, \big\|z_2 \mid \widetilde{W}_2^{(|2r|-|\alpha|)}\big\|.$$

To the summand related to the remainder of the Taylor series, there corresponds the sum over α, $|\alpha| = [|2r|] + 1$, of the integrals

$$\int_0^1 \frac{(1-t)^{[|2r|]}}{\alpha!} \sum_{k,s} (D^\alpha(|2\pi(s+t(k-s))|^2+1)^r)(k-s)^\alpha z_1[k-s]z_2[s]e^{i2\pi kx}\,dt.$$

Using the Parseval identity, find an upper bound for the norm of this sum in $\widetilde{L}_2$. It is easy to see that a sought dominant is as follows

$$\mathrm{I} = \max_{0\le t\le 1} \sum_{|\alpha|=[|2r|]+1} \left(\sum_k \left| \sum_s z_1[k-s](k-s)^\alpha z_2[s]\psi(t,\alpha,k,s) \right|^2 \right)^{1/2},$$

with

$$\psi(t,\alpha,k,s) = D_\xi^\alpha(|2\pi\xi|^2+1)^r, \quad \xi = s+t(k-s), \quad |\alpha| = [|2r|]+1.$$

Clearly, the function ψ is bounded uniformly in t, α, k, and s. Move the l_2 norm over k under the summation sign over s in the expression I. Estimating $|\psi|$ with a constant, obtain

$$\begin{aligned} \mathrm{I} &\le C_2 \sum_{|\alpha|=[|2r|]+1} \sum_s |z_2[s]| \left(\sum_k |z_1[k]k^\alpha|^2 \right)^{1/2} \\ &\le C_3 \sum_{|\alpha|=[|2r|]+1} \max_x |D^\alpha z_1(x)| \sum_s |z_2[s]|. \end{aligned}$$

Since $m - \varepsilon > n/2$; therefore,

$$\begin{aligned} \sum_s |z_2[s]| &\le \left\{ \sum_s |z_2[s]|^2(1+|2\pi s|^2)^{m-\varepsilon} \right\}^{1/2} \left\{ \sum_s \frac{1}{(1+|2\pi s|^2)^{m-\varepsilon}} \right\}^{1/2} \\ &\le C_\varepsilon \| z_2 \mid \widetilde{W}_2^{(m-\varepsilon)} \|. \end{aligned}$$

Thus, (2.2) is established.

It is easy that the norm of the left side of (2.1) is less than or equal to the sum of two terms each of which is a $\widetilde{L}_2$ norm similar to the left side of (2.2). In more detail, those are the $\widetilde{L}_2$ norms as follows: first, with

$$r = m_1 + m_2, \quad z_1 = \zeta, \quad z_2 = (1-\Delta)^{-m_3} l_h^\omega, \quad m = 2r - 1$$

and, second, with

$$r = m_1, \quad z_1 = \zeta, \quad z_2 = (1-\Delta)^{m_2-m_3} l_h^\omega, \quad m = 2r-1.$$

In each of the two cases, obtain the dominant

$$K|\log h|^{2(m_1+m_2)+1} \| l_h^\omega \mid \widetilde{W}_2^{(2(m_3-m_1-m_2)+1)*}\|$$

which is $o(h^{2(m_3-m_1-m_2)})$ by Corollary 4.2.2. Using this, we estimate the norm of the commutator in $\widetilde{L}_2$. The proof of Lemma 6.7 is complete.

Lemma 6.8. *For every $m \in (n, M)$ the inequality holds*

$$\|l_h^{h_0} \mid \widetilde{C}^{m*}\| \le h_0^n \|l_\infty \mid \widetilde{C}^{m*}\|(1+o(1)) \tag{2.3}$$

as $h \to 0$.

PROOF. Taking a function $\varphi(\tau) \in C^{(\infty)}(\mathbb{R})$ equal 0 for $\tau < 0$ and equal 1 for $\tau > 1$, arrange a logarithmic truncator $\zeta_h^{h_0}(x)$ by letting

$$\zeta_h^{h_0}(x) = \prod_{j=1}^{n} \varphi(x_j|\log h|)\varphi((h_0 - x_j)|\log h|).$$

Clearly, $\zeta_h^{h_0}(x)$ equals 0 beyond h_0Q, and at the interior points of h_0Q at a distance greater than $1/|\log h|$ from the boundary, the function $\zeta_h^{h_0}(x)$ equals 1.

Show that the norm of

$$l_h^{h_0}(x) - (1-\Delta)^m \zeta_h^{h_0}(x)(1-\Delta)^{-m} l_\infty(x)$$

is negligibly small as compared with the right side of (2.3).

In further proving there will appear several summands that are $o(h^m)$. We neglect them since

$$\|l_\infty \mid \widetilde{C}^m\| \ge K h^m,$$

with K independent of h. We now validate the above estimate for l_∞.

To this end, take some infinitely differentiable function $\zeta(x)$ supported in Q and put

$$\psi(x) = \sum_{hk \in Q} \zeta\left(\frac{x}{h} - k\right).$$

We then have the inequality $\|\psi \mid \widetilde{C}^m\| \le K_3 h^m$ and hence the estimate

$$\|l_\infty \mid \widetilde{C}^m\| \ge \frac{\left|\int_Q \psi(x)\,dx\right|}{\|\psi \mid \widetilde{C}^m\|} \ge K h^m \int \zeta(y)\,dy.$$

We now have

$$\begin{aligned}&\big\|l_h^{h_0}(x)-(1-\Delta)^m\zeta_h^{h_0}(x)(1-\Delta)^{-m}l_\infty(x)\mid \widetilde{C}^{m*}\big\|\\ &\le K\big\|(1-\Delta)^{-m/2}l_h^{h_0}(x)-(1-\Delta)^{m/2}\zeta_h^{h_0}(x)(1-\Delta)^{-m}l_\infty(x)\mid \widetilde{L}_2\big\|.\end{aligned}$$

Replace the second term of the difference under the $\widetilde{L}_2$ norm sign with

$$\zeta_h^{h_0}(x)(1-\Delta)^{-m/2}l_\infty(x),$$

on commuting $(1-\Delta)^{m/2}$ and $\zeta_h^{h_0}(x)$. Then bound the whole norm of the difference by

$$\begin{aligned}&\big\|(1-\zeta_h^{h_0}(x))(1-\Delta)^{-m/2}l_h^{h_0}(x)\mid \widetilde{L}_2\big\|\\ &+\big\|\zeta_h^{h_0}(x)(1-\Delta)^{-m/2}\big(l_\infty(x)-l_h^{h_0}(x)\big)\mid \widetilde{L}_2\big\|.\end{aligned}\tag{2.4}$$

Both norms in (2.4) are estimated similarly. We thus estimate the first.

Given an arbitrary $\varepsilon>0$, split the error $l_h^{h_0}(x)$ into the two summands

$$l_h^{h_0}(x)=l_h^{\omega_\varepsilon}(x)+l_h^{\Gamma_\varepsilon}(x).$$

The first of them only contains local errors $l_0(x/h-k)$ such that the point hk is at a distance greater than ε from the support of $1-\zeta_h^{h_0}(x)$. Thus, for h sufficiently small, the support of $l_0(x/h-k)$ is at a distance less than $\varepsilon/2$ from the support of $1-\zeta_h^{h_0}(x)$. The rest of the local errors $l_0(x/h-k)$ sum up to the generalized function $l_h^{\Gamma_\varepsilon}(x)$. It is clear that, as $\varepsilon\to 0$, the volume of the domain Γ_ε also vanishes. Considering this, we see that the first summand of (2.4) is bounded by

$$\big\|l_h^{\Gamma_\varepsilon}\mid \widetilde{W}_2^{(m)*}\big\|+\sup_f\frac{\big|\big(l_h^{\omega_\varepsilon}(x),(1-\Delta)^{-m/2}(1-\zeta_h^{h_0}(x))f(x)\big)\big|}{\|f\mid \widetilde{L}_2\|}.\tag{2.5}$$

By Remark 3 on Theorem 6.2, there is a constant K such that for all $\varepsilon>0$ the inequality holds

$$\big\|l_h^{\Gamma_\varepsilon}\mid \widetilde{W}_2^{(m)*}\big\|\le K\sqrt{\varepsilon}\,h^m\big(1+o(1)\big).$$

Consider the second summand of (2.5). Denote by ω the set of x such that

$$1-\zeta_h^{h_0}(x)=0.$$

Given a function $f\in\widetilde{L}_2$, put

$$u(x)=(1-\Delta)^{-m/2}\big(1-\zeta_h^{h_0}(x)\big)f(x).$$

In ω this function is a solution to the equation

$$(1-\Delta)^{m/2}u(x)=0.$$

Consequently, $u(x)$ is infinitely differentiable in ω. Considering that ω is at a distance at least $\varepsilon/2$ from the support of the generalized function $l_h^{\omega_\varepsilon}$ in question, obtain the estimate

$$\|u \mid \widetilde{W}_2^{(M)}(\operatorname{supp} l_h^{\omega_\varepsilon})\| \le K_2(\varepsilon)\|u \mid \widetilde{W}_2^{(m)}\| \le K_2(\varepsilon)\|f \mid \widetilde{L}_2\|.$$

Thus, the total of (2.5) does not exceed

$$K_1\varepsilon^{1/2}h^m + K_2(\varepsilon)\|l_h^{\omega_\varepsilon} \mid \widetilde{W}_2^{(M)*}(\operatorname{supp} l_h^{\omega_\varepsilon})\| = K_1\sqrt{\varepsilon}\,h^m + K_3(\varepsilon)h^M.$$

By the arbitrariness of ε, this sum is $o(h^m)$.

To estimate the second summand in (2.4), let $\omega = Q\setminus h_0Q$, and take $l_h^{\omega_\varepsilon}$ as the sum of elementary errors $l_0(x/h-k)$ such that the points hk lie at a distance greater than ε from the support of $\zeta_h^{h_0}(x)$. As a result, infer that (2.4) is $o(h^m)$.

Thus, to obtain (2.3) it suffices to write down the norm in $\widetilde{C}^{m*}$ of the following generalized function

$$(1-\Delta)^m\zeta_h^{h_0}(x)(1-\Delta)^{-m}l_\infty(x).$$

Subtracting from the latter the generalized function

$$(1-\Delta)^{m-\gamma/2}\zeta_h^{h_0}(x)(1-\Delta)^{-(m-\gamma)/2}l_\infty(x),$$

observe that the $\widetilde{C}^{m*}$ norm of this difference is bounded by

$$\|(1-\Delta)^{-\gamma/2}\left[(1-\Delta)^{m+\gamma/2},\zeta_h^{h_0}(x)\right](1-\Delta)^{-m}l_\infty(x) \mid \widetilde{L}_2\|,$$

and so, by Lemma 6.5, it is $o(h^m)$. Put

$$\mathrm{I} = \|(1-\Delta)^{m-\gamma/2}\zeta_h^{h_0}(x)(1-\Delta)^{-(m-\gamma)/2}l_\infty(x) \mid \widetilde{C}^{m*}\|$$

and continue estimating this norm. By definition, we have

$$\mathrm{I} = \sup_{f\in\widetilde{C}^\gamma}\frac{|(\zeta_h^{h_0}(x)(1-\Delta)^{-(m-\gamma)/2}l_\infty(x), f(x))|}{\max(|f[0]|, \max_x \mathscr{H}^\gamma f(x))},$$

with

$$\mathscr{H}^\gamma f(x) = \sup_y \frac{|f(x+y)-f(x)|}{|y|^\gamma}.$$

Take a sequence of functions $f^{(j)}(x)$ with the $\widetilde{C}^\gamma$ norm equal 1. Assume further that this sequence has the property

$$\mathrm{I} = \lim_{j\to\infty} \left| \left(\zeta_h^{h_0}(x)(1-\Delta)^{-(m-\gamma)/2} l_\infty(x), f^{(j)}(x) \right) \right|.$$

The generalized function for which we examine the limit of its values at $f^{(j)}(x)$ is symmetric with respect to the coordinate planes passing through the point $(h_0/2, \ldots, h_0/2)$. Consequently, we may also assume that the sequence $f^{(j)}(x)$ has the same property.

Show that $f^{(j)}(x)$ may be replaced with functions which in addition to the above-listed properties of $f^{(j)}(x)$ are periodic with period matrix $h_0 I$.

Let $\Phi^{(j)}(x) = f^{(j)}(x - h_0 k)$ for $x \in h_0(Q+k)$. This is a periodic function with period matrix $h_0 I$. In view of the symmetry of $f^{(j)}(x)$, the function $\Phi^{(j)}(x)$ is continuous. Since $f^{(j)}(x)$ satisfies the Hölder condition with exponent $\gamma < 1$ and Hölder constant $\max_x \mathscr{H}^\gamma f^{(j)}(x)$; therefore, $\Phi^{(j)}(x)$ meets the same Hölder condition and, moreover,

$$\max_x \mathscr{H}^\gamma \Phi^{(j)}(x) \le \max_x \mathscr{H}^\gamma f^{(j)}(x).$$

Denote the integral $\int_Q \Phi^{(j)}(x)\,dx$ by $\Phi_0^{(j)}$ and consider $\Phi^{(j)}(x) - \Phi_0^{(j)}$. It is clear that

$$\max_x \mathscr{H}^\gamma (\Phi^{(j)}(x) - \Phi_0^{(j)}) \le \max_x \mathscr{H}^\gamma f^{(j)}(x).$$

Consequently, the estimate holds

$$\left\| \Phi^{(j)}(x) - \Phi_0^{(j)} \mid \widetilde{C}^\gamma \right\| \le \| f^{(j)} \mid \widetilde{C}^\gamma \| = 1.$$

The action of the generalized function $\zeta_h^{h_0}(x)(1-\Delta)^{-(m-\gamma)/2} l_\infty(x)$ under consideration at the difference $\Phi^{(j)}(x) - \Phi_0^{(j)}$ is defined by the equality

$$\begin{aligned} &\left(\zeta_h^{h_0}(x)(1-\Delta)^{-(m-\gamma)/2} l_\infty(x), \Phi^{(j)}(x) - \Phi_0^{(j)} \right) \\ &= \left(\zeta_h^{h_0}(x)(1-\Delta)^{-(m-\gamma)/2} l_\infty(x), f^{(j)}(x) \right) - \Phi_0^{(j)} \left(\zeta_h^{h_0}(x)(1-\Delta)^{-(m-\gamma)/2} l_\infty(x), 1 \right). \end{aligned}$$

Show that the last summand is $o(h^m)$ uniformly in j. First, observe that

$$|\Phi_0^{(j)}| \le \max_x |\Phi^{(j)}(x)| \le \max_x |f^{(j)}(x)| \le K \| f^{(j)} \mid \widetilde{C}^\gamma \| = K,$$

with K independent of j. Second, we have the relations

$$\begin{aligned} &\left| \left(\zeta_h^{h_0}(x)(1-\Delta)^{-(m-\gamma)/2} l_\infty(x), 1 \right) \right| = \left| \left(l_\infty(x), (1-\Delta)^{-(m-\gamma)/2} \zeta_h^{h_0}(x) \right) \right| \\ &\le \left\| l_\infty \mid \widetilde{W}_2^{(m+2-\gamma)*} \right\| \left\| \zeta_h^{h_0} \mid \widetilde{W}_2^{(2)} \right\| \le K h^{m+2-\gamma} |\log h|^2 = o(h^m). \end{aligned}$$

Thus, we have shown that

$$\lim_{j\to\infty}\left(\zeta_h^{h_0}(x)(1-\Delta)^{-(m-\gamma)/2}l_\infty(x), \Phi^{(j)}(x)-\Phi_0^{(j)}\right) = \mathrm{I} + o(h^m)$$

Now, arrange the function

$$g^{(j)}(x) = \frac{\Phi^{(j)}(x)-\Phi_0^{(j)}}{\|\Phi^{(j)}(x)-\Phi_0^{(j)} \mid \widetilde{C}^\gamma\|}.$$

The sequence $g^{(j)}$ possesses the following properties

$$\|g^{(j)} \mid \widetilde{C}^\gamma\| = 1, \quad \lim_{j\to\infty}\left(\zeta_h^{h_0}(x)(1-\Delta)^{-(m-\gamma)/2}l_\infty(x), g^{(j)}(x)\right) = \mathrm{I} + o(h^m).$$

In this event, $g^{(j)}(x)$ are periodic functions with period matrix $h_0 I$.

Extend the function $\zeta_h^{h_0}(x)$ from $h_0 Q$ to the whole space with the same period. In other words, define the function $z_h^{h_0}(x)$ as

$$z_h^{h_0}(x) = \zeta_h^{h_0}(x - h_0 k) \quad \text{for } x - h_0 k \in Q.$$

Then the expression I satisfies the following estimate

$$\begin{aligned}
\mathrm{I} &= \lim_{j\to\infty}\left(\zeta_h^{h_0}(x)(1-\Delta)^{-(m-\gamma)/2}l_\infty(x), g^{(j)}(x)\right) + o(h^m) \\
&= h_0^n \lim_{j\to\infty}\left(z_h^{h_0}(x)(1-\Delta)^{-(m-\gamma)/2}l_\infty(x), g^{(j)}(x)\right) + o(h^m) \\
&\le h_0^n \|z_h^{h_0}(x)(1-\Delta)^{-(m-\gamma)/2}l_\infty(x) \mid \widetilde{C}^{\gamma *}\| + o(h^m).
\end{aligned}$$

We may estimate the last norm through the sum of the norms of the two generalized functions

$$(1-\Delta)^{-(m-\gamma)/2}l_\infty(x), \quad (1-z_h^{h_0}(x))(1-\Delta)^{-(m-\gamma)/2}l_\infty(x).$$

For the first of them, the equality holds

$$h_0^n\|(1-\Delta)^{-(m-\gamma)/2}l_\infty \mid \widetilde{C}^{\gamma *}\| = h_0^n\|l_\infty \mid \widetilde{C}^{m*}\|,$$

while the second is bounded by

$$\mathrm{II} = K\|(1-z_h^{h_0}(x))(1-\Delta)^{-(m-\gamma)/2}l_\infty(x) \mid \widetilde{W}_2^{(\gamma)*}\|.$$

Transposing the operator of multiplication by $\left(1 - z_h^{h_0}(x)\right)$ and the operator $(1-\Delta)^{\gamma/2}$ in II, from Lemma 6.5 infer the inequality

$$\mathrm{II} \leq \mathrm{III} + o(h^m),$$

with

$$\mathrm{III} = \left\|(1 - z_h^{h_0}(x))(1-\Delta)^{-m/2} l_\infty(x) \mid \tilde{L}_2\right\|.$$

To estimate the norm of III, repeat the scheme of the reasoning involved in estimating (2.4).

Let Γ_ε be the union of cubes $h(k+Q)$ such that the point hk is at a distance less than ε from the support of $1 - z_h^{h_0}(x)$. Put

$$l_h^{\Gamma_\varepsilon}(x) = \sum_{hk \in \Gamma_\varepsilon} l_0\left(\frac{x}{h} - k\right), \quad l_h^{\omega_\varepsilon}(x) = l_\infty(x) - l_h^{\Gamma_\varepsilon}(x).$$

Obviously, we have the inequalities

$$\mathrm{III} \leq K\left\|l_h^{\Gamma_\varepsilon} \mid \widetilde{W}_2^{(m)*}\right\| + \left\|(1 - z_h^{h_0}(x))(1-\Delta)^{-m/2} l_h^{\omega_\varepsilon}(x) \mid \widetilde{L}_2\right\|;$$

$$\left\|l_h^{\Gamma_\varepsilon} \mid \widetilde{W}_2^{(m)*}\right\| \leq K\sqrt{\varepsilon}\, h^m.$$

Estimate the second summand in the penultimate inequality as follows

$$\sup_f \frac{\left|\left(l_h^{\omega_\varepsilon}(x), (1-\Delta)^{-m/2}(1 - z_h^{h_0}(x)) f(x)\right)\right|}{\|f \mid \widetilde{L}_2\|}$$

$$\leq Kh^M \sup_f \frac{\left\|(1-\Delta)^{-m/2}(1 - z_h^{h_0}(x)) f(x) \mid W_2^{(M)}(\operatorname{supp} l_h^{\omega_\varepsilon})\right\|}{\|f \mid \widetilde{L}_2\|}.$$

Arrange the set ω, the complement to the cube Q of the support of $1 - z_h^{h_0}(x)$. Let $f(x)$ be an arbitrary function in $\widetilde{L}_2$. Put

$$u(x) = (1-\Delta)^{-m/2}(1 - z_h^{h_0}(x)) f(x).$$

Then $u(x)$ is a solution to the equation

$$(1-\Delta)^{m/2} u(x) = 0$$

in ω. Noting that the support of $l_h^{\omega_\varepsilon}(x)$ is compactly embedded in the interior of ω, from Lemma 6.2 we obtain the estimate

$$\left\|u \mid \widetilde{W}_2^{(M)}(\operatorname{supp} l_h^{\omega_\varepsilon})\right\| \leq K(\varepsilon)\|f \mid \widetilde{L}_2\|,$$

with $K(\varepsilon)$ a constant independent of f. We may thus dominate II that was introduced earlier as follows

$$\mathrm{II} \leq o(h^m) + K_1\sqrt{\varepsilon}\, h^m + K_2(\varepsilon) h^M.$$

By the arbitrariness of ε, this expression is $o(h^m)$. The proof of Lemma 6.8 is complete.

Theorem 6.5. *Assume that, for some p, $1 < p < 2$, and $m = m_1, m_3$ with $n < m_1 < m_3$, the* BBL*-error $l_h^\Omega(x)$ admits the estimates*

$$\| l_h^\Omega \mid \widetilde{W}_p^{(m)*} \| \leq K h^m.$$

Then $l_h^\Omega(x)$ is asymptotically optimal on every space $\widetilde{C}^m$ for $m \in [m_1, m_3)$.

PROOF. Theorem 4.2 yields a lower bound on the norm of l_h^Ω in $\widetilde{C}^m$

$$|\Omega| \, \| l_\infty \mid \widetilde{C}^{m*} \| (1 + o(1)) \leq \| l_h^\Omega \mid \widetilde{C}^{m*} \|.$$

It thus suffices to obtain the same upper bound on this norm. Put

$$\Omega_{h_0} = \bigcup_{\rho(h_0 k, \mathbb{R}^n \setminus \Omega) > \sqrt{n}\, h_0} h_0(Q + k).$$

Obviously, $\Omega_{h_0} \subset \Omega$ and $|\Omega \setminus \Omega_{h_0}| \to 0$ as $h_0 \to 0$. Decompose $l_h^\Omega(x)$ in the sum of two homogeneous errors of degree M by putting

$$l_h^\Omega(x) = l_h^{\Omega_{h_0}}(x) + l_h^{\Omega \setminus \Omega_{h_0}}(x), \quad \text{with} \quad l_h^{\Omega_{h_0}}(x) = \sum_{hk \in \Omega_{h_0}} l_0 \left(\frac{x}{h} - k \right).$$

By Corollary 4.2.2, the inequality holds

$$\| l_h^{\Omega_{h_0}} \mid \widetilde{W}_p^{(m)*} \| \leq K_1 h^m + o(h^m), \quad m = m_1, m_3,$$

with K_1 a constant independent of h and h_0.

Since the norm of $l_h^{\Omega \setminus \Omega_{h_0}}$ is at most the sum of the norms of l_h^Ω and $l_h^{\Omega_{h_0}}$; therefore,

$$\| l_h^{\Omega \setminus \Omega_{h_0}} \mid \widetilde{W}_p^{(m)*} \| \leq K_2 h^m + o(h^m), \quad m = m_1, m_3.$$

By Theorem 4.5, for $m \in [m_1, m_3)$, we have

$$\| l_h^{\Omega \setminus \Omega_{h_0}} \mid \widetilde{W}_2^{(m)*} \| \leq K_2 \mid \Omega \setminus \Omega_{h_0} |^{1/p - 1/2} h^m + o(h^m).$$

In view of the embedding $\widetilde{C}^m$ in $\widetilde{W}_2^{(m)}$ this estimate is also valid for the $\widetilde{C}^{m*}$ norm of the error. In proving Lemma 6.8, we observe the estimate $\| l_\infty \mid \widetilde{C}^m \| \geq K h^m$. Consequently, the norm of $l^{\Omega \setminus \Omega_{h_0}}$ may be estimated as follows

$$\| l_h^{\Omega \setminus \Omega_{h_0}} \mid \widetilde{C}^{m*} \| \leq K |\Omega \setminus \Omega_{h_0}|^{1/p - 1/2} \| l_\infty \mid \widetilde{C}^{m*} \| (1 + o(1)). \tag{2.6}$$

Now, estimate the $\widetilde{C}^{m*}$ norm of $l_h^{\Omega_{h_0}}(x)$. Applying to the generalized function $l_h^{\Omega_{h_0}}(x)$ the triangle inequality and Lemma 6.8, obtain

$$\left\| l_h^{\Omega_{h_0}} \mid \widetilde{C}^{m*} \right\| \le |\Omega_{h_0}| h_0^{-n} \left\| l_h^{h_0} \mid \widetilde{C}^{m*} \right\| \le |\Omega| \left\| l_\infty \mid \widetilde{C}^{m*} \right\| (1 + o(1)). \tag{2.7}$$

Combining (2.6) and (2.7), write

$$\left\| l_h^{\Omega} \mid \widetilde{C}^{m*} \right\| \le \left\| l_\infty \mid \widetilde{C}^{m*} \right\| \left\{ |\Omega| + K |\Omega \setminus \Omega_{h_0}|^{1/p-1/2} + o(1) \right\}.$$

Dividing both sides of the last inequality by $\left\| l_\infty \mid \widetilde{C}^{m*} \right\|$ and letting h vanish, come to the inequality

$$\overline{\lim_{h \to 0}} \left\{ \frac{\left\| l_h^{\Omega} \mid \widetilde{C}^{m*} \right\|}{\left\| l_\infty \mid \widetilde{C}^{m*} \right\|} \right\} \le |\Omega| + K |\Omega \setminus \Omega_{h_0}|^{1/p-1/2}.$$

Now, letting h_0 vanish, arrive at what was required. The proof of Theorem 6.5 is complete.

An analogous assertion holds also in the case of a lattice with arbitrary matrix H.

§3. Constructing Universal Asymptotically Optimal Formulas

The integration domain Ω under study is always a bounded domain with piecewise-smooth boundary. Let a sequence of cubature formulas be such that each of their errors $l_{hH}^{\Omega}(x)$, have a bounded boundary layer and possess the norm that is $O(h^m)$ in $W_p^{(m)}$ for some $p < 2$, $p \ge 1$ and $m = M_1, M_2$, with $n < M_1 < M_2$. Then these cubature formulas are asymptotically optimal in the collection of spaces $\widetilde{W}_2^{(m)}$, and $(\widetilde{C}^{\gamma})^m$ for $m \in [M_1, M_2)$, as well as in $\widetilde{H}_2^{\mu}$ with the function μ of Theorem 6.1.

Formulas with regular boundary layer satisfy these sufficient conditions of universal asymptotic optimality. Computer tasks for calculating integrals by using such formulas were written by L. V. Voĭtishek [292] for rational polyhedra and by N. I. Blinov [30–31] for plane domains with smooth boundary.

We describe the construction of cubature formulas of another class which are also instances of a universal asymptotically optimal formula. These are formulas with bounded boundary layer. Indeterminacy of the constants entering the definition of BBL-formula enables us to use in constructing these formulas the conventional operations of analysis. In particular, available are smooth changes of variables and bounded linear operations in the corresponding spaces.

We assume that H is the identity matrix I and so the lattice is cubic. Consider Ω with smooth boundary Γ. Given a point $x \in \overline{\Omega}$, we may indicate a neighborhood $U(x)$ such that the part of the boundary $\Gamma \cap U(x)$ may be rectified on one of the coordinate planes.

By the compactness of Γ, we may choose a finite cover $U_t = U(x^t)$, $t = 1, \dots, T$, and a *partition of unity* $\{\varphi_t\}_{t=1}^T$ *subject to this cover*, i.e.,

$$\forall \varphi_t \in C_0^{(\infty)}, \quad \operatorname{supp} \varphi_t \subset U_t, \quad \sum_t \varphi_t(x) = 1.$$

Construct the error $l_h^{\Omega \cap U_t}(x)$ in each neighborhood U_t, and obtain a cubature formula for the entire domain Ω by summing the found local formulas with the weight $\varphi_t(x)$. In other words, we assume further that

$$l_h^{\Omega}(x) = \sum_{t=1}^{T} \varphi_t(x) l_h^{\Omega \cap U_t}(x). \tag{3.1}$$

If $c_k^t(h)$ are the weights of the cubature formula corresponding to $l_h^{\Omega \cap U_t}(x)$ then $l_h^{\Omega}(x)$ has the weights $c_k^a(h)$ subject to the condition

$$c_k^a(h) = \sum_{t=1}^{T} \varphi_t(hk) c_k^t(h). \tag{3.2}$$

We are left with finding out how to construct the local errors $l_h^{\Omega \cap U_t}(x)$.

Designate $x = (x_1, \dots, x_{n-1}, x_n) = (x', x_n)$. In general, let the prime stand for the $(n-1)$-dimensional version of a notation. For instance, $\delta'(x)$ is the $(n-1)$-dimensional Dirac delta function,

$$(\delta'(x'), \varphi(x)) = \varphi(0', x_n).$$

Also, let $\delta_n(x_n)$ stand for the singly-indexed Dirac delta function acting along the nth coordinate, and let $[\alpha]$ and $\{\alpha\}$ stand for the integral and fractional parts of a real α.

For definiteness, take a neighborhood $U = U_t$ such that

$$\overline{\Gamma \cap U} = \{x : x_n = \gamma(x'),\ \gamma \in C^{(M)}\}.$$

Let $\Phi : \mathbb{R}^n \to \mathbb{R}^n$ be a smooth one-to-one mapping of U which rectifies $\Gamma \cap U$ and is given in U by equalities

$$y' = \Phi'(x') = x', \quad y_n = \Phi_n(x) = x_n - \gamma(x').$$

It is natural to assume that $\Omega \cap U$ lies to the one side of the boundary $\Gamma \cap U$. Assume further that for all $x \in \Omega \cap U$ we have $x_n > \gamma(x')$.

We obtain the error $l_h^{\Omega \cap U}(x)$ by the change of variables $y = \Phi(x)$ from $l^\omega(y)$, with $\omega = \Phi(\overline{\Omega} \cap U)$ and

$$l^\omega(y) = \chi_\omega(y) - h^n \sum_{hk \in \Omega \cap U} c_k^t \delta'(y' - hk')\delta_n(y_n - hk_n + \gamma(hk'))$$
$$= \chi_\omega(y) - h^n \sum_{hk \in \omega} c_{k',k_n+[\gamma(hk')/h]}^t \delta'(y' - hk')\delta_n(y_n - hk_n + h\{\gamma(hk')/h\}).$$

It is convenient to have the weights c_k defined for all k in $\mathbb{Z}^n$. To this end, let $c_k = 0$, for $hk \notin \omega$. In this event the error $l^\omega(y)$ takes the shape

$$l^\omega(y) = \chi_\omega(y) - \sum_{k, k_n \geq 1} c_{k',k_n+[\gamma(hk')/h]}^t \delta(y' - hk')\delta(y_n - hk_n + h\{\gamma(hk')/h\}).$$

Determining the weights of $l^\omega(y)$ we should bear in mind two circumstances. First, the nodes of $l^\omega(y)$ must lie on the "curved lattice," i.e., at the points of

$$\{(hk', hk_n - h\{\gamma(hk')/h\}) : k \in \mathbb{Z}^n\}.$$

Second, in the last summation over k, we may substitute the condition

$$k_n > \{\gamma(hk')/h\}$$

for the condition $hk \in \omega$; since only the weights of delta functions with support in ω are automatically preserved under multiplication by the truncator $\varphi_t(\Phi^{-1}(y))$.

Moreover, in the formula for $l^\omega(y)$ we may replace ω by an arbitrary bounded subset $\hat{\omega}$ in the half-space $y_n \geq 0$, if only $\Phi^{-1}(\hat{\omega})$ includes $\overline{\Omega} \cap U$. Instead of the condition $k_n > \{\gamma(hk')/h\}$ require the condition $k_n \geq 1$, which guarantee the validity of the preceding inequality because $0 \leq \{\gamma(hk')/h\} < 1$.

Define the coefficients $b_k^t = c_{k',k_n+[\gamma(hk')/h]}^t$ so as to guarantee the asymptotic equality

$$\|l^\omega \mid \widetilde{W}_p^{(m)*}\| = O(h^m).$$

After the change of variables $x = \Phi^{-1}(y)$, multiplication by $\varphi_t(x)$, and summation over t, this property, obviously translates to the norm $\|l_h^\Omega \mid \widetilde{W}_p^{m*}\|$.

To this end construct an auxiliary homogeneous error on the cubic lattice in the variables y as follows

$$q_h^\omega(y) = \sum_{hk \in \omega} l_0\left(\frac{y}{h} - k\right) = \sum_{k'} \sum_{k_n \geq 0} l_0\left(\frac{y}{h} - k\right). \tag{3.3}$$

Take a local error $l_0(y)$ of order $M+1$ in (3.3) such that the support of the discrete part of $l_0(y)$ lie in the half-space $y_n > 0$, i.e., put

$$l_0(y) = \chi_Q(y) - \sum_{k,k_n \geq 1} a_k\delta(y-k); \quad (l_0(y), y^\alpha) = 0, \quad |\alpha| \leq M. \tag{3.4}$$

Here a_k vanishes for every vector k beyond some ball with center the origin and radius $R = R(M)$. If $R(M)$ is sufficiently large for a given M then the local error (3.4) exists.

Now, summation in (3.3) is taken over k such that $k_n \geq 1$, namely,

$$q_h^\omega(y) = \chi_\omega(y) - h^n \sum_{k,k_n \geq 1} b_k\delta(y-hk). \tag{3.5}$$

From (3.3)–(3.4) it follows that b_k in (3.5) vanishes for each k beyond some ball. Assume further that $b_k = 0$ for $k_n < 1$. We now arrange the expression

$$\Delta(y) \equiv l^\omega(y) - q_h^\omega(y)$$

which, if summed up with $q_h^\omega(y)$, yields $l^\omega(y)$. The generalized function $\Delta(y)$ is a linear combination of delta functions. By choosing $\Delta(y)$, we will ensure the replacement of the delta functions supported at the nodes of an auxiliary cubic lattice by delta functions supported at the nodes of the "curved lattice"

$$\{(hk', hk_n - h\{\gamma(hk')/h\}) : k \in \mathbb{Z}^n\}.$$

We perform this operation separately on each half-line

$$\{y : y' = hk',\, y_n \geq 0\}$$

with the help of a homogeneous generalized function

$$\Delta_n(y_n) = \delta_n(y_n) - \sum_{j=1}^{J} v_j\delta_n(y_n - h(j-\eta)), \tag{3.6}$$

with $\eta = \{\gamma(hk')/h\}$.

Using (3.6), it is easy to see that

$$\sum_{k_n \geq 1} b_k\delta_n(y_n - hk_n) = \sum_{k_n \geq 1} b_k\Delta_n(y_n - hk_n)$$
$$+ \sum_{k_n \geq 1}\Bigg(\sum_{j=1}^{\min\{k_n-1,J\}} b_{k',k_n-j}v_j\Bigg)\delta(y - hk_n + h\eta).$$

Consequently, the shape of $\Delta(y)$ for suitable c_k^t is as follows

$$\Delta(y) = h^n \sum_{k,k_n \geq 1} \{b_k \delta(y - hk) - c^t_{k',k_n+[\gamma(hk')/h]} \delta'(y' - hk')\delta_n(y_n - h(k_n - \eta))\}$$
$$= h^{n-1} \sum_{k'} \delta'(y' - hk')h \sum_{k_n \geq 1} b_k \Delta_n(y_n - hk_n).$$

Thus, the initial unknown weights c_k are linear combinations of the new unknowns v_j depending on η.

The main requirement we have to meet when choosing v_j is as follows: The error $l^\omega(y)$ must have an optimal mode of convergence in $\widetilde{W}_p^{(m)}$. To ensure this, it suffices that the $\widetilde{W}_p^{(m)*}$ norm of the generalized function $\Delta(y)$ be $O(h^m)$. We may achieve the latter if $(\Delta_n(y_n), f(y_n))$ is a finite difference of high order with a variable mesh-size. The node $y_n = 0$ of such finite difference does not belong to the set of nodes $\{h(j - \eta)\}_{j=1}^J$.

We subject the coefficients v_j to the conditions

$$(\Delta_n(\tau), \tau^\alpha) = 0, \quad |\alpha| \leq M,$$

i.e., we require the following

$$\sum_{j=1}^J v_j j^\alpha = \eta^\alpha, \quad |\alpha| \leq M. \tag{3.7}$$

Obviously, for $J \geq M + 1$, these simultaneous linear equations for the unknowns v_j, $j = 1, \ldots, J$, are solvable. Take $J = M + 1$. Then v_j are determined uniquely.

The collection of the conditions (3.3)–(3.7) enables us to construct a sought functional $\varphi_t(x) l_h^{\Omega \cap U}(x)$. We justify this claim below, and now we give as an example the final formula for the weight c_k^t of the cubature formula at the node hk of a given neighborhood U_t such that $\Omega \cap U_t \subset \{x : x_n > \gamma(x')\}$.

We see that a_k, the weights of the n-dimensional error $l_0(x)$, may be expressed through the weights of its one-dimensional analogs

$$a_{(k_1,\ldots,k_n)} = a_{k_1} a_{k_2} \ldots a_{k_n},$$

with

$$\sum_{j=1}^{M+1} a_j j^\alpha = \frac{1}{\alpha + 1}, \quad |\alpha| \leq M.$$

This manner of prescribing a_k is convenient because the matrix B of the algebraic system determining a_j coincides with the matrix of (3.7). Denoting by a_{ij} the entries of the inverse matrix B^{-1}, see

$$v_i = \sum_{j=1}^{M+1} a_{ij}\eta^{j-1}, \quad a_i = \sum_{j=1}^{M+1} a_{ij}/j, \quad i = 1, 2, \ldots, M+1. \tag{3.8}$$

It is of service to set the coefficients a_i, $i \notin [1, M+1]$, equal 0. Inserting (3.4) in (3.3) and collecting similar terms, calculate b_k to obtain

$$b_k = \sum_{t'} \sum_{t_n=1}^{\min(k_n, M+1)} a_{t', t_n} = \sum_{t'} a_{t_1} \ldots a_{t_{n-1}} \sum_{t_n=1}^{\min(k_n, M+1)} a_{t_n}.$$

Since $\sum_{j=1}^{M+1} a_j = 1$; therefore,

$$b_k = \sum_{j=1}^{\min(k_n, M+1)} a_j.$$

In other words, the value of b_k is determined only by the nth component of the index k, i.e., $b_k = b_{k_n}$.

The final formula for the error $l^\omega(y)$ results from substitution of b_k and v_j in the relations defining Δ_k and Δ. The weights c_k^t in the case under study take the form

$$c_k^t = \sum_{p=1}^{M+1} \frac{1}{p} \sum_{q=1}^{M+1} \eta^{q-1} \sum_{r=1}^{\min(k_n - \xi - 1, M+1)} a_{rp} \sum_{s=1}^{\min(k_n - \xi - r, M+1)} a_{sq}, \tag{3.9}$$

with $\xi = [\gamma(hk')/h]$ and $\eta = \{\gamma(hk')/h\}$. The formula (3.9) is valid for $k_n > \xi + 1$ and $hk \in \omega$. For $k_n \leq \xi + 1$, put $c_k^t = 0$.

Check fulfillment of the conditions (1.1)–(1.2) for $l_h^\Omega(x)$ with the weights $c_k^a(h)$ calculated by (3.2). Obviously, $c_k^a(h)$ are bounded uniformly in k and h. Observe also that in (3.9) all sums are taken up to $M+1$, if only the node hk lies far from the boundary, i.e., for $hk_n > \gamma(hk') - 2h(M+1)$. In this event, the equality holds

$$c_k^t = \sum_{r=1}^{M+1} \sum_{p=1}^{M+1} a_{rp} \frac{1}{p} \sum_{s=1}^{M+1} \sum_{q=1}^{M+1} a_{sq}\eta^{q-1} = \sum_{r=1}^{M+1} a_r \sum_{s=1}^{M+1} v_s = 1.$$

An analogous equality holds for the weights $c_k^a(h)$ provided by (3.2). This is easy to check by using the properties of the functions $\varphi_t(x)$, $t = 1, \ldots, T$. Indeed, if $c_k^t = 1$ in (3.2) then

$$c_k^a(h) = \sum_{t=1}^{T} \varphi_t(hk) = 1.$$

Thus, our construction yields BBL-formulas.

Theorem 6.6. *The* BBL-*error* $l_h^\Omega(x)$ *with weights* $c_k^a(h)$ *defined by* (3.2) *has an optimal mode of convergence in* $W_p^{(m)}$ *for all* $p \geq 1$ *and* $m \in [M_1, M_2]$ *with* M_1 *and* M_2 *integers subject to the condition* $n < M_1 \leq M_2 \leq M$.

PROOF. From the preceding reasoning we see that it suffices to obtain the following estimate

$$\|\varphi(\Phi^{-1}(y))\Delta(y) \mid \widetilde{W}_p^{(m)*}\| \leq Kh^m$$

with $\varphi \in \{\varphi_1, \dots, \varphi_T\}$. Applying Theorem 4.4 to $l(y,h) = \varphi(\Phi^{-1}(y))\Delta(y)$, reduce the problem to the two extreme cases $m = M_1, M_2$, i.e., m an integer.

Given an arbitrary function $f \in \widetilde{W}_p^{(m)}$, put $g(y) = \varphi(\Phi^{-1}(y))f(y)$. Then the equality holds

$$(\varphi(\Phi^{-1}(y))\Delta(y), f(y)) = h^n \sum_{k'} \sum_{k_n \geq 1} b_k \Big\{ g(hk', hk_n)$$
$$- \sum_{j=1}^{M+1} v_j g(hk', hk_n + hj - h\{\gamma(hk')/h\}) \Big\}. \tag{3.10}$$

Introduce the notation

$$q(y', k, h) = g(y', hk_n) - \sum_{j=1}^{M+1} v_j g(y', hk_n + hj - h\{\gamma(hk')/h\}).$$

The braces in (3.10) contain the value of $q(hk', k, h)$. Express it through the integrals over

$$\omega = h(Q' + k') = \{y' : 0 \leq y_j - k_j < h, \ j = 1, 2, \dots, n-1\}.$$

It is obvious that

$$h^{n-1} q(hk', k, h) = \int_\omega q(y', k, h)\, dy' + \int_\omega (q(hk', k, h) - q(y', k, h))\, dy'$$
$$= \int_\omega q(y', k, h)\, dy' + \int_\omega \int_0^1 D_t q(y' + t(hk' - y'), k, h)\, dt dy'$$
$$= \int_\omega q(y', k, h)\, dy' + \int_0^1 \sum_{i=1}^{n-1} \int_\omega hD_i q(y' + t(hk' - y'), k, h)(k_i - y_i)\, dy' dt.$$

Inserting this expression in (3.10), obtain

$$
\begin{aligned}
&(\varphi(\Phi^{-1}(y))\Delta(y), f(y)) \\
&= \sum_{k'} \int\limits_{h(Q'+k')} h \sum_{k_n \geq 1} b_k \Big\{ g(y', hk_n) - \sum_{j=1}^{M+1} v_j g(y', hk_n + hj - h\eta) \Big\} dy' \\
&+ \int\limits_0^1 \sum_{i=1}^{n-1} \Big\{ \sum_{k'} \int\limits_{h(Q'+k')} h^2 \sum_{k_n \geq 1} b_k \Big[D_i g(y' + t(hk' - y'), hk_n) \\
&- \sum_{j=1}^{M+1} v_j D_i g(y' + t(hk' - y'), hk_n + hj - h\eta) \Big] (k_i - y_i) dy' \Big\} dt,
\end{aligned}
$$

with $\eta = \{\gamma(hk')/h\}$.

Substitute for g and $D_i g$ their Taylor expansions in the last argument at the point hk_n with the remainders of degree m and $m-1$ respectively. In virtue of the conditions on v_j all terms of the expansion, except for the remainder, vanish.

Coarsening the estimates, in particular, replacing $b_k = b_{k_n}$ and v_j by the same constant, obtain

$$
\begin{aligned}
&|(\varphi(\Phi^{-1}(y))\Delta(y), f(y))| \\
&\leq Kh^m \Big\{ \sum_{k'} \int\limits_{h(Q'+k')} \Big(\sum_{j=1}^{M+1} \int\limits_{hk_n}^{hk_n+(j-\eta)h} |D_n^m(g)(y', s)| \, ds \Big) dy' \\
&+ \int\limits_0^1 \sum_{i=1}^{n-1} \sum_{k'} \int\limits_{h(Q'+k')} \int\limits_{hk_n}^{hk_n+(j-\eta)h} |D_n^{m-1} D_i(g)(y' + t(hk' - y'), s)| \, ds dy dt \Big\} \\
&\leq Kh^m \|g \mid \widetilde{W}_p^{(m)}\|.
\end{aligned}
$$

This actually means that

$$
\|\varphi(\Phi^{-1}(y))\Delta(y) \mid \widetilde{W}_p^{(m)*}\| \leq Kh^m.
$$

The proof of Theorem 6.6 is complete.

The above algorithm for constructing universal asymptotically optimal formulas was implemented by I. Umarkhanov as a program for calculating multiple integrals over bounded domains with smooth boundary. He also generalized the algorithm to bounded domains with piecewise-smooth boundary. We now describe his results.

To begin with, we formulate the definition of a domain Ω with piecewise-smooth boundary Γ rigorously. For such a domain in $\mathbb{R}^n$, to each point x of its closure, $x \in \overline{\Omega}$, there exist a neighborhood U_x and a $C^{(M)}$-diffeomorphism $y = \Phi(x)$ mapping this neighborhood to the open unit ball with center the origin. The image of the set $\Omega \cap U_x$ under this diffeomorphism is given by one of the relations

$$V_0 = \{y : y_1^2 + \cdots + y_n^2 < 1\},$$
$$V_j = \{y : y_1 \geq 0, \ldots, y_j \geq 0;\ y_1^2 + \cdots + y_n^2 < 1\}, \quad 1 \leq j \leq n.$$

Observe that this definition is not satisfied, for instance, by tetrahedral pyramid in 3-dimensional space. A neighborhood with the desired property fails to exist about the apex, the vertex which all four faces are contiguous to.

Further, using a partition of unity, obtain a sought cubature formula by summing the local formulas that are already-constructed for individual neighborhoods U_t comprising some finite cover $\overline{\Omega}$. We explain how it is possible to construct such local formulas.

If $\Phi(\overline{\Omega} \cap U)$ coincides with the entire unit ball in $\mathbb{R}^n$ then we put

$$l^U(x) = \chi_U(x) - h^n \sum_{hk \in U} \delta\left(\frac{x}{h} - k\right).$$

If the image $\Phi(\overline{\Omega} \cap U)$ coincides with V_1, i.e., in the neighborhood under consideration the boundary Γ is smooth, then we perform the construction as before. We are left with addressing the cases $\Phi(\overline{\Omega} \cap U) = V_j$, $j = 2, \ldots, n$. We elaborate the construction for the case of 2-dimensional space when $n = 2$ and the sets $\Phi(\overline{\Omega} \cap U)$ and V_2 coincide, i.e.,

$$\Phi(\overline{\Omega} \cap U) = V_2 = \{y : y_1 \geq 0,\ y_2 \geq 0,\ y_1^2 + y_2^2 < 1\}.$$

Consider the particular case in which one of the two smooth pieces of the boundary which lies in U is a coordinate axis, i.e.,

$$\begin{gathered} \Gamma \cap U = \Gamma_1 \cup \Gamma_2, \quad \Gamma_1 = \{x : x_1 = 0\}, \quad \Gamma_2 = \{x : x_2 = \gamma(x_1)\}, \\ \overline{\Omega} \cap U = \{x : x_1 \geq 0,\ x_2 \geq \gamma(x_1),\ x \in U\}. \end{gathered} \tag{3.11}$$

We perform the construction of a local cubature formula as it was described above for a neighborhood about the smooth boundary $x_2 = \gamma(x_1)$. We need only two minor changes. First, it is convenient to assume that the function $\gamma(x_1)$ is defined not only for $x_1 \geq 0$, but also for all x_1 of the set resulting from projecting U to the

x_1 axis. Second, on the support of the local error $l_0(y)$ to be involved we should impose the conditions

$$\operatorname{supp} l_0 \subset \{y : y_2 \geq 0\}, \quad \operatorname{supp} l_0 \subset \{y : y_1 \geq 0\}.$$

It is easy to check that such properties of an error as possession of a bounded boundary layer and an optimal mode of convergence are perfectly preserved, so are the equalities (3.2) and (3.9) for the weights of a cubature formula. We have thus completely solved the problem of constructing universal asymptotically optimal formulas for the plane domains of the shape (3.11).

We reduce the general case to the particular case of (3.11). Denote the initial variables by $z = (z_1, z_2)$; a neighborhood, by W; and the boundary, by G. In this event, we have

$$G \cap W = G_1 \cup G_2, \quad G_1 = \{z : z_1 = g_1(z_2)\}, \quad G_2 = \{z : z_2 = g_2(z_1)\}.$$

Assume that G_1 and G_2 meets at the coordinate origin at the angle α. If this angle is acute then transform it into an obtuse angle by a change of variables $z' = Az$ with a unimodular matrix A. The entries of A are integers and its determinant equals 1. Such transformation preserves the nodes and so the class of the cubature formulas under consideration.

We thus assume that the angle α is obtuse, i.e., $\pi/2 < \alpha < \pi$. Obtain a sought formula from another one by the change of variables $y_1 = z_1 - g_1(z_2)$ and $y_2 = z_2$. We suppose that $g_1(z_2)$ is a smooth function defined on the entire projection of W to the z_2 axis. In the variables y we encounter the problem of constructing a BBL-formula over (3.11) having an optimal mode of convergence but with nodes on a "curved lattice." In Theorem 6.6 we justified the procedure of obtaining such formula from an auxiliary formula with nodes forming a cubic lattice. Consequently, it suffices to construct a lattice formula for the case of (3.11). We have just solved this problem.

If a boundary comprises two smooth pieces, then the construction consists in double application of the algorithm for the case of a smooth boundary.

This reasoning translates to the case of an n-dimensional domain. In this event a k-hedral piece of the boundary requires k rectifications of faces and reevaluations of delta functions we have described for a smooth boundary.

Chapter 7

Cubature Formulas of Infinite Order

We speak about a *cubature formula of infinite order* whenever the error of the formula is $O(h^m)$ for all integer m and all functions in the space under study. Here h is the mesh-size of the lattice of integration. The Mean Value Theorem for harmonic functions provides the simplest example of a formula of such kind. In § 6 and § 8 of the current section we consider not so obvious examples of cubature and quadrature formulas of infinite order. The starting point of our study is the next observation. The estimate of accuracy of a formula of approximate integration by means of the norm of its error is unimprovable as long as we consider the whole $L_2^{(m)}$ space. However, for every individual function φ in $L_2^{(m)}$ the vanishing of (l_h, φ) as $h \to 0$, i.e., weak convergence, is always faster. This circumstance prompts us to seek for formulas of infinite order on various subspaces of $\widetilde{L}_2^{(m)}$. As such we naturally take classes of infinitely differentiable functions, in particular, the Gevrey classes. The properties of the latter are scrutinized in § 2–§ 5 and § 7 of the current chapter. We thus address the formulas whose rate of convergence exceeds the rate we may expect from the formulas of a given polynomial degree. A similar phenomenon is referred to as *superconvergence* in [221].

§1. Weak Convergence of Cubature Formulas

In Chapter 1 we found the exact value of the norm of the optimal error of a cubature formula in the $\widetilde{L}_2^{(m)}(H)$ space of periodic functions. Recall that this error has the shape

$$l_h(x) = \chi_{\Omega_0}(x) - \sum_{hH\gamma\in\Omega_0} h^n \delta(x - hH\gamma), \tag{1.1}$$

with Ω_0 the fundamental parallelepiped of the orthogonal matrix H. The appropriate candidates for the numbers h are only the quantities $h = 1/N$, with N an integer, since the lattice of nodes $hH\gamma$ must also be periodic with matrix H.

For every h, the error attains its maximal value at some element $u_h(x)$ of the unit ball of $\widetilde{L}_2^{(m)}(H)$. The function $u_h(x)$ is $\widetilde{L}_2^{(m)}$-extremal for the error $l_h(x)$. In this event $u_h(x) = N^{-m}u_1(x/h)$. Assuming the volume of the fundamental parallelepiped equal to 1, namely, $|\Omega_0| = 1$, we have

$$(l_h, u_h) = \|l_h \mid \widetilde{L}_2^{(m)*}\| \, \|u_h \mid \widetilde{L}_2^{(m)}\| = \left(\frac{h}{2\pi}\right)^m \sqrt{\zeta(H^{-1*} \mid 2m)}.$$

The set of the $\widetilde{L}_2^{(m)}$-extremal functions corresponding to all admissible N is noncompact. Moreover, this set has no condensation point. The reader may check it by simple estimation of the norm of the difference between two elements of the set. Consequently, equality in the inequality

$$|(l_h, \varphi)| \le \|l_h \mid \widetilde{L}_2^{(m)*}\| \, \|\varphi \mid \widetilde{L}_2^{(m)}\| \tag{1.2}$$

is not attained as $h \to 0$. Moreover, we now show that at every fixed element $\varphi \in \widetilde{L}_2^{(m)}$ the expression $h^{-m}(l_h(x), \varphi(x))$ (denote it by $g_\varphi[N]$) vanishes faster than it follows from (1.2), with the rate of convergence dependent on m.

We have the following

Theorem 7.1. *If $m > n$ then the sequence $g_\varphi[N]$, $N = 1, 2, \ldots$, belongs to l_2 and the next estimate for the norm holds*

$$\|g_\varphi[N] \mid l_2\| = \left\{\sum_{N=1}^{\infty} |g_\varphi[N]|^2\right\}^{1/2} \le K(m)\|\varphi \mid \widetilde{L}_2^{(m)}\|, \tag{1.3}$$

with $K(m)$ independent of φ.

PROOF. To simplify calculations, restrict our consideration to the case in which the period matrix H is the identity matrix. This leads to no loss of generality.

A function $\varphi(x)$ expands in the following Fourier series

$$\varphi(x) = \sum_{\beta \ne 0} \frac{c[\beta]}{|\beta|^m} e^{i2\pi\beta x} + c[0]. \tag{1.4}$$

Summation here is taken over all integer multi-indices β. The $\widetilde{L}_2^{(m)}(H)$ norm of the function $\varphi(x)$ of (1.4) is expressed as

$$\|\varphi \mid \widetilde{L}_2^{(m)}\| = (2\pi)^m \left\{\sum_{\beta \ne 0} |c[\beta]|^2\right\}^{1/2}. \tag{1.5}$$

Obviously, convergence of the series on the right side of (1.5) amounts to the containment of φ in $\widetilde{L}_2^{(m)}(H)$.

Calculate the error of an optimal cubature formula at the function $\varphi(x)$ expanded like (1.4). Along with this, it is convenient to consider the error of a slightly more general cubature formula when the mesh-size varies with coordinates. Let

$$l(x \mid \Gamma) = \chi_Q(x) - \Phi_0(\Gamma x)\chi_Q(x),$$

with Q the unit cube in $\mathbb{R}^n$ and Γ a diagonal matrix with natural diagonal entries $\gamma_1, \ldots, \gamma_n$. For the generalized function $\Phi_0(\Gamma x)$ by definition we have

$$\Phi_0(\Gamma x) = \sum_\alpha \delta(\Gamma x - \alpha) = \frac{1}{\gamma_1\gamma_2\ldots\gamma_n}\sum_\alpha \delta(x - \Gamma^{-1}\alpha),$$

with α an integer vector.

To find the value of the error $l(x \mid \Gamma)$ at $\varphi(x)$, calculate $(l(x \mid \Gamma), e^{i2\pi\beta x})$ and use the expansion (1.4). For $\beta \neq 0$, the equality holds

$$L_\Gamma[\beta] \equiv (l(x \mid \Gamma), e^{i2\pi\beta x})$$

$$= -\frac{1}{\gamma_1\ldots\gamma_n}\sum_{\alpha_j=0,\ldots,\gamma_j-1} e^{i2\pi\sum\limits_{j=1}^{n}\beta_j\alpha_j/\gamma_j} = -\prod_{j=1}^{n}\frac{1}{\gamma_j}\sum_{\alpha_j=0}^{\gamma_j-1} e^{i2\pi\beta_j\alpha_j/\gamma_j}.$$

The inner sum equals 0, if β_j is not a multiple of γ_j, and equals γ_j otherwise. Whence obtain

$$(l(x \mid \Gamma), e^{i2\pi\beta x}) = \begin{cases} -1, & \text{if } \beta_j \text{ is a multiple of } \gamma_j \text{ for every } j, \\ 0, & \text{otherwise.} \end{cases}$$

Returning to the value of $l(x \mid \Gamma)$ at $\varphi(x)$, find

$$(l(x \mid \Gamma), \varphi(x)) = -\sum_{\beta\neq 0}\frac{c[\Gamma\beta]}{|\Gamma\beta|^m}. \tag{1.6}$$

In the case when $\Gamma = NI$, with I the identity matrix of order n, the equality holds

$$g_\varphi[N] = -\sum_{\beta\neq 0}\frac{c[N\beta]}{|\beta|^{m/2}}\frac{1}{|\beta|^{m/2}}. \tag{1.7}$$

Applying the Cauchy inequality to the right side of (1.7), obtain

$$|g_\varphi[N]|^2 \leq \sum_{\beta\neq 0}\frac{|c[N\beta]|^2}{|\beta|^m}\sum_{\beta\neq 0}\frac{1}{|\beta|^m}.$$

By hypothesis $m > n$. Consequently, the second factor on the right side of the last inequality is finite:

$$\sum_{\beta\neq 0} 1/|\beta|^m = K_1(m) < \infty.$$

Whence and from the definition of the norm of l_2 it follows that

$$\|g_\varphi \mid l_2\|^2 = \sum_{N=1}^{\infty} |g_\varphi[N]|^2 \leq K_1(m) \sum_{N=1}^{\infty} \sum_{\beta\neq 0} \frac{|c[N\beta]|^2}{|\beta|^m}.$$

Changing the order of summation, infer

$$\|g_\varphi \mid l_2\|^2 \leq K_1^2(m) \sum_{\lambda\neq 0} |c[\lambda]|^2 \leq K_1^2(m) \|\varphi \mid \widetilde{L}_2^{(m)}\|^2/(2\pi)^{2m}$$

(in the second inequality we used (1.5)).

Thus, we arrive at (1.3), which completes the proof of Theorem 7.1.

The estimate (1.3) is unimprovable in the sense that for no $q > 2$ we may assert that the norm $\|g_\varphi \mid l_q\|$ is finite.

Theorem 7.2. *If $n/2 < m \leq n$, then for all $q_* > q = 2n/(2m-n)$ the inequality holds*

$$\|g_\varphi \mid l_{q_*}\| = \left\{ \sum_{N=1}^{\infty} |g_\varphi[N]|^{q_*} \right\}^{1/q_*} \leq K \|\varphi \mid \widetilde{L}_2^{(m)}\|, \tag{1.8}$$

with K independent of φ.

PROOF. Let $0 < \varepsilon < m - n/2$ and $0 < \mu < 1$. Recall that for a function $\varphi \in \widetilde{L}_2^{(m)}$ we have (1.7). Replacing therein the coefficient $c[N\beta]$ by its modulus and properly collecting factors, come to the inequality

$$|g_\varphi[N]| \leq \sum_{\beta\neq 0} |c[N\beta]|^{1-\mu} \left(\frac{|c[N\beta]|}{|\beta|^{(m-n/2-\varepsilon)/\mu}} \right)^{\mu} \frac{1}{|\beta|^{n/2+\varepsilon}}.$$

Applying to the sum on the right side the Hölder inequality for the product of three functions with exponents $1/p_1 = (1-\mu)/2$, $1/p_2 = \mu/2$, and $1/p_3 = 1/2$ and considering that $1/p_1 + 1/p_2 + 1/p_3 = 1$, obtain

$$|g_\varphi[N]| \leq \left(\sum_{\beta\neq 0} |c[N\beta]|^2 \right)^{(1-\mu)/2} \left(\sum_{\beta\neq 0} \frac{|c[N\beta]|^2}{|\beta|^{2\varepsilon_*}} \right)^{\mu/2} \left(\sum_{\beta\neq 0} \frac{1}{|\beta|^{n+2\varepsilon}} \right)^{1/2}. \tag{1.9}$$

Here ε_* stands for $(m - n/2 - \varepsilon)/\mu$. Observe now that we have the relations

$$\sum_{\beta \neq 0} |c[N\beta]|^2 \leq \|\varphi \mid \widetilde{L}_2^{(m)}\|^2/(2\pi)^{2m}, \quad \sum_{\beta \neq 0} 1/|\beta|^{n+2\varepsilon} = K_1(\varepsilon) < +\infty.$$

Inserting them in (1.9), come to the estimate

$$|g_\varphi[N]|^2 \leq (2\pi)^{-2m(1-\mu)} \|\varphi \mid \widetilde{L}_2^{(m)}\|^{2(1-\mu)} K_1(\varepsilon) \Big(\sum_{\beta \neq 0} \frac{|c[N\beta]|^2}{|\beta|^{2\varepsilon_*}} \Big)^{\mu}. \tag{1.10}$$

Let $\varepsilon = ((m - n/2)q_* - n)/(1 + q_*)$ and $\mu = 2/q_*$. Using the condition $q_* > q$, we readily validate the following relations

$$0 < \varepsilon < m - n/2, \quad 0 < \mu < 1, \quad 2\varepsilon_* = n + \varepsilon_1, \quad \varepsilon_1 > 0.$$

Raising both sides of (1.10) to the power $1/\mu$ and summing the result over all natural N, obtain the estimate

$$\sum_{N=1}^{\infty} |g_\varphi[N]|^{q_*} \leq K \|\varphi \mid \widetilde{L}_2^{(m)}\|^{2(1/\mu - 1)} \sum_{N=1}^{\infty} \sum_{\beta \neq 0} \frac{|c[N\beta]|^2}{|\beta|^{n+\varepsilon_1}}.$$

Inserting in this inequality the variable $\lambda = N\beta$ and transposing sums, derive the inequality

$$\sum_{N=1}^{\infty} |g_\varphi[N]|^{q_*} \leq K \|\varphi \mid \widetilde{L}_2^{(m)}\|^{q_*}.$$

Thus (1.8) is established, and so the proof of Theorem 7.2 is complete.

Associating with each function $\varphi \in \widetilde{L}_2^{(m)}$ the sequence of errors $g_\varphi[N]$, we thus define some embedding operator with codomain in l_2 (for $m > n$) or in l_{q_*} (for $n/2 < m \leq n$). In Theorems 7.1 and 7.2 it is proven that such operator is continuous. Of interest is the question as to whether or not the above embedding is one-to-one and invertible.

Leaving aside solution of this question in a general case, show that (1.6) enables us in principle to construct all Fourier coefficients $a[\gamma]$ of a function $\varphi(x)$ from a given system of errors.

If $\varphi(x) = \sum_\gamma a[\gamma] e^{i2\pi\gamma x}$ then we let

$$b_\varphi[\gamma] \equiv \sum_{\beta \neq 0} a[\Gamma\beta] = -(l(x \mid \Gamma), \varphi(x)). \tag{1.11}$$

Here Γ is a diagonal matrix with natural diagonal entries $|\gamma_j|$. For $\Gamma = NI$ and $\mathbf{e} = (1, \dots, 1)$ the equality holds

$$b_\varphi[N\mathbf{e}] = g_\varphi[N].$$

Express the Fourier coefficients $a[\gamma]$ of $\varphi(x) \in \widetilde{L}_2^{(m)}$ through the errors $b_\varphi[\gamma]$, i.e., invert the operator $\mathscr{E}$ that is given by (1.11) on the set of functions $a[\gamma]$ of sufficient decay.

We need the following formula

$$\chi\left(\bigcup_{j=1}^{n} \mathscr{E}_j\right) = \sum_{j_1=1}^{n} \chi(\mathscr{E}_{j_1}) - \sum_{j_1 \neq j_2} \chi(\mathscr{E}_{j_1} \cap \mathscr{E}_{j_2})$$
$$+ \cdots + (-1)^{s+1} \sum_{j_m \neq j_l} \chi\left(\bigcap_{t=1}^{s} \mathscr{E}_{j_t}\right) + \cdots + (-1)^{n+1} \chi\left(\bigcap_{t=1}^{n} \mathscr{E}_t\right) \tag{1.12}$$

for the indicator of the union of finitely many sets $\mathscr{E}_j$ belonging to some σ-algebra. (The indicator $\chi_A(a) \equiv \chi(a \mid A)$ of A is a numerical function. The argument a is omitted in (1.12) for the sake of brevity.)

To prove (1.12), consider the set E_k whose every element belongs precisely to k of the sets $\mathscr{E}_1, \dots, \mathscr{E}_n$. Over the set E_k, we readily see that

$$\sum_{j_m \neq j_l} \chi\left(\bigcap_{t=1}^{s} \mathscr{E}_{j_t}\right) = \binom{k}{s}.$$

Thus, on E_k the indicator of the right side of (1.12) takes the form

$$\binom{k}{1} - \binom{k}{2} + \binom{k}{3} + \cdots + (-1)^{k+1}\binom{k}{k} = 1,$$

i.e., coincides with the left side.

Denote the union of the sets $\mathscr{E}_j$, $j = 1, \dots, n$, by $\mathscr{E}_0$. It is not hard to see that

$$\mathscr{E}_0 = \bigcup_{k=1}^{n} E_k, \quad E_i \cap E_j = \varnothing \quad \text{for} \quad i \neq j.$$

The equality (1.12), valid over the set E_k, holds also on their union $\mathscr{E}_0$. At the points not belonging to $\mathscr{E}_0$, both sides of (1.12) equal 0. Thus, (1.12) is proven.

Consider the tuples of numbers $\vec{L} = \{l_1, \dots, l_t\}$ composed of distinct elements of the set of indices $\{1, 2, \dots, n\}$. To each tuple we assign its *size*, the number $t = t(\vec{L})$. For a fixed n the set of tuples $\vec{L}$ is finite. Enumerate its elements by means of the index j to obtain

$$\vec{L}_1, \vec{L}_2, \dots, \vec{L}_j, \dots, \vec{L}_{2^n},$$

Require also that $t_j = t(\vec{L}_j)$ be not greater than t_{j+1}. Using the above notation, rewrite (1.12) as

$$\chi(\mathscr{E}_0) = \sum_{\vec{L}} (-1)^{t(\vec{L})+1} \chi\Big(\bigcap_{l \in \vec{L}} \mathscr{E}_l\Big) = \sum_{j=1}^{2^n} (-1)^{t_j+1} \chi\Big(\bigcap_{l \in \vec{L}_j} \mathscr{E}_l\Big). \tag{1.13}$$

Revert to the problem of expressing the Fourier coefficients $a[\gamma]$ of a function $\varphi \in \widetilde{L}_2^{(m)}$ through the errors $b_\varphi[\gamma]$. Enumerate all primes in conventional increasing order

$$p_0 = 1; \quad p_1 = 2; \quad p_2 = 3; \quad p_3 = 5, \dots.$$

Consider the set $E(\gamma, s)$ of nonzero vectors $\beta \neq \gamma$ such that each component β_j is a multiple of at least one of the products $\gamma_j p_{t_j}$ with $t_j = 0, 1, \dots, s$ and $|t| > 0$. Arrange the following difference

$$b_\varphi[\gamma] - \sum_{\beta \in E(\gamma, s)} a[\beta]. \tag{1.14}$$

Obviously, the sum in (1.14) as s increases becomes however close to the sum of $a[\beta]$ taken over all $\beta \neq 0$ each of which is a multiple of γ but differs from γ. Whence and from (1.11) it follows that

$$\lim_{s \to \infty} \Big\{ b_\varphi[\gamma] - \sum_{\beta \in E(\gamma, s)} a[\beta] \Big\} = a[\gamma]. \tag{1.15}$$

Transform the sum in the preceding equality under the limit sign by using the next formula

$$\sum_{\beta \in E(\gamma, s)} a[\beta] = \sum_{\beta} \chi(\beta \mid E(\gamma, s))\, a[\beta],$$

with $\chi(\cdot \mid E(\gamma, s))$ the indicator of the subset $E(\gamma, s)$ in the lattice $\mathbb{Z}^n$ of all integer vectors β. Fix some s and consider the system of the sets

$$B_j(\gamma \mid l), \quad 1 \leq j \leq n, \quad 1 \leq l \leq s.$$

Each set $B_j(\gamma \mid l)$ comprises vectors $\beta \in E(\gamma, s)$ such that the component β_j divides into $\gamma_j p_l$. Clearly, $E(\gamma, s)$ is the union of the sets $B_j(\gamma \mid l)$ over all l and j under consideration

$$E(\gamma, s) = \bigcup_{\substack{1 \le l \le s \\ 1 \le j \le n}} B_j(\gamma \mid l).$$

Consequently, the equality takes place

$$\sum_{\beta \in E(\gamma,s)} a[\beta] = \sum_{\beta} \chi\Big(\beta \mid \bigcup_{\substack{1 \le j \le n \\ 1 \le l \le s}} B_j(\gamma \mid l)\Big) a[\beta]. \tag{1.16}$$

It is essential that each of the sets $B_j(\gamma \mid l)$ and their intersections comprises integer vectors β whose every component β_j divides into the same factor. We explain what this factor is by the example of the intersection

$$\bigcap_{k \in \vec{M}} \bigcap_{l \in \vec{L}_j} B_k(\gamma \mid l),$$

with $\vec{L}_j = \{l_1^{(j)}, l_2^{(j)}, \ldots, l_{t_j}^{(j)}\}$ some collection of nonnegative integers each of which is at most s and $\vec{M}$ a subset of $\{1, 2, \ldots, n\}$. By definition, this intersection consists of all multi-indices $\beta \in E(\gamma, s)$ whose component β_k divides into $\gamma_k p_{l_1^{(j)}} p_{l_2^{(j)}} \cdots p_{l_{t_j}^{(j)}}$ for all $k \in \vec{M}$. In other words, every vector β of the intersection may be written as

$$\beta = \Gamma Q \vec{\alpha}. \tag{1.17}$$

Here $\vec{\alpha}$ is an nonzero integer multi-index and ΓQ is the diagonal matrix with diagonal entries

$$(\gamma Q)_k = \gamma_k p_{l_1^{(j)}} p_{l_2^{(j)}} \cdots p_{l_{t_j}^{(j)}}, \ k \in \vec{M}; \quad (\gamma Q)_k = \gamma_k, \ k \notin \vec{M}. \tag{1.18}$$

Each multi-index of the shape (1.17) belongs to the intersection of the sets $B_k(\gamma \mid l)$ under consideration.

Applying (1.13) to the indicator of the union $E(\gamma, s)$ of the sets $B_k(\gamma \mid l)$, from (1.16), obtain

$$\sum_{\beta \in E(\gamma,s)} a[\beta] = \sum_{\vec{L}, \vec{M}} (-1)^{t(\vec{L}) + t(\vec{M}) + 1} \sum_{\beta \in \bigcap_{k \in \vec{M}} \bigcap_{l \in \vec{L}} B_k(\gamma \mid l)} a[\beta]. \tag{1.19}$$

Summation on the right side of this equality is taken over all collections $\vec{L}$ lying in $\{1, 2, \ldots, s\}$ and all subsets $\vec{M}$ of $\{1, 2, \ldots, n\}$. The inner summation on the right side of (1.19) is easy to perform on using (1.17) and (1.11), which yields

$$\sum_{\beta \in \bigcap_{k \in \vec{M}} \bigcap_{l \in \vec{L}} B_k(\gamma \mid l)} a[\beta] = b_\varphi[\gamma Q], \tag{1.20}$$

with the integer entries $(\gamma Q)_j$ of the vector γQ defined by (1.18). Inserting (1.20) in (1.19) and putting the result in (1.15), finally obtain

$$a[\gamma] = b_\varphi[\gamma] - \lim_{s \to \infty} \sum_{\vec{L} = \vec{L}(s)} \sum_{\vec{M}} (-1)^{t(\vec{L}) + t(\vec{M}) + 1} b_\varphi[\gamma Q(\vec{L}, \vec{M})].$$

We found this expansion on assuming that every entry of the multi-index γ is nonzero. When this assumption is rejected, an analogous formula remains valid but needs clarification. If there is a zero entry of γ then $b_\varphi[\gamma]$ is the value at $\varphi(x)$ of the functional $l(x \mid \Gamma)$ with the singular matrix Γ. For the sake of definiteness, let $\gamma_{k+1} = \ldots = \gamma_n = 0$, and for the remaining $\gamma_j \neq 0$, define $l(x \mid \Gamma))$ as

$$(l(x \mid \Gamma), \varphi(x)) = \int_Q \varphi(x) dx$$

$$- \frac{1}{|\gamma_1 \ldots \gamma_k|} \sum_{\alpha_1 = 0}^{|\gamma_1| - 1} \cdots \sum_{\alpha_k = 0}^{|\gamma_k| - 1} \int_{Q_k} \varphi\left(\frac{\alpha_1}{|\gamma_1|}, \ldots, \frac{\alpha_k}{|\gamma_k|}, x_{k+1}, \ldots, x_n\right) dx_{k+1} \ldots dx_n,$$

with Q_k the unit cube in the space of $x_{k+1}, \ldots, x_n$.

Thus, the coefficients $a[\gamma]$ are expressed through $b_\varphi[\gamma]$, what was required.

It is difficult to verify the conditions for a function $\varphi(x)$ with the Fourier coefficients $a[\gamma]$ expressed in terms of the errors $b_\varphi[\gamma]$ to belong to $\widetilde{L}_2^{(m)}$.

We examined convergence of cubature formulas at fixed functions in $\widetilde{L}_2^{(m)}$. Now we are interested in the question of how such formulas converge at infinitely differentiable functions in $\widetilde{L}_2^{(m)}$.

For every given periodic infinitely differentiable function $\varphi(x)$ there are various estimates for the error given by (1.1). These estimates depend on the choice of m, namely,

$$|(l_h, \varphi)| \leq K_m h^m \|\varphi \mid \widetilde{L}_2^{(m)}\|. \tag{1.21}$$

Advancing various hypotheses about the rate of growth of the norm $\|\varphi \mid \widetilde{L}_2^{(m)}\|$ as m increases, we obtain the succession of inequalities (1.21). Knowing the behavior of K_m, for each h we may choose the strongest of these inequalities and establish the estimate

$$|(l_h, \varphi)| < \inf_m \{K_m \|\varphi \mid \widetilde{L}_2^{(m)}\| h^m\} = \eta(h, \varphi). \tag{1.22}$$

Constructing $\eta(h, \varphi)$ is an important problem. The inverse $\eta^{-1}(\varepsilon, \varphi)$ of $\eta(h, \varphi)$, i.e., the function satisfying the relation $\eta^{-1}(\eta(h, \varphi), \varphi) = h$, gives the mesh-size h that we must take so as to achieve the required accuracy $\varepsilon = \eta(h, \varphi)$.

Since, for infinitely differentiable functions, the number m may be arbitrary; it is obvious that the error (l_h, φ) of a cubature formula for each of these functions vanishes faster than every power of h.

As a rule, at a fixed h, among the estimates (1.21) corresponding to various m's, there is a best one and the index of this estimate $m_0(h)$ depends on h. In the opposite case, the formula is absolutely exact for h sufficiently small. Denote

$$K(m, \varphi) = K_m \|\varphi \mid \widetilde{L}_2^{(m)}\|.$$

Lemma 7.1. *If for some function φ at a given $h = h_0$ there is an infinite set of $m = m_j$ such that $K(m, \varphi) h_0^m$ remains bounded, then for such φ and all $h < h_0$ the cubature formula under study is exact, i.e.,*

$$(l_h, \varphi) = 0.$$

This lemma is immediate from the estimates

$$|(l_h, \varphi)| < K(m_j, \varphi) h^{m_j} = (h/h_0)^{m_j} \left[K(m_j, \varphi) h_0^{m_j}\right],$$

making it clear that (l_h, φ) is less than every given number for $h < h_0$.

§2. The Function Classes $H(\varkappa, A, \lambda)$ and $C(\varkappa, A, \lambda)$

Consider some special spaces of infinitely differentiable functions defined by the growth rate of their derivatives which depends on the order of the derivatives.

Say that a function $\varphi(x)$, given in a domain Ω and infinitely differentiable in the closure $\overline{\Omega}$, *belongs to the class* $H(\varkappa, A, \lambda)$, if for every $m \geq 1$ the following inequalities are satisfied

$$\|\varphi \mid L_2^{(m)}(\Omega)\| \equiv \left(\int_\Omega \sum_{|\alpha|=m} \frac{m!}{\alpha!} |D^\alpha \varphi(x)|^2 \, dx \right)^{1/2} \leq K m^{\varkappa m} A^m m^\lambda,$$

with K a constant independent of m. It is not hard to see that the set $H(\varkappa, A, \lambda)$ is a linear space. Moreover, we have

Lemma 7.2. *For all real* λ_1, λ_2, A_1, A_2, *and* $\varepsilon > 0$, *the following embeddings are valid*

$$H(\varkappa, A, \lambda_1 - \varepsilon) \subset H(\varkappa, A, \lambda_1); \quad H(\varkappa, A - \varepsilon, \lambda_1) \subset H(\varkappa, A, \lambda_2); \\ H(\varkappa - \varepsilon, A_1, \lambda_1) \subset H(\varkappa, A_2, \lambda_2). \tag{2.1}$$

The relations (2.1) follow from the definition of $H(\varkappa, A, \lambda)$.
We use the natural notation

$$H(\varkappa, A) = \bigcup_{\lambda} H(\varkappa, A, \lambda), \quad H(\varkappa) = \bigcup_{A} H(\varkappa, A).$$

Alongside the class $H(\varkappa, A, \lambda)$, consider the set $C(\varkappa, A, \lambda)$ comprising functions such that, for all $m \geq 1$, the inequality holds

$$\|\varphi \mid C^{(m)}(\Omega)\| = \sup_{x \in \overline{\Omega}} \left\{ \sum_{|\alpha|=m} \frac{m!}{\alpha!} |D^\alpha \varphi(x)|^2 \right\}^{1/2} \leq K m^{\varkappa m} A^m m^\lambda. \tag{2.2}$$

The spaces $C(\varkappa, A, \lambda)$ satisfy the embedding relations (2.1). The classes $C(\varkappa, A)$ and $C(\varkappa)$ are defined as follows

$$C(\varkappa, A) = \bigcup_{\lambda} C(\varkappa, A, \lambda), \quad C(\varkappa) = \bigcup_{A} C(\varkappa, A).$$

The inequality (2.2) enables us to obtain an obvious estimate for each mth order derivative of φ at an arbitrary point $x \in \overline{\Omega}$. Given $\varphi \in C(\varkappa, A, \lambda)$, for $m \geq 1$ observe

$$\sqrt{\frac{m!}{\alpha!}}\, |D^\alpha \varphi(x)| \leq K m^{\varkappa m} A^m m^\lambda. \tag{2.3}$$

Conversely, for every function $\varphi(x)$ given in $\overline{\Omega}$ and satisfying (2.3), elementarily infer

$$\|\varphi \mid C^{(m)}(\Omega)\| \leq K m^{\varkappa m} A^m m^\lambda \left(\sum_{|\alpha|=m} 1 \right)^{1/2},$$

with

$$\sum_{|\alpha|=m} 1 = \frac{(m+n-1)!}{m!(n-1)!}$$

a polynomial in m of degree $n-1$. Consequently, every $\varphi(x)$ satisfying (2.3) belongs to $C(\varkappa, A, \lambda + (n-1)/2)$.

Theorem 7.3. *Let $k = [n/2] + 1$. Then the following embeddings are valid*

$$C(\varkappa, A, \lambda - (n-1)/2) \subset H(\varkappa, A, \lambda) \subset C(\varkappa, A, \lambda + k\varkappa). \tag{2.4}$$

PROOF. The first of the embeddings (2.4) it is not hard to derive from the definition of the norm $\|\varphi \mid L_2^{(m)}(\Omega)\|$ on using the estimates for derivatives like (2.3) which follow from the containment of φ in $C(\varkappa, A, \lambda - (n-1)/2)$.

Show the second of the embeddings (2.4). Transforming the expression of the norm of $\varphi \in H(\varkappa, A, \lambda)$, obtain

$$\|\varphi \mid L_2^{(m+k)}(\Omega)\|^2 = \int\limits_\Omega \sum_{\substack{1\le j_s \le n\\ s=1,2,\dots,m+k}} \left(\frac{\partial^{m+k}\varphi}{\partial x_{j_1}\partial x_{j_2}\dots\partial x_{j_{m+k}}}\right)^2 dx$$

$$= \sum_{\substack{1\le j_s \le n\\ s=1,2,\dots,m}} \int\limits_\Omega \sum_{\substack{1\le j_s \le n\\ m+1\le s\le m+k}} \left[\frac{\partial^k}{\partial x_{j_{m+1}}\dots\partial x_{j_{m+k}}}\left(\frac{\partial^m\varphi}{\partial x_{j_1}\dots\partial x_{j_m}}\right)\right]^2 dx$$

$$= \sum_{\substack{1\le j_s \le n\\ s=1,2,\dots,m}} \left\|\frac{\partial^m\varphi}{\partial x_{j_1}\dots\partial x_{j_m}} \mid L_2^{(k)}(\Omega)\right\|^2 = \sum_{|\alpha|=m} \frac{m!}{\alpha!}\|D^\alpha\varphi \mid L_2^{(k)}(\Omega)\|^2. \tag{2.5}$$

By hypothesis, the derivative $D^\alpha\varphi(x)$ is continuous in $\overline{\Omega}$ and belongs to $L_2^{(k)}(\Omega)$. Applying to it the First Embedding Theorem of [265], deduce

$$|D^\alpha\varphi(x)|^2 \le K\left[\|D^\alpha\varphi \mid L_2(\Omega)\| + \|D^\alpha\varphi \mid L_2^{(k)}(\Omega)\|\right]^2,$$

with $x \in \overline{\Omega}$ and K a constant independent of m. Whence and from (2.5), for all integer $m \ge 1$, we have

$$\|\varphi \mid C^{(m)}(\Omega)\|^2 \le \sup_{x\in\overline{\Omega}} \sum_{|\alpha|=m} \frac{m!}{\alpha!}|D^\alpha\varphi(x)|^2$$

$$\le K(m+k)^{2\varkappa(m+k)}A^{2(m+k)}(m+k)^{2\lambda}. \tag{2.6}$$

Estimate the expression on the right side of this inequality. Obviously, $(m+k)^{2\lambda} \le Km^{2\lambda}$. Furthermore,

$$(m+k)^{\varkappa(m+k)} = m^{\varkappa(m+k)}\left[(1 - k/(m+k))^{(m+k)/k}\right]^{-\varkappa k} \le m^{\varkappa(m+k)}e^{k\varkappa(1+\eta)},$$

with η however small at m large. Whence and from (2.6) it follows that φ actually belongs to $C(\varkappa, A, \lambda + k\varkappa)$. The proof of Theorem 7.3 is complete.

Theorem 7.3 entails the coincidence of the classes $H(\varkappa, A)$ and $C(\varkappa, A)$ as well as the coincidence of the classes $H(\varkappa)$ and $C(\varkappa)$. Observe that the set $H(\varkappa) = C(\varkappa)$ with $\varkappa \geq 1$ is known in the literature as the *class of Gevrey functions* or simply the *Gevrey class* [274]. An important property of this class is the fact that $H(\varkappa)$ is an algebra. The sum of two functions in $H(\varkappa)$, as well as the product of a function in $H(\varkappa)$ and a complex number, is again a member of the class $H(\varkappa)$. Moreover, we have

Theorem 7.4. *The product of every two functions of $H(\varkappa)$ is a member of this class.*

PROOF. Let the functions $\varphi_1(x)$ and $\varphi_2(x)$ belong to $C(\varkappa, A, \lambda)$. The derivative of order α of their product may be written as

$$D^\alpha(\varphi_1\varphi_2) = \sum_{0\leq\beta\leq\alpha} \frac{\alpha!}{\beta!(\alpha-\beta)!} D^\beta\varphi_1(x) D^{\alpha-\beta}\varphi_2(x).$$

Using the estimates for $D^\beta\varphi_1(x)$ and $D^{\alpha-\beta}\varphi_2(x)$ that are valid by hypothesis, at every point $x \in \overline{\Omega}$ observe

$$|D^\alpha(\varphi_1\varphi_2)(x)| \leq \sum_{0\leq\beta\leq\alpha} \frac{\alpha!}{\beta!(\alpha-\beta)!} |\beta|^{\varkappa|\beta|} A^{|\beta|} |\beta|^\lambda |\alpha-\beta|^{\varkappa|\alpha-\beta|} A^{|\alpha-\beta|} |\alpha-\beta|^\lambda.$$

Substituting $|\alpha|$, for $|\beta|$ and $|\alpha-\beta|$, for the arguments of the powers on the right side of this inequality, which may only increase the latter, obtain

$$\begin{aligned} |D^\alpha(\varphi_1\varphi_2)(x)| &\leq \sum_{0\leq\beta\leq\alpha} \frac{\alpha!}{\beta!(\alpha-\beta)!} (|\alpha|^\varkappa A)^{|\beta|+|\alpha-\beta|} |\alpha|^{2\lambda} \\ &= |\alpha|^{\varkappa|\alpha|} A^{|\alpha|} |\alpha|^{2\lambda} \Big(\sum_{0\leq\beta\leq\alpha} \frac{\alpha!}{\beta!(\alpha-\beta)!} \Big), \end{aligned}$$

with the sum in parentheses equal to $2^{|\alpha|}$. Thus, the product $\varphi_1\varphi_2$ belongs to $H(\varkappa)$. The proof of Theorem 7.4 is complete.

The change of variable $x = Ky$, i.e., *scaling*, sends the class $C(\varkappa, A, \lambda)$ with respect to x to the class $C(\varkappa, A_1, \lambda)$ with respect to y, where $A_1 = KA$. The functions of the class $C(\varkappa, A, \lambda)$ satisfy some invariance condition under change of variables. We have

Theorem 7.5. *For $\varkappa \geq 1$, each change of variables $x = \chi(y)$ analytic up to the boundary of a domain Ω sends a function $\varphi(x)$ of the class $C(\varkappa, A, \lambda)$ to a function $\psi(y)$ of the class $C(\varkappa, A_2, \lambda_2)$ with the same value of $\varkappa$.*

PROOF. To simplify calculations, site the coordinate origin at a chosen point x of the space. Also assume that this point corresponds to the coordinate origin of y, i.e., $\chi(0) = 0$. Consider the polynomial

$$T_m(x) = \sum_{|\alpha| \leq m} \frac{D^\alpha \varphi(0) x^\alpha}{\alpha!}$$

which is a part of the Taylor series for $\varphi(x)$. Put $Q_m(y) = T_m(\chi(y))$. The function $Q_m(y)$ at the coordinate origin has the same derivatives with respect to y up to order m as the function $\psi(y) = \varphi(\chi(y))$. This is easy to validate, writing $\psi(y)$ as the sum of $Q_m(y)$ and the remainder $R_m(\chi(y))$ of the Taylor series for $\varphi(x)$. In this event the function $R_m(x)$ has a root of multiplicity $m+1$ at the coordinate origin. Since $\chi(0) = 0$, it follows that $R_m(\chi(y))$ also has a root of multiplicity $m+1$ at $y = 0$. Thus, it suffices to estimate the derivatives up to order m of $Q_m(y)$, with the aim of obtaining a dominant for the corresponding derivatives of $\chi(y)$. We proceed with such estimation.

To begin with, construct a dominant for all components of the vector-function $\chi(y) = (\chi_1(y), \ldots, \chi_n(y))$, i.e., a function $Y(y)$ such that, for all $j = 1, \ldots, n$ and every α, in a neighborhood of the coordinate origin the inequality holds

$$|D^\alpha \chi_j(y)| \leq D^\alpha Y(|y_1|, \ldots, |y_n|).$$

It is possible to prove by using the analiticity of $\chi(y)$ that as a common dominant at M sufficiently large and R sufficiently small we may take the following ratio

$$Y(y) = M / \left[1 - \left(\sum_{i=1}^n y_i\right) / R\right].$$

Let $\xi = y_1 + \cdots + y_n$ and $f(\xi) = RM/(R - \xi)$. Then $Y(y) = f(\xi)$. It is not hard to calculate that for $|\alpha| = m$ we have

$$D^\alpha Y(y) = \frac{d^m f}{d\xi^m} = \frac{RMm!}{(R-\xi)^{m+1}}. \tag{2.7}$$

All derivatives of $(\chi(y))^\alpha$ are polynomials in the derivatives of $\chi_j(y)$ with positive coefficients. Since at the coordinate origin the derivatives of $Y(y)$ are positive and greater than the corresponding derivatives of $\chi_j(y)$, the function

$$M_m(y) = \sum_{l=0}^m \left[\sum_{|\alpha|=l} \frac{D^\alpha \varphi(0)}{\alpha!} (Y(y))^{|\alpha|}\right] = \sum_{l=0}^m b_l \left(\frac{RM}{R-\xi}\right)^l \equiv \Lambda_m(\xi)$$

with $b_l = \sum_{|\alpha|=l} D^\alpha\varphi(0)/\alpha!$ is a dominant for $Q_m(y)$. Considering that all derivatives of order m with respect to y of $M_m(y)$ are equal to the derivatives of the same order of $\Lambda_m(\xi)$ with respect to ξ, for $|\alpha| = m$ obtain

$$|D^\alpha\psi(y)|\Big|_{y=0} = |D^\alpha Q_m(y)|\Big|_{y=0} \le \frac{d^m\Lambda_m}{d\xi^m}(\xi)\Big|_{\xi=0}. \tag{2.8}$$

Using the Cauchy inequality for sums and the fact that by hypothesis $\varphi(x)$ belongs to $C(\varkappa)$, obtain

$$|b_l| \le \sum_{|\alpha|=l}\frac{1}{\alpha!}|D^\alpha\varphi(0)| \le \frac{1}{l!}\Big\{\sum_{|\alpha|=l}\frac{l!}{\alpha!}|D^\alpha\varphi(0)|^2\Big\}^{1/2}\Big\{\sum_{|\alpha|=l}\frac{l!}{\alpha!}\Big\}^{1/2} \le \frac{Kl^{\varkappa l}A^l l^\lambda n^{l/2}}{l!}.$$

This, together with the Stirling formula, entails the inequality

$$\frac{d^m\Lambda_m}{d\xi^m}(\xi)\Big|_{\xi=0} \le K\sum_{l=0}^{m} l^{(\varkappa-1)l}R^{-m}(e\sqrt{n}\,AM)^l l^{\lambda-1/2}\frac{(m+l-1)!}{(l-1)!}.$$

The summands on the right side of this inequality grow with l in virtue of the hypothesis that $\varkappa \ge 1$. Replacing all of them by the greatest, i.e., that with $l = m$, deduce

$$\frac{d^m\Lambda_m}{d\xi^m}\Big|_{\xi=0} \le Km^{(\varkappa-1)m}A_1^m m^{\lambda+1/2}\frac{(2m-1)!}{(m-1)!},$$

where $A_1 = e\sqrt{n}\,AM/R$. Applying the Stirling formula to the factorials in this inequality, obtain the sought estimate

$$\frac{d^m\Lambda_m}{d\xi^m}\Big|_{\xi=0} \le Km^{\varkappa m}A_2^m m^{\lambda_2}.$$

Inserting it in (2.8), we see that $\psi(y) \in C(\varkappa, A_2, \lambda_2)$. The proof of Theorem 7.5 is complete.

§3. The Properties of $H(\varkappa, A)$ and $C(\varkappa, A)$ for $\varkappa \ge 1$

Consider a domain Ω in $\mathbb{R}^n$ with boundary a smooth analytic surface $\mathscr{F}(x) = 0$. We assume that $\mathscr{F}(x) < 0$ in Ω and $\mathscr{F}(x) > 0$ in the exterior of Ω. Let $\varphi(x)$ belong to $H(\varkappa, A, \lambda)$ in a domain Ω_1 whose interior includes $\overline{\Omega}$. Then we have the following

Theorem 7.6. *Each $\varphi(x)$ in $H(\varkappa, A, \lambda)$ for $\varkappa > 1$ extends from the domain Ω to an arbitrary domain $\Omega_2 \supset \overline{\Omega}_1$ to a compactly-supported function belonging to some space $H(\varkappa, A, \lambda_1)$ in Ω_2.*

This theorem implies that the values of a function $\varphi(x)$ in $H(\varkappa)$ in two analytic disjoint domains ω_1 and ω_2 may be prescribed independently.

Before launching into the proof of Theorem 7.6, we state the following

Lemma 7.3. *The function $\varphi_m(y) = e^{-1/y^m}$, given on the semiaxis $0 \le y < \infty$, belongs to the space $C(1 + 1/m, A, \lambda)$.*

PROOF. The derivative of $\varphi_m(y)$ of order k has the form

$$\frac{d^k \varphi_m}{dy^k}(y) = \left(\sum_{s=1}^{k} A_s^{(k)} y^{-sm-k} \right) e^{-1/y^m}. \tag{3.1}$$

This is obvious for $k = 1$, and the differentiation of (3.1) shows that this formula, when valid for k, remains valid for $k + 1$. Thus, by the induction principle, (3.1) holds for every k.

Differentiating (3.1) once, obtain a recurrent formula that relates two consecutive derivatives of $\varphi_m(y)$, namely,

$$\begin{aligned} e^{1/y^m} \frac{d^{k+1}\varphi_m}{dy^{k+1}}(y) &= -\{(m+k)A_1^{(k)} y^{-m-k-1} \\ &+ \cdots + (sm+k)A_s^{(k)} y^{-sm-k-1} + \cdots + (mk+k)A_k^{(k)} y^{-mk-k-1}\} \\ &+ \{mA_1^{(k)} y^{-2m-k-1} + \cdots + mA_s^{(k)} y^{-(s+1)m-k-1} \\ &+ \cdots + mA_k^{(k)} y^{-(k+1)m-k-1}\}. \end{aligned} \tag{3.2}$$

Whence we see that the coefficients $A_s^{(k)}$ are solutions to the difference equation

$$A_s^{(k+1)} = -(sm+k)A_s^{(k)} + mA_{s-1}^{(k)} \tag{3.3}$$

subject to the conditions

$$A_s^{(k)} = 0 \quad \text{for } s \le 0; \quad A_s^{(k)} = 0 \quad \text{for } s > k; \quad A_1^{(1)} = m. \tag{3.4}$$

Performing the change of variables $A_s^{(k)} = (-1)^{k+s} m^k C_s^{(k)}$ in (3.3) and taking account of (3.4), obtain

$$C_s^{(k+1)} = \left(s + \frac{k}{m}\right) C_s^{(k)} + C_{s-1}^{(k)}; \quad C_1^{(1)} = 1; \quad C_s^{(k)} = 0 \quad \text{for } s \le 0 \text{ or } s > k.$$

Show that for $0 < s \le k$ the following formula is valid for a solution to this problem

$$C_s^{(k)} = \sum_{0<j_1<\cdots<j_{s-1}<k} \left[\prod_{0<t<j_1} \left(1+\frac{t}{m}\right) \cdots \prod_{j_{l-1}<t<j_l} \left(l+\frac{t}{m}\right) \cdots \prod_{j_{s-1}<t<k} \left(s+\frac{t}{m}\right) \right]. \tag{3.5}$$

Each summand on the right side is a product of $k - s$ factors. Furthermore, the variable t of multiplication ranges over all indices from 1 to $k-1$ with the omission of $j_1, j_2, \ldots, j_{s-1}$. At each omitted value of t, the number l in the factors like $l+t/m$ becomes greater by 1.

The validity of (3.5) is established by induction on k.

Assume (3.5) valid for all $C_{s-1}^{(k)}$ and $C_s^{(k)}$. Calculate the sum

$$\left(s+\frac{k}{m}\right)C_s^{(k)} + C_{s-1}^{(k)}, \tag{3.6}$$

inserting in it the values of $C_s^{(k)}$ and $C_{s-1}^{(k)}$ taken from (3.5). The quantity $C_{s-1}^{(k)}$ equals the sum of the products of $k+1-s$ factors like $l+t/m$ with the omission of $s-2$ factors with the indices $j_1, j_2, \ldots, j_{s-2}$. This corresponds to the sum of the products of the same factors with the omission of the $(s-1)$th factor of $j_1, j_2, \ldots, j_{s-1}$, where $j_{s-1} = k < k+1$. In the sum of the shape (3.5), there are also $k+1-s$ factors like $l+t/m$ following in succession as before with the omission of an arbitrary $(s-1)$th number, namely, $j_1, j_2, \ldots, j_{s-1}$. This is clear from the fact that, for $j_{s-1} < k-1$, the new factor $s+k/m$ lengthens the last product by one factor; and, for $j_s = k-1$, it introduces this product which is omitted in $C_s^{(k)}$. Thus,

$$\left(s+\frac{k}{m}\right)C_s^{(k)} + C_{s-1}^{(k)} = \sum_{0<j_1<\cdots<j_{s-1}<k+1} \left\{ \prod_{l=1}^{s} \prod_{j_{l-1}<t<j_l} \left(l+\frac{t}{m}\right) \right\}.$$

So, the validity of (3.5) is established.

Estimate the magnitude of the coefficients $C_s^{(k)}$ from above. Replacing $l+t/m$ by $s+k/m$ in all factors on the right side of (3.5) and noting that the total number of summed products in (3.5) equals the number of combinations selecting $j_1, j_2, \ldots, j_{s-1}$ out of $1, 2, \ldots, k-1$, i.e., it equals $(k-1)!/((s-1)!(k-s)!) < 2^k$; obtain

$$C_s^{(k)} \le 2^k \left(s+\frac{k}{m}\right)^{k-s}, \quad |A_s^{(k)}| = m^k C_s^{(k)} \le (2m)^k \left(s+\frac{k}{m}\right)^{k-s}. \tag{3.7}$$

Return to estimating the derivatives of $\varphi_m(y)$. It is easy to check that

$$\max_{y\ge 0} y^{-k-sm} e^{-1/y^m} = \left(\frac{sm+k}{me}\right)^{(sm+k)/m} = \left(s+\frac{k}{m}\right)^{s+k/m} e^{-(s+k/m)}.$$

Whence, using (3.1) and (3.7), obtain

$$\left|\frac{d^k\varphi_m}{dy^k}(y)\right| \le \sum_{s=1}^{k}(2m)^k\left(s+\frac{k}{m}\right)^{k+k/m} e^{-(s+k/m)}$$
$$\le k(2m)^k\,[k\,(1+1/m)]^{k(1+1/m)}\,e^{-k(1+1/m)}.$$

This, however, means that $\varphi_m(y)$ belongs to $C(\varkappa, A, \lambda)$ for

$$\varkappa = 1+\frac{1}{m}, \quad A = 2m\left[\left(1+\frac{1}{m}\right)/e\right]^{1+1/m}, \quad \lambda = 1.$$

The proof of Lemma 7.3 is complete.

Corollary 3.1. *The function $\varphi_{0,m}(y)$, $[0,1]$ and agreeing with the product $\varphi_m(y)\varphi_m(1-y)$ in the interval, belongs to $C(1+1/m)$ on the whole axis $-\infty < y < \infty$.*

From (3.2) it is immediate that $\varphi_{0,m}(y)$ is an infinitely differentiable function with all derivatives at $y=0$ and $y=1$ equal 0. Using Theorem 7.4 and Lemma 7.3, we now see that $\varphi_{0,m}(y)$ indeed belongs to the class $C(1+1/m)$.

Corollary 3.2. *The function $\psi_m(y)$, given by the equality*

$$\psi_m(y) = \frac{1}{c_{m,0}}\int_y^\infty \varphi_{0,m}(\xi)\,d\xi, \tag{3.8}$$

with $c_{m,0} = \int_0^1 \varphi_{0,m}(\xi)\,d\xi$, belongs to the space $C(1+1/m)$, equals 0 for $y \ge 1$, and equals 1 for $y \le 0$.

Corollary 3.3. *Let Ω be a bounded domain with an analytic boundary defined by the equation $\mathscr{F}(x) = 0$. Assume that $\mathscr{F}(x) < 0$ in Ω and $\mathscr{F}(x) > 0$ in the exterior of Ω. Assume further that the system of equalities $\mathscr{F}(x) = \varepsilon$ at small ε determines an analytic family of surfaces close to the boundary of Ω. Then for all $m > 0$ there is a function that is supported in a bounded domain Ω_1 including $\overline{\Omega}$, equals 1 in Ω and belongs to the class $C(1+1/m)$.*

It is not hard to see that as such function we may take $\psi_m(\mathscr{F}(x)/\varepsilon)$, with $\varepsilon > 0$ and $\psi_m(y)$ defined by (3.8). The fact that this function belongs to the class $C(1+1/m)$ follows from Theorem 7.5.

PROOF OF THEOREM 7.6. Assume that the function $\varphi(x)$ belongs to the class $C(\varkappa)$ with $\varkappa = 1+1/m$ and Ω is bounded with the analytic surface $\mathscr{F}(x) = 0$. Consider the product $\psi_m(\mathscr{F}(x)/\varepsilon)\varphi(x)$. By Theorem 7.4 this product is again a member of the class $C(\varkappa)$. By Corollary 3.2 it equals $\varphi(x)$ in Ω and equals 0 outside Ω_2 for sufficiently small $\varepsilon > 0$. The proof is complete.

Consider the case of $\varkappa = 1$.

***Theorem* 7.7.** *The class $H(1, A)$ consists of analytic functions on Ω.*

PROOF. Take $x_0 \in \Omega$. Estimate the term of degree m of the Taylor series for φ in $H(1, A)$ at the point x_0. Without loss of generality, we may assume that x_0 coincides with the coordinate origin. Using the Cauchy inequality, deduce

$$\left| \sum_{|\alpha|=m} \frac{D^\alpha \varphi(0)}{\alpha!} x^\alpha \right| \leq \left\{ \sum_{|\alpha|=m} \frac{m!}{\alpha!} |D^\alpha \varphi(0)|^2 \right\}^{1/2} \frac{1}{m!} \left\{ \sum_{|\alpha|=m} \frac{m!}{\alpha!} x^{2\alpha} \right\}^{1/2}$$
$$= \frac{|x|^m}{m!} \left\{ \sum_{|\alpha|=m} \frac{m!}{\alpha!} |D^\alpha \varphi(0)|^2 \right\}^{1/2}.$$

The function φ belongs to $H(1, A)$. Consequently, the last inequality may be continued as follows

$$\left| \sum_{|\alpha|=m} \frac{D^\alpha \varphi(0)}{\alpha!} x^\alpha \right| \leq \frac{|x|^m}{m!} m^m A^m m^\lambda. \tag{3.9}$$

By the Stirling formula there are constants K_1 and K_2 such that, for all $m \geq 1$, the inequality holds

$$K_1 (m/e)^m \sqrt{m} < m! < K_2 (m/e)^m \sqrt{m+1}. \tag{3.10}$$

Using the lower bound for $m!$, from (3.9) infer the inequality

$$|\varphi(x)| = \left| \sum_{\alpha} \frac{D^\alpha \varphi(0)}{\alpha!} x^\alpha \right| \leq K \sum_{m=0}^{\infty} (A|x|e)^m \, m^{\lambda - 1/2}. \tag{3.11}$$

Thus, the Taylor series for $\varphi(x)$ converges for $|x| < e/A$. The proof of Theorem 7.7 is complete.

In a similar fashion it is not hard to prove the following

***Theorem* 7.8.** *For $\varkappa < 1$ the class $C(\varkappa, A, \lambda)$ consists of entire functions.*

The properties of the classes $C(\varkappa, A)$ with $\varkappa < 1$ are inspected in more detail in the next section.

§4. The Function Classes $\Psi(\rho, \sigma, \mu)$

Estimates for the growth rate of functions in a single independent variable which belong to the space $C(\varkappa, A)$ with $\varkappa < 1$ are obtained in many classical books [122, 123, 135]. Analogous estimates are readily available for functions in many variables. We expose them estimates below.

Say that an *entire function* $\varphi(x)$ *in* n *variables belongs to class* $\Psi(\rho,\sigma,\mu)$, if there is a constant K such that, for all complex x, the inequality holds

$$|\varphi(x)| \le K|x|^{\mu}e^{\sigma|x|^{\rho}}. \tag{4.1}$$

The quantities ρ and σ are customarily called the *order* and the *type of an entire function* $\varphi(x)$.

Denote by $\Psi(\rho,\sigma)$ the union of the classes $\Psi(\rho,\sigma,\lambda)$ over all λ. We have the following

Theorem 7.9. *If an entire function $\varphi(x)$ belongs to the class $\Psi(\rho,\sigma,\mu)$, with ρ and σ positive; then in every bounded domain Ω this function belongs to $C(\varkappa,A,\lambda)$, with*

$$\varkappa = 1-1/\rho, \quad A=(\sigma\rho e)^{1/\rho}/e, \quad \lambda = 5n/4-1/4+\mu/\rho. \tag{4.2}$$

Before launching into the proof of the theorem, one lemma is in order.

Lemma 7.4. *Let θ be a unit vector with nonnegative components $\theta_1,\dots,\theta_n$. For the function*

$$\Phi(\theta) = \theta^{y} = \theta_1^{y_1}\theta_2^{y_2}\dots\theta_n^{y_n},$$

with $y_j>0$ and $\sum_{j=1}^{n} y_j = s$, the equality holds

$$\max_{|\theta|=1}\Phi(\theta) = \frac{y^{y/2}}{s^{s/2}} = \frac{y_1^{y_1/2}y_2^{y_2/2}\dots y_n^{y_n/2}}{s^{s/2}}. \tag{4.3}$$

PROOF. Seek for a maximum of $\Phi(\theta)$ by the Lagrange method of multipliers. Put

$$\Psi(\theta) = \Phi(\theta) - \frac{\mu}{2}\sum_{j=1}^{n}\theta_j^2.$$

Differentiating $\Psi(\theta)$ with respect to θ_j and equating the corresponding derivatives with 0, obtain the system of equations

$$y_j\frac{\Phi(\theta)}{\theta_j} - \mu\theta_j = 0, \quad j=1,2,\dots,n.$$

Now, by the properties of proportions, come to the equality

$$\Phi(\theta) = \mu\frac{\theta_j^2}{y_j} = \mu\frac{\sum\limits_{j=1}^{n}\theta_j^2}{\sum\limits_{j=1}^{n}y_j} = \frac{\mu}{s}.$$

Thus, the entries θ_j of the maximum points of $\Phi(\theta)$ satisfy the conditions

$$\theta_j^2 = \frac{y_j}{s} \quad \text{or} \quad \theta_j^{y_j} = \frac{y^{y_j/2}}{s^{y_j/2}}.$$

These equalities imply (4.3). The proof of Lemma 7.4 is complete.

In particular, for $y_1 = y_2 = \cdots = y_n = 1$ we have

$$\max_{|\theta|=1} (\theta_1 \theta_2 \dots \theta_n) = \frac{1}{n^{n/2}}.$$

Finally, for arbitrary nonnegative $\alpha_1, \alpha_2, \dots, \alpha_n$, putting $\theta_j^2 = \alpha_j / \sum_{j=1}^n \alpha_j$, obtain

$$(\alpha_1 \alpha_2 \dots \alpha_n)^{1/2} \leq \left(\sum_{j=1}^n \frac{\alpha_j}{n} \right)^{n/2}$$

This is the celebrated inequality between the arithmetic and geometric means.

Proof of Theorem 7.9. Estimate the value of $D^\alpha \varphi(0)/\alpha!$ for $|\alpha| = s$. Represent $\varphi(x)$ as a contour integral taken in every plane of a complex variable x over the circle $\Gamma_k = \{x_k : |x_k| = r\theta_k,\ \theta_k \geq 0\}$, namely,

$$\varphi(z) = \frac{1}{(i2\pi)^n} \int_{\Gamma_1} \cdots \int_{\Gamma_n} \frac{\varphi(x)\, dx_1 \dots dx_n}{(x_1 - z_1) \dots (x_n - z_n)}. \tag{4.4}$$

Choose the circles $\Gamma_1, \dots, \Gamma_n$ so that the normalization condition be met

$$\sum_{j=1}^n \theta_j^2 = |\theta|^2 = 1. \tag{4.5}$$

Differentiating both sides of (4.4) and putting $z = 0$, obtain

$$\left| \frac{D^\alpha \varphi(0)}{\alpha!} \right| \leq \frac{r^{-s}}{\theta^\alpha} \frac{1}{(2\pi)^n} \int_{\Gamma_1} \cdots \int_{\Gamma_n} |\varphi(x)| \frac{|dx_1|}{r\theta_1} \cdots \frac{|dx_n|}{r\theta_n}.$$

The function $\varphi(x)$ by hypothesis belongs to the class $\Psi(\rho, \sigma, \mu)$. Consequently, the last estimate may be continued as follows

$$\left| \frac{D^\alpha \varphi(0)}{\alpha!} \right| \leq K r^{\mu - s} e^{\sigma r^\rho} \theta^{-\alpha}. \tag{4.6}$$

Here $\theta = (\theta_1, \dots, \theta_n)$ is an arbitrary vector with nonnegative components θ_j satisfying (4.5). Choose θ so that the right side of (4.6) be minimal. Then in accord with Lemma 7.4 obtain the inequality

$$\left|\frac{D^\alpha \varphi(0)}{\alpha!}\right| \leq K\alpha^{-\alpha/2} s^{s/2} r^{\mu-s} e^{\sigma r^\rho}, \tag{4.7}$$

with K a constant independent of α and r. Assume in what follows that $s = |\alpha| > \mu$, and minimize over $r > 0$ the right side of (4.7). It is easy to check that for $\lambda > 0$ the minimum of $z^{-\lambda} e^z$ is attained at $z = \lambda$ and equals $\lambda^{-\lambda} e^\lambda$. Putting $z = \sigma r^\rho$, i.e., $r = (z/\sigma)^{1/\rho}$, obtain

$$\min_{r>0} r^{\mu-s} e^{\sigma r^\rho} = \min_{z>0} \left(\frac{z}{\sigma}\right)^{(\mu-s)/\rho} e^z = (\sigma\rho e)^{(s-\mu)/\rho} (s-\mu)^{(\mu-s)/\rho}.$$

This, together with (4.7), implies the inequality

$$\left|\frac{D^\alpha \varphi(0)}{\alpha!}\right| \leq K\alpha^{-\alpha/2} (\sigma\rho e)^{(s-\mu)/\rho} s^{s/2+(\mu-s)/\rho}, \quad |\alpha| = s. \tag{4.8}$$

From (3.10) it follows that

$$s! \leq K_2 (s/e)^s \sqrt{s+1}, \quad \alpha! \leq K_2 \alpha^\alpha e^{-s} \prod_{j=1}^n \sqrt{\alpha_j + 1}.$$

Whence and from (4.8) by simple calculations obtain the inequality

$$\sqrt{\frac{s!}{\alpha!}}\, |D^\alpha \varphi(0)| \leq K s^{(1-1/\rho)s} \left[\frac{1}{e} (\sigma\rho e)^{1/\rho}\right]^s s^{(n+1)/4+\mu/\rho}.$$

Squaring both sides, summing over α and noting that the number of different α, $|\alpha| = s$, equals $(s+n-1)!/s!(n-1)!$, i.e., is at most Ks^{n-1}, come to the estimate

$$\left\{\sum_{|\alpha|=s} \frac{s!}{\alpha!} |D^\alpha \varphi(0)|^2\right\}^{1/2} \leq K s^{(1-1/\rho)s} \left(\frac{1}{e} (\sigma\rho e)^{1/\rho}\right)^s s^{5n/4-3/4+\mu/\rho}.$$

The proof of Theorem 7.9 is complete.

We also have the following theorem in a sense converse to the preceding theorem.

***Theorem* 7.10.** *An entire function $\varphi(x)$ belonging to the class $C(\varkappa, A, \lambda)$ for $\varkappa < 1$ and $\lambda \leq -1/2$ is also a member of the class $\Psi(\rho, \sigma, \mu)$, with*

$$\varkappa = 1 - 1/\rho, \quad (\sigma\rho e)^{1/\rho} = Ae, \quad \mu = (\lambda + 1/2)\,\rho. \tag{4.9}$$

Observe that $\varkappa$ and A relate with ρ and σ by the same dependence as in Theorem 7.9, whereas $(5n+1)/4$ is subtracted from the exponent λ.

PROOF. Take $\varphi(x) \in C(\varkappa, A, \lambda)$. Suppose that $\varphi(0) = 0$. This leads to no loss of generality, since by hypothesis $\mu \leq 0$. Consequently, every constant function belongs to $\Psi(\rho, \sigma, \mu)$. At the coordinate origin we arrange the Taylor series for $\varphi(x)$, namely,

$$\varphi(x) = \sum_{s=1}^{\infty} \sum_{|\alpha|=s} \frac{D^\alpha \varphi(0)}{\alpha!} x^\alpha. \tag{4.10}$$

Estimate the homogeneous polynomial of degree s in the last expression. For $s \geq 1$ we have the inequality

$$\begin{aligned}\left|\sum_{|\alpha|=s} \frac{D^\alpha \varphi(0)}{\alpha!} x^\alpha\right| &\leq \left\{\sum_{|\alpha|=s} \frac{s!}{\alpha!} |D^\alpha \varphi(0)|^2\right\}^{1/2} \left\{\sum_{|\alpha|=s} \frac{1}{s!\alpha!} x^{2\alpha}\right\}^{1/2} \\ &\leq K \frac{s^{\varkappa s}(A|x|)^s s^\lambda}{s!}.\end{aligned}$$

Thus, the sum

$$\sum_{s=1}^{\infty} g(s) \equiv K \sum_{s=1}^{\infty} \frac{s^{\varkappa s}(A|x|)^s s^\lambda}{s!} \tag{4.11}$$

dominates (4.10). Find the ratio of two consecutive values of $g(s)$, namely,

$$\frac{g(s+1)}{g(s)} = \frac{A|x|}{s^{1-\varkappa}} \left(1 + \frac{1}{s}\right)^{\varkappa s + \varkappa + \lambda - 1}.$$

For the last quantity, when $\varkappa \geq 0$ and $\lambda \geq 1 - \varkappa$, we have the inequality

$$2^\varkappa \frac{A|x|}{s^{1-\varkappa}} \leq \frac{g(s+1)}{g(s)} \leq e^\varkappa 2^{\varkappa+\lambda-1} \frac{A|x|}{s^{1-\varkappa}}.$$

If, on the other hand, $\varkappa \geq 0$ and $\lambda < 1 - \varkappa$ then

$$2^\varkappa 2^{\varkappa+\lambda-1} \frac{A|x|}{s^{1-\varkappa}} \leq \frac{g(s+1)}{g(s)} \leq e^\varkappa \frac{A|x|}{s^{1-\varkappa}}.$$

Thus, for all λ and $\varkappa$, $0 \le \varkappa < 1$, there are two positive numbers $\eta_1 < \eta_2$ such that

$$g(s+1)/g(s) > 2 \quad \text{for} \quad s < \eta_1(A|x|)^{1/(1-\varkappa)},$$
$$g(s+1)/g(s) < 1/2 \quad \text{for} \quad s > \eta_2(A|x|)^{1/(1-\varkappa)}.$$

It is not hard to see that analogous constants η_1 and η_2 also exist for $\varkappa \le 0$. With this in mind, estimate the sum of (4.11). Observe first that each term of the series with $s > s_2 = [\eta_2(A|x|)^{1/(1-\varkappa)} + 1]$ decreases faster than a geometric progression with ratio 1/2. Consequently, the sum of these terms is less than $2g(s_2)$. Similarly, when $s < s_1 = [\eta_1(A|x|)^{1/(1-\varkappa)}]$ each of the respective terms decreases faster than a geometric progression with the same ratio as s goes from s_1 to 1. Thus, the sum of the respective terms is less than $2g(s_1)$. Consequently, instead the sum of s_1 first terms and the whole remainder of (4.11), starting with the term with index s_2, we may take the sum of the remaining summands, thus arriving at the inequality

$$\sum_{s=1}^{\infty} g(s) \le 2\sum_{s=s_1}^{s_2} g(s). \tag{4.12}$$

Estimate the last sum in (4.12), replacing each summand by the maximal one

$$\sum_{s=s_1}^{s_2} g(s) \le K(s_2 - s_1) \max_{s_1 \le s \le s_2} \frac{s^{\varkappa s}(A|x|)^s s^{\lambda}}{s!}.$$

Using now the estimate (3.10) for $s!$, obtain

$$\begin{aligned}\sum_{s=s_1}^{s_2} g(s) &\le K(s_2 - s_1) \max_{s_1 \le s \le s_2} s^{(\varkappa-1)s}(A|x|e)^s s^{\lambda-1/2} \\ &\le K(s_2 - s_1) \max_{s_1 \le s \le s_2} s^{\lambda-1/2} \max_{s_1 \le s \le s_2} s^{(\varkappa-1)s}(A|x|e)^s.\end{aligned} \tag{4.13}$$

The definition of s_1 and s_2 entails the next obvious inequality

$$(s_2 - s_1) \max_{s_1 \le s \le s_2} s^{\lambda-1/2} \le K|x|^{(\lambda+1/2)/(1-\varkappa)}. \tag{4.14}$$

To estimate the second maximum in (4.13), perform the change of variables

$$s = z(Ae|x|)^{1/(1-\varkappa)} \quad \text{or} \quad z = s(Ae|x|)^{-1/(1-\varkappa)}.$$

Then

$$\max_{s_1 \le s \le s_2} s^{(\varkappa-1)s}(A|x|e)^s \le \max_{z \ge 0} (z^{-z})^{(1-\varkappa)(Ae|x|)^{1/(1-\varkappa)}}.$$

The maximum of z^{-z} is attained at $z = 1/e$. Consequently, the last inequality may be continued as follows

$$\max_{s_1 \leq s \leq s_2} s^{(\varkappa-1)s}(A|x|e)^s \leq e^{(1-\varkappa)e^{-1}(Ae|x|)^{1/(1-\varkappa)}}. \tag{4.15}$$

Inserting the inequalities (4.14) and (4.15) in (4.13) and putting the result in (4.12), infer

$$\sum_{s=1}^{\infty} g(s) \leq K|x|^{(\lambda+1/2)/(1-\varkappa)} e^{(1-\varkappa)e^{-1}(Ae)^{1/(1-\varkappa)}|x|^{1/(1-\varkappa)}}.$$

Setting $\rho = 1/(1-\varkappa)$, $\sigma\rho e = (Ae)^\rho$, and $\mu = (\lambda + 1/2)\rho$ and recalling that the series on the left side is a dominant for $\varphi(x)$, we see that $\varphi(x)$ does belong to the class $\Psi(\rho, \sigma, \mu)$. The proof of Theorem 7.10 is complete.

§5 The Classes of Periodic Functions $\widetilde{H}(\varkappa, A, \lambda)$

By now in the current chapter we do not assume a function $\varphi(x)$ in $H(\varkappa, A, \lambda)$ periodic. Now we agree to consider this condition fulfilled. Moreover, the domain of definition Ω of $\varphi(x)$ is the fundamental parallelepiped of an orthogonal matrix H.

The set of periodic functions with period matrix H is a subspace in $H(\varkappa, A, \lambda)$ which we denote by $\widetilde{H}(\varkappa, A, \lambda)$. Inspect its structure in more detail.

First, consider the case $\varkappa = 0$. Take a trigonometric polynomial $P(x)$ in n complex variables $x = (x_1, \dots, x_n)$, namely,

$$P(x) = \sum_{\xi \in \Xi} a[\xi] e^{i2\pi\xi x}, \tag{5.1}$$

with Ξ some finite set in $\mathbb{Z}^n$ on which the function $a[\xi]$ is defined. We call the set

$$N_P = \{\xi \in \Xi : a[\xi] \neq 0\}$$

the *support of the coefficients of* $P(x)$ and the number

$$R_P = \max_{\xi \in N_P} |\xi|,$$

the *coefficient support radius* of P. The trigonometric polynomial $P(x)$ may obviously be written as

$$P(x) = \sum_{\xi \in N_P} a[\xi] e^{i2\pi\xi x}. \tag{5.2}$$

The radius of the support of the coefficients of a trigonometric polynomial is analogous to the degree of an ordinary algebraic polynomial.

Lemma 7.5. *A trigonometric polynomial $P(x)$ with coefficient support radius R_P is an entire function of order 1 and type $2\pi R_P$.*

PROOF. Estimate each summand of (5.2) separately. Let $x_k = u_k + iv_k$. Then

$$|a[\xi]e^{i2\pi\xi x}| = |a[\xi]|e^{-2\pi\sum_{j=1}^{n}\xi_j v_j} \le e^{2\pi|\xi||v|}|a[\xi]| \le |a[\xi]|e^{2\pi R_P|x|}.$$

Thus, $|P(x)| \le (\sum_{\xi\in N_P}|a[\xi]|)e^{2\pi R_P|x|}$. This readily entails the inequality (4.1) with $\rho = 1$, $\sigma = 2\pi R_P$, and $\mu = 0$. The proof of Lemma 7.5 is complete.

Lemma 7.6. *The support of the coefficients of a trigonometric polynomial (5.2) of order $\rho = 1$ and type σ lies in the ball of radius $R_P = \sigma/2\pi$.*

PROOF. All summands composing $P(x)$ are linearly independent. Consequently, it suffices to show that every function $e^{i2\pi\xi x}$ of order $\rho = 1$ may be of type σ only if $|\xi| \le \sigma/2\pi$.

Assume the contrary, namely, $|\xi| > \sigma/2\pi$. Let $v > 0$ and $x = -iv\xi$. Then we would have the equalities

$$|x| = v|\xi|, \quad e^{i2\pi\xi x} = e^{2\pi|\xi||x|} = e^{(\sigma+\varepsilon_0)|x|},$$

with $\varepsilon_0 = 2\pi|\xi| - \sigma$ positive. Consequently, the function $e^{i2\pi\xi x}$ might fail to satisfy (4.1) with $\rho = 1$ for whatever μ. Consequently, this function and the trigonometric polynomial $P(x)$ do not belong to $\Psi(\rho,\sigma)$. The proof is complete.

Theorem 7.11. *Each periodic function $\varphi(x)$ with period matrix H of order $\rho = 1$ and type σ, taking real values for $x \in \mathbb{R}^n$, is a trigonometric polynomial like*

$$\varphi(x) = \sum_{|H^{-1*}\alpha|\le\sigma/2\pi} a[\alpha]e^{i2\pi\alpha H^{-1}x}. \tag{5.3}$$

PROOF. To simplify calculations, let $\varphi(x)$ be a periodic function whose period matrix is the identity matrix and, moreover, $|\varphi(x)| \le e^{\sigma|x|}$. Introduce the new variables

$$x_k = \frac{1}{i2\pi}\log z_k, \quad z_k = e^{i2\pi x_k}, \quad z = (z_1,\dots,z_n), \tag{5.4}$$

and put

$$\psi(z) = \varphi\left(\frac{1}{i2\pi}\log z_1,\dots,\frac{1}{i2\pi}\log z_n\right) = \sum a[\alpha]z^\alpha.$$

This change of variables maps the domain $\{x : -1/2 \le \operatorname{Re} x_k < 1/2,\ k = 1,\dots,n\}$, i.e., the fundamental domain of the period matrix in the space $\mathbb{C}^n_x$ of complex variables x, to the whole space $\mathbb{C}_z^n$ of complex variables z. By periodicity of $\varphi(x)$, the function $\psi(z)$ is one-to-one in $\mathbb{C}_z^n$. It is clear also that $\psi(z)$ is regular everywhere at finitely remote points, except possibly, those at which at least one of the variables z_k vanishes.

Consider the product

$$(z_1 z_2 \ldots z_n)^{[\sigma/2\pi]+1}\psi(z) = \Psi(z)$$

and show that $\Psi(z)$ is bounded on the cube $\{z : |z_k| < 1,\ k = 1, \ldots, n\}$. Observe that

$$(z_1 z_2 \ldots z_n)^{[\sigma/2\pi]+1}\psi(z) = \varphi(x)e^{i2\pi([\sigma/2\pi]+1)\sum\limits_{j=1}^{n} x_j}. \tag{5.5}$$

Estimate the right side of (5.5) in the domain $\{x : \operatorname{Im} x_k > 0,\ k = 1, \ldots, n\}$.

Let $x_k = u_k + iv_k$ with $v_k > 0$. Use the fact that the sum of n positive numbers, i.e., the length of a polygonal line in n-dimensional space, is greater than the length of vector joining the endpoints of the line

$$\sum_{k=1}^{n} v_k > |v| .$$

Bearing in mind that $2\pi[\sigma/2\pi] + 2\pi \geq \sigma$ obtain the estimate

$$\begin{aligned} |\Psi(z)| &\leq |z_1 z_2 \ldots z_n|^{[\sigma/2\pi]+1} |\psi(z)| \\ &\leq Ke^{\sigma|x| - 2\pi([\sigma/2\pi]+1)\sum\limits_{j=1}^{n} v_j} \leq Ke^{\sigma(|x|-|v|)}. \end{aligned} \tag{5.6}$$

Further, in the strip $\{x : -1/2 \leq u_k = \operatorname{Re} x_k \leq 1/2,\ k = 1, \ldots, n\}$ we have the relations

$$|x| \leq \left(\sum_{j=1}^{n} v_j^2 + n/4\right)^{1/2} = \frac{n/4}{\left(\sum\limits_{j=1}^{n} v_j^2 + n/4\right)^{1/2} + \left(\sum\limits_{j=1}^{n} v_j^2\right)^{1/2}} + \left(\sum_{j=1}^{n} v_j^2\right)^{1/2},$$

$$|x| - \left(\sum_{j=1}^{n} v_j^2\right)^{1/2} \leq \frac{n}{4|v| + 2\sqrt{n}}. \tag{5.7}$$

Inserting (5.7) in (5.6), we see that $\Psi(z)$ is bounded for $|z_k| < 1,\ k = 1, \ldots, n$.

The function $\psi(z)$ is real provided that $|z_k| = 1$, $k = 1, 2, \ldots, n$. Consequently, its values at the points conjugate to one another with respect to the unit circle are also conjugate

$$\overline{\psi}(1/\overline{z}_1, 1/\overline{z}_2, \ldots, 1/\overline{z}_n) = \psi(z_1, z_2, \ldots, z_n).$$

From the last equality and the boundedness of

$$\Psi(w) = (w_1 \ldots w_n)^{[\sigma/2\pi]+1}\psi(w)$$

in the domain $\{w : |w_k| < 1,\ k = 1, \ldots, n\}$ it follows that the product

$$(z_1 z_2 \ldots z_n)^{-([\sigma/2\pi]+1)}\ \psi(z)$$

is bounded in the domain $\{z : |z_k| > 1,\ k = 1, \ldots, n\}$.

Thus, $\psi(z)$ is an entire function growing at infinity not faster than a polynomial. As is known from function theory, in this case it is a polynomial, namely,

$$\psi(z) = \sum_{0 \le |\alpha| \le M} a[\alpha] z^{\alpha}.$$

Inserting the expression of z through x by (5.4) we see that $\varphi(x)$ is a trigonometric polynomial,

$$\varphi(x) = \sum_{|\beta| \le M_1} a[\beta] e^{i2\pi\beta x}. \tag{5.8}$$

By Lemma 7.6 the coefficient support of $\varphi(x)$ lies in the ball of radius $\sigma/2\pi$; namely, in (5.8) we may set $M_1 \le \sigma/2\pi$.

If $\varphi(x)$ is such that

$$|\varphi(x)| \le K|x|^{\mu} e^{\sigma|x|^{\rho}},$$

with $\mu \ne 0$, then the above proof should be modified in an obvious manner.

The case in which the period matrix H of $\varphi(x)$ is not necessarily the identity matrix reduces to that settled above by the change of variables $x = Hy$. The proof of Theorem 7.11 is complete.

Corollary 5.1. *An entire real function $\varphi(x)$ with orthogonal period matrix H of order $\rho = 1$ and type $\sigma < 2\pi$ is a constant function.*

Corollary 5.2. *An entire real periodic function $\varphi(x)$ of order $\rho < 1$ is a constant function.*

Indeed, such function has order $\rho = 1$ and type σ however small. By Corollary 5.1 it is a constant function.

Corollary 5.3. *A function $\varphi(x)$ in the space $\widetilde{H}(\varkappa, A, \lambda)$ with orthogonal period matrix H for $\varkappa = 0$ and $\lambda \leq -1/2$ is a trigonometric polynomial,*

$$\varphi(x) = \sum_{|\alpha| \leq A/2\pi} a[\alpha] e^{i2\pi\alpha H^{-1}x}.$$

If, on the other hand, $\varkappa < 0$ then the class $\widetilde{H}(\varkappa, A, \lambda)$ consists of constant functions.

This claim is easy to justify on successively applying Theorems 7.3, 7.10 and 7.11.

Take $\varkappa > 0$. Expand $\varphi(x)$ in $\widetilde{H}(\varkappa, A)$ with period matrix the identity matrix in the Fourier series

$$\varphi(x) = \sum_{\beta} c_{\varphi}[\beta] e^{i2\pi\beta x}. \tag{5.9}$$

The general case reduces to this by the change of variable $x = Hy$. Below we formulate a condition on the coefficients $c_{\varphi}[\beta]$ of the expansion (5.9) whose fulfillment is necessary and sufficient for $\varphi(x)$ to be a member of the class $\widetilde{H}(\varkappa, A)$.

Introduce some notation. Let the sequence of numbers a_m be defined by the equalities

$$a_0 = 1, \quad a_m = m^{\varkappa m} A^m m^{\lambda} \quad \text{for} \quad m \geq 1.$$

For every function $\varphi(x) \in \widetilde{H}(\varkappa, A, \lambda)$ the following quantity is finite

$$[\varphi]_H = \sup_{m \geq 0} \frac{\left\| \varphi \mid \widetilde{L}_2^{(m)} \right\|}{a_m}. \tag{5.10}$$

If the entries of the multi-index $\beta = (\beta_1, \beta_2, \ldots, \beta_n) \neq 0$ range independently over the set of all integers, then the square of its Euclidean norm $|\beta|^2 = \beta_1^2 + \cdots + \beta_n^2$ ranges obviously over some subset of the set of naturals. Denote the members of this subset by N_j, assuming in addition that $N_j < N_{j+1}$. Obviously, every N_j is greater than j. We have the following

Theorem 7.12. *The Fourier coefficient $c_{\varphi}[\beta]$ of a function $\varphi(x)$ belonging to $\widetilde{H}(\varkappa, A, \lambda)$ for $\varkappa > 0$ satisfies the estimate*

$$\left(\sum_{|\beta|^2 = N_j} |c_{\varphi}[\beta]|^2 \right)^{1/2} \leq C[\varphi]_H N_j^{v_0/2} e^{-s N_j^{1/(2\varkappa)}}, \tag{5.11}$$

with C a constant independent of φ and j, where the numbers s, $\varkappa$, A, λ, and v_0 are related as follows

$$s = \left(\frac{\varkappa}{e}\right) \left(\frac{2\pi}{A}\right)^{1/\varkappa}, \quad v_0 = \frac{\lambda}{\varkappa}. \tag{5.12}$$

PROOF. Let $\varphi(x) \in \widetilde{H}(\varkappa, A, \lambda)$ for $\varkappa > 0$. Using the periodicity of $\varphi(x)$, we readily see that the coefficient $c_\varphi[\beta]$ of the expansion (5.9) decreases with the growth of $|\beta|$ faster than every negative power of $|\beta|$. Consequently, we may differentiate (5.9) arbitrarily many times, obtaining uniformly convergent series. Using this remark, calculate the norm of $\varphi(x)$. By the Parseval identity, we have

$$\|\varphi \mid \widetilde{L}_2^{(m)}\|^2 = \sum_{\beta\neq 0}\Big[\sum_{|\alpha|=m} \frac{m!}{\alpha!}(2\pi\beta)^{2\alpha}\Big]|c_\varphi[\beta]|^2.$$

The sum in brackets by the well-known Multinomial Theorem may be written as $(2\pi|\beta|)^{2m}$. Considering this, rearrange the sum over β as follows

$$\|\varphi \mid \widetilde{L}_2^{(m)}\|^2 = \sum_{j=1}^{\infty}(2\pi\sqrt{N_j}\,)^{2m}\Big(\sum_{|\beta|^2=N_j} |c_\varphi[\beta]|^2\Big). \tag{5.13}$$

Whence and from the definition of $[\varphi]_H$ the following inequality is immediate

$$\Big(\sum_{|\beta|^2=N_j} |c_\varphi[\beta]|^2\Big)^{1/2} \leq \frac{\|\varphi \mid \widetilde{L}_2^{(m)}\|}{(2\pi\sqrt{N_j}\,)^m} \leq [\varphi]_H \frac{a_m}{(2\pi\sqrt{N_j}\,)^m}. \tag{5.14}$$

Minimize over m the sequence on the right side of (5.14), assuming N_j sufficiently large. The ratio of the $(m+1)$th entry of the sequence to the preceding entry is $a_{m+1}/(2\pi\sqrt{N_j}\,a_m)$. Using the explicit form of a_{m+1} and a_m, we readily see that this ratio is at most 1 when $1 \leq m \leq m_0(N_j)$, after that is becomes greater than 1. In this event $m_0(N_j)$ is equivalent to the sequence $(s/\varkappa)N_j^{1/(2\varkappa)}$ as $j \to \infty$. Thus, the minimum of the right side of (5.14) is attained at $m = m_0(N_j)$ and equals $m_0^\lambda(N_j)e^{-\varkappa m_0(N_j)}$. Inserting here the principal term of the asymptotic expansion of $m_0(N_j)$, arrive at the required estimate (5.11). The proof is complete.

The following theorem is in a sense a converse of the preceding theorem.

Theorem 7.13. *Let the coefficients $c_\varphi[\beta]$ of the expansion (5.9) of a periodic function $\varphi(x)$ satisfy the condition*

$$\Big(\sum_{|\beta|^2=N_j} |c_\varphi[\beta]|^2\Big)^{1/2} \leq C N_j^{(v_0+v_1)/2} e^{-sN_j^{1/2\varkappa}}, \tag{5.15}$$

with the constants s, $\varkappa$, A, λ, and v_0 related by (5.12) and $v_1 = \min\{0, 1/(4\varkappa)-1\}$. Then $\varphi(x)$ belongs to $\widetilde{H}(\varkappa, A, \lambda)$.

PROOF. Use (5.13) to estimate the $\widetilde{L}_2^{(m)}$ norm of $\varphi(x)$. By (5.15)

$$\left\|\varphi \mid \widetilde{L}_2^{(m)}\right\|^2 \leq C(2\pi)^{2m} \sum_{j=1}^{\infty} N_j^{m+v} e^{-2sN_j^{1/2\varkappa}}, \tag{5.16}$$

with $v = v_0 + v_1$. Examine the series on the right side of (5.16). Recall that N_j is natural. Hence, the sum in question does not diminish if we add to it the values of the function being summed at all other points of the set of naturals. Doing so, obtain the relation

$$\sum_{j=1}^{\infty} N_j^{m+v} e^{-2sN_j^{1/2\varkappa}} \leq \sum_{j=1}^{\infty} j^{m+v} e^{-2sj^{1/2\varkappa}}. \tag{5.17}$$

For $n \geq 4$ this relation becomes an equality, which follows from the celebrated *Lagrange Theorem* on decomposition of each natural number into a sum of four squares.

Examine how the sum of (5.17) changes with the growth of m. Put $\mu = m + v$ and $q = 2s$ and consider the function $\psi(r) = r^{\mu} e^{-qr^{1/\varkappa}}$ for $\mu > 0$. The series (5.17) is the sum of the values of $\psi(r)$ at all points of the set of naturals. It is not hard to calculate the derivative of $\psi(r)$, namely,

$$\psi'(r) = \left(\mu - \frac{q}{2\varkappa} r^{1/(2\varkappa)}\right) \frac{\psi(r)}{r}.$$

Thus, $\psi(r)$ increases for $0 \leq r \leq r_0(\mu)$ with $r_0(\mu) = (2\varkappa\mu/q)^{2\varkappa}$, and decreases for $r \geq r_0(\mu)$. The integral of $\psi(r)$ taken over the positive semiaxis is expressed through the *Euler gamma function*

$$\int_0^{\infty} \psi(r)\, dr = \frac{2\varkappa}{q^{2\varkappa(\mu+1)}} \int_0^{\infty} t^{2\varkappa(\mu+1)-1} e^{-t}\, dt = \frac{2\varkappa\Gamma(2\varkappa(\mu+1))}{q^{2\varkappa(\mu+1)}}.$$

Split it into the sum of integrals over the intervals $[j, j+1]$ and estimate $\psi(r)$ from below on each of the intervals. Considering separately the case in which $r_0(\mu)$ is natural, come to the inequality

$$\sum_{j=1}^{\infty} \psi(j) \leq \frac{2\varkappa\Gamma(2\varkappa(\mu+1))}{q^{2\varkappa(\mu+1)}} + \psi(r_0(\mu)). \tag{5.18}$$

At large μ, use the Stirling formula and establish the asymptotic expansion of the right side of (5.18). If $\varkappa > 1/4$, then the first summand prevails in the asymptotic expansion of the sum; whereas, for $0 < \varkappa \leq 1/4$, so does the second summand on the right side of (5.18). The inequality suitable in both cases may be written as

$$\sum_{j=1}^{\infty} j^{\mu} e^{-qj^{1/2\varkappa}} \leq C \left(\frac{2\varkappa}{qe}\right)^{2\varkappa\mu} \mu^{2\varkappa(\mu - v_1')}, \tag{5.19}$$

with $v_1' = \min\{0, 1/(4\varkappa) - 1\}$. Inserting (5.19) in (5.17) and putting the result in (5.16), obtain the estimate

$$\|\varphi \mid \widetilde{L}_2^{(m)}\|^2 \leq C(2\pi)^{2m} \left(\frac{\varkappa}{se}\right)^{2\varkappa m} m^{2\varkappa(m+v_0)}.$$

Recalling now the relations (5.12) between the parameters s, v, $\varkappa$, A, and λ, we see that $\varphi(x)$ does belong to $\widetilde{H}(\varkappa, A, \lambda)$. The proof is complete.

Corollary 5.4. *If $0 < \varkappa \leq 1/4$, then the space $\widetilde{H}(\varkappa, A, \lambda)$ coincides with the class comprising the Fourier series* (5.9) *whose coefficients satisfy* (5.11).

The proof is immediate on observing that, for the values of $\varkappa$ under consideration, the parameter v_1 in (5.15) equals 0.

If $\varkappa > 1/4$, then (5.11) and (5.15) do not coincide. It is of interest to find out whether this discrepancy is essential and whether either of the estimates (5.11) and (5.15) may be sharpened.

Corollary 5.5. *A function $\varphi(x)$ in a single variable belongs to $\widetilde{H}(\varkappa, A, \lambda)$ provided that its Fourier coefficients meet the condition* (5.15) *with*

$$v_1 = \min\{1/(4\varkappa) - 1/2, 0\}.$$

The singleness of the independent variable enables us to specify the sequence of N_j. Obviously, N_j equals j^2. Consequently, the estimate (5.16) for the norm of $\|\varphi \mid \widetilde{L}_2^{(m)}\|$ takes the form

$$\|\varphi \mid \widetilde{L}_2^{(m)}\|^2 \leq C(2\pi)^{2m} \sum_{j=1}^{\infty} j^{2(m+v)} e^{-2sj^{1/\varkappa}},$$

with $v = \lambda/\varkappa + v_1$. Further apply (5.19) with appropriate parameters μ, q, and $\varkappa$.

Corollary 5.6. *The function*

$$\varphi_*(x) = \sum_{\beta \neq 0} |\beta|^v e^{-s|\beta|^{1/\varkappa}} a[\beta] e^{i2\pi\beta x} \tag{5.20}$$

belongs to $\widetilde{H}(\varkappa, A, \lambda)$, *if the following conditions are fulfilled*

$$s = \left(\frac{\varkappa}{e}\right)\left(\frac{2\pi}{A}\right)^{1/\varkappa}, \quad v = \frac{\lambda}{\varkappa} + v_1, \quad \sum_{|\beta|^2 = N_j} |a[\beta]|^2 = 1. \tag{5.21}$$

Recall that the parameter v_1 is defined in accordance with the dimension of the space of independent variables. The assertion under proof is immediate from Theorem 7.13, on observing that

$$\sum_{|\beta|^2 = N_j} |c_{\varphi_*}[\beta]|^2 = N_j^{\nu} e^{-2sN_j^{1/2\varkappa}}.$$

§6. Convergence of Cubature Formulas in $\widetilde{H}(\varkappa, A, \lambda)$

In § 1 of the current chapter we inspected behavior of the sequence of the errors (l_h, φ) corresponding to an arbitrary function $\varphi \in \widetilde{L}_2^{(m)}$. It turned out that, as N increases, the sequence vanishes faster in comparison with what is implied by the estimates of the $\widetilde{L}_2^{(m)*}$ norm of l_h. It is worthwhile to find what properties of $\varphi(x)$ essentially influence the rate of this convergence. We now demonstrate that the periodicity of $\varphi(x)$, its infinite differentiability and certain growth of derivatives with respect to the order of differentiation guarantee exponential rather than polynomial decay of errors with the growth of N.

Theorem 7.14. *Let a function* $\varphi(x)$ *belong to* $\widetilde{H}(\varkappa, A, \lambda)$. *Then the inequality holds*

$$|(l_h, \varphi)| \le K[\varphi]_H N^{v_0} e^{-sN^{1/\varkappa}}, \tag{6.1}$$

with $N = 1/h$, s *and* v_0 *the constants defined in* (5.12) *and* K *a constant independent of* φ *and* N.

PROOF. We may apply the estimate (1.22) to (l_h, φ). Moreover, we may specify the latter as follows

$$|(l_h, \varphi)| \le K[\varphi]_H \inf_{m \ge 0} \left\{ \frac{a_m}{(2\pi N)^m} \right\}, \tag{6.2}$$

with K a constant independent of φ, m, and N. To derive (6.2), appeal, first, to the inequality

$$\|\varphi \mid \widetilde{L}_2^{(m)}\| \le [\varphi]_H a_m,$$

ensuing from (5.10) and, second, to the familiar expansion of the norm square of the error (1.1)

$$\|l_h \mid \widetilde{L}_2^{(m)*}\|^2 = \left(\frac{h}{2\pi}\right)^{2m} \sum_{\beta \ne 0} \frac{1}{|H^{-1*}\beta|^{2m}}.$$

As was mentioned in the proof of Theorem 7.12, the sequence $a_m/(2\pi N)^m$ attains its minimal value for m equal to $m_0(N) \sim (s/\varkappa)N^{1/\varkappa}$. Therefore, the minimum on the right side of (6.2) equals $m_0^\lambda(N)e^{-\varkappa m_0(N)}$ for large N. Inserting here the principal term of the asymptotic expansion of $m_0(N)$, come to the sought estimate (6.1). The proof is complete.

The function $\varphi(x)$ in (6.1) is an arbitrary member of $\widetilde{H}(\varkappa, A, \lambda)$. Consequently,

$$\sup_{\varphi\in\widetilde{H}} \frac{(l_h,\varphi)}{[\varphi]_H} \leq KN^{v_0}e^{-sN^{1/\varkappa}}. \tag{6.3}$$

The left side of (6.3) presents the $\widetilde{H}^*(\varkappa, A, \lambda)$ norm of the error (1.1). It is thus shown that the sequence of norms $\|l_h \mid \widetilde{H}(\varkappa, A, \lambda)^*\|$ vanishes exponentially as $h \to 0$. Examine the extend to which (6.1) and (6.3) are unimprovable, restricting consideration to the case in which H is the identity matrix.

Consider the function $\varphi_*(x)$ given by the Fourier series (5.20). The coefficients $a[\beta]$ must satisfy (5.21). By Corollary 5.6 $\varphi_*(x)$ belongs to $\widetilde{H}(\varkappa, A, \lambda)$. The norm of $[\varphi_*]_H$ is, obviously, other than 0.

Calculate the value of the error (1.1) with $H = I$ at φ_*. We have

$$(l_h,\varphi_*) = \sum_{\beta\neq 0} |\beta|^v e^{-s|\beta|^{1/\varkappa}} a[\beta]L_N[\beta], \tag{6.4}$$

with $L_N[\beta]$ the value of l_h at the exponential $e^{i2\pi\beta x}$. As mentioned in § 1, $L_N[\beta]$ equals 0 if the multi-index β is not a multiple of N and equals -1 otherwise. Consequently,

$$(l_h,\varphi_*) = -\sum_{\alpha\neq 0} N^v|\alpha|^v e^{-sN^{1/\varkappa}|\alpha|^{1/\varkappa}} a[N\alpha]. \tag{6.5}$$

Let the Euclidean norm square of β equal N_j. We set one of the coefficients $a[\beta]$ with such β equal 1 and the rest of them, equal 0. It is clear that this does not violate (5.21). Then (6.5) may be continued as follows

$$(l_h,\varphi_*) = -N^v \sum_{j=1}^{\infty} N_j^{v/2} e^{-sN^{1/\varkappa}N_j^{1/2\varkappa}}. \tag{6.6}$$

At sufficiently large N the main contribution to the error is given by the first term of the series on the right side of (6.6). Considering that $N_1 = 1$, obtain the inequality

$$|(l_h,\varphi_*)| \geq CN^v e^{-sN^{1/\varkappa}}, \tag{6.7}$$

with C a constant independent of N, where s and v are the parameters defined by (5.21). Whence, it is not hard to infer the following lower bound on the norm of l_h in the space $\widetilde{H}^*(\varkappa, A, \lambda)$, namely,

$$\|l_h \mid \widetilde{H}^*(\varkappa, A, \lambda)\| \geq C N^v e^{-sN^{1/\varkappa}}. \tag{6.8}$$

Compare the inequalities (6.1) and (6.7), as well as (6.3) and (6.8), pairwise. We see that the exponents of the exponentials in all these inequalities coincide. Difference is incorporated only in the powers of N. Recall that the difference between v and v_0 was denoted by v_1 and, moreover,

$$v_1 = \begin{cases} \min\left\{\frac{1}{4\varkappa} - 1, 0\right\}, & \text{if } \ n \geq 2, \\ \min\left\{\frac{1}{4\varkappa} - \frac{1}{2}, 0\right\}, & \text{if } \ n = 1. \end{cases}$$

In particular, the difference v_1 is nonpositive, and for certain $\varkappa$s is exactly equal to 0. In this case, (6.1) and (6.7), as well as (6.3) and (6.8), differ only in the constants K and C.

Thus, the maximal value of l_h on the unit ball of the Banach space $\widetilde{H}(\varkappa, A, \lambda)$ is almost attained at $\varphi_*/[\varphi_*]_H$. In other words, this function is "almost extremal" simultaneously for all errors l_h.

We now find out to what extend the results for periodic functions may be translated to the class $H(\varkappa, A, \lambda)$ containing aperiodic members. If in this case we also want to use the collection of inequalities analogous to (1.21), then as a cubature formula we should take such of them whose error is exact for polynomials of degree less than m. This leads to the necessity of considering double sequences of errors $l_{m,N}(x)$, with the first index standing for the order and second, for the number of nodes of a cubature formula.

Assume that the space $\mathbb{R}^n$ of independent variables x is one-dimensional, i.e., $n = 1$. As a double sequence consider the errors $l_{m,N}^{EM}$ corresponding to the celebrated *Euler–Maclaurin formula*

$$(l_{m,N}^{EM}, \varphi) = \int_0^1 \varphi(x)\, dx - T_N(\varphi) + \sum_{k=1}^{\nu-1} h^{2k} \frac{B_{2k}}{(2k)!} \left[\varphi^{(2k-1)}(1) - \varphi^{(2k-1)}(0)\right]. \tag{6.9}$$

Here B_{2k} is the Bernoulli number with index k; also ν is an integer, $m = 2\nu$, and

$$T_N(\varphi) = \frac{h[\varphi(0) + \varphi(1)]}{2} + \sum_{\alpha=1}^{N-1} h\varphi(\alpha h).$$

Let φ be defined on the unit interval and belong to $H(\varkappa, A, \lambda)$. Then we have a collection of estimates analogous to (1.21)

$$\begin{aligned}|(l_{m,N}^{EM}, \varphi)| &\le \| l_{m,N}^{EM} \mid L_2^{(m)}[0,1]^* \| \, \| \varphi \mid L_2^{(m)}[0,1] \| \\ &\le [\varphi]_H \| l_{m,N}^{EM} \mid L_2^{(m)}[0,1]^* \| a_m. \end{aligned} \tag{6.10}$$

The norm square of the error (6.9) is known to admit the expansion

$$\| l_{m,N}^{EM} \mid L_2^{(m)}[0,1]^* \|^2 = h^{2m} \left\{ \left(\sum_{\beta \neq 0} \frac{1}{(2\pi\beta)^m} \right)^2 + \sum_{\beta \neq 0} \frac{1}{(2\pi\beta)^{2m}} \right\}. \tag{6.11}$$

Inserting this equality in (6.10), obtain

$$|(l_{m,N}^{EM}, \varphi)| \le K[\varphi]_H \frac{a_m}{(2\pi N)^m}. \tag{6.12}$$

The minimum of the right side of (6.12) is attained at an even m equal to

$$m_0(N) \sim \left(\frac{s}{\varkappa}\right) N^{1/\varkappa},$$

and coincides with $m_0^{\lambda}(N) e^{-\varkappa m_0(N)}$. Thus, for every function $\varphi \in H(\varkappa, A, \lambda)$ the estimate holds

$$|(l_{m_0(N),N}^{EM}, \varphi)| \le K[\varphi]_H N^{v_0} e^{-sN^{1/\varkappa}}. \tag{6.13}$$

The constants s and v_0 are defined in (5.12), and K is a constant independent of N.

Thus, when the order is chosen properly, the vanishing of the error (6.9) is exponential. Show that in this event the exponent of the exponential on the right side of (6.13) is impossible to diminish.

Calculate the value of the error (6.9) at the function φ_* given by the Fourier series (5.20). Choose the coefficients $a[\beta]$ of this expansion so that (6.6) be valid. Observe that the errors $l_{m,N}^{EM}(x)$ and $l_h(x)$ agree at every periodic infinitely differentiable function. Consequently, (6.6) holds also for the error of the Euler–Maclaurin formula at φ_*, namely,

$$(l_{m,N}^{EM}, \varphi_*) = -N^v \sum_{j=1}^{\infty} N_j^{v/2} e^{-sN^{1/\varkappa} N_j^{1/2\varkappa}}. \tag{6.14}$$

In the one-dimensional case, $N_j = j^2$ and at large N the first summand prevails in the sum (6.14). This yields a lower bound on the norm $l_{m,N}^{EM}$, namely,

$$\| l_{m,N}^{EM} \mid H^*(\varkappa, A, \lambda) \| \ge \frac{|(l_{m,N}^{EM}, \varphi_*)|}{[\varphi_*]_H} \ge K N^v e^{-sN^{1/\varkappa}}. \tag{6.15}$$

In particular, for $m = m_0(N)$ we have the two-sided estimate for the quantity $\|l^{EM}_{m_0(N),N} \mid H^*(\varkappa, A, \lambda)\|$ which results from (6.13) and (6.15). The question arises at this juncture of how essential is the fact that the order of the error equals $m_0(N)$, and whether the same property is shared with the errors possessing another much larger m. It turns out that the norm $\|l^{EM}_{m,N} \mid H^*(\varkappa, A, \lambda)\|$ tends to infinity as $m \to \infty$ at a fixed N, and its growth is controlled by the estimate

$$\|l^{EM}_{m,N} \mid H^*(\varkappa, A, \lambda)\| \geq C m^\gamma N^2 \frac{a_m}{(2\pi N)^m}, \tag{6.16}$$

with C and γ constants independent of m and N. This, in particular, implies that the increase of the order of a formula for approximate integration at a fixed number of nodes N does not diminish the norm of the error but, on the contrary, it increases the norm unboundedly.

The estimate (6.16) yields a theoretic explanation of an effect known from the practice of calculating definite integrals. This effect is as follows: from some $m = m_0(N)$ on, the accuracy of the result does not increase as m increases.

To derive (6.16) and specify the coefficient γ, we need more detailed information on the space $H(\varkappa, A, \lambda)$.

§7. Gevrey Classes of Functions in a Single Independent Variable

Let Ω be a bounded domain in $\mathbb{R}^n$ with analytic boundary. Say that a function φ *belongs to the Gevrey class* $G_\varkappa(\Omega)$, $\varkappa \geq 1$, if φ is infinitely differentiable in a neighborhood about Ω and for some L the estimate holds [274]

$$\|D^\alpha \varphi \mid L_2(\Omega)\| \leq L^{|\alpha|+1}(|\alpha|!)^\varkappa, \quad |\alpha| \geq 0. \tag{7.1}$$

For $\varkappa > 1$ the class $G_\varkappa(\Omega)$ coincides with the space $H(\varkappa)$ presenting the union of $H(\varkappa, A, \lambda)$ over all A and λ. Indeed, the embedding $G_\varkappa(\Omega) \subset H(\varkappa)$ follows from (7.1) and the Stirling formula. Conversely, if a function φ belongs to the class $H(\varkappa)$ in Ω; then in accord with Theorem 7.6 we may extend it to a function of the same class in a neighborhood about Ω. Obviously, (7.1) remains valid in this event.

Assume that $n = 1$, and Ω coincides with the unit interval of the real axis. The error of a quadrature formula at every function $\varphi \in G_\varkappa[0,1]$ may be estimated according to the above scheme, using the fact that φ belongs to the Banach space $H(\varkappa, A, \lambda)$ with some A and λ. However, this leaves obscure the extend to which such estimates are unimprovable. We are thus impelled to pursue a slightly different approach.

As we have already seen, it is more convenient to study the convergence of cubature formulas in a Hilbert space. Each bounded linear functional is now presented by means of the extremal function whose norm is easy to estimate from above as well as from below, if we know some orthogonal basis for the initial space. To use the advantages of this approach, reduce the problem we address to estimating the norm of an error acting in a Hilbert space.

Say that a *function* φ *belongs to* $\overline{H}(\varkappa, A, \lambda)$ on the unit interval, if φ is infinitely differentiable on this interval and, moreover, the following series converges

$$\sum_{m=0}^{\infty}\left\{\frac{\left\|\varphi \mid L_2^{(m)}[0,1]\right\|^2}{c_m^2}+\frac{|D^m\varphi|_0^1|^2}{b_m^2}\right\}<\infty, \tag{7.2}$$

with

$$D^m\varphi\big|_0^1 = D^m\varphi(1) - D^m\varphi(0).$$

The sequences b_m and c_m on the left side are similar to the sequence a_m, i.e.,

$$b_0 = c_0 = 1, \quad b_m = m^{\varkappa m}A^m m^{\mu_2}, \quad c_m = m^{\varkappa m}A^m m^{\mu_1}. \tag{7.3}$$

In this event the exponents μ_1 and μ_2 depend on $\varkappa$ and λ, namely,

$$\mu_1 = \lambda + \varkappa + 2, \quad \mu_2 = \lambda + 3\varkappa/2 + 1/2. \tag{7.4}$$

Observe right away that μ_1 and μ_2 satisfy the following conditions of service in the sequel

$$\mu_1 - \lambda + \varkappa/2 > \mu_1 - \lambda > 1/2, \quad \mu_2 - \lambda - \varkappa > 1/2, \quad \mu_1 - \mu_2 > 1 - \varkappa/2. \tag{7.5}$$

Conditions like (7.2) are often used for defining classes of infinitely differentiable functions. For instance, if we eliminate from (7.2) the summands with the increments of the derivatives of φ, then we obtain the class that is known as the *Sobolev space of infinite order* [58].

We define the inner product of $\varphi, \psi \in \overline{H}(\varkappa, A, \lambda)$ by the equality

$$\langle \varphi, \psi \rangle_1 = \sum_{m=0}^{\infty}\left\{\int_0^1 \frac{D^m\varphi\overline{D^m\psi}}{c_m^2}\,dx + \frac{\left(D^m\varphi\big|_0^1\right)\left(\overline{D^m\psi\big|_0^1}\right)}{b_m^2}\right\}. \tag{7.6}$$

The corresponding norm is denoted by $\langle\cdot\rangle_1$. It is not hard to see that the resultant normed space $\overline{H}(\varkappa, A, \lambda)$ is complete and so it is a Hilbert space.

Let $\varphi \in H(\varkappa, A, \lambda)$. Then, for every $m \geq 0$, we have

$$|D^m\varphi(1) - D^m\varphi(0)| \leq \left\|\varphi \mid L_2^{(m+1)}[0,1]\right\| \leq [\varphi]_H\, a_{m+1}.$$

Using these estimates, dominate the norm $\langle \varphi \rangle_1$, namely,

$$\langle \varphi \rangle_1 \leq [\varphi]_H \left\{ \sum_{m=0}^{\infty} \left(\frac{a_m}{c_m}\right)^2 + \left(\frac{a_{m+1}}{b_m}\right)^2 \right\}^{1/2}.$$

As follows from (7.5), the series on the right side of the last inequality converges. Thus, the Banach space $H(\varkappa, A, \lambda)$ is continuously embedded in the Hilbert space $\overline{H}(\varkappa, A, \lambda)$.

Let $\varphi \in \overline{H}(\varkappa, A, \lambda)$. For every $m \geq 0$, it is obvious that

$$\left\| \varphi \mid L_2^{(m)}[0,1] \right\| \leq \langle \varphi \rangle_1 \, c_m. \tag{7.7}$$

The sequence c_m is defined by (7.3). Consequently, from (7.7) we may deduce that $\overline{H}(\varkappa, A, \lambda)$ is continuously embedded in $H(\varkappa, A, \mu_1)$. Thus, the union of $\overline{H}(\varkappa, A, \lambda)$ over all A and λ coincides with $G_\varkappa[0,1]$.

We now examine the structure of the space $\overline{H}(\varkappa, A, \lambda)$, fixing $\varkappa$, A, and λ. Describe it by constructing a special basis for this space.

To begin with, consider the set of periodic elements of $\overline{H}(\varkappa, A, \lambda)$, i.e., functions φ with the increments of derivatives of every order equal 0 on the unit interval. In $\overline{H}(\varkappa, A, \lambda)$, these functions constitute a closed subspace. We denote the latter by $\approx{H}(\varkappa, A, \lambda)$. The exponentials $e^{i2\pi\beta x}$, $\beta = 0, \pm 1, \ldots$, are examples of the functions belonging to $\approx{H}(\varkappa, A, \lambda)$. From the definition of the inner product in $\overline{H}(\varkappa, A, \lambda)$ it is immediate that the exponentials corresponding to different indices β are mutually orthogonal. The norm square of the exponential with exponent β is the value of some entire function at the point $z = 2\pi\beta$, namely,

$$\langle e^{i2\pi\beta x} \rangle_1^2 = \sum_{m=0}^{\infty} \frac{(2\pi\beta)^{2m}}{c_m^2} \equiv A_0[\beta]. \tag{7.8}$$

Expand an arbitrary $\varphi(x) \in \approx{H}(\varkappa, A, \lambda)$ in the Fourier series

$$\varphi(x) = \sum_{\beta} c_\varphi[\beta] e^{-i2\pi\beta x}. \tag{7.9}$$

As in the case of $\widetilde{H}(\varkappa, A, \lambda)$, we may differentiate this series however many times, obtaining a series that converges uniformly on the unit interval. It is not hard to prove that (7.9) converges to $\varphi(x)$ in the normed space $\overline{H}(\varkappa, A, \lambda)$. This means that the exponentials $e^{i2\pi\beta x}$, $\beta = 0, \pm 1, \ldots$, constitute an orthogonal basis for $\approx{H}(\varkappa, A, \lambda)$.

Theorem 7.15. *A function $\varphi(x)$ in a single variable belongs to $\overset{\approx}{H}(\varkappa, A, \lambda)$ if and only if the Fourier series (7.9) converges in the norm $\langle\cdot\rangle_1$, i.e., the Fourier coefficients $c_\varphi[\beta]$ satisfy the inequality*

$$\sum_{\beta} A_0[\beta]|c_\varphi[\beta]|^2 < +\infty, \tag{7.10}$$

with the function $A_0[\beta]$ defined by (7.8).

PROOF. Take $\varphi(x) \in \overset{\approx}{H}(\varkappa, A, \lambda)$. Its Fourier series (7.9), as mentioned above, converges in the norm of $C[0,1]$. Denote by φ_M the partial sum of (7.9) in which summation is carried out over all integer β, $|\beta| \leq M$. Using the orthogonality of the system of exponentials, as well as (7.8), deduce that

$$\langle \varphi_M \rangle_1^2 = \sum_{|\beta|\leq M} |c_\varphi[\beta]|^2 A_0[\beta]. \tag{7.11}$$

Considering the obvious relation $\langle \varphi_M \rangle_1 \leq \langle \varphi \rangle_1$ and passing to the limit in (7.11) as $M \to \infty$, conclude that (7.10) is indeed valid.

If the Fourier coefficients of φ satisfy the condition (7.10); then, using (7.11), we readily see that the sequence of the partial sums of (7.9) is a Cauchy sequence. The limit of this sequence in the norm $\langle\cdot\rangle_1$ belongs to $\overline{H}(\varkappa, A, \lambda)$ and coincides with φ. The proof is complete.

Corollary 7.1. *The Fourier coefficients of a function $\varphi(x)$ in $\overset{\approx}{H}(\varkappa, A, \lambda)$ may be written as the following ratio*

$$c_\varphi[\beta] = \frac{\langle \varphi, e^{-i2\pi\beta x} \rangle_1}{A_0[\beta]}, \quad \beta = 0, \pm 1, \ldots. \tag{7.12}$$

Every *Bernoulli polynomial* $B_{m+1}(x)$ of degree $m+1$ belongs obviously to the class $\overline{H}(\varkappa, A, \lambda)$. Dividing this polynomial by $(m+1)!$, find the projection $\widetilde{\varphi}_m(x)$ of the resultant function to the subspace $\overset{\approx}{H}(\varkappa, A, \lambda)$. Denote by $\widetilde{c}_m[\beta]$ the coefficients of the Fourier series for $\widetilde{\varphi}_m(x)$. In virtue of (7.12),

$$\widetilde{c}_m[\beta] = \frac{1}{(m+1)!} \frac{\langle B_{m+1}(x), e^{-i2\pi\beta x} \rangle_1}{A_0[\beta]}. \tag{7.13}$$

Calculate the inner product on the right side of this equality. It is well known that the sequence of increments of the derivatives of $B_{m+1}(x)$ over the unit interval satisfies the conditions

$$\frac{D^\alpha B_{m+1}(1) - D^\alpha B_{m+1}(0)}{(m+1)!} = \delta_m^\alpha. \tag{7.14}$$

In the unit interval, we have the expansion

$$\frac{1}{(m+1)!} B_{m+1}(x) = (-1)^m \sum_{\beta \neq 0} \frac{1}{(i2\pi\beta)^{m+1}} e^{-i2\pi\beta x}. \tag{7.15}$$

Now, using (7.14) and (7.15) and integration by parts, we readily calculate explicitly each integral

$$\frac{1}{(m+1)!} \int_0^1 D^\alpha B_{m+1}(x) D^\alpha (e^{i2\pi\beta x})\, dx. \tag{7.16}$$

Inserting the so-obtained values of (7.16) in the definition of $\langle B_{m+1}(x), e^{-i2\pi\beta x} \rangle_1$ and putting the result in (7.13), come to the equality

$$\widetilde{c}_m[\beta] = \frac{(-1)^m}{(i2\pi\beta)^{m+1}} \frac{A_0^m[\beta]}{A_0[\beta]}, \quad \beta \neq 0. \tag{7.17}$$

Here, by $A_0^m[\beta]$ we denote the partial sum of (7.8)

$$A_0^m[\beta] = \sum_{\alpha=0}^{m} \frac{(2\pi\beta)^{2\alpha}}{c_\alpha^2}. \tag{7.18}$$

Split the space $\overline{H}(\varkappa, A, \lambda)$ into the direct sum of the subspace $\overset{\approx}{H}(\varkappa, A, \lambda)$ and the corresponding orthogonal complement $\overset{\approx}{H}{}^\perp(\varkappa, A, \lambda)$. By $\varphi_m(x)$ we denote the projection of $B_{m+1}(x)/(m+1)!$ to the complement. It is clear that the polynomial under projection is the sum of the mutually orthogonal functions $\varphi_m(x)$ and $\widetilde{\varphi}_m(x)$. The Fourier coefficients $c_m[\beta]$ of $\varphi_m(x)$, as follows from (7.15) and (7.17), may be found by the formulas

$$c_m[\beta] = \frac{(-1)^m}{(i2\pi\beta)^{m+1}} \frac{A_{m+1}[\beta]}{A_0[\beta]}, \quad \beta \neq 0, \tag{7.19}$$

with $A_{m+1}[\beta]$ standing for the difference between $A_0[\beta]$ and $A_0^m[\beta]$. The coefficient $c_m[0]$, corresponding to the function $\varphi_m(x)$, equals 0 by orthogonality between $\varphi_m(x)$ and every constant function

$$\int_0^1 \varphi_m(x)\, dx = \langle \varphi_m, 1 \rangle_1 = 0, \quad m = 0, 1, \ldots. \tag{7.20}$$

The integral of the derivative $D^{\alpha+1}\varphi_m$ over the unit interval in view of (7.14) may be calculated by the formula

$$\int_0^1 D^{\alpha+1}\varphi_m(x)\,dx = D^\alpha\varphi_m(1) - D^\alpha\varphi_m(0) = \delta_m^\alpha, \quad \alpha = 0, 1, \ldots. \tag{7.21}$$

Theorem 7.16. *Each function φ, a member of the orthogonal complement to $\tilde{\overline{H}}(\varkappa, A, \lambda)$, expands into a series in $\varphi_m(x)$. Moreover, the coefficient of $\varphi_m(x)$ equals the increment on the unit interval of the derivative of φ of order m, namely,*

$$\varphi(x) = \sum_{m=0}^{\infty} (D^m\varphi(1) - D^m\varphi(0))\,\varphi_m(x). \tag{7.22}$$

The series (7.22) converges in the norm of $\overline{H}(\varkappa, A, \lambda)$.

Before launching into the proof of Theorem 7.16, we state a few lemmas.

We introduce an auxiliary inner product in the space $\overline{H}(\varkappa, A, \lambda)$, by letting for all functions φ and ψ

$$\langle \varphi, \psi \rangle = \sum_{m=0}^{\infty} \frac{1}{c_m^2} \int_0^1 D^m\varphi(x)\overline{D^m\varphi(x)}\,dx. \tag{7.23}$$

Lemma 7.7. *The norm of φ_m corresponding to the inner product (7.23) may be calculated by the formula*

$$\langle \varphi_m \rangle^2 = \frac{1}{c_{m+1}^2} + 2\sum_{\beta=1}^{\infty} \frac{A_0^m[\beta]A_{m+1}[\beta]}{(2\pi\beta)^{2(m+1)}A_0[\beta]}. \tag{7.24}$$

PROOF. Using the equalities (7.19)–(7.21), it is easy to express the norm square of φ_m in $L_2^{(k)}[0,1]$ through the values of the functions $A_m[\beta]$, $A_0^m[\beta]$, and $A_0[\beta]$. Inserting the result in the definition of the norm of $\langle \varphi_m \rangle$ and performing easy calculations, come to the sought formula (7.24). The proof is complete.

Inspect the behavior of the norm of $\langle \varphi_m \rangle$ as m tends to infinity. Considering (7.24) observe that it suffices to do this only for the series on the right side as $m \to \infty$. Moreover, it is more convenient to deal with $A_m[\beta]$, $A_0^m[\beta]$, and some other similar functions.

Introduce into consideration a numerical sequence d_m, by putting

$$d_0 = 1, \quad d_m = L^m[(m-1)!]^\varkappa m^\mu, \quad m \geq 1. \tag{7.25}$$

If $L = Ae^{\varkappa}$ and $\mu = \mu_1 + \varkappa/2$, then the sequences d_m and c_m are equivalent by the Stirling formula. Associate with the sequence d_m the entire function in the variable $z = 2\pi\beta$ by letting

$$\widetilde{A}_0[\beta] = \sum_{m=0}^{\infty} \frac{(2\pi\beta)^{2m}}{d_m^2}. \tag{7.26}$$

By $\widetilde{A}_0^{\alpha}[\beta]$ denote the partial sum of (7.26) in which m ranges from 0 to α. Denote the difference between $\widetilde{A}_0[\beta]$ and $\widetilde{A}_0^{\alpha}[\beta]$ by $\widetilde{A}_{\alpha+1}[\beta]$.

The sequences d_m and c_m are equivalent. Consequently, there are positive contants K_1 and K_2 such that, for all α and β, we have the simultaneous inequalities

$$K_1\widetilde{A}_\alpha[\beta] \leq A_\alpha[\beta] \leq K_2\widetilde{A}_\alpha[\beta], \quad K_1\widetilde{A}_0^{\alpha}[\beta] \leq A_0^{\alpha}[\beta] \leq K_2\widetilde{A}_0^{\alpha}[\beta].$$

Thus, to estimate the norm of $\langle \varphi_\alpha \rangle_1$ from above it suffices to inspect the behavior as $\alpha \to \infty$ of the sum of the following series

$$\widetilde{S}[\alpha] = \sum_{\beta=1}^{\infty} \frac{\widetilde{A}_0^{\alpha}[\beta]\widetilde{A}_{\alpha+1}[\beta]}{(2\pi\beta)^{2(\alpha+1)}\widetilde{A}_0[\beta]}. \tag{7.27}$$

We settle this problem by paying more attention to the ratio of the consecutive terms of (7.26). It is not hard to calculate that the ratio of the mth term to the $(m+1)$th term equals the square of $\beta_\mu(m)/\beta$, with

$$\beta_\mu(m) = \frac{L}{2\pi} m^{\varkappa} \left(1 + \frac{1}{m}\right)^{\mu}. \tag{7.28}$$

Lemma 7.8. *For every positive q there is an increasing function $\alpha_{\mu,q}(\beta)$ defined on the interval $(\beta_q, +\infty)$ and possessing the following properties. The ratio of the mth term of (7.26) to the consecutive term is at most q if $m \in [1, \alpha_{\mu,q}(\beta)]$, and this ratio is at least $1/q$ if $m \in (\alpha_{\mu,1/q}(\beta), +\infty)$. The asymptotic behavior of $\alpha_{\mu,q}(\beta)$ at large β is given by*

$$\alpha_{\mu,q}(\beta) = \left(2\pi q \frac{\beta}{L}\right)^{1/\varkappa} \left(1 + O\left(\beta^{-1/\varkappa}\right)\right), \quad \beta \to \infty. \tag{7.29}$$

Proof. Set the function $\beta_{\mu,q}(\alpha)$ equal $\beta_\mu(\alpha)/q$ and calculate its logarithmic derivative

$$\frac{d}{d\alpha}[\log \beta_{\mu,q}(\alpha)] = \frac{\varkappa - \mu}{\alpha} + \frac{\mu}{\alpha+1} = \frac{1}{\alpha}\left(\varkappa - \frac{\mu}{\alpha+1}\right).$$

This derivative is nonnegative for $\alpha \geq \alpha_0$, with α_0 the maximum of the two numbers, 0 and $\mu/\varkappa - 1$. Thus, for $\alpha \geq \alpha_0$ the function $\beta_{\mu,q}(\alpha)$ is an increasing function.

Consequently, there exists an inverse of this function which we denote by $\alpha_{\mu,q}(\beta)$. The latter function is certainly defined and monotone on the interval $(\beta_q, +\infty)$, with $\beta_q = \beta_{\mu,q}(\alpha_0)$. Check that $\alpha_{\mu,q}(\beta)$ possesses the required properties.

The ratio of the mth term of (7.26) to the consecutive term, as mentioned above, equals the square of

$$\frac{\beta_\mu(m)}{\beta} = q\frac{\beta_{\mu,q}(m)}{\beta}. \tag{7.30}$$

If for a given β the argument m is at most $\alpha_{\mu,q}(\beta)$, then by monotonicity of $\beta_{\mu,q}(m)$ the ratio in the right side of (7.30) does not exceed the quantity $\beta_{\mu,q}(\alpha_{\mu,q}(\beta))/\beta = 1$. If, on the other hand, m is at least $\alpha_{\mu,1/q}(\beta)$, then we are to use the following estimate

$$\frac{\beta_\mu(m)}{\beta} = \frac{1}{q}\frac{\beta_{\mu,1/q}(m)}{\beta} \geq \frac{1}{q}\frac{\beta_{\mu,1/q}(\alpha_{\mu,1/q}(\beta))}{\beta} = \frac{1}{q}.$$

Finally, we find the asymptotic expansion of $\alpha_{\mu,q}(\beta)$. By definition, the equality holds

$$\alpha_{\mu,q}(\beta) = \left(\frac{2\pi q\beta}{L}\right)^{1/\varkappa}\left(1 + \frac{1}{\alpha_{\mu,q}(\beta)}\right)^{-\mu/\varkappa}. \tag{7.31}$$

Since the function $\alpha_{\mu,q}(\beta)$ tends to infinity at large β, (7.31) entails the sought relation (7.29). The proof is complete.

Corollary 7.2. *The functions $\widetilde{A}_0^\alpha[\beta]$ and $\widetilde{A}_\alpha[\beta]$ grow at large β not faster than this is allowed by the following estimates*

$$\widetilde{A}_\alpha[\beta] \leq K\alpha_\mu(\beta) \max_{\alpha\leq m<\infty} \frac{(2\pi\beta)^{2m}}{d_m^2}, \tag{7.32}$$

$$\widetilde{A}_0^\alpha[\beta] \leq K\alpha_\mu(\beta) \max_{0\leq m\leq \alpha} \frac{(2\pi\beta)^{2m}}{d_m^2}. \tag{7.33}$$

Here K is a constant independent of α and β, and the function $\alpha_\mu(\beta) = \alpha_{\mu,1}(\beta)$ is defined in Lemma 7.8.

PROOF. We may estimate the growth of the modulus of an entire function through the maximal term of the corresponding Taylor series, leaning on the basic facts of entire function theory (see, for instance, [121] or [135]). However, we do the same on using Lemma 7.8.

First, derive (7.32) for $\alpha = 0$. Decompose the function $\widetilde{A}_0[\beta]$ into three summands. The first of them is the partial sum of (7.26) with the index m ranging from 1 to $\alpha_{\mu,1/2}(\beta)$. The maximal summand at that has index $m_* = [\alpha_{\mu,1/2}(\beta)]$.

Factoring the summand out of the partial sum sign and using Lemma 7.8, obtain a dominant in the shape

$$\frac{(2\pi\beta)^{2m_*}}{d_{m_*}^2} \sum_{m=1}^{\infty} \left(\frac{1}{2}\right)^m . \tag{7.34}$$

Analogous reasoning applied to the third summand with summation over all m greater than $\alpha_{\mu,2}(\beta)$, allows us to dominate the summand with

$$\frac{(2\pi\beta)^{2m_{**}}}{d_{m_{**}}^2} \sum_{m=1}^{\infty} \left(\frac{1}{2}\right)^m , \tag{7.35}$$

where m_{**} stands for the integer nearest to $\alpha_{\mu,2}(\beta)$. Summation in the second, central, summand is taken over the integers m in the interval $(\alpha_{\mu,1/2}(\beta), \alpha_{\mu,2}(\beta))$, and the maximum is attained at m equal to $\alpha_\mu(\beta)$. Thus, a dominant for the central summand takes the form

$$(\alpha_{\mu,2}(\beta) - \alpha_{\mu,1/2}(\beta)) \max_{0\leq m<\infty} \frac{(2\pi\beta)^{2m}}{d_m^2} . \tag{7.36}$$

The last of the quantities (7.34)–(7.36) is the greatest among them.

Recalling the asymptotic expansion of $\alpha_{\mu,q}(\beta)$ which is given by (7.29), we see that $\widetilde{A}_0[\beta]$ indeed satisfies (7.32). In an analogous fashion we may dominate $\widetilde{A}_0^\alpha[\beta]$ and $\widetilde{A}_\alpha[\beta]$. The proof of Corollary 7.2 is complete.

Some remarks on maxima and minima in (7.32) and (7.33) are now in order. A. Ostrowski used the function

$$T(z) = \max_{0\leq m\leq\infty} \frac{z^m}{d_m}$$

in study of various classes of quasianalytic functions [211]. In our case when the sequence d_m satisfies (7.25), at large z the *Ostrowski function* $T(z)$ is equivalent to the function $\beta^\nu e^{s\beta^{1/\varkappa}}$, with $\nu = 1/2 - \mu/\varkappa$ and $\beta = z/2\pi$. Therefore, as follows from (7.32), the entire function $\widetilde{A}_0[\beta]$ has order $\beta = 1/\varkappa$ and type $\sigma = 2s$ in a neighborhood about the point at infinity.

Lemma 7.9. *There is a constant K such that the inequality*

$$\langle \varphi_\alpha \rangle \leq K \frac{\alpha^{(1-\varkappa)/2}}{c_\alpha} \tag{7.37}$$

holds uniformly in α.

PROOF. Use the expansion (7.24) of $\langle\varphi_\alpha\rangle$. As follows from the definition of c_α and (7.5), the first summand in (7.24) is dominated by the function $K\alpha^{-2\varkappa}/c_\alpha^2$. Thus, it suffices to examine at large α the behavior of the second summand in (7.24) given as a series. As mentioned, this problem amounts to estimating the function $\widetilde{S}[\alpha]$ that is defined by (7.27). Split the corresponding series into two summands.

In the first of them β ranges from 1 to $\beta_{\mu_1}(\alpha)$, the ratio of $\widetilde{A}_0^\alpha[\beta]$ to $\widetilde{A}_0[\beta]$ is at most 1, and the function $\widetilde{A}_{\alpha+1}[\beta]$ admits the estimate (7.32). In this case, $\alpha_{\mu_1}(\beta)\le\alpha$ and the inequality (7.32) becomes simpler

$$\widetilde{A}_{\alpha+1}[\beta]\le K\frac{(2\pi\beta)^{2(\alpha+1)}}{d_{\alpha+1}^2}\alpha_{\mu_1}(\beta_{\mu_1}(\alpha))\le K\frac{(2\pi\beta)^{2(\alpha+1)}}{d_{\alpha+1}^2}\alpha.$$

Thus, the first summand we selected in $\widetilde{S}[\alpha]$ admits the dominant

$$K\sum_{\beta=1}^{\beta_{\mu_1}(\alpha)}\frac{\alpha}{d_{\alpha+1}^2}=K\frac{\alpha\beta_{\mu_1}(\alpha)}{d_{\alpha+1}^2}\sim K\frac{\alpha^{1-\varkappa}}{c_\alpha^2}. \tag{7.38}$$

In the second summand we selected in $\widetilde{S}[\alpha]$ the summation index β is greater than $\beta_{\mu_1}(\alpha)$. Moreover, the ratio of $\widetilde{A}_{\alpha+1}[\beta]$ to $\widetilde{A}_0[\beta]$ is at most 1, and the function $\widetilde{A}_0^\alpha[\beta]$ satisfies (7.33). In this case, $\alpha\le\alpha_{\mu_1}(\beta)$ and so (7.33) becomes simpler

$$\widetilde{A}_0^\alpha[\beta]\le K\frac{(2\pi\beta)^{2\alpha}}{d_\alpha^2}\beta^{1/\varkappa}. \tag{7.39}$$

At large α we may sharpen (7.39). Take an arbitrary natural $\gamma\le\alpha$ and split the function $\widetilde{A}_0^\alpha[\beta]$ into two parts

$$\widetilde{A}_0^\alpha[\beta]=\widetilde{A}_0^{\alpha-\gamma}[\beta]+\sum_{m=\alpha-\gamma+1}^{\alpha}\frac{(2\pi\beta)^{2m}}{d_m^2}. \tag{7.40}$$

Applying (7.39) to the first summand, obtain

$$\widetilde{A}_0^{\alpha-\gamma}[\beta]\le K\frac{(2\pi\beta)^{2(\alpha-\gamma)}}{d_{\alpha-\gamma}^2}\beta^{1/\varkappa}\sim K\frac{(2\pi\beta)^{2\alpha}}{d_\alpha^2}\beta^{1/\varkappa-2\gamma}\alpha^{2\varkappa\gamma}. \tag{7.41}$$

The second sum on the right side of (7.40) comprises exactly γ summands, the maximal among which corresponds to $m=\alpha$ by the condition $\alpha\le\alpha_{\mu_1}(\beta)$. Consequently,

$$\sum_{m=\alpha-\gamma+1}^{\alpha}\frac{(2\pi\beta)^{2m}}{d_m^2}\le\gamma\frac{(2\pi\beta)^{2\alpha}}{d_\alpha^2}. \tag{7.42}$$

From (7.40)–(7.42) it follows that for $\gamma \leq \alpha \leq \alpha_{\mu_1}(\beta)$ the estimate holds

$$\widetilde{A}_0^\alpha[\beta] \leq K\frac{(2\pi\beta)^{2\alpha}}{d_\alpha^2}\alpha^{2\varkappa\gamma}\beta^{1/\varkappa-2\gamma},$$

with K independent of α and β. Thus, the second summand we selected in $\widetilde{S}[\alpha]$ admits the following dominant

$$K\sum_{\beta\geq\beta_{\mu_1}(\alpha)}^{\infty}\frac{\alpha^{2\varkappa\gamma}}{\beta^{2+2\gamma-1/\varkappa}d_\alpha^2}. \tag{7.43}$$

The series in (7.43) converges provided that $m = 2+2\gamma-1/\varkappa$ is greater than 1. In this event, the sum of the series (7.43) may be estimated from above by the integral

$$\int_{\beta_{\mu_1}(\alpha)-1}^{\infty} x^{-m}\,dx \sim \frac{\beta_{\mu_1}^{1-m}(\alpha)}{m-1}.$$

Therefore, the expression (7.43) at large α admits an estimate from above by the function

$$K\frac{\alpha^{2\varkappa\gamma}(\beta_{\mu_1}(\alpha))^{1-m}}{d_\alpha^2} \sim K\frac{\alpha^{1-\varkappa}}{c_\alpha^2}. \tag{7.44}$$

This means that a dominant for the function $\widetilde{S}[\alpha]$ takes the form $\alpha^{1-\varkappa}/c_\alpha^2$ at large α. The proof is complete.

Using (7.37), we readily estimate the inner product $\langle\varphi_\alpha, \varphi_\beta\rangle$ as follows

$$|\langle\varphi_\alpha,\varphi_\beta\rangle| \leq \langle\varphi_\alpha\rangle\langle\varphi_\beta\rangle \leq K\frac{(\alpha\beta)^{(1-\varkappa)/2}}{c_\alpha c_\beta}. \tag{7.45}$$

PROOF OF THEOREM 7.16. Let φ belong to $\overline{\overline{H}}(\varkappa, A, \lambda)$. Arrange the partial sum $S_N(x)$ of (7.22) corresponding to φ,

$$S_N(x) = \sum_{m=0}^{N}\left(D^m\varphi\Big|_0^1\right)\varphi_m(x).$$

Prove that the sequence of S_N is fundamental in $\overline{H}(\varkappa, A, \lambda)$. Let $N_2 > N_1$. Then, using (7.37), we readily estimate the norm of the difference between $S_{N_2}(x)$ and $S_{N_1}(x)$,

$$\langle S_{N_1} - S_{N_1}\rangle \leq K\sum_{m=N_1+1}^{N_2}\frac{m^{(1-\varkappa)/2}}{c_m}\left|D^m\varphi\Big|_0^1\right|.$$

Applying the Cauchy inequality for sums to the right side, obtain

$$\langle S_{N_2} - S_{N_1} \rangle \le K \left\{ \sum_{m=N_1+1}^{N_2} \frac{|D^m\varphi|_0^1|^2}{b_m^2} \right\}^{1/2} \left\{ \sum_{m=N_1+1}^{N_2} m^{1-\varkappa} \left(\frac{b_m}{c_m}\right)^2 \right\}^{1/2}. \quad (7.46)$$

By (7.3), the ratio b_m/c_m is equivalent to $1/m^{\mu_1-\mu_2}$ as $m \to \infty$. As mentioned in (7.5), $2(\mu_1 - \mu_2) + \varkappa - 1 > 1$. Consequently, to the last factor on the right side of (7.46) there corresponds a convergent series. By the definition of the norm $\langle \cdot \rangle_1$ of the space, we further have

$$\langle S_{N_2} - S_{N_1} \rangle_1^2 = \langle S_{N_2} - S_{N_1} \rangle^2 + \sum_{m=N_1+1}^{N_2} \frac{|D^m\varphi|_0^1|^2}{b_m^2}.$$

This and (7.46) imply that the sequence of the partial sums $S_N(x)$ is fundamental in $\overline{H}(\varkappa, A, \lambda)$. Its limit $S(x)$ belongs obviously to the orthogonal complement $\tilde{\overline{H}}^{\perp}(\varkappa, A, \lambda)$.

Check that the initial function $\varphi(x)$ agrees with $S(x)$.

To begin with, show that in our case the Fourier coefficients $c_\varphi[\beta]$ satisfy the equality

$$c_\varphi[\beta] = \sum_{m=0}^{\infty} \left(D^m\varphi\Big|_0^1\right) c_m[\beta], \quad (7.47)$$

with $c_m[\beta]$ the Fourier coefficients of $\varphi_m(x)$ defined by (7.19). The series on the right side of (7.47) converges absolutely. This may be proven on using the inequality

$$|c_m[\beta]|^2 \le \int_0^1 \varphi_m^2\, dx \le \langle \varphi_m \rangle^2 \le K \frac{m^{1-\varkappa}}{c_m^2},$$

and reasoning further by the scheme analogous to validating that the sequence S_N is fundamental. The inner product of φ and the exponential $e^{-i2\pi\beta x}$ is expressed through the coefficient $c_\varphi[\beta]$ and the increments of the derivatives of φ as follows

$$\langle \varphi, e^{-i2\pi\beta x} \rangle_1 = A_0[\beta] \left\{ c_\varphi[\beta] - \sum_{m=0}^{\infty} c_m[\beta] D^m\varphi\Big|_0^1 \right\}.$$

This, together with the orthogonality of φ to exponentials, entails the validity of (7.47).

Each Fourier coefficient of $S(x)$ is obviously the limit of the sequence of the Fourier coefficients of the partial sums $S_N(x)$, namely, it is also expressed as the series on the right side of (7.47). This is possible only in the case when $S(x)$ and $\varphi(x)$ agree. The proof is complete.

Corollary 7.3. *The set comprising the functions $\varphi_m(x)$ and the exponentials $e^{i2\pi\beta x}$, $\beta = 0, \pm 1, \dots$, is a basis for the space $\overline{H}(\varkappa, A, \lambda)$.*

The inner product of the basis functions φ_m and φ_n defined by (7.23) we denote by $A_{m,n}$; the same quantity multiplied by $b_m b_n$, we denote by $A^*_{m,n}$; namely,

$$A^*_{m,n} = b_m b_n A_{m,n} = b_m b_n \langle \varphi_m, \varphi_n \rangle.$$

The infinite matrix $(A^*_{m,n})$ determines a Hilbert–Schmidt selfadjoint bounded linear operator in the l_2 space of square summable numerical sequences. This assertion is easy on using the next estimate which follows from (7.45)

$$\left(\sum_{\alpha,\gamma=0}^{\infty} \left| A^*_{\alpha,\gamma} \right|^2 \right)^{1/2} \le C \sum_{\alpha=1}^{\infty} 1/\alpha^{2(\mu_1-\mu_2)+\varkappa-1}. \tag{7.48}$$

The series on the right side converges by (7.5).

For every function φ in the orthogonal complement $\overline{\overline{H}}{}^{\perp}(\varkappa, A, \lambda)$ it is possible to express $\langle \varphi \rangle^2$ also through the increments of the derivatives of this function by the formula

$$\langle \varphi \rangle^2 = \sum_{m,n=0}^{\infty} A^*_{m,n} \{ D^m \varphi \big|_0^1 \overline{D^n \varphi} \big|_0^1 \} / (b_m b_n). \tag{7.49}$$

To justify (7.49), it suffices to expand the function φ in the series (7.22) and use the fact that the series converges to φ in the norm $\langle \cdot \rangle_1$ and, consequently, in the norm $\langle \cdot \rangle$.

Not all functions of the basis we constructed for the space $\overline{H}(\varkappa, A, \lambda)$ are mutually orthogonal. This shortcoming may be repaired by endowing $\overline{H}(\varkappa, A, \lambda)$ with a new inner product. We have the following

Theorem 7.17. *The functions $\varphi_m(x)$, $m = 0, 1, \dots$, and the exponentials $e^{i2\pi\beta x}$, $\beta = 0, \pm 1, \dots$, constitute a **Riesz basis** for the space $\overline{H}(\varkappa, A, \lambda)$. In other words, there is an inner product in this such that, first, it generates the norm which is equivalent to the initial norm and, second, the basis functions are mutually orthogonal.*

PROOF. Given arbitrary functions $\varphi, \psi \in \overline{H}(\varkappa, A, \lambda)$, define a new bilinear form by letting

$$\langle \varphi, \psi \rangle_2 = \langle \varphi, \psi \rangle_1 - \sum_{m,n=0}^{\infty} A^*_{m,n} \frac{D^m \varphi \big|_0^1 \overline{D^n \psi} \big|_0^1}{b_m b_n}. \tag{7.50}$$

Denote the orthoprojections of φ and ψ to $\tilde{\overline{H}}(\varkappa, A, \lambda)$ by $\widetilde{\varphi}$ and $\widetilde{\psi}$. Using (7.49), it is not hard to validate the formula

$$\langle \varphi, \psi \rangle_2 = \langle \widetilde{\varphi}, \widetilde{\psi} \rangle_1 + \sum_{m=0}^{\infty} \frac{1}{b_m^2} \{ D^m \varphi \big|_0^1 \overline{D^m \psi \big|_0^1} \}. \tag{7.51}$$

Whence and from (7.21) it follows that (7.50) really determines in $\overline{H}(\varkappa, A, \lambda)$ an inner product such that the functions of the basis we constructed are mutually orthogonal.

Check that the norm generated by this new inner product is equivalent to the initial norm of $\overline{H}(\varkappa, A, \lambda)$. The norm square of φ may be expressed as

$$\langle \varphi \rangle_1^2 = \langle \widetilde{\varphi} \rangle_1^2 + \sum_{m=0}^{\infty} \frac{1}{b_m^2} |D^m \varphi \big|_0^1|^2 + \langle \varphi - \widetilde{\varphi} \rangle^2. \tag{7.52}$$

Comparing the expansions (7.51) and (7.52) of the squared norms of the same function φ, come to the conclusion that

$$\langle \varphi \rangle_2 \leq \langle \varphi \rangle_1. \tag{7.53}$$

Establish a reverse estimate to (7.53). Using (7.51) and (7.52), rearrange the difference of the norm squares of φ to obtain

$$\langle \varphi \rangle_1^2 - \langle \varphi \rangle_2^2 = \langle \varphi - \widetilde{\varphi} \rangle^2. \tag{7.54}$$

Expand $\langle \varphi - \widetilde{\varphi} \rangle^2$ in a series by (7.49) and, using (7.48), infer the inequality

$$\langle \varphi - \widetilde{\varphi} \rangle^2 \leq C \sum_{m=0}^{\infty} \frac{1}{b_m^2} |D^m \varphi \big|_0^1|^2 \leq C \langle \varphi \rangle_2^2.$$

This, together with (7.54), entails the sought estimate for $\langle \varphi \rangle_1$ via $\langle \varphi \rangle_2$. The proof is complete.

§8. Convergence of Euler–Maclaurin and Gregory Quadrature Formulas on Gevrey Classes

Revert to examining the sequences of quadrature formulas at a function belonging to the class $G_\varkappa[0,1]$. In § 6 we considered the error corresponding to the classical Euler–Maclaurin formula and checked that for every function φ in $H(\varkappa, A, \lambda)$ the error of such formula vanishes exponentially as the number of nodes N tends to infinity. Moreover, we claimed that it is not always reasonable to increase the order m at a fixed N. To justify the claim, we exhibit a lower bound (6.16) of the $H^*(\varkappa, A, \lambda)$ norm of $l_{m,N}^{EM}$ without proof. Now we proceed with demonstration.

Consider the Hilbert space $\overline{H}(\varkappa, A, \mu)$ with inner product $\langle \cdot, \cdot \rangle_2$. This space has an orthogonal basis containing the functions φ_α. The norm of each of these functions is easy to calculate by the formula

$$\langle \varphi_\alpha \rangle_2 = \frac{1}{b_\alpha}.$$

If $\mu = \lambda - \varkappa - 2$ then the space $\overline{H}(\varkappa, A, \mu)$ is embedded in $H(\varkappa, A, \lambda)$, with the embedding operator bounded. Consequently, the ratio of the norms of the error $l_{m,N}^{EM}$ in the dual spaces to $H(\varkappa, A, \lambda)$ and $\overline{H}(\varkappa, A, \mu)$ is bounded from below by a positive constant K independent of m and N, namely,

$$\left\| l_{m,N}^{EM} \mid H^*(\varkappa, A, \lambda) \right\| \geq K \left\| l_{m,N}^{EM} \mid \overline{H}^*(\varkappa, A, \mu) \right\|.$$

We may continue the estimate. Observe that

$$\left\| l_{m,N}^{EM} \mid \overline{H}^*(\varkappa, A, \mu) \right\| \geq \frac{\left| \left(l_{m,N}^{EM}, \varphi_\alpha \right) \right|}{\langle \varphi_\alpha \rangle_2} = b_\alpha \left| \left(l_{m,N}^{EM}, \varphi_\alpha \right) \right|.$$

Calculate the error of the Euler–Maclaurin formula at a basis function φ_α. Let $\alpha \leq m - 2$. Then the error $l_{m,N}^{EM}$ is exact for the polynomial $B_{\alpha+1}(x)/(\alpha+1)!$. Hence, its values at the projections φ_α and $\widetilde{\varphi}_\alpha$ differ only in sign

$$\left(l_{m,N}^{EM}, \varphi_\alpha \right) = -\left(l_{m,N}^{EM}, \widetilde{\varphi}_\alpha \right).$$

The value of the error $l_{m,N}^{EM}$ at every infinitely differentiable periodic function $\widetilde{\varphi}$ is $(l_h(x), \widetilde{\varphi})$, with the error l_h defined by (1.1). Considering this and expanding $\widetilde{\varphi}_\alpha$ in the Fourier series, infer

$$\left(l_{m,N}^{EM}, \varphi_\alpha \right) = -\sum_{\beta \neq 0} \widetilde{c}_\alpha[\beta] (l_h, e^{-i2\pi\beta x}).$$

The value of l_h at the exponential $e^{-i2\pi\beta x}$, as mentioned above, equals 0 if β is not a multiple of N and equals -1 otherwise. Considering this and using the formula (7.17) for the Fourier coefficients of $\widetilde{\varphi}_\alpha$, obtain

$$\left(l_{m,N}^{EM}, \varphi_\alpha \right) = \sum_{k \neq 0} \frac{(-1)^\alpha}{(i2\pi kN)^{\alpha+1}} \frac{A_0^\alpha[kN]}{A_0[kN]}.$$

Taking in the last equality α odd and equal to $m-3$, come to the following expansion of the modulus of the error

$$\left|\left(l_{m,N}^{EM}, \varphi_{m-3}\right)\right| = 2h^{m-2}\sum_{k=1}^{\infty}\frac{1}{(2\pi k)^{m-2}}\frac{A_0^{m-3}[kN]}{A_0[kN]}.$$

The series on the right side consists of positive terms. Hence, its sum is surely greater than the term with index $k=1$. Consequently,

$$\left\|l_{m,N}^{EM} \mid \overline{H}^*(\varkappa, A, \lambda)\right\| \geq 2b_{m-3}\left(\frac{h}{2\pi}\right)^{m-2}\frac{A_0^{m-3}[N]}{A_0[N]}.$$

Assume now that the order m is greater than $\alpha_{\mu,2}(N)+3$. Then the ratio of $A_0^{m-3}[N]$ and $A_0[N]$ is bounded from below by a positive constant independent of N.

We have thus demonstrated the existence of a constant K such that, for all m greater than $\alpha_{\mu,2}(N)+3$, the inequality holds

$$\left\|l_{m,N}^{EM} \mid H^*(\varkappa, A, \lambda)\right\| \geq KN^2\frac{b_{m-3}}{(2\pi N)^m}.$$

Recall that at large m the sequence b_m behaves like $m^{\varkappa m+\mu_2}A^m$, with $\mu_2 = \lambda + \varkappa/2 - 3/2$. Consequently, as $m\to\infty$ the right side of the our estimate is equivalent to the sequence

$$K\frac{a_m}{(2\pi N)^m}N^2 m^{\gamma},$$

with $\gamma = -(5\varkappa+3)/2$.

Thus, for $m > \alpha_{\mu,2}(N)+3$ we indeed see that the all inequalities (6.16) hold.

Now we examine the behavior of the double sequence of errors corresponding to the *Gregory quadrature formulas* on the functions of a Gevrey class. Put

$$\left(l_{N,m}^{G}, \varphi\right) = \int_0^1 \varphi(x)\,dx - T_N(\varphi) + \sum_{\alpha=1}^{m-2} a[\alpha]h\{\Delta^\alpha\varphi_{N-\alpha} + (-1)^\alpha\Delta^\alpha\varphi_0\}. \quad (8.1)$$

Here $\varphi(x)$ is an arbitrary continuous function on the unit interval. The product of a natural N and a real h equals 1, and a natural m must be an even number at most N. The finite differences $\Delta^\alpha\varphi_k$ are as usual determined by the equality

$$\Delta^\alpha\varphi_k = \sum_{j=0}^{\alpha}(-1)^{\alpha-j}\binom{\alpha}{j}\varphi(kh+jh).$$

The denotation $T_N(\varphi)$ is introduced in § 6.

The error $l^G_{N,m}$ has order m if and only if the weights $a[\alpha]$ may be written as the following integrals of the Newtonian powers

$$a[\alpha] = \frac{(-1)^\alpha}{(\alpha+1)!} \int_0^1 t^{[\alpha+1]}\, dt, \quad \alpha = 0, 1, \ldots, m-2. \tag{8.2}$$

For the function $\varphi \in H(\varkappa, A, \lambda)$ we have a collection of inequalities analogous to (6.10)

$$\left|(l^G_{N,m}, \varphi)\right| \le K[\varphi]_H \| l^G_{N,m} \mid L_2^{(m)}[0,1]^* \| a_m. \tag{8.3}$$

The sequence a_m is the same as in § 6. To minimize the right side of (8.3) over all even m, it is necessary to know the behavior of the $L_2^{(m)}[0,1]^*$ norms of $l^G_{N,m}$, with N fixed and m varying. If $m \le N/2$, then the norm square may be calculated by the formula

$$\| l^G_{N,m+2} \mid L_2^{(m)}[0,1]^* \|^2 = h^{2m} \frac{|B_{2m}|}{(2m)!}$$

$$+ h^{2m+1} a^2[m+1] \frac{32}{\pi} \int_0^\infty \left(\frac{\sin t}{t}\right)^{2m} \sin^4 t\, P^2_{m-2}(\cos t)\, dt. \tag{8.4}$$

Here B_{2m} is the Bernoulli number, the quantity $a[m+1]$ is defined by (8.2) for $\alpha = m+1$, and the polynomial $P_{m-2}(w)$ of degree $m-2$ is independent of N and differs on the unit interval from the Chebyshev polynomial of the second kind $U_{m-2}(w)$ only by a quantity that is $O(1/m)$ as $m \to \infty$. We will prove (8.4) in the last section of this chapter.

From (8.4), in particular, it is easy to infer the estimate

$$\| l^G_{N,m+2} \mid L_2^{(m)}[0,1]^* \| \le K \frac{h^m}{m}, \tag{8.5}$$

with K independent of N and m. Inserting (8.5) in (8.3), come to the following relation

$$\left|(l^G_{N,m+2}, \varphi)\right| \le K[\varphi]_H m^{\varkappa m + \lambda - 1} \left(\frac{A}{N}\right)^m.$$

The minimum of $m^{\varkappa m+\lambda-1}(A/N)^m$ is attained at m and equals $m_0(N)$, the even number nearest to $(s/\varkappa)(N/2\pi)^{1/\varkappa}$. This minimal value is $m_0^{\lambda-1} e^{-\varkappa m_0}$. Therefore, for an arbitrary function $\varphi \in H(\varkappa, A, \lambda)$ the error may be dominated as follows

$$\left|(l^G_{N,m_0(N)+2}, \varphi)\right| \le K[\varphi]_H N^{(\lambda-1)/\varkappa} e^{-s(N/2\pi)^{1/\varkappa}}, \tag{8.6}$$

with s defined by (5.12) and K a constant independent of N.

Thus, in much the same way as in the case of Euler–Maclaurin formulas, the vanishing of the errors of Gregory formulas is exponential for a suitable choice of order. Moreover, the norm of the error, as follows from (8.6), satisfies the inequality

$$\left\| l^G_{N,m_0(N)+2} \mid H^*(\varkappa, A, \lambda) \right\| \leq K N^{(\lambda-1)/\varkappa} e^{-s(N/2\pi)^{1/\varkappa}}. \tag{8.7}$$

Derive a lower bound on the norm in the left side of (8.7). As in the case of the Euler–Maclaurin formulas we are to know the explicit form of the function

$$L_N^{m,G}[\beta] = \left(l^G_{N,m}, e^{-i2\pi\beta x} \right). \tag{8.8}$$

If β is not a multiple of the number of nodes N, then the Fourier coefficients $L_N^{m+2,G}[\beta]$ may be expressed through the polynomial $P_{m-2}(w)$ of (8.4). We have the formula

$$L_N^{m+2,G}[\beta] = h 2^{m+2} (-1)^{m/2+1} (\sin \pi\beta h)^{m+2} a[m+1] P_{m-2}(\cos \pi\beta h). \tag{8.9}$$

The number $a[m+1]$ is defined by (8.2) as before. We will derive (8.9) in Theorem 7.18.

From (8.9) we infer the following expansion

$$L_N^{m+2,G}[\beta] = h 2^{m+2} (-1)^{m/2+1} (\sin \pi\beta h)^{m+1} a[m+1] \times \left(\sin (m-1)\pi\beta h + O\left(\frac{1}{m}\right) \right) \tag{8.10}$$

which is asymptotically exact as $m \to \infty$.

Consider the periodic function

$$\varphi_*(x) = \sum_{\beta=1}^{\infty} \beta^{\nu_0} e^{-s\beta^{1/\varkappa}} \operatorname{sgn}\left(L_N^{m+2,G}[\beta] \right) e^{-i2\pi\beta x}. \tag{8.11}$$

For some ν_0, the above function belongs to $H(\varkappa, A, \lambda)$. This is easy to prove by the same scheme as in § 6. The error of the Gregory formula at φ_* is at most the product of the $H(\varkappa, A, \lambda)$ norm of φ_* and the $H^*(\varkappa, A, \lambda)$ norm of $l^G_{N,m+2}$, namely,

$$\left| \left(l^G_{N,m+2}, \varphi_* \right) \right| \leq [\varphi_*]_H \left\| l^G_{N,m+2} \mid H^*(\varkappa, A, \lambda) \right\|.$$

Using (8.11), calculate the lower bound of this inequality

$$\left\| l^G_{N,m+2} \mid H^*(\varkappa, A, \lambda) \right\| \geq K \sum_{\beta=1}^{\infty} \beta^{\nu_0} e^{-s\beta^{1/\varkappa}} \left| L_N^{m+2,G}[\beta] \right|.$$

Preserving in the series on the right side the only term corresponding to the index β nearest to $N/6$, and using the expansion (8.10), obtain a sought lower bound for the norm of $l^G_{N,m+2}$

$$\left\| l^G_{N,m+2} \mid H^*(\varkappa, A, \lambda) \right\| \geq K N^{\nu_0 - 1} e^{-s(N/6)^{1/\varkappa}}, \tag{8.12}$$

with K a constant independent of N.

Observe that the exponents of the exponentials in (8.7) and (8.12) do not coincide. The question arises of finding out which of these estimates, (8.7) or (8.12), we may sharpen. It turns out that the exponent $s(N/2\pi)^{1/\varkappa}$ in (8.7) may be increased up to $s(N/6)^{1/\varkappa}$. The proof of this claim grounds on the representation of the norm of the error $l^G_{N,m+2}$ in the Hilbert space $\overline{H}(\varkappa, A, \lambda)$ as the sum of the squared errors at the functions composing the orthogonal basis that was constructed in § 7. The relevant calculations from [281] are cumbersome and thus omitted.

§9. The Sequence of the Fourier Coefficients of an Error

Studying the question of convergence of Gregory quadrature formulas in the space $H(\varkappa, A, \lambda)$, we essentially use two important results. The first of them, (8.4), explicitly expresses the norm of the error of a Gregory formula as a function of m and h. The second, (8.9), presents the error for trigonometric functions which is asymptotically exact as $m \to \infty$. This and two subsequent sections are devoted to derivation of these formulas.

We find it convenient to consider in what follows the errors $l^m_N(x)$ of a more general form than $l^G_{N,m}(x)$. We define the value of $l^m_N(x)$ at an arbitrary continuous function $\varphi(x)$ by the equality analogous to (8.1), but with summation on the right side performed over all α from 1 to N rather than to m, as in the case of a Gregory formula. The weights $a[\alpha]$ of this expansion of the error for all $\alpha = 1, \dots, m-2$ are given by the integrals (8.2) of the Newtonian powers, whereas for $\alpha > m-2$ they are arbitrary.

Agree on one more denotation. Let a natural M be at most N. Arrange the function of a discrete variable α, defined on the semiaxis $\alpha \geq 1$, equal to 0 for $\alpha > M$, and coincident with the weights $a[\alpha]$ for $\alpha \leq M$. We denote this function by $a_M[\alpha]$. Thus, for every continuous function φ, the equality holds

$$(l^m_N, \varphi) = \int_0^1 \varphi(x)\, dx - T_N(\varphi) + \sum_{\alpha=1}^{\infty} a_N[\alpha] h \{\Delta^\alpha \varphi_{N-\alpha} + (-1)^\alpha \Delta^\alpha \varphi_0\}. \tag{9.1}$$

Observe that an arbitrary error (9.1) is even with respect to the midpoint of the interval $[0, 1]$, i.e.,

$$l^m_N(x) = l^m_N(1-x). \tag{9.2}$$

The error (9.1) may be written down differently as

$$(l_N^m, \varphi) = \int_0^1 \varphi(x)\,dx - \sum_{\alpha=0}^{N} c[\alpha] h \varphi(\alpha h). \tag{9.3}$$

By (9.2) the weights $c[\alpha]$ in this expansion are symmetric with respect to the midpoint of the interval $[0, N]$.

Obviously, the error $l_N^m(x)$ is a continuous linear functional in the space $\mathscr{S}$ of tempered functions. Thus, on the set of functionals like (9.1), which we denote by $EF_N^{(m)}$, there are defined the following operations: addition, multiplication by a scalar, linear changes of the independent variable, generalized differentiation, and the Fourier transform. The restriction of the Fourier transform of l_N^m to the set of integers is the function of a discrete variable whose values may be found by the formula

$$L_N^m[\beta] = \left(l_N^m(x), e^{i2\pi\beta x}\right). \tag{9.4}$$

We call $L_N^m[\beta]$ the *sequence of the Fourier coefficients of* l_N^m or, simply, the *Fourier coefficient* of l_N^m.

Denote by $CF_N^{(m)}$ the set of all functions (9.4) corresponding to the errors $l_N^m(x) \in EF_N^{(m)}$. The equality (9.4) establishes a one-to-one correspondence between $EF_N^{(m)}$ and $CF_N^{(m)}$.

Note the simplest properties of Fourier coefficients (9.4).

For every $l_N^m(x) \in EF_N^{(m)}$ the function $L_N^m[\beta]$ is real and even, i.e.,

$$L_N^m[\beta] = \left(l_N^m(x), \cos 2\pi\beta x\right), \quad L_N^m[-\beta] = L_N^m[\beta]. \tag{9.5}$$

This remark is easy from (9.2). Further, by (9.4) and (9.3) obtain

$$L_N^m[\beta] = \delta[\beta] - \sum_{\alpha=0}^{N} c[\alpha] h e^{-i2\pi\beta\alpha h} = \delta[\beta] - \sum_{\alpha=0}^{N} c[\alpha] h \cos 2\pi\beta\alpha h. \tag{9.6}$$

Here the function $\delta[\beta]$ is 1 at $\beta = 0$ and vanishes at the other points. An easy check shows that $\delta[\beta]$ is the sequence of the Fourier coefficients of the indicator of the unit interval.

It is not hard to calculate the sequence of the Fourier coefficients of $T_N(\varphi)$. This is the function $T_N[\beta]$ equal to 0 at all integer points but the multiples of N at which it equals 1.

The difference between $L_N^m[\beta]$ and $\delta[\beta]$, as follows from (9.6), is the function of period N equal to -1 at all integer multiples of N, i.e.,

$$L_N^m[\beta + N] = L_N^m[\beta], \quad L_N^m[\beta N] = -1, \quad \beta \neq 0. \tag{9.7}$$

It also follows from (9.6) that, for all β such that $0 < \beta < N$, the equality holds

$$L_N^m[N-\beta] = L_N^m[\beta]. \tag{9.8}$$

Denote by $\widetilde{X}_N^{(0)}$ the space of even real functions of a discrete variable β having period N, vanishing at every β that is a multiple of N, and symmetric with respect to the midpoint of the unit interval. Every function $\widetilde{L}_N^m[\beta]$ of this space is uniquely determined by its values at positive βs not exceeding the integral part of $N/2$. To each function $L_N^m[\beta] \in CF_N^{(m)}$, as follows from (9.5)–(9.8), we may assign the function $\widetilde{L}_N^m[\beta] \in \widetilde{X}_N^0$ such that

$$L_N^m[\beta] = \delta[\beta] - T_N[\beta] + \widetilde{L}_N^m[\beta]. \tag{9.9}$$

The functions $\widetilde{L}_N^m[\beta]$, corresponding by (9.9) to all possible errors $l_N^m(x) \in EF_N^{(m)}$, comprise the subset $\widetilde{CF}_N^{(m)}$ of $\widetilde{X}_N^{(0)}$.

Lemma 7.10. *An arbitrary function in $\widetilde{CF}_N^{(m)}$ is an even polynomial in the variable $w = \cos \pi\beta h$ of degree at most $2N$ vanishing for $w = 1$.*

PROOF. By the definition of Δ^α, we have

$$\Delta^\alpha (e^{\mp i2\pi\beta x})_k = (-1)^\alpha e^{\mp i2\pi\beta kh}(1 - e^{\mp i2\pi\beta h})^\alpha. \tag{9.10}$$

From (9.1) and (9.9) it follows that

$$\widetilde{L}_N^m[\beta] = \sum_{\alpha=1}^{\infty} a_N[\alpha]h\{\Delta^\alpha(e^{-i2\pi\beta x})_{N-\alpha} + (-1)^\alpha \Delta^\alpha (e^{-i2\pi\beta x})_0\}.$$

Inserting (9.10), after easy manipulations obtain

$$\widetilde{L}_N^m[\beta] = \sum_{\alpha=1}^{m-2} a[\alpha]h\, 2^{\alpha+1}(\sin \pi\beta h)^\alpha \cos \pi(\beta h - 1/2)\alpha$$
$$+ 2^{m-1}(\sin \pi\beta h)^{m-2}(-1)^{m/2+1} \sum_{\alpha=0}^{\infty} a_N[\alpha + m - 1]\, h$$
$$\times \operatorname{Re}\left[e^{-i(m-2)\pi\beta h}(1 - e^{-i2\pi\beta h})^{\alpha+1}\right]. \tag{9.11}$$

By hypothesis $\cos \pi\beta h = w$ and $\sin \pi\beta h = (1 - w^2)^{1/2}$. Moreover,

$$\cos \pi(\beta h - \frac{1}{2})\alpha = \begin{cases} (-1)^{\alpha/2} T_\alpha(w) & \text{for even } \alpha, \\ (-1)^{(\alpha-1)/2}(1 - w^2)^{1/2} U_{\alpha-1}(w) & \text{for odd } \alpha. \end{cases}$$

Here by $T_n(w)$ ($U_n(w)$) we as usual denote a Chebyshev polynomial of the first (second) kind. Using these equalities, transform (9.11).

Split each sum on the right side of (9.11) into two: one, over the even αs and the other, over the odd αs. Inserting therein the expressions of trigonometric functions through Chebyshev polynomials, infer

$$\widetilde{L}_N^m[\beta] = h\{Q_{2(m-2)}^s(w) + 2^m(w^2-1)^{m/2}Q_{2N-m}^f(w)\}. \tag{9.12}$$

The polynomials $Q_{2(m-2)}^s(w)$ and $Q_{2N-m}^f(w)$ of degree $2(m-2)$ and $2N-m$ are determined by the following equalities

$$Q_{2(m-2)}^s(w) = \sum_{k=1}^{m/2-1} 2^{2k}(w^2-1)^k\{2a[2k]T_{2k}(w) - a[2k-1]U_{2(k-1)}(w)\},$$

$$Q_{2N-m}^f(w) = \sum_{k=0}^{\infty} 2^{2k}(w^2-1)^k$$

$$\times\{2a_N[2k+m]T_{2k+m}(w) - a_N[2k+m-1]U_{2(k-1)+m}(w)\}. \tag{9.13}$$

Since all m are even by hypothesis, the polynomials $Q_{2(m-2)}^s(w)$ and $Q_{2N-m}^f(w)$ are also even. Moreover, $Q_{2(m-2)}^s(1) = 0$. Thus, on the right side of (9.12) there is a polynomial of the desired shape. The proof of Lemma 7.10 is complete.

The equality (9.12) may serve as a criterion for an arbitrary function $\widetilde{L}_N^m[\beta] \in \widetilde{X}_N^{(0)}$ to belong to $\widetilde{CF}_N^{(m)}$.

Consider a special *local error* $l_m^{(0)}(x)$. Given an arbitrary continuous function $\varphi(x)$, put

$$(l_m^{(0)}(x), \varphi(x)) = \int_0^1 \varphi(x)\,dx - \varphi(0) - \sum_{k=0}^{\infty}(-1)^k a_N[k]\Delta_1^{k+1}\varphi_0. \tag{9.14}$$

The coefficients $a_N[k]$ here are the same as in the expansion (9.1) of the error $l_N^m(x)$ and the finite differences $\Delta_1^{k+1}\varphi_0$ are taken with the mesh-size $h=1$. The local error $l_m^{(0)}(x)$ has order m. This is easy to show by integrating over $[0,1]$ the Newton interpolant of the function $\varphi(x)$ which is constructed for the nodes $\alpha = 0, 1, \dots, m-1$.

Using the change of variable $x = hy$, $h > 0$, and given $l_m^{(0)}(y)$, obtain the error $l_m^{(0)}(x/h)$. For an arbitrary continuous function $\varphi(x)$, the equality holds

$$\left(l_m^{(0)}\left(\frac{x}{h}\right), \varphi(x)\right) = \int_0^h \varphi(x)\,dx - h\varphi(0) - \sum_{k=0}^{\infty} h(-1)^k a_N[k]\Delta^{k+1}\varphi_0. \tag{9.15}$$

Denote the sequence of the Fourier coefficients of the error (9.15) by

$$\lambda_0^{m,h}[\beta] = \left(l_m^{(0)}\left(\frac{x}{h}\right), e^{-i2\pi\beta x}\right). \tag{9.16}$$

We have

***Lemma* 7.11.** *The following algebraic identity is valid*

$$i2\sin\pi\beta h\,\widetilde{L}_N^m[\beta] = e^{i\pi\beta h}\lambda_0^{m,h}[\beta] - e^{-i\pi\beta h}\overline{\lambda_0^{m,h}}[\beta]. \tag{9.17}$$

PROOF. It suffices to check two intermediate equalities. The first of them

$$\lambda_0^{m,h}[\beta] = \int_0^h e^{-i2\pi\beta x}\,dx - h + \sum_{\alpha=0}^{\infty} a_N[\alpha]h(1 - e^{-i2\pi\beta h})^{\alpha+1}$$

results from (9.15) and (9.10). By easy calculations, derive the second equality

$$e^{i\pi\beta h}\lambda_0^{m,h}[\beta] - e^{-i\pi\beta h}\overline{\lambda_0^{m,h}}[\beta]$$
$$= i2\sin\pi\beta h\sum_{\alpha=1}^{\infty} a_N[\alpha]h2^{\alpha+1}(\sin\pi\beta h)^{\alpha}\cos\pi\left(\beta h - \frac{1}{2}\right)\alpha.$$

Comparing the right side of the result with (9.11), obtain (9.17). The proof of Lemma 7.11 is complete.

Thus, the function $\widetilde{L}_N^m[\beta]$ belongs to $\widetilde{CF}_N^{(m)}$ if and only if there is a local error (9.15) whose Fourier coefficients relate with $\widetilde{L}_N^m[\beta]$ by the equality

$$\widetilde{L}_N^m[\beta] = \frac{1}{\sin\pi\beta h}\operatorname{Im}\left[e^{i\pi\beta h}\lambda_0^{m,h}[\beta]\right], \quad \beta \neq kN. \tag{9.18}$$

If all weights $a[\alpha]$ in (9.14) equal 0 for $\alpha \geq m-1$ then (9.17) becomes simpler and reduces to the following

$$Q_{2(m-2)}^s(\cos\pi\beta h) = \frac{N}{\sin\pi\beta h}\operatorname{Im}\left[e^{i\pi\beta h}\lambda_0^{m,h}[\beta]\right]. \tag{9.19}$$

The function $\lambda_0^{m,h}[\beta]$ in (9.19) is the sequence of the Fourier coefficients of the local error $l_m^G(x/h)$. The value of this error at a continuous function $\varphi(x)$ is given by the formula

$$\left(l_m^G\left(\frac{x}{h}\right), \varphi(x)\right) = \int_0^h \varphi(x)\,dx - \sum_{\alpha=0}^{m-1} hc_m[\alpha]\varphi(\alpha h). \tag{9.20}$$

The error $l_m^G(x/h)$ has order m, and the weights $c_m[\alpha]$ of $l_m^G(x/h)$ may be written down as the integrals

$$c_m[\alpha] = (-1)^\alpha \int_{-1}^{0} \frac{t(t+1)\dots(t+m-1)}{(t+\alpha)\alpha!(m-\alpha-1)!}\,dt. \tag{9.21}$$

The equalities (9.21) are not hard to obtain by integrating over the interval $[-1,0]$ the Lagrange interpolation formula of degree $m-1$ constructed for the function $\varphi(-x)$ on the set of nodes $\alpha = 0,\dots,-m+1$.

§10. The Fourier Transform of a Local Error

The Fourier transform of the error $l_m^G(x/h)$ defined by (9.20) is a function of a real variable ξ which is given by the equality

$$\lambda_0^{m,h}(\xi) = \left(l_m^G\left(\frac{x}{h}\right), e^{-i2\pi\xi x}\right). \tag{10.1}$$

Its restriction to the set of the integer points β is the sequence $\lambda_0^{m,h}[\beta]$ of the Fourier coefficients of $l_m^G(x/h)$.

We further consider the function $\lambda_0^m(\xi)$ related to $\lambda_0^{m,h}(\xi)$ as follows

$$\lambda_0^m(\xi) = \lambda_0^{m,1}(\xi) = \left(\frac{1}{h}\right)\lambda_0^{m,h}\left(\frac{\xi}{h}\right) = (l_m^G(y), e^{-i2\pi\xi y}). \tag{10.2}$$

We thus obtain an analytic description for $\lambda_0^m(\xi)$ which enables us, in particular, to expand $\lambda_0^m(\xi)$ in a series of a special shape.

Lemma 7.12. *For every real ξ, the Fourier transform of a local error may be written as the integral*

$$\lambda_0^m(\xi) = \int_{-1}^{0} \frac{t(t+1)\dots(t+m-1)}{(m-1)!}\psi_t^m(2\pi\xi)\,dt. \tag{10.3}$$

Moreover, the function $\psi_t^m(x)$ may be found by the formula

$$\psi_t^m(x) = i\int_{0}^{x} e^{it(x-y)}(1-e^{-iy})^{m-1}\,dy. \tag{10.4}$$

PROOF. By the definition of $l_m^G(x)$ we have

$$\lambda_0^m(\xi) = \frac{1}{i2\pi\xi}(1 - e^{-i2\pi\xi}) - \sum_{\alpha=0}^{m-1} c_m[\alpha] e^{-i2\pi\xi\alpha}.$$

Whence, using the formula (9.21) for the weights $c_m[\alpha]$, obtain

$$\lambda_0^m(\xi) = \frac{(1 - e^{-i2\pi\xi})}{i2\pi\xi} - \int_{-1}^{0} \frac{t(t+1)\dots(t+m-1)}{(m-1)!} \varphi_t^m(2\pi\xi)\, dt. \tag{10.5}$$

The integrand $\varphi_t^m(x)$ in (10.5) may be written as the following linear combination

$$\varphi_t^m(x) = \sum_{\alpha=0}^{m-1} (-1)^\alpha \binom{m-1}{\alpha} \frac{1}{t+\alpha} e^{-ix\alpha}. \tag{10.6}$$

Differentiating both sides of (10.6) with respect to x, come to the differential equation

$$\frac{d}{dx}\varphi_t^m(x) = -i(1 - e^{-ix})^{m-1} + it\varphi_t^m(x).$$

Solving it at a fixed t, find that

$$\varphi_t^m(x) = e^{itx}\varphi_t^m(0) - i\int_0^x e^{it(x-y)}(1 - e^{-iy})^{m-1}\, dy. \tag{10.7}$$

Considering that the subtrahend on the right side of (10.7) is denoted by $\psi_t^m(x)$ and inserting (10.7) in (10.5), obtain

$$\lambda_0^m(\xi) - \int_{-1}^{0} \frac{t(t+1)\dots(t+m-1)}{(m-1)!} \psi_t^m(2\pi\xi)\, dt$$

$$= \frac{1}{i2\pi\xi}(1 - e^{-i2\pi\xi}) - \int_{-1}^{0} \frac{t(t+1)\dots(t+m-1)}{(m-1)!} e^{it2\pi\xi}\varphi_t^m(0)\, dt. \tag{10.8}$$

Check that the right side of (10.8) equals 0. By the definition of $\varphi_t^m(x)$, we have

$$\int_{-1}^{0} \frac{t(t+1)\dots(t+m-1)}{(m-1)!} e^{it2\pi\xi}\varphi_t^m(0)\, dt$$

$$= \int_{-1}^{0} e^{it2\pi\xi} \left[\sum_{\alpha=0}^{m-1} \frac{t(t+1)\dots(t+m-1)}{(t+\alpha)\alpha!(m-1-\alpha)!}(-1)^\alpha \right] dt.$$

However, the sum in brackets under the integral sign is identically 1. To prove this, it suffices apply the Lagrange interpolation formula of degree $m-1$ constructed for a constant function over the set of nodes $\alpha = 0, -1, \ldots, -m+1$. Consequently, the right side of (10.8) equals 0. The proof of Lemma 7.12 is complete.

Introducing the notation $u(x) = 1-e^{-ix}$ and carrying out the change of variable $\tau = (e^{-iy}-1)/(e^{-ix}-1)$ in (10.4), obtain

$$\psi_t^m(x) = e^{itx}(1-e^{-ix})^m \int_0^1 \frac{\tau^{m-1}}{(1-\tau u)^{1-t}}\, d\tau. \tag{10.9}$$

Using this formula, find an asymptotically exact expansion of $\psi_t^m(x)$ as $m \to \infty$.

Lemma 7.13. *For all real t and x, the function $\psi_t^m(x)$ expands in a series in the Newtonian powers of t, namely,*

$$\frac{1}{(m-1)!}\psi_t^m(x) = (1-e^{-ix})^{m-1}$$
$$\times\left\{\sum_{j=0}^{n} a_j^m(x)\frac{(t-1)\ldots(t-j)}{j!} + \frac{1}{(m+n)!}A_n^m(x,t)\right\}. \tag{10.10}$$

The coefficients $a_j^m(x)$ of this expansion are determined by

$$a_j^m(x) = \frac{j!}{(m+j)!}(e^{ix}-1)^{j+1}. \tag{10.11}$$

The function $A_n^m(x,t)$ is uniformly bounded when x varies over $[-\pi,\pi]$ and t varies over $[-1,0]$. Moreover,

$$\sup_{\substack{-\pi\le x\le\pi\\ -1\le t\le 0}} \left|A_n^m(x,t)\right| \le \frac{a_n}{m+n}\sin^{n+2}\left(\frac{x}{2}\right), \tag{10.12}$$

with a_n a constant independent of m.

PROOF. Rewrite (10.9) as

$$\psi_t^m(x) = e^{itx}(1-e^{-ix})^{m+t-1} \int_0^1 \frac{\tau^{m-1}}{(z(x)-\tau)^{1-t}}\, d\tau. \tag{10.13}$$

Here $z(x)$ stands for the ratio $1/u(x)$. Integrating by parts $n+1$ times in the right side of (10.13), obtain

$$\int_0^1 \frac{\tau^{m-1}}{(z-\tau)^{1-t}}\,d\tau = \frac{1}{m}\frac{1}{(z-1)^{1-t}}\left\{1+\frac{t-1}{m+1}\frac{1}{z-1}\right.$$

$$\left.+\cdots+\frac{m!}{(m+n)!}\frac{(t-1)\dots(t-n)}{(z-1)^n}+\frac{m!}{(m+n)!}(z-1)A_n^m(x,t)\right\}. \tag{10.14}$$

Write down the remainder of this expansion as

$$A_n^m(x,t)=(z-1)^{-t}(t-1)\dots(t-n-1)\int_0^1 \frac{\tau^{m+n}}{(z-\tau)^{n+2-t}}\,d\tau. \tag{10.15}$$

Inserting (10.14) in (10.13), obtain (10.10).

Check now that, for $x\in[-\pi,\pi]$ and $t\in[-1,0]$, the estimate (10.12) holds. If $|x|\le 2\pi/3$, then for all $\tau\in[0,1]$ we have the inequality

$$|z(x)-\tau|^2=\left(\frac{1}{2}-\tau\right)^2+\frac{1}{4}\cot^2\frac{x}{2}\ge\frac{1}{16}\frac{1}{\sin^2 x/2}.$$

Consequently,

$$\left|A_n^m(x,t)\right|\le\frac{(n+2)!}{m+n}2^{2n+7}\sin^{n+2}\left(\frac{x}{2}\right).$$

The function $A_n^m(x,t)$ is thus bounded for $|x|\le 2\pi/3$ and $-1\le t\le 0$ by a quantity of the desired shape.

Let x belong to the interval $[2\pi/3,\pi]$ or $[-\pi,-2\pi/3]$. Then the integrand in (10.15) has singularities at $x=\pm\pi$, which to some extend complicates the derivation of the required estimate. Rewrite (10.15) in a more convenient form by letting

$$J_n^m(x,t)=\int_0^1\frac{\tau^{n+m}}{(z(x)-\tau)^{n-t}}\,d\tau.$$

This, together with (10.15), yields

$$A_n^m(x,t)=(z-1)^{-t}(t-1)\dots(t-n-1)J_{n+2}^{m-2}(x,t).$$

Thus, we readily obtain a dominant for the function $A_n^m(x,t)$, knowing an upper bound on the integral $J_{n+2}^{m-2}(x,t)$. Assume that $m=2$, and prove by induction on n that, for x and t real, the inequality holds

$$\left|J_n^0(x,t)\right|\le|z(x)|^{t+1}2^n|x|. \tag{10.16}$$

Indeed, the function $J_0^0(x,t)$ is easy to write down explicitly, namely,

$$J_0^0(x,t) = z^{t+1}(1 - e^{-ix(t+1)})/(t+1),$$

which readily entails (10.16). The induction step is immediate on using the recurrent relation

$$J_{n+1}^0(x,t) = zJ_n^0(x,t-1) - J_n^0(x,t).$$

Assume now that $m > 0$. Inducting on n, we then prove existence of a numerical sequence c_n such that, for all $t \geq -1$ and every integer m the inequality holds

$$\left|J_n^m(x,t)\right| \leq \frac{|z(x)|^{t+1} c_n}{m+1}. \tag{10.17}$$

If $n = 0$ then

$$J_0^m(x,t) = z^m J_0^0(x,t) - \sum_{\alpha=0}^{m-1} z^{m-1-\alpha} \int_0^1 \tau^\alpha (z-\tau)^{1+t}\, d\tau. \tag{10.18}$$

By hypothesis, $1 + t \geq 0$. Consequently,

$$\left|\int_0^1 (z-\tau)^{1+t} \tau^\alpha\, d\tau\right| \leq \frac{|z|^{1+t}}{\alpha+1}. \tag{10.19}$$

Inserting (10.16) and (10.19) in (10.18), obtain

$$\left|J_0^m(x,t)\right| \leq q^{m+t+1}|x| + \sum_{\alpha=0}^{m-1} \frac{q^{m-\alpha+t}}{\alpha+1}. \tag{10.20}$$

Here $q = |z(x)|$ is at most $1/\sqrt{3}$, i.e., less than 1. Consequently, there is a constant c_0 independent of m, x, and n and such that the ratio $q^{t+1}c_0/(m+1)$ dominates the right side of (10.20). Therefore, (10.17) is proven for $n = 0$ and $t \geq -1$.

The induction step is easy on using the following recurrent relations

$$J_{n+1}^m(x,t) = z^m J_{n+1}^0(x,t) - \sum_{\alpha=0}^{m-1} z^{m-1-\alpha} J_n^{\alpha+1}(x,t)$$

together with (10.16). Define the constant c_{n+1} as

$$c_{n+1} = \sup_m \left\{ (m+1)q^m 2^{n+1}\pi + c_n(m+1) \sum_{\alpha=1}^{m} \frac{q^{m-\alpha}}{\alpha} \right\}.$$

By hypothesis $q < 1$. Consequently, c_{n+1} is finite. Thus, (10.17) is proven completely. Using it, we readily validate (10.12). The proof of Lemma 7.13 is complete.

We now address in more detail the question of convergence of the sequence of the partial sums on the right side of (10.10). The limit of the sequence, if exists, is called the *Newton series*. As is known [74], the Newton series converges uniformly on every bounded set in the half-plane $\mathrm{Re}\, t \geq \lambda + \varepsilon$. Here ε is positive, and λ is a numerical parameter called the *abscissa of convergence* of the series under consideration.

In the case when $|x| \leq \pi/3$, the modulus of the complex number $e^{ix} - 1$ is at most 1. Consequently, we have the inequalities

$$\left|\sum_{k=n}^{\infty}(-1)^k a_k^m(x)\right| \leq \sum_{k=n}^{\infty}\frac{1}{k^m} \leq \int_{n-1}^{\infty} x^{-m}\, dx = \frac{(n-1)^{1-m}}{m-1}. \tag{10.21}$$

Moreover, we may determine the abscissa of convergence λ for the Newton series from the following relations (see [74])

$$\lambda = \varlimsup_{n\to\infty} \frac{\log\left|\sum\limits_{k=n}^{\infty}(-1)^k a_k^m(x)\right|}{\log n}.$$

Inserting (10.21) in the preceding formula, infer that $\lambda \leq -m+1 < -1$. Consequently, for all $t > -m+1$ and x with modulus at most $\pi/3$, we may pass to the limit in (10.10) as $n \to \infty$.

Inserting the Newton series that corresponds to the expansion (10.10) in (10.3) and performing easy calculations, expand $\lambda_0^m(\xi)$ as follows

$$\lambda_0^m(\xi) = (1 - e^{-i2\pi\xi})^{m-1} \sum_{\alpha=0}^{\infty} b_m[\alpha](e^{i2\pi\xi} - 1)^{\alpha+1}, \tag{10.22}$$

where $|\xi| \leq 1/6$, and the coefficients of the series may be written as

$$b_m[\alpha] = \frac{1}{(m+\alpha)!}\int_{m-2}^{m-1} t^{[m+\alpha]}\, dt. \tag{10.23}$$

Observe that the coefficient $b_m[0]$ is the negative of the coefficient $a[m-1]$ defined by (8.2).

In the case when $\pi/3 < |x| \leq \pi$ and $-1 \leq t \leq 0$, the series corresponding to (10.10) diverges.

Return to the sequence of the Fourier coefficients corresponding to the errors of a Gregory formula. We have

Theorem 7.18. *For β other than every multiple of N, the function of a discrete argument, defined by (8.8), admits the following factorization*

$$L_N^{m,G}[\beta] = h\,2^m(-1)^{m/2+1}(\sin\pi\beta h)^m b_m[0]P_{m-4}(\cos\pi\beta h). \tag{10.24}$$

The coefficient $b_m[0]$ is an integral like (10.23), and the polynomial $P_{m-4}(w)$ of degree $m-4$ is independent of N and differs on the unit interval from the Chebyshev polynomial of the second kind $U_{m-4}(w)$ only by a summand that is $O(1/m)$ as $m \to \infty$.

PROOF. The Fourier coefficients of the errors of a Gregory quadrature formula and the corresponding local error are related by (9.17). Rearrange the right side of (9.17). Inserting the expansion (10.10) in (10.3), obtain

$$\operatorname{Im}\left[e^{i\pi\xi}\lambda_0^m(\xi)\right] = 2^{m-1}(-1)^{(m-2)/2}(\sin\pi\xi)^{m-1}$$
$$\times\left\{\sum_{\alpha=0}^{n} b_m[\alpha]\operatorname{Re}\left[e^{-i\pi\xi(m-2)}(e^{i2\pi\xi}-1)^{\alpha+1}\right] + \Lambda_n^m(\xi)/(m+n)!\right\}. \tag{10.25}$$

The function characterizing the remainder of the series in the last equality is written down as

$$\Lambda_n^m(\xi) = \int_{-1}^{0} t(t+1)\ldots(t+m-1)\operatorname{Re}\left[e^{-i(m-2)\pi\xi}A_n^m(2\pi\xi,t)\right]dt. \tag{10.26}$$

It is not hard to dominate the function on using (10.12).

Assume that β is not a multiple of N and $w = \cos\pi\beta h$. Then from (9.9) and (9.12) it follows that

$$L_N^{m,G}[\beta] = hQ_{2(m-2)}^s(w). \tag{10.27}$$

The polynomial $Q_{2(m-2)}(w)$ is defined by (9.13) and does not depend on h. Show that it divides by the polynomial $(1-w^2)^{m/2}$.

Study the behavior of the rational function

$$R_{m-4}(w) = -Q_{2(m-2)}^s(w)/(2^m b_m[0](w^2-1)^{m/2}) \tag{10.28}$$

in a neighborhood about $w = \pm 1$.

Using (9.19) and (10.2), infer

$$R_{m-4}(\cos\pi\beta h) = \frac{\operatorname{Im}\left[e^{i\pi\beta h}\lambda_0^m(\beta h)\right]}{2^m b_m[0](-1)^{m/2}(\sin\pi\beta h)^{m+1}}.$$

Inserting (10.25) with $\xi = \beta h$ into the right side of the preceding equality, come to the relation

$$R_{m-4}(\cos\pi\beta h) = \left\{\sum_{\alpha=0}^{n} \frac{b_m[\alpha]}{b_m[0]} \operatorname{Re}\left[\frac{e^{-i(m-2)\pi\beta h}(e^{i2\pi\beta h}-1)^{\alpha+1}}{2\sin^2\pi\beta h}\right] + \frac{1}{(m+n)!}\frac{\Lambda_n^m(\beta h)}{(2b_m[0]\sin^2\pi\beta h)}\right\}. \quad (10.29)$$

From (10.26) and (10.12) it follows that the function $\Lambda_n^m(\beta h)$ admits the estimate

$$\sup_{|\beta|\le N/2} |\Lambda_n^m(\beta h)| \le \frac{a_n|b_m[0]|m!}{m+n}\sin^{n+2}\pi\beta h. \quad (10.30)$$

For $\alpha = 2k$ even, the real part of the quantity in brackets in (10.29) is the product

$$2^{2k}(\cos^2\pi\beta h - 1)^k U_{m-4-2k}(\cos\pi\beta h).$$

For $\alpha = 2k+1$ odd, this real part coincides with the value of the polynomial

$$-2^{2k+1}(\cos^2\pi\beta h - 1)^k T_{m-4-2k}(\cos\pi\beta h).$$

Here $T_n(w)$ ($U_n(w)$) remains to be a Chebyshev polynomial of the first (second) kind.

Putting $\beta = 1$ in (10.29), let h vanish. The limit of the expression on the right side of exist and is finite, as follows from (10.30). Consequently, the rational function $R_{m-4}(w)$ is impossible to have a pole at $w = 1$. The function $R_{m-4}(w)$, as follows from (10.28), is even. Therefore, the point $w = -1$ is impossible to be a pole of $R_{m-4}(w)$ either. This may happen only of $R_{m-4}(w)$ is a polynomial of degree $m-4$.

Putting $n = 0$ in (10.29), we further obtain

$$R_{m-4}(\cos\pi\beta h) - U_{m-4}(\cos\pi\beta h) = \frac{1}{b_m[0]m!}\Lambda_0^m(\beta h)/(2\sin^2\pi\beta h).$$

This, together with (10.30), yields

$$\sup_{|w|\le 1} |R_{m-4}(\cos\pi\beta h) - U_{m-4}(\cos\pi\beta h)| \le \frac{a_0}{2m}, \quad (10.31)$$

with a_0 the constant of (10.12). Consequently, the polynomial $P_{m-4}(w) = R_{m-4}(w)$ possesses all desired properties. The proof of Theorem 7.18 is complete.

***Corollary* 10.1.** *If N is even and $m \geq a_0$, with a_0 the constant of* (10.12); *the sum of the moduli of the weights of a Gregory quadrature formula of degree $m - 1$ cannot be less than $h^{2m-1}|h_m[0]|$*

PROOF. Observe that the sum of the moduli values of the weights of a quadrature formula is always at least the modulus of each of the Fourier coefficients of the respective error. Bearing this in mind, apply (10.24) with $\beta = N/2$ and use (10.31) to arrive at what is desired. The proof of Corollary 10.1 is complete.

§11. The Norm of the Error of a Gregory Quadrature Formula

We derive a formula for the $L_2^{(m)}[0,1]^*$ norm of the error of a Gregory quadrature formula. Our preliminary reasoning applies to a general error (9.1).

Use the available one-to-one correspondence between the set of the errors $EF_N^{(m)}$ and the set of their Fourier coefficients $CF_N^{(m)}$. Associate with the $L_2^{(m)}[0,1]^*$ norm of a functional in $EF_N^{(m)}$ the respective norm defined on the functions of a discrete argument in $CF_N^{(m)}$. So, instead of comparing two errors in $EF_N^{(m)}$, we may compare their images in $CF_N^{(m)}$.

Given functions $f[\beta]$ and $g[\beta]$ of a discrete argument, define the bilinear form

$$\langle f, g \rangle_m = \sum_{\beta \neq 0} \frac{f[\beta] g[\beta]}{(2\pi\beta)^{2m}} + \left(\sum_{\beta \neq 0} \frac{f[\beta]}{(2\pi\beta)^m} \right) \left(\sum_{\beta \neq 0} \frac{g[\beta]}{(2\pi\beta)^m} \right). \tag{11.1}$$

Recall that m is an even natural. Denote by $\|f\|_m$ the value $\langle f, f \rangle_m^{1/2}$ and consider only functions $f[\beta]$ such that $\|f\|_m$ is finite. It is not hard to see that $\|\cdot\|_m$ is a nonnegative, homogeneous, and convex functional, i.e., it defines a norm in $\widetilde{X}_N^{(0)}$.

***Lemma* 7.14** (V. I. Polovinkin [151]). *The norm of an arbitrary error l_N^m in $EF_N^{(m)}$ in the dual of $L_2^{(m)}[0,1]$ coincides with the norm of the corresponding sequence of the Fourier coefficients, namely,*

$$\left\| l_N^m \mid L_2^{(m)}[0,1]^* \right\| = \left\| L_N^m \right\|_m. \tag{11.2}$$

PROOF. By the Riesz Theorem on the general form of a bounded linear functional in a Hilbert space, there is an element $u_N^m(x)$ in $L_2^{(m)}[0,1]$ such that, for all φ in the same space, the equality holds

$$(l_N^m, \varphi) = \int_0^1 D^m \varphi \, D^m u_N^m \, dx. \tag{11.3}$$

The norm of l_N^m in the dual of $L_2^{(m)}[0,1]$ coincides with the norm of the extremal function $\overline{u}_N^m$, namely,

$$\left\| l_N^m \mid L_2^{(m)}[0,1]^* \right\|^2 = \left(l_N^m, u_N^m \right) = \int_0^1 \left| D^m u_N^m \right|^2 dx. \tag{11.4}$$

Taking as $\varphi(x)$ in (11.3) consecutively the functions x^m, $\cos 2\pi\beta x$, and $\sin 2\pi\beta x$, obtain the Fourier series for the mth order derivative of the extremal function

$$D^m u_N^m(x) = \frac{1}{m!}\left(l_N^m, x^m \right) + 2\sum_{\beta=1}^{\infty} \frac{L_N^m[\beta]}{(2\pi\beta)^m} \cos 2\pi\beta x.$$

Whence, by the Parseval identity, infer

$$\int_0^1 \left| D^m u_N^m \right|^2 dx = \left| \frac{\left(l_N^m, x^m \right)}{m!} \right|^2 + \sum_{\beta \neq 0} \frac{\left| L_N^m[\beta] \right|^2}{(2\pi\beta)^{2m}}. \tag{11.5}$$

The power function x^m splits into two summands: one, the Bernoulli polynomial $B_m(x)$, and the other, a polynomial of degree at most $m-1$, namely,

$$x^m = B_m(x) + P_{m-1}(x).$$

Considering that the error l_N^m is exact for polynomials of degree $m-1$ and using (7.15), deduce

$$\frac{\left(l_N^m, x^m \right)}{m!} = (-1)^{(m-2)/2} \sum_{\beta \neq 0} \frac{L_N^m[\beta]}{(2\pi\beta)^m}. \tag{11.6}$$

Whence and from (11.1) we infer that the right side of (11.5) is the norm square of $L_N^m[\beta]$. Recalling (11.4), we finish the proof of Lemma 7.14.

The further reasoning aims at translating the norm of $L_N^m[\beta]$ to the shape in which the variables m and h are separated. The manner of translating the norm we use is applicable to the case of an arbitrary error like (9.1). However, we use it only for the error of the corresponding Gregory quadrature formula. This enables us to shorten the final formula slightly.

Lemma 7.15. *The error of a Gregory quadrature formula of degree $m-1$ at the polynomial $x^m/m!$ may be written as the sum*

$$\frac{\left(l_{N,m}^G, x^m \right)}{m!} = -h^m \frac{B_m}{m!} + (-1)^{m/2} h^{m+1} b_m[0] P_{m-4}(1)$$
$$+(-1)^{m/2+1} h^m b_m[0] \frac{2}{\pi} \int_0^\infty \left(\frac{\sin t}{t} \right)^m P_{m-4}(\cos t)\, dt, \tag{11.7}$$

with the coefficient $b_m[0]$ defined by (10.23) and $P_{m-4}(w)$ the same polynomial as in (10.24).

PROOF. Using (11.6), (9.9) and (10.21), obtain

$$\frac{(l^G_{N,m}, x^m)}{m!}$$

$$= -h^m \frac{B_m}{m!} - h^m \left\{ 2(-1)^{m/2} b_m[0] h \sum_{\beta=1}^{\infty} \left(\frac{\sin \pi\beta h}{\pi\beta h} \right)^m P_{m-4}(\cos \pi\beta h) \right\}. \quad (11.8)$$

Express the right side of (11.8) through some integral on involving the familiar *Poisson summation formula* in accord with the scheme proposed in [43]. The formula in question reads

$$\sqrt{\alpha} \left\{ \frac{1}{2} f(0) + \sum_{n=1}^{\infty} f(n\alpha) \right\} = \sqrt{\gamma} \left\{ \frac{1}{2} g(0) + \sum_{n=1}^{\infty} g(n\gamma) \right\}. \quad (11.9)$$

Here $\alpha \geq 0$, $\alpha\gamma = 2\pi$, and the functions f and g are related as follows

$$g(\theta) = \sqrt{\frac{2}{\pi}} \int_0^{\infty} f(t) \cos \theta t \, dt. \quad (11.10)$$

If the function $f(t)$ we sum in (11.9) coincides with $(\sin t/t)^m$, then the Fourier cosine transform $g(\theta)$ of $f(t)$ vanishes for $\theta \geq m$. The relevant formula is, for instance, in [56]. Write down the integral

$$\int_0^{\infty} f(t) P_{m-4}(\cos t) \cos \theta t \, dt \quad (11.11)$$

as a linear combination of the elementary summands

$$\int_0^{\infty} f(t) \cos \beta t \cos \theta t \, dt, \quad (11.12)$$

with $0 \leq \beta \leq m-4$. Expanding in a sum the product of cosines under the integral sign in (11.12), we see that for $\theta - \beta \geq m$ the quantity (11.12) equals 0. Consequently, for all $\theta \geq 2(m-2)$ the integral (11.11) is also equal to 0.

Applying now the Poisson formula with $\alpha = \pi h$ and $\gamma = 2N$ to the function $(\sin t/t)^m P_{m-4}(\cos t)$ infer

$$\pi h\left\{\frac{1}{2}P_{m-4}(1) + \sum_{\beta=1}^{\infty}\left(\frac{\sin\pi\beta h}{\pi\beta h}\right)^m P_{m-4}(\cos\pi\beta h)\right\} = \int_0^{\infty}\left(\frac{\sin t}{t}\right)^m P_{m-4}(\cos t)\,dt. \tag{11.13}$$

Expressing the sum of the series over β through the remaining terms and inserting the result in (11.8), obtain (11.7). The proof of Lemma 7.15 is complete.

We analogously transform the second summand in the definition of the norm of $L_N^{m,G}[\beta]$. By (9.9) and (10.24), for $m \leq N$ we have

$$\sum_{\beta\neq 0}\frac{\left|L_N^{m,G}[\beta]\right|^2}{(2\pi\beta)^{2m}} = h^{2m}\left\{\frac{|B_{2m}|}{(2m)!} + 2|b_m[0]|^2\right.$$
$$\left.\times h^2\sum_{\beta=1}^{\infty}\left(\frac{\sin\pi\beta h}{\pi\beta h}\right)^{2m}|P_{m-4}(\cos\pi\beta h)|^2\right\}. \tag{11.14}$$

In the Poisson formula (11.9) we further set $\alpha = \pi h$ and $\gamma = 2N$ and take

$$f(t) = \left(\frac{\sin t}{t}\right)^{2m}|P_{m-4}(\cos t)|^2,$$

as a function to be summed. We thus obtain

$$\int_0^{\infty}\left(\frac{\sin t}{t}\right)^{2m}|P_{m-4}(\cos t)|^2\left(\frac{1}{2} + \cos 2Nt\right)dt$$
$$= \frac{\pi h}{2}\left\{\frac{1}{2}|P_{m-4}(1)|^2 + \sum_{\beta=1}^{\infty}\left(\frac{\sin\pi\beta h}{\pi\beta h}\right)^{2m}|P_{m-4}(\cos\pi\beta h)|^2\right\}. \tag{11.15}$$

The expression on the left side splits naturally into the sum of two integrals. Easy analysis shows that the integral whose integrand has the factor $\cos 2Nt$, equals 0 for $m \leq (N+4)/2$. Assume the last condition met. Then, expressing from (11.15) the sum of the series over β through the remaining terms and inserting the result in (11.14), come to the some presentation of the $L_2^{(m)}[0,1]^*$ norm of $l_{N,m}^G$.

The formula for the norm of the error $l_{N,m+2}^G$ in $L_2^{(m)}[0,1]^*$ is derived similarly. This formula is precisely the equality (8.4) which we have already used in estimating the error of Gregory formulas for the functions in the class $G_{\varkappa}[0,1]$.

Chapter 8

Functions of a Discrete Variable

In this chapter, the domain of definition of the functions under study is the multidimensional integer lattice of nodes

$$\mathbb{Z}^n = \{\beta : \beta = (\beta_1, \dots, \beta_n),\ \beta_j = 0, \pm 1, \pm 2, \dots\}.$$

Consider the set of complex-valued functions $f[\beta]$, on agreeing to put a discrete argument in brackets. For brevity we sometimes call a function of a discrete argument a *discrete function*. Clearly, the set of discrete functions gives rise to an infinite-dimensional linear space.

It is impossible to apply the operations of mathematical analysis to studying discrete functions immediately. However, analogous operations are available. The analogy grounds on the fact that $\varphi[\beta]$ results often from restricting some function $\varphi(x)$ defined on the whole of $\mathbb{R}^n$ to the set $\{hH\beta : \beta \in \mathbb{Z}^n\}$. The latter thickens as the scalar parameter h vanishes. Obviously, under this interpretation we may routinely apply the methods of continuous mathematical analysis to examination of the properties of $\varphi(hH\beta)$. However, it is more convenient to ignore this possibility and to define needed operations over discrete functions axiomatically.

§1. Operations over Discrete Functions

Let the *support of a function* $\varphi[\beta]$ be the set of points at which it is other than 0. When this set is bounded, $\varphi[\beta]$ is called *compactly-supported*.

The *inner product* $[\varphi[\beta], \psi[\beta]]$ of $\varphi[\beta]$ and $\psi[\beta]$ is the sum of the series

$$\sum_{\beta} \varphi[\beta]\overline{\psi}[\beta]$$

provided that the latter converges absolutely. The inner product exists of every two functions either of which is compactly-supported.

The *convolution* $\varphi[\beta] * \psi[\beta]$ of $\varphi[\beta]$ and $\psi[\beta]$ is the inner product

$$\chi[\beta] = \varphi[\beta] * \psi[\beta] = [\varphi[\gamma], \overline{\psi}[\beta - \gamma]], \tag{1.1}$$

with γ the summation index. This definition implies that the series defining $\chi[\beta]$ converges absolutely for all β.

Commutativity of convolution, i.e., the equality

$$\varphi[\beta] * \psi[\beta] = \psi[\beta] * \varphi[\beta],$$

is easy to check by letting $\beta - \gamma = \delta$ and $\gamma = \beta - \delta$ in (1.1). Moreover, we clearly see that

$$\varphi[\beta + \varepsilon] * \psi[\beta + \delta] = [\varphi[\gamma + \varepsilon], \overline{\psi}[\beta + \delta - \gamma]] = \chi[\beta + \delta + \varepsilon].$$

In other words, if the argument of one of the convoluted factors translates then the same translation applies to the argument of the convolution. If the arguments of both factors are independently translated then the argument of the convolution undergoes the composition of the translations of the factors. The convolution of two functions by means of (1.1) was defined through an inner product. Conversely, the inner product is expressed through convolution as follows

$$[\varphi[\beta], \psi[\beta]] = \varphi[\beta] * \overline{\psi}[-\beta]\Big|_{\beta=0} = \varphi[-\beta] * \overline{\psi}[\beta]\Big|_{\beta=0} = \varphi[\beta+\varepsilon] * \overline{\psi}[\delta-\beta]\Big|_{\beta=\delta-\varepsilon}. \tag{1.2}$$

Introduce the notation

$$\overset{\frown}{\psi}[\beta] = \psi[-\beta]$$

and rewrite the inner product of $\varphi[\beta]$ and $\psi[\beta]$ as

$$[\varphi[\beta], \psi[\beta]] = \varphi[\beta] * \overset{\frown}{\overline{\psi}}[\beta]\Big|_{\beta=0} = \overset{\frown}{\varphi}[\beta] * \overline{\psi}[\beta]\Big|_{\beta=0} = \varphi[\beta + \varepsilon] * \overset{\frown}{\overline{\psi}}[\beta - \delta]\Big|_{\beta=\delta-\varepsilon}. \tag{1.3}$$

The formulas (1.2) and (1.3) make sense if so does at least one of the expressions on their right sides. In this event all remaining expressions also make sense. It is worthy to list a few cases in which the convolution of two functions exists.

Introducing various norms, we distinguish classes of functions of a discrete argument which constitute Banach spaces. Define l_p, $1 < p < \infty$, as the linear space of discrete functions with finite norm

$$\|f \mid l_p\| = \left\{\sum_{\beta} | f[\beta]|^p\right\}^{1/p}.$$

Theorem 8.1. *Let $\varphi[\beta]$ and $\psi[\beta]$ belong to l_p and $l_{p'}$ with $1/p + 1/p' = 1$, in which case the spaces are dual to one another. Then the convolution*

$$\varphi[\beta] * \psi[\beta] = \xi[\beta]$$

of $\varphi[\beta]$ and $\psi[\beta]$ exists and vanishes as $|\beta| \to \infty$.

PROOF. The existence of the convolution is obvious. We are only to prove that it vanishes. Given a positive a, split $\varphi[\beta]$ into the sum

$$\varphi[\beta] = \varphi_a[\beta] + \widetilde{\varphi}_a[\beta],$$

with

$$\varphi_a[\beta] = \begin{cases} \varphi[\beta], & \text{if } |\beta| \le a, \\ 0, & \text{if } |\beta| > a. \end{cases}$$

By analogy, decompose the function $\psi[\beta]$. Then

$$\|\varphi \mid l_p\|^p = \|\varphi_a \mid l_p\|^p + \|\widetilde{\varphi}_a \mid l_p\|^p.$$

For all $\varepsilon > 0$ and a sufficiently large, $a > N(\varepsilon)$, we have

$$\|\widetilde{\varphi}_a \mid l_p\| \le \varepsilon, \quad \|\widetilde{\psi}_a \mid l_{p'}\| \le \varepsilon.$$

At sufficiently large $|\gamma|$, the supports of $\varphi_a[\beta]$ and $\overline{\psi}_a[\gamma - \beta]$ are disjoint. Hence,

$$[\varphi[\beta], \overline{\psi}[\gamma - \beta]] = [\varphi_a[\beta], \overline{\widetilde{\psi}}_a[\gamma - \beta]] + [\widetilde{\varphi}_a[\beta], \overline{\psi}_a[\gamma - \beta]] + [\widetilde{\varphi}_a[\beta], \overline{\widetilde{\psi}}_a[\gamma - \beta]].$$

Every summand on the right side of this equality is easy to estimate:

$$\begin{aligned} |[\varphi_a[\beta], \overline{\widetilde{\psi}}_a[\gamma - \beta]]| &\le \|\varphi \mid l_p\| \, \|\widetilde{\psi}_a \mid l_{p'}\|, \\ |[\widetilde{\varphi}_a[\beta], \overline{\psi}_a[\gamma - \beta]]| &\le \|\widetilde{\varphi}_a \mid l_p\| \, \|\psi \mid l_{p'}\|, \\ |[\widetilde{\varphi}_a[\beta], \overline{\widetilde{\psi}}_a[\gamma - \beta]]| &\le \|\widetilde{\varphi}_a \mid l_p\| \, \|\widetilde{\psi}_a \mid l_{p'}\|. \end{aligned}$$

Consequently, for all γ, the convolution $\xi[\gamma]$ does not exceed in magnitude a quantity that is proportional to ε. Whence Theorem 8.1 follows.

The next theorem is obvious.

Theorem 8.2. *The convolution of functions φ and ψ exists provided that, for all γ, the intersection of the supports of $\varphi[\beta]$ and $\psi[\gamma - \beta]$ is finite.*

The hypothesis of the claim holds, for instance, if either of the functions φ and ψ has finite support. However, it is easy to exhibit the examples in which the intersection of the supports is finite whereas the support of each of the functions is unbounded. Assume that the supports of φ and ψ are some sets composed of β whose every component is bounded from below, namely,

$$\beta_j \geq \beta_j^{(0)}, \quad j = 1, 2, \ldots, n. \tag{1.4}$$

We write such system of inequalities as $\beta \geq \beta^{(0)}$. Then the intersection of the supports of $\varphi[\beta]$ and $\psi[\gamma - \beta]$ is the parallelepiped

$$\beta^{(0)} \leq \beta \leq \gamma - \beta^{(0)}.$$

Hence, this intersection is finite for every γ.

Consider the functions $\varphi[\beta]$, $\psi[\beta]$, and $\chi[\beta]$ of a discrete argument. If all pairwise convolutions are commutative then, for instance, we have

$$\varphi * (\psi * \chi) = \varphi * (\chi * \psi) = (\psi * \chi) * \varphi = (\chi * \psi) * \varphi.$$

So, the twelve possible triple convolutions reduce in fact only to three of them, namely,

$$\varphi * (\psi * \chi), \quad \psi * (\chi * \varphi), \quad \chi * (\varphi * \psi).$$

The next theorem yields some conditions under which all three possible versions of triple convolution coincide and, consequently, the operation of convolution is associative.

***Theorem* 8.3.** *Assume existent the three pairwise convolutions*

$$|\varphi[\beta]| * |\psi[\beta]|, \quad |\psi[\beta]| * |\chi[\beta]|, \quad |\chi[\beta]| * |\varphi[\beta]|$$

and at least one triple convolution, for definiteness,

$$|\varphi[\beta]| * (|\psi[\beta]| * |\chi[\beta]|).$$

Then all three triple convolutions

$$\varphi * (\psi * \chi), \quad \psi * (\chi * \varphi), \quad \chi * (\varphi * \psi)$$

exist and coincide.

PROOF. The theorem is immediate from the absolute convergence of the double series

$$\sum_{\gamma}\sum_{\delta}\varphi[\gamma]\,\psi[\beta-\gamma-\delta]\,\chi[\delta]. \tag{1.5}$$

Consequently, it is unconditionally convergent and we may sum it in whatever order. Collecting terms with a fixed $\chi[\delta]$, obtain an absolutely convergent series, which yields $\chi*(\varphi*\psi)$. Collecting terms in another order, obtain the convolution $\varphi*(\psi*\chi)$ and so on. Thus,

$$\varphi*(\psi*\chi)=\psi*(\chi*\varphi)=\chi*(\varphi*\psi). \tag{1.6}$$

The proof of Theorem 8.3 is complete.

Observe that if, given each of the functions φ, ψ and χ, we may find at least one point at which it is other than 0, then absolute convergence of the series determining one of the triple convolutions entails existence for all pairwise convolutions mentioned in the theorem.

Corollary 1.1. *Assume existent the convolution*

$$|\varphi_1|*(\,|\varphi_2|*[\,|\varphi_3|*\cdots*\{\,|\varphi_{n-1}|*|\varphi_n|\,\}\ldots]) \tag{1.7}$$

and all convolutions with $n-1$ factors out of those entering (1.7)*. Then there exist all convolutions in arbitrary order*

$$\varphi_{s_1}*\varphi_{s_2}*\cdots*\varphi_{s_n}$$

independent of the order and combination of factors.

As before, if given each of the functions $\varphi_i[\beta]$ we may find at least one point at which it is other than 0, then the convergence of (1.7) entails existence of all convolutions with $n-1$ factors out of those mentioned in Corollary 1.1.

The convolutions of the shape (1.7) are called *repeated.*

Corollary 1.2. *Convolutions iterated in arbitrary order exist and coincide provided that all but possibly one of the factors are compactly-supported. In this event the sum like* (1.5) *contains only finitely many nonzero terms.*

Corollary 1.3. *Convolutions iterated in arbitrary order exist and coincide provided that the support of each function φ_k lies in the set $\beta\geq\beta^{(0,k)}$, with $\beta^{(0,k)}$ a vector varying with k. In this event the sum like* (1.5) *contains only finitely many terms.*

Corollary 1.4. *If the variables β_j split into two collections so that one of them satisfies the hypotheses of Corollary 1.2 whereas the other satisfies the hypotheses of Corollary 1.3, then the hypotheses of Theorem 8.3 are also satisfied.*

The class of compactly-supported functions, being closed under convolution, is a convolution algebra. Also, the class of functions with support bounded below in the sense of (1.4) is a convolution algebra.

The functions of exponential decay at infinity again comprise a convolution algebra. The last claim ensues from the following property.

***Theorem* 8.4.** *The convolution of functions in a discrete variable of exponential decay is also a function of exponential decay.*

PROOF. Suffice it to check that, for $\eta_1 > 0$ and $\eta_2 > 0$, the estimates

$$|\varphi[\beta]| \leq K_1 e^{-\eta_1|\beta|}, \quad |\psi[\beta]| \leq K_2 e^{-\eta_2|\beta|} \tag{1.8}$$

imply

$$|\varphi[\beta] * \psi[\beta]| \leq K_3 e^{-\eta_3|\beta|}.$$

We establish this inequality only in the case of a single independent variable, since for many variables the proof is analogous.

Equating the exponents of the exponentials on the right side of (1.8), i.e., letting $\eta = \min(\eta_1, \eta_2)$, for $\beta > 0$ find

$$\left|\sum_{\gamma} \varphi[\gamma]\overline{\psi}[\beta - \gamma]\right| \leq K_1 K_2 \sum_{\gamma} e^{-\eta|\gamma| - \eta|\beta-\gamma|}$$
$$= K_1 K_2 \left\{ \sum_{0 \leq \gamma \leq \beta} e^{-\eta|\beta|} + \sum_{\gamma > \beta} e^{-2\eta|\gamma| + \eta|\beta|} + \sum_{\gamma < 0} e^{2\eta\gamma - \eta|\beta|} \right\}.$$

Estimate each term on the right side separately. Given $\eta > \eta_3$, observe

$$\sum_{0 \leq \gamma \leq \beta} e^{-\eta|\beta|} = e^{-\eta\beta}(\beta + 1) \leq K e^{-\eta_3\beta},$$
$$\sum_{\beta < \gamma} e^{-2\eta\gamma + \eta\beta} = e^{-\eta(\beta+2)} \frac{1}{1 - e^{-2\eta}} \leq K e^{-\eta\beta},$$
$$\sum_{\gamma < 0} e^{2\eta\gamma - \eta\beta} = e^{-\eta\beta} \frac{1}{1 - e^{-2\eta}} \leq K e^{-\eta\beta}.$$

Analogous estimation for β negative is carried out similarly. The proof of Theorem 8.4 is complete.

Consider some important discrete functions. As for a continuous variable, introduce the *monomials* $[\beta]^\alpha$ and the *polynomials* $P[\beta] = \sum_\alpha a_\alpha[\beta]^\alpha$. Here, as before, $[\beta]^\alpha = \beta_1^{\alpha_1}\beta_2^{\alpha_2}\dots\beta_n^{\alpha_n}$. Then the convolution of the polynomial $P[\beta]$ of degree m with an arbitrary compactly-supported function is again a polynomial of degree at most m.

Indeed,

$$P[\beta-\gamma] = \sum_k a_k[\gamma]P_k[\beta],$$

with $P_k[\beta]$ some polynomial of degree at most m. Consequently,

$$\varphi[\beta] * P[\beta] = \sum_k P_k[\beta][\varphi[\gamma], \bar{a}_k[\gamma]],$$

which proves our claim.

Define a discrete analog of the Dirac delta function. Put

$$\delta[\beta] = \begin{cases} 1, & \text{if } \beta = 0, \\ 0, & \text{if } \beta \neq 0. \end{cases}$$

The convolution of $\delta[\beta]$ with an arbitrary function $\varphi[\beta]$ does not change the latter, namely,

$$\delta[\beta] * \varphi[\beta] = [\,\delta[\gamma], \overline{\varphi}[\beta-\gamma]\,] = \varphi[\beta].$$

Thus, $\delta[\beta]$ plays the role of unity for convolution. The functions δ in fewer variables are related as follows

$$\delta[\beta] = \delta[\beta_1]\delta[\beta_2]\dots\delta[\beta_n].$$

The delta function of arbitrary coordinates is constructed on using the above formula. Namely,

$$\delta[\beta_{j_1}, \beta_{j_2}, \dots, \beta_{j_k}] = \delta[\beta_{j_1}]\delta[\beta_{j_2}]\dots\delta[\beta_{j_k}].$$

The following formulas hold

$$\delta[\beta_j - \gamma_j]\,\varphi[\beta] = \varphi[\beta_1, \dots, \beta_{j-1}, \gamma_j, \beta_{j+1}, \dots, \beta_n]\,\delta[\beta_j - \gamma_j],$$
$$\delta[\beta-\gamma]\,\varphi[\beta] = \delta[\beta-\gamma]\,\varphi[\gamma].$$

In other words, the argument β_j of φ is fixed in multiplication of $\varphi[\beta]$ by $\delta[\beta_j - \gamma_j]$. For $\beta_j \neq \gamma_j$, the product is equal to 0. In multiplication of $\varphi[\beta]$ by $\delta[\beta-\gamma]$, fixed is the whole n-dimensional argument of φ, whereas at the other points the product vanishes.

In the sequel, we need the notation

$$\widehat{\beta}_j = (\beta_1, \beta_2, \dots, \beta_{j-1}, \beta_{j+1}, \dots, \beta_n).$$

Introduce into consideration the function

$$\Delta_j[\beta] = \delta[\widehat{\beta}_j](\delta[\beta_j + 1] - \delta[\beta_j]). \tag{1.9}$$

The *partial difference* of an arbitrary function φ we call the convolution of $\Delta_j[\beta]$ with $\varphi[\beta]$ which takes the form

$$\Delta_j[\beta] * \varphi[\beta] = \varphi[\beta + \delta_j] - \varphi[\beta],$$

with δ_j standing for the vector $\delta_j = (\underbrace{0, \ldots, 0}_{j-1}, 1, 0, \ldots, 0)$.

For discrete functions, partial difference plays an analogous role to that of partial derivative for functions of a continuous argument. We define an analog of the differential operator of an arbitrary order on using the notation

$$\Delta_j^{[\alpha_j]} = \overbrace{\Delta_j * \Delta_j * \cdots * \Delta_j}^{\alpha_j}, \quad \Delta^{[\alpha]} = \Delta_1^{[\alpha_1]} * \Delta_2^{[\alpha_2]} * \cdots * \Delta_n^{[\alpha_n]},$$

with $\Delta_i^{[0]}[\beta] = \delta[\beta]$.

The taking of convolution with $\Delta^{[\alpha]}$ is the taking of the difference of order α, namely,

$$(\Delta^{[\alpha]} * \varphi)[\beta] = \sum_{0 \le k \le \alpha} (-1)^{|\alpha|-|k|} \frac{\alpha!}{k!(\alpha - k)!} \varphi[\beta + k]$$
$$= \sum_{k_1=0}^{\alpha_1} \cdots \sum_{k_n=0}^{\alpha_n} \left\{ \prod_{j=1}^{n} \frac{(-1)^{\alpha_j - k_j} \alpha_j!}{k_j!(\alpha_j - k_j)!} \right\} \varphi\left[\beta + \sum_{j=1}^{n} k_j \delta_j\right],$$

with $0 \le k \le \alpha$ implying the simultaneous inequalities

$$0 \le k_j \le \alpha_j, \quad j = 1, 2, \ldots, n.$$

The differences $\Delta^{[\alpha]} * \varphi$ are analogs of partial derivatives of higher order.

In dealing with functions of a discrete variable it is convenient to use Newtonian powers. For one variable y and an integer $k \ge 0$, put

$$y^{[0]} = 1, \quad y^{[k]} = y(y-1)\ldots(y-k+1), \; k \ge 1.$$

We may expand in the Newtonian powers every ordinary polynomial

$$P(y) = \sum_{k=0}^{m} a_k y^k$$

of degree m in y as follows

$$P(y) = \sum_{k=0}^{m} b_k y^{[k]}.$$

This ensues from the fact that the transition matrix from the system $1, y, y^{[2]}, \dots, y^{[k]}, \dots$ to the system $1, y, y^2, \dots, y^k, \dots$ is triangular with 1s on the principal diagonal and thus nonsingular.

Let $y = (y_1, \dots, y_n)$ and assume that $\alpha = (\alpha_1, \alpha_2, \dots, \alpha_n)$ is an integer vector with nonnegative entries. Put

$$y^{[\alpha]} = y_1^{[\alpha_1]} \dots y_n^{[\alpha_n]}.$$

Every polynomial of degree m in n variables may be expanded in the Newtonian powers by what was explained above.

One of the main advantages of the Newtonian powers consists in the elementary formula

$$\Delta * y^{[k]} = k y^{[k-1]}$$

entailing the general formula

$$\Delta^{[\alpha]} * y^{[\gamma]} = \frac{\gamma!}{(\gamma - \alpha)!} y^{[\gamma - \alpha]} \tag{1.10}$$

for $\gamma \geq \alpha$ and

$$\Delta^{[\alpha]} * y^{[\gamma]} = 0 \tag{1.11}$$

provided that $\gamma \leq \alpha$ and $\gamma \neq \alpha$. Hence, we arrive again at the assertion that the convolution of a polynomial P of degree m with $\Delta^{[\alpha]}$ diminishes the degree of the former by $|\alpha|$ units.

The formulas (1.10) and (1.11) lead to the *biorthogonality condition*

$$\Delta^{[\alpha]} * \left. \frac{[\beta]^{[\gamma]}}{\gamma!} \right|_{\beta=0} = \delta_\alpha^\gamma = \begin{cases} 1, & \text{if } \alpha = \gamma, \\ 0, & \text{if } \alpha \neq \gamma. \end{cases}$$

From this condition we readily infer the general formula for expansion of a polynomial $P[\beta]$ of degree l in the Newtonian powers

$$P[\beta] = \sum_{|\alpha| \leq l} \Delta^{[\alpha]} * P[\beta] \Big|_{\beta=0} \frac{[\beta]^{[\alpha]}}{\alpha!}.$$

In particular, obtain

$$[\beta + \gamma]^{[\alpha]} = \sum_{\varepsilon + \eta = \alpha} \frac{\alpha!}{\varepsilon! \eta!} [\beta]^{[\varepsilon]} [\gamma]^{[\eta]},$$

an analog of the *Newton binomial.*

Revert to Theorem 8.3 on associativity of convolution and demonstrate by example that the hypotheses of this theorem are essential and cannot be rejected.

EXAMPLE. Consider the three functions in a single variable

$$\varphi[\beta] = \beta, \quad \psi[\beta] = \delta[\beta+1] - 2\delta[\beta] + \delta[\beta-1], \quad \chi[\beta] = |\beta|.$$

The function $\psi[\beta]$ is compactly-supported. Hence, both convolutions $\varphi * \psi$ and $\psi * \chi$ make sense. It is easy that

$$\varphi * \psi = 0, \quad \psi * \chi = 2\delta[\beta],$$

which implies the relations

$$\varphi * (\psi * \chi) = 2\beta, \quad (\varphi * \psi) * \chi = 0.$$

Hence, (1.6) fails. More exactly, two of the triple convolutions in (1.6) make sense but fail to coincide whereas the third is senseless.

Taking in this example the function $\varphi[\beta] = \beta^2$ instead of $\varphi[\beta] = \beta$, we see that existent is only one of the three triple convolutions.

For calculating the partial difference of the product of two functions there is a formula analogous to the differential product rule, namely,

$$\Delta_j[\beta] * (\varphi[\beta]\psi[\beta]) = \varphi[\beta + \delta_j](\Delta_j[\beta] * \psi[\beta]) + \psi[\beta](\Delta_j[\beta] * \varphi[\beta]). \tag{1.12}$$

This formula ensues elementarily from the identity

$$\begin{aligned}&\varphi(x+1)\psi(x+1) - \delta(x)\psi(x)\\ &= \varphi(x+1)[\psi(x+1) - \psi(x)] + \psi(x)[\varphi(x+1) - \varphi(x)].\end{aligned}$$

Transposing φ and ψ in (1.12), obtain

$$\Delta_j * (\varphi[\beta]\psi[\beta]) = \psi[\beta + \delta_j](\Delta_j[\beta] * \varphi[\beta]) + \varphi[\beta](\Delta_j[\beta] * \psi[\beta]).$$

Now, introduce into consideration a function of a single variable analogous to the Heaviside theta function. Namely, put

$$\Sigma[\beta_j] = \begin{cases} 0, & \text{if } \beta_j \le 0,\\ 1, & \text{if } \beta_j > 0.\end{cases}$$

For $j = 1, \ldots, n$ and every $\beta \in \mathbb{Z}^n$, assign

$$\Sigma_j[\beta] = \delta[\widehat{\beta}_j]\Sigma[\beta_j].$$

The functions $\Sigma[\beta_j]$ and $\Sigma_j[\beta]$ satisfy the obvious relations

$$\Sigma[-\beta_j] = 1 - \Sigma[\beta_j + 1], \quad \Sigma_j[-\beta] = \delta[\widehat{\beta}_j] - \Sigma_j[\beta + \delta_j].$$

The function $\Sigma_j[\beta]$ is inverse to $\Delta_j[\beta]$ with respect to convolution. The converse is also valid

$$\Sigma_j[\beta] * \Delta_j[\beta] = \Delta_j[\beta] * \Sigma_j[\beta] = \delta[\beta].$$

The proof is easy from (1.9) and the equality

$$\Delta[\beta_j] * \Sigma[\beta_j] = \delta[\beta_j]$$

which may be checked straightforward.

Clearly, the convolution of $\varphi[\beta]$ with the function $\Sigma_j[\beta]$ is the summation of the values of $\varphi[\beta']$ over all β' such that $\widehat{\beta}'_j = \widehat{\beta}_j$, and β'_j ranges over all indices less than β_j. In other words, the convolution with $\Sigma_j[\beta]$ is the ***partial summation*** over the coordinate β_j. Indeed,

$$\varphi[\beta] * \Sigma_j[\beta] = \sum_{\beta'_j=-\infty}^{\beta_j-1} \varphi[\widehat{\beta}_j, \beta'_j].$$

The pair $\widehat{\beta}_j, \beta'_j$ stands here for the vector β with the jth entry replaced by β'_j. An analogous agreement is effective in the sequel.

The function $\Sigma_j[\beta]$ is not compactly-supported. The convolution with it may fail to exist sometimes. The convolution with $\sum_j[\beta]$ exists provided that the support of φ has the coordinate β_j bounded from below for all fixed $\widehat{\beta}_j$, i.e., if there is a function $k[\widehat{\beta}_j]$ such that $\operatorname{supp}\varphi \subset \{\beta : \beta_j \geq k[\widehat{\beta}_j]\}$. The convolution of $\Sigma_j[\beta]$ with $\varphi[\beta]$ is a discrete analog of integration with respect to the variable β_j over the half-line form $-\infty$ to $\beta_j - 1$. Integration with finite limits is analogous to another operation, summation with finite limits, which is performed by the rule

$$\Sigma_j[\beta] * \left(\varphi[\beta]\Sigma_j[\beta - \gamma]\right) = \sum_{\beta'_j=\gamma_j+1}^{\beta_j-1} \varphi[\widehat{\beta}_j, \beta'_j]. \tag{1.13}$$

Observe that an inverse of $\Delta_j[\beta]$ is defined nonuniquely. Clearly, the convolution of $\Delta_j[\beta]$ with an arbitrary function ψ independent of β_j, i.e., with an arbitrary function only in $\widehat{\beta}_j$, is zero, namely, $\Delta_j[\beta] * \psi[\widehat{\beta}_j] = 0$. Consequently, the sum of the shape $\Sigma_j[\beta] + \psi[\widehat{\beta}_j]$ is also inverse to $\Delta_j[\beta]$ in the sense that

$$\Delta_j[\beta] * \left(\Sigma_j[\beta] + \psi[\widehat{\beta}_j]\right) = \left(\Sigma_j[\beta] + \psi[\widehat{\beta}_j]\right) * \Delta_j[\beta] = \delta[\beta].$$

Given a function $\varphi[\beta]$ with support bounded in β_j from above, i.e.,

$$\operatorname{supp}\varphi \subset \{\beta : \beta_j \le k[\widehat{\beta}_j]\}, \tag{1.14}$$

as an inverse with respect to convolution of $\Delta_j[\beta]$ we may take the function

$$\Sigma_j^{(-)}[\beta] = -\Sigma_j[\widehat{\beta}_j, 1-\beta_j] = \delta[\widehat{\beta}_j](-\Sigma_j[-\beta_j+1])$$
$$= \delta[\widehat{\beta}_j](\Sigma_j[\beta_j]-1) = \delta[\widehat{\beta}_j]\Sigma_j[\beta_j] - \delta[\widehat{\beta}_j] = \Sigma_j[\beta] - \delta[\widehat{\beta}_j].$$

The convolution of $\varphi[\beta]$ with $\Sigma_j^{(-)}[\beta]$, calculated by the formula

$$\varphi[\beta] * \Sigma_j^{(-)}[\beta] = -\sum_{\gamma_j=\beta_j}^{\infty} \varphi[\widehat{\beta}_j, \gamma_j],$$

is an analog of the integral

$$-\int_{\beta_j}^{\infty} \varphi(\widehat{x}_j, \xi_j)\, d\xi_j.$$

For functions satisfying (1.14), we have

$$\Sigma_j^{(-)}[\beta] * (\Delta_j[\beta] * \varphi[\beta]) = \varphi[\beta], \quad \Delta_j[\beta] * \left(\Sigma_j^{(-)}[\beta] * \varphi[\beta]\right) = \varphi[\beta].$$

Consider a single discrete independent variable t and introduce the function $t^{[k]}\Sigma[t+1]$ which is 0 for t negative and agrees with $t^{[k]}$ for other t. In particular, for $k=0$ this function is given by the equality

$$t^{[0]}\Sigma[t+1] = \begin{cases} 0, & \text{if } t<0, \\ 1, & \text{if } t \ge 0. \end{cases}$$

We call $t^{[k]}\Sigma[t+1]$ a ***truncated Newtonian power.***

The validity of the formula

$$\Sigma[t] * \left(\frac{t^{[k]}}{k!}\Sigma[t+1]\right) = \frac{t^{[k+1]}}{(k+1)!}\Sigma[t+1]$$

is easy from (1.13) and (1.10).

In the theory of differential equations, ordinary and partial, the derivatives of some function of a continuous argument are often replaced with divided differences of this function at the points of some cubic lattice with side h, i.e., by partial differences of the function $\varphi_h[\beta] = \varphi(h\beta)$. Such replacement is not unique. The same derivative $\partial\varphi/\partial x_1$ may be replaced, for instance, by the "forward" difference

$$\frac{\partial\varphi}{\partial x_1} \approx \frac{1}{h}\Delta_1[\beta] * \varphi_h[\beta],$$

or by the "backward" difference

$$\frac{\partial\varphi}{\partial x_1} \approx \frac{1}{h}(\varphi_h[\beta] - \varphi_h[\beta - h\delta_1]).$$

As an approximation to $\partial\varphi/\partial x_1$ we may also take the *symmetric difference*, i.e., the half-sum of the "forward" and "backward" differences or in general every combination of the values of $\Delta_1 * \varphi_h$ at several points

$$\frac{\partial\varphi}{\partial x_1} \approx \frac{1}{h}\sum_{\varepsilon}\chi[\varepsilon](\Delta_1[\beta] * \varphi_h[\beta - \varepsilon]) = \frac{1}{h}\chi * \Delta_1 * \varphi_h,$$

with $\chi[\varepsilon]$ a compactly-supported function and $\sum_\varepsilon \chi[\varepsilon] = 1$. We similarly approximate the value of the γth order derivative of a given function

$$D^\gamma\varphi \approx \frac{1}{h^{|\gamma|}}\sum_{\varepsilon}\chi[\varepsilon]\left(\Delta^{[\gamma]}[\beta] * \varphi_h[\beta - \varepsilon]\right) = \frac{1}{h^{|\gamma|}}\chi * \Delta^{[\gamma]} * \varphi_h.$$

The argument of $\Delta^{[\gamma]}$ on the right side is absent. From now on we agree to omit the argument of $\Delta^{[\gamma]}$ when this leads to no confusion.

Let a discrete function $\psi[\beta]$ be compactly-supported. The operator of convolution with $\psi[\beta]$, i.e. the operator $\psi[\beta]*$, is referred to as a *difference operator*. The domain of definition of such operator is the entire space of discrete functions.

Take $\gamma \in \mathbb{Z}^n$ and a compactly-supported discrete function $\chi_\gamma[\beta]$. We call the operator

$$\chi_\gamma[\beta] * \Delta^{[\gamma]}[\beta]*$$

a *difference operator of order* γ.

Let l be a natural. Assume given a compactly-supported function $\chi_\gamma[\beta]$ for every $\gamma \in \mathbb{Z}^n$. By a *difference operator of scalar order* l we mean a sum of the shape

$$\sum_{|\gamma|=l}\chi_\gamma[\beta] * \Delta^{[\gamma]} * .$$

Clearly, for an arbitrary l times continuously differentiable function $\varphi(x)$ we have the relation

$$\lim_{h\to 0}\frac{1}{h^l}\sum_{|\gamma|=l}\chi_\gamma[\beta] * \Delta^{[\gamma]}[\beta] * \varphi_h[\beta] = \sum_{|\gamma|=l} a_\gamma D^\gamma\varphi(0)$$

with $a_\gamma = \sum_\beta \chi_\gamma[\beta]$. Consequently, in passage to the limit as $h \to 0$ a difference operator of order l becomes a differential operator of order l. We do not exclude the possibilities in which some or even all a_γ vanish, since we may assume to treat the zero operator as a differential operator of every order l.

By definition, a difference operator of order l is simultaneously a difference operator of order each of the numbers $l-1, l-2, \ldots, 1, 0$. To the operator

$$\psi[\beta]* = \left(\sum_{j=0}^{k} \sum_{|\gamma|=l} \chi_\gamma[\beta] * \Delta^{[\gamma]}\right) *,$$

written as the sum of operators of order one of the integers from 0 to k, we respectively ascribe an arbitrary order from 0 to k. This stands to reason since for an arbitrary k times continuously differentiable function $\varphi(x)$ there are limits

$$\lim_{h\to 0} \frac{1}{h^j} \psi[\beta] * \varphi_h[\beta], \quad j = 0, 1, \ldots, k,$$

which are possibly equal to 0.

Observe that this definition diverges with an analogous definition of the order of a differential operator. We agree that the order of the expression

$$\sum_{j=k}^{m} a_j \frac{d^j}{dx^j}, \quad a_k \neq 0,\ a_m \neq 0,$$

is by definition equal to the highest order $j = m$ of the involved derivatives. For difference operators, the order of the expression

$$\sum_{j=k}^{m} \psi_j[\beta]*$$

with the order of $\psi_j[\beta]$ being k or at least k is k, i.e., the least of the orders of $\psi_j[\beta]$.

A formally natural continuous analog of the convolution operator acting on a discrete function $\psi[\beta]$ by the formula

$$\psi[\beta] = L\varphi[\beta] = \chi_\gamma[\beta] * \Delta^{[\gamma]}[\beta] * \varphi[\beta]$$

is the integro-differential convolution operator acting on a smooth function $\varphi(y)$ as

$$\psi(x) = L\varphi(x) = \int \chi_\gamma(x-y) D^\gamma \varphi(y)\, dy.$$

However, for the compactly-supported coefficients $\chi_\gamma[\beta]$, we consider such integro-differential operator does not result from a difference operator by passage to the limit. On the other hand, as was mentioned, analogs of an ordinary differential operator are constructed in a nonunique fashion.

This explains in part why it is convenient to use nonsimilar definitions of order for differential and difference operators.

As is easy to see, a compactly-supported difference operator of order l is orthogonal with respect to convolution to every polynomial of degree less than l. In other words, for a compactly-supported function

$$\varphi[\beta] = \sum_{|\gamma|=l} \chi_\gamma[\beta] * \Delta^{[\gamma]}$$

we have the system of relations

$$\varphi[\beta] * [\beta]^{[\alpha]} = 0, \quad |\alpha| < l. \tag{1.15}$$

As we prove below, this property is characteristic of such operators. It is worth observing that (1.15) amounts to the condition

$$[\varphi[\beta], [\beta]^{[\alpha]}] = 0, \quad |\alpha| < l. \tag{1.16}$$

Indeed, (1.16) is immediate from (1.15) on recalling (1.2). Conversely, assuming (1.16) fulfilled, by an analog of the Newton binomial formula for Newtonian powers obtain

$$\varphi[\beta] * [\beta]^{[\alpha]} = \sum_\gamma \varphi[\gamma][\beta - \gamma]^{[\alpha]} = \sum_{0\le\varepsilon\le\alpha} \sum_\gamma \varphi[\gamma] c_{\alpha,\varepsilon}[\beta]^{[\alpha-\varepsilon]}[\gamma]^{[\varepsilon]},$$

with $c_{\alpha,\varepsilon}$ easily calculable constants. Further derive

$$\varphi[\beta] * [\beta]^{[\alpha]} = \sum_{0\le\varepsilon\le\alpha} c_{\alpha,\varepsilon}[\beta]^{[\alpha-\varepsilon]}[\varphi[\gamma], [\gamma]^{[\varepsilon]}] = 0.$$

Hence, (1.16) entails (1.15).

We now demonstrate that (1.15) is a characteristic property of the operators under consideration. We have

Theorem 8.5. *Let $\varphi[\beta]$ be a compactly-supported function orthogonal with respect to convolution to every polynomial of degree less than l, in symbols,*

$$\varphi[\beta] * [\beta]^{[\gamma]} = 0, \quad |\alpha| < l, \tag{1.17}$$

$$\operatorname{supp}\varphi[\beta] \subset \{\beta : |\beta_j| < L\}. \tag{1.18}$$

Then $\varphi[\beta]$ is a compactly-supported difference operator of scalar order l; i.e., it may be written in divergence form

$$\varphi[\beta] = \sum_{|\gamma|=l} \chi_\gamma[\beta] * \Delta^{[\gamma]}[\beta], \quad \operatorname{supp} \chi_\gamma[\beta] \subset \{\beta : |\beta_j| < L\}, \tag{1.19}$$

thus presenting a compactly-supported difference operator of scalar order l.

We precede the proof of Theorem 8.5 with two lemmas.

Lemma 8.1. *Let a discrete function $\varphi[\beta]$ satisfy (1.17) and (1.18). Then the convolution*

$$\Phi_n[\beta] = \varphi[\beta] * \Sigma_n[\beta] = \sum_{\beta_n'=-\infty}^{\beta_n-1} \varphi[\widehat{\beta}_n, \beta_n']$$

possesses the following properties:

1. *All entries of the vectors $\beta \in \operatorname{supp} \Phi_n[\beta]$ but the entry β_n are bounded in magnitude, whereas the entry β_n is in general bounded only from below*

$$\operatorname{supp} \Phi_n[\beta] \subset \{\beta : |\beta_j| < L, j < n;\ \beta_n > -L\}.$$

2. *For $\beta_n > L$, the function $\Phi_n[\beta]$ depends only on $\widehat{\beta}_n$.*

3. *For $\beta_n > L$, the function $\widehat{\Phi}_n[\widehat{\beta}_n] = \Phi_n[\beta]$ is orthogonal in the space of discrete functions of the variable $\widehat{\beta}_n \in \mathbb{Z}^{(n-1)}$ to every polynomial of degree less than l, namely,*

$$[\widehat{\Phi}_n[\widehat{\beta}_n], [\widehat{\beta}_n]^{[\widehat{\alpha}_n]}] = 0 \quad \textit{for } |\widehat{\alpha}_n| < l.$$

4. *For the truncation $\chi_n[\beta; k]$ of $\Phi_n[\beta]$ with respect to the coordinate β_n which is defined as*

$$\chi_n[\beta; k] = \Phi_n[\beta]\Sigma[k - \beta_n] = \begin{cases} \Phi_n[\beta], & \textit{if } \beta_n < k; \\ 0, & \textit{if } \beta_n \geq k; \end{cases} \tag{1.20}$$

the equality holds

$$[\chi_n[\beta; k], [\beta]^{[\alpha]}] = 0 \quad \textit{for } k > L, \quad |\alpha| < l - 1. \tag{1.21}$$

PROOF. The first two properties of $\Phi_n[\beta]$ are obvious by definition. Show the property 3.

The inner product of $\widehat{\Phi}_n[\widehat{\beta}_n]$ and $[\widehat{\beta}_n]^{[\widehat{\alpha}_n]}$ in the space of discrete functions in the argument $\widehat{\beta}_n \in \mathbb{Z}^{n-1}$ is easily seen to maintain the following relations

$$[\widehat{\Phi}_n[\widehat{\beta}_n], [\widehat{\beta}_n]^{[\widehat{\alpha}_n]}] = \sum_\beta \varphi[\beta][\widehat{\beta}_n]^{[\widehat{\alpha}_n]} = \sum_\beta \varphi[\beta][\widehat{\beta}_n]^{[\widehat{\alpha}_n]}[\beta_n]^{[0]}$$

$$= \sum_\beta \varphi[\beta][\beta]^{[\alpha]} = [\varphi[\beta], [\beta]^{[\alpha]}] = 0.$$

Here α stands for the multi-index $(\widehat{\alpha}_n, 0)$.

To proof the property 4, rewrite the left side of (1.21) as

$$\left[\chi_n[\beta;k],[\beta]^{[\alpha]}\right] = \left[\Phi_n[\beta]\Sigma[k-\beta_n],[\beta]^{[\alpha]}\right] = \left[\Phi_n[\beta],[\beta]^{[\alpha]}\Sigma[k-\beta_n]\right]. \qquad (1.22)$$

The second factor of the last inner product may be written as convolution of $\Delta_n[\beta]$ and some discrete function.

As it is easy to see, the function

$$\lambda_\alpha[\beta] = \begin{cases} 0, & \text{if } \beta_n \geq k, \\ \dfrac{[\widehat{\beta}_n]^{[\widehat{\alpha}_n]}}{\alpha_n+1}\{[\beta_n]^{[\alpha_n+1]} - k^{[\alpha_n+1]}\}, & \text{if } \beta_n < k \end{cases}$$

serves as a solution to the equation

$$\Delta_n * \lambda[\beta] = [\beta]^{[\alpha]}\Sigma[k-\beta_n].$$

Clearly, $\lambda_\alpha[\beta]$ for $\beta_n \leq k$ is a polynomial of degree less than l. The definition of $\lambda_\alpha[\beta]$ and the property 1 of $\Phi_n[\beta]$ imply that the product $\lambda_\alpha[\beta]\Phi_n[\beta]$ is a compactly-supported function. Using this, readily derive the equality

$$\sum_\beta \Delta_n * (\Phi_n[\beta]\lambda_\alpha[\beta]) = 0. \qquad (1.23)$$

Applying (1.12) with $j = n$, transform the sum on the left side of (1.23) as follows

$$\Delta_n * (\varphi\psi) = (\Delta_n * \varphi[\beta])\psi[\beta] + (\Delta_n * \psi[\beta])\varphi[\beta+\delta_n].$$

Letting $\varphi[\beta] = \lambda_\alpha[\beta]$ and $\psi[\beta] = \Phi_n[\beta]$ and inserting the result in (1.23), obtain

$$\sum_\beta \Phi_n[\beta](\Delta_n * \lambda_\alpha[\beta]) + \sum_\beta \lambda_\alpha[\beta+\delta_n]\Delta_n * \Phi_n[\beta]$$
$$= \sum_\beta \Phi_n[\beta][\beta]^{[\alpha]}\Sigma[k-\beta_n] + \sum_\beta \lambda_\alpha[\beta+\delta_n]\varphi[\beta] = 0.$$

Since $\lambda_\alpha[\beta+\delta_n]$ for $k > L$ is a polynomial of degree less than l at every point of the support of $\varphi[\beta]$, by (1.17) the second summand in the last equality equals 0. Consequently, the inner product (1.22) also equals 0. The proof of Lemma 8.1 is complete.

Lemma 8.2. *Let a discrete function $\varphi[\beta]$ satisfy the hypotheses* (1.17) *and* (1.18) *of Theorem* 8.5. *Then $\varphi[\beta]$ may be written as*

$$\varphi[\beta] = \sum_{j=1}^{n} (\Delta_j * \chi_j)[\beta],$$

with $\chi_j[\beta]$ compactly-supported functions orthogonal to every polynomial of degree less than $l-1$, i.e.,

$$\operatorname{supp} \chi_j[\beta] \subset \{\beta : |\beta_j| < L\}, \quad [\chi_j[\beta], [\beta]^{[\alpha]}] = 0, \ |\alpha| < l-1. \tag{1.24}$$

PROOF. Induct on the number of independent variables. Prove that for a single variable the lemma is valid. Indeed, for β_1 sufficiently large by the hypotheses of the lemma we have

$$\Sigma[\beta_1] * \varphi[\beta_1] = \psi[\beta_1] = \sum_{\gamma_1 < \beta_1} \varphi[\gamma_1] = (\varphi[\beta_1], 1) = 0.$$

Consequently, the function $\psi[\beta_1]$ is compactly-supported and agrees with $\Phi_1[\beta]$ of Lemma 8.1. By Lemma 8.1 this function is orthogonal to every polynomial of degree less than $l-1$. However,

$$\varphi[\beta_1] = \Delta_1 * \psi[\beta_1],$$

which proves our claim for $n = 1$.

Show now that the validity of Lemma 8.2 for $n-1$ variables entails its validity for n variables.

Again consider the function $\chi_n[\beta; k]$ defined by (1.20). For $k > L$ by the definition of $\Delta_n[\beta]$ and (1.20) we have

$$\Delta_n * \chi_n[\beta; k] = \varphi[\beta] - \delta[\beta_n - k + 1]\widehat{\Phi}_n[\widehat{\beta}_n].$$

Hence,

$$\varphi[\beta] = \Delta_n * \chi_n[\beta; k] + \delta[\beta_n - k + 1]\widehat{\Phi}_n[\widehat{\beta}_n].$$

The function $\chi_n[\beta; k]$ is compactly-supported and orthogonal to every polynomial of degree less than $l-1$ by Lemma 8.1. By the induction hypothesis the function $\widehat{\Phi}_n[\widehat{\beta}_n]$, being a function in $n-1$ variables, may be written as

$$\widehat{\Phi}_n[\widehat{\beta}_n] = \sum_{j=1}^{n-1} \Delta_j * \chi_j[\widehat{\beta}_n].$$

Hence,

$$\varphi[\beta] = \sum_{j=1}^{n-1} \Delta_j * (\delta[\beta_n - k + 1]\chi_j[\widehat{\beta}_n]) + \Delta_n * \chi_n[\beta] \tag{1.25}$$

which is the sought decomposition of $\varphi[\beta]$. The proof of Lemma 8.2 is complete.

PROOF OF THE MAIN THEOREM 8.5. Apply induction, this time on the degree of a polynomial.

For the polynomials of degree 0, the theorem is trivial. A compactly-supported function $\varphi[\beta]$ serves itself in (1.19) with $l = 0$.

Assume the theorem valid for compactly-supported functions orthogonal to every polynomial of degree less than $l-1$. Show that it is valid also for the functions orthogonal to the polynomials of degree less than l.

Indeed, from Lemma 8.2 it follows that a function $\varphi[\beta]$ satisfying (1.17) and (1.18) admits the expansion

$$\varphi[\beta] = \sum_{j=1}^{n} \Delta_j * \chi_j\,[\beta], \tag{1.26}$$

with $\chi_j\,[\beta]$ compactly-supported and orthogonal with respect to convolution to every polynomial of degree $l-1$, i.e., satisfying (1.24). Each function $\chi_j\,[\beta]$ by the induction hypothesis may be written in divergence form as

$$\chi_j\,[\beta] = \sum_{|\gamma|=l-1} \chi_{j,\gamma}[\beta] * \Delta^{[\gamma]}[\beta].$$

Inserting this equality in (1.26), obtain (1.19). The proof of Theorem 8.5 is complete.

We further trace an analogy between the difference operators and differentiation operators.

Consider the set of all partial differences of order l of a given function of a discrete argument

$$\{\omega_\alpha[\beta] = \Delta^{[\alpha]} * \varphi[\beta] : \ |\alpha| = l\}.$$

Obviously, the functions $\omega_\alpha[\beta]$ maintain the relation

$$\Delta_k * \omega_{\alpha+\delta_j} = \Delta_j * \omega_{\alpha+\delta_k}, \tag{1.27}$$

with α an arbitrary vector such that $|\alpha| = l-1$. This relation is an analog of the formula $\partial w_j/\partial x_k = \partial w_k/\partial x_j$.

We call a *difference gradient of order* l an arbitrary system of $\{w_\alpha, |\alpha| = l\}$ satisfying (1.27).

Theorem 8.6. *Each difference gradient w_α of order l is the set of partial differences of order l of some function in a discrete variable.*

PROOF. We check the claim only for a gradient of order 1 since the case of a general order l has no specific features as compared with the case of $l = 1$.

Let φ_k possess the property

$$\Delta_j * \varphi_k = \Delta_k * \varphi_j. \tag{1.28}$$

Show that there is a function $\varphi_0[\beta]$ such that

$$\varphi_k = \Delta_k * \varphi_0, \quad k = 1, \dots, n. \tag{1.29}$$

This is easy to achieve by using consecutive partial summation over each discrete variable in much the same way as in classical calculus for reconstructing a function from its gradient.

Let $\varphi_0[0]$ be an arbitrary constant and let $\beta = (\beta_1, \dots, \beta_n)$ belong to $\mathbb{Z}^n$. Define the value $\varphi_0[\beta]$ by recursion.

For $m = 1, 2, \dots, n$ put

$$\varphi_0\left[\sum_{k=1}^{m-1} \beta_k\delta_k + \beta_m\delta_m\right] = \begin{cases} \varphi_0\left[\sum\limits_{k=1}^{m-1} \beta_k\delta_k\right] + \sum\limits_{\gamma_m=0}^{\beta_m-1} \varphi_m[\gamma_m\delta_m], & \beta_m > 0, \\ \varphi_0\left[\sum\limits_{k=1}^{m-1} \beta_k\delta_k\right] - \sum\limits_{\gamma_m=0}^{-\beta_m+1} \varphi_m[\gamma_m\delta_m], & \beta_m \le 0. \end{cases}$$

A routine check shows that in the case of (1.28) the function

$$\varphi_0[\beta] = \varphi_0\left[\sum_{k=1}^{n} \beta_k\delta_k\right]$$

satisfies (1.29). The proof of Theorem 8.6 is complete.

We call a function $\varphi_0[\beta]$ corresponding to a difference gradient of order l by Theorem 8.6 a ***primitive*** of this gradient.

A primitive $\varphi_0[\beta]$ satisfying (1.29) is determined from the values of a difference gradient of first order to within an arbitrary constant summand. With a gradient of order l available, construct a gradient of order $l-1$ that consists of the functions φ_α, $|\alpha| = l-1$, each of which defined to within a constant summand. Repeating the process, come to a gradient of order $l-2$, and so on. Show that the function $\varphi_0[\beta]$ we construct finally is defined to within an arbitrary polynomial summand of degree $l-1$. In other words, we establish

Lemma 8.3. *A primitive of an arbitrary difference gradient of order l is determined to within a polynomial summand of degree at most $l-1$.*

PROOF. Check that, if for all α such that $|\alpha| = l$ the equality holds

$$\Delta^{[\alpha]} * \varphi[\beta] = 0,$$

then $\varphi[\beta]$ is a polynomial of degree at most $l-1$.

We prove by induction on the number of independent variables. For $n = 1$ the assertion is easy by induction on l. Assume that it is valid for $n-1$ variables. Since all differences of order l of φ with respect to the variables $\beta_1, \beta_2, \dots, \beta_{n-1}$ vanish, the values of $\varphi[\beta]$ on each plane $\beta_n = \text{const}$ coincide with the values of some polynomials in $\widehat{\beta}_n$, i.e.,

$$\varphi[\beta] = \sum_{|\widehat{\alpha}_n|<l} a_{\widehat{\alpha}_n}[\beta_n][\widehat{\beta}_n]^{[\widehat{\alpha}_n]}.$$

Applying to both sides of this equality the convolution operator $\Delta^{[\widehat{\gamma}_n]}*$, obtain

$$\Delta^{[\widehat{\gamma}_n]} * \varphi = \sum_{|\widehat{\alpha}_n|<l} a_{\widehat{\alpha}_n}[\beta_n]\Delta^{[\widehat{\gamma}_n]} * [\widehat{\beta}_n]^{[\widehat{\alpha}_n]}.$$

The difference in β_n of order $\gamma_n = l - |\widehat{\gamma}_n|$ of this polynomial is 0 by hypothesis. Consequently, all coefficients $a_{\widehat{\alpha}_n}[\beta_n]$ are polynomials in β_n of degree $l - [\widehat{\gamma}_n] - 1$. Whence the claim follows. The proof of Lemma 8.3 is complete.

The difference between two primitive functions $\varphi_0^{(1)}$ and $\varphi_0^{(2)}$ of the same difference gradient of order l, i.e., between two solutions to the simultaneous equations

$$\Delta^{[\alpha]} * \varphi_0 = \omega_\alpha, \quad |\alpha| = l,$$

satisfies the corresponding homogeneous equations and by Lemma 8.3 is a polynomial of degree at most $l-1$.

§2. Spaces of Discrete Functions

The spaces of functions of a discrete argument are constructed by means of differences in the same manner as this is carried out for the functions of a continuous argument by means of derivatives.

The functions of a single discrete argument, i.e., *sequences*, have already been examined in thorough detail. For them, available are the celebrated l_p *spaces*, $1 \le p < \infty$, comprising the functions $\varphi[\beta]$ of an integer argument β with the finite norm

$$\|\varphi\| = \left\{ \sum_{\beta=-\infty}^{\infty} |\varphi[\beta]|^p \right\}^{1/p}.$$

We now consider their natural abstraction, the l_p space of functions of several discrete variables with the norm

$$\|\varphi \mid l_p\| = \left[\sum_{\beta} |\varphi[\beta]|^p\right]^{1/p}.$$

It is worthy to consider also the space $\tilde{l}_{p,\tau}$ of periodic functions of a discrete argument. By definition, this is the linear space of functions $\varphi[\beta]$ periodic with period τ, $\tau \in \mathbb{Z}^n$, and possessing the finite norm

$$\|\varphi \mid \tilde{l}_{p,\tau}\| = \left\{\sum_{0 \leq \beta < \tau} |\varphi[\beta]|^p\right\}^{1/p}.$$

Alongside these spaces, consider the set $l_p^{(l)}$ comprising the classes of functions $\varphi[\beta]$ whose differences of order l are summable to the power p, namely,

$$\|\varphi \mid l_p^{(l)}\| = \left\{\sum_{|\alpha|=l} \sum_{\beta} |\Delta^{[\alpha]} * \varphi[\beta]|^p\right\}^{1/p} < \infty.$$

The members of each class in $l_p^{(l)}$ differ from one another in polynomial summands of degree at most $l-1$.

For the $l_p^{(l)}$ spaces we have the embedding theorems analogous to those valid for the $L_p^{(l)}$ spaces of functions of a continuous argument whose definition is in exact analogy with that given in §2 of Chapter 1 in the case of $p = 2$. Also valid is a theorem on the denseness of compactly-supported functions (see [265]).

The relevant section of theory of the spaces of discrete functions bases on several principal theorems relating $l_p^{(l)}$ and $L_p^{(l)}$.

Assume given a continuous function $\varphi(x)$ in $\mathbb{R}^n$, and let

$$\Gamma = \{A\beta + \beta^{(0)} : \beta \in \mathbb{Z}^n\}$$

be an integer lattice in $\mathbb{R}^n$. We call the function

$$\varphi[\beta] = \varphi(A\beta + \beta^{(0)})$$

of a discrete variable the *trace* or *restriction* of $\varphi(x)$ on Γ. Given a function $\varphi[\beta]$, we may somehow define a function $\varphi(x)$ on $\mathbb{R}^n$ that is continuous together with several derivatives and whose trace equals $\varphi[\beta]$. We call such function an *interpolation function* or *interpolant*.

Theorem 8.7 (of tracing). *If $\varphi(x)$ in $L_p^{(l)}$ is continuous in $\mathbb{R}^n$ then the trace of φ on Γ is a member of $l_p^{(l)}$. The norms of $\varphi(x)$ and $\varphi[\beta]$ in the corresponding spaces satisfy the inequality*

$$\|\varphi[\beta] \mid l_p^{(l)}\| \leq K\|\varphi(x) \mid L_p^{(l)}\|$$

with K a constant independent of φ.

PROOF. Let $\varphi(x)$ be a smooth function in $L_p^{(l)}$, and let $\varphi[\beta]$ be its trace on Γ. To simplify calculations, assume that Γ is a cubic lattice with vertex the coordinate origin, i.e., $\beta^{(0)} = 0$ and $A = I$. Express the differences of $\varphi[\beta]$ through the integrals of the derivatives of $\varphi(x)$.

As usual, let $\Theta(\xi)$ stand for the *Heaviside function*, i.e.,

$$\Theta(\xi) = \frac{1}{2}(1 + \operatorname{sgn} \xi) = \begin{cases} 1, & \text{if } \xi \geq 0, \\ 0, & \text{if } \xi < 0. \end{cases}$$

Given a natural m, put

$$\psi_m(\xi) = \sum_{k=0}^{m} (-1)^{m-k} \frac{m!}{k!(m-k)!} \frac{(\xi + k)^{m-1}}{(m-1)!} \Theta(\xi + k).$$

Clearly, for $\xi < -m$ the function $\psi_m(\xi)$ is identically 0. For $\xi > 0$ the sum $\psi_m(\xi)$ is the mth derivative of the monomial $\xi^{m-1}/(m-1)!$, and so it equals 0 too.

Given a multi-index $\alpha \in \mathbb{Z}^n$ with natural entries α_j, consider the product

$$\psi_\alpha(y) = \prod_{j=1}^{n} \psi_{\alpha_j}(y_j).$$

The function $\psi_\alpha(y)$ has compact support included in the cube

$$\{y : -\alpha_j \leq y_j \leq 0, j = 1, \dots, n\}.$$

For $|\alpha| \leq l$ elementary calculations lead to the formula

$$\int \psi_\alpha(x - y) D^\alpha \varphi(y)\, dy = \Delta^{[\alpha]} * \varphi(x). \tag{2.1}$$

Letting $x = \beta$ in (2.1), rewrite the latter as

$$\Delta^{[\alpha]} * \varphi[\beta] = \int \psi_\alpha(\beta - y) \chi_\sigma(\beta - y) D^\alpha \varphi(y)\, dy.$$

Here $\chi_\sigma(y)$ is the indicator of the support σ of $\psi_\alpha(y)$. It is worth observing that the product $\psi_\alpha(\beta-y)\chi_\sigma(\beta-y)$ is a function compactly-supported in y whose support lies entirely in the cube

$$\{y : \beta_j \le y_j \le \beta_j + \alpha_j,\ j = 1,\dots,n\}.$$

The found representation of the difference $\Delta^{[\alpha]} * \varphi[\beta]$ allows us to write the norm of $\varphi[\beta]$ as follows

$$\|\varphi[\beta] \mid l_p^{(l)}\| = \left\{\sum_{|\alpha|=l}\sum_{\beta}\left|\int \psi_\alpha(\beta-y)\chi_\sigma(\beta-y)D^\alpha\varphi(y)\,dy\right|^p\right\}^{1/p}.$$

Estimate the inner integral by the Hölder inequality

$$\left|\int \psi_\alpha(\beta-y)\chi_\sigma(\beta-y)D^\alpha\varphi(y)\,dy\right|$$
$$\le \|\chi_\sigma(\beta-y)D^\alpha\varphi(y) \mid L_p\|\,\|\psi_\alpha(\beta-y) \mid L_{p'}\|.$$

Here $1/p + 1/p' = 1$. Since the second factor on the right side is independent of β, the inequality holds

$$\|\varphi[\beta] \mid l_p^{(l)}\| \le \max_{|\alpha|=l}\|\psi_\alpha \mid L_{p'}\|\left\{\sum_{|\alpha|=l}\sum_{\beta}\int |\chi_\sigma(\beta-y)D^\alpha\varphi(y)|^p\,dy\right\}^{1/p}.$$

Each integral on the right side admits the estimate

$$\int |\chi_\sigma(\beta-y)D^\alpha\varphi(y)|^p\,dy \le \int\limits_{\beta\le y\le\beta+\alpha} |D^\alpha\varphi(y)|^p\,dy.$$

For every $\beta \in \mathbb{Z}^n$ the cube

$$\{y : \beta_j \le y_j \le \beta_j + \alpha_j,\ j = 1,\dots,n\}$$

is the union of some finite number c_α of elementary meshes like $\{y : \gamma \le y < \gamma+1\}$. Furthermore, the number c_α depends on β in no way. Considering this and summing over $\beta \in \mathbb{Z}^n$ the obtained inequalities, arrive at the relations

$$\sum_{\beta}\int \left|\chi_\sigma(\beta-y)D^\alpha\varphi(y)\right|^p dy \le c_\alpha\int |D^\alpha\varphi(y)|^p\,dy, \quad |\alpha| = l.$$

Summing them over α, $|\alpha| = l$, finally derive

$$\|\varphi[\beta] \mid l_p^{(l)}\| \le C_l\max_{|\alpha|=l}\|\psi_\alpha \mid L_{p'}\|\sum_{|\alpha|=l}\|D^\alpha\varphi \mid L_p\| \le K\|\varphi \mid L_p^{(l)}\|.$$

The proof of Theorem 8.7 is complete.

Corollary 2.1. *If $lp > n$ then the assertion of Theorem 8.7 remains valid.*

The preceding claim is easy from the Embedding Theorem of $L_p^{(l)}$ into the space of continuous functions on a bounded domain.

Extend a discrete function $\varphi[\beta]$ to the whole of $\mathbb{R}^n$ smoothly, i.e., construct an interpolation function $\varphi(x)$ for $\varphi[\beta]$. To implement this interpolation, we use a special linear operator which was suggested by V. S. Ryaben'kiĭ [189, 190].

We describe how it is constructed.

STEP 1. Consider an auxiliary interpolation problem: *Given two $(l+1)$-dimensional vectors*

$$a^{(0)} = \left(a_0^{(0)}, a_1^{(0)}, \ldots, a_l^{(0)}\right), \quad a^{(1)} = \left(a_0^{(1)}, a_1^{(1)}, \ldots, a_l^{(1)}\right),$$

find a polynomial $P(y)$ of degree $2l+1$ satisfying

$$\left.\frac{d^s P(y)}{dy^s}\right|_{y=0} = a_s^{(0)}, \quad \left.\frac{d^s P(y)}{dy^s}\right|_{y=1} = a_s^{(1)}, \quad s = 0, 1, \ldots, l. \tag{2.2}$$

Show that (2.2) always define a unique polynomial $P(y)$ of degree $2l+1$. Indeed, the knowledge of all derivatives of $P(y)$ up to order l at the points $y = 0$ and $y = 1$ enables us to find in a unique fashion the expansion

$$\frac{P(y)}{y^{l+1}(y-1)^{l+1}} = \frac{b_{l+1}}{y^{l+1}} + \frac{b_l}{y^l} + \cdots + \frac{b_1}{y} + \frac{c_{l+1}}{(y-1)^{l+1}} + \frac{c_l}{(y-1)^l} + \cdots + \frac{c_1}{y-1}.$$

This relation completely characterizes the polynomial $P(y)$ which may be determined from the formula

$$\begin{aligned} P(y) &= (y-1)^{l+1}[b_{l+1} + b_l y + b_{l-1} y^2 + \cdots + b_1 y^l] \\ &+ y^{l+1}[c_{l+1} + c_l(y-1) + c_{l-1}(y-1)^2 + \cdots + c_1(y-1)^l]. \end{aligned} \tag{2.3}$$

Calculation of the coefficients b_s and c_s is as usual implemented by differentiating (2.3) and substituting the values $y = 0$ and $y = 1$ for y.

The coefficients of the polynomial $P(y)$, as well as its values, are not necessarily numbers. They may be elements of an arbitrary nature. What we need is only a possibility of summing them and multiplying them by scalars.

Let $\mathbf{X}$ be an arbitrary vector space (say, the space of polynomials in some auxiliary variable z, etc.). Consider the vectors $a = (a_0, a_1, \ldots, a_l)$ whose entries are members of $\mathbf{X}$. Addition and scalar multiplication are determined for these vectors in a natural coordinatewise manner. Clearly, in this case the problem (2.2) is also uniquely solvable for all vectors $a^{(0)}$ and $a^{(1)}$ of the indicated type. The coefficients of the polynomial $P(y)$ are also elements of $\mathbf{X}$.

Assume that an $(l+1)$-dimensional vector field $a[\beta]$ is given on the one-dimensional lattice of β, i.e., to each integer β there is assigned some vector

$$a[\beta] = (a_0[\beta], a_1[\beta], \dots, a_l[\beta]), \quad a_s[\beta] \in \mathbf{X}.$$

In every interval $\beta \leq y \leq \beta + 1$, given $a[\beta]$ and $a[\beta+1]$, construct the polynomial $R(y \mid \beta]$ that solves (2.2). In result, obtain a piecewise-polynomial function $R(y)$ with codomain the vector space $\mathbf{X}$ which is continuous with all derivatives up to order l. Moreover, the value of $R(y)$ at an integer point $y = \beta$ agrees with $a_0[\beta]$.

Were the exact values of the derivatives of the sought function $\varphi(y)$ known at integer points, we would construct it by solving the interpolation problem (2.2) piecewise. However, the values of the derivatives are unknown, and as $a_s[\beta]$ we are compelled to choose some difference approximations. We show how this can be carried out.

STEP 2. Take a function $\varphi[\beta]$ given on a lattice and construct from the values of $\varphi[\beta]$ at the points $\beta, \beta+1, \dots, \beta+l$ the Newton interpolant

$$Q(y \mid \beta] = \sum_{k=0}^{l} \frac{\Delta^{[k]} * \varphi[\beta]}{k!} (y-\beta)^{[k]}$$

of degree l. To this polynomial we in turn assign the vector of its derivatives at the point β,

$$\left. \frac{d^k Q(y \mid \beta]}{dy^k} \right|_{y=\beta} = a_k[\beta], \quad k = 0, 1, \dots, l. \tag{2.4}$$

We thus obtain a vector field $a[\beta]$. This field is given on the lattice and defines some linear operator that sends the difference $\Delta^{[k]} * \varphi[\beta]$ to $a_k[\beta]$, $k = 0, 1, \dots, l$.

The quantities $a_k[\beta]$ are the coefficients of the expansion of $Q(y \mid \beta]$ in the conventional powers, namely,

$$Q(y \mid \beta] = \sum_{k=0}^{l} a_k[\beta] \frac{y^k}{k!}.$$

The connection between the Newtonian and ordinary powers is revealed by the formulas

$$y^0 = y^{[0]}; \quad y^n = \sum_{k=1}^{n} S(n,k) y^{[k]}, \quad n = 1, 2, \dots,$$

$$y^{[0]} = y^0; \quad y^{[n]} = \sum_{k=1}^{n} s(n,k) y^k, \quad n = 1, 2, \dots,$$

with $s(n,k)$ ($S(n,k)$) the so-called *Stirling numbers* of the first (second) kind [94].

STEP 3. The Ryaben'kiĭ interpolator is a simple combination of the operators considered at Steps 1 and 2. Having constructed the vector field $a[\beta]$ over a function $\varphi[\beta]$, we take as the ***Ryaben'kiĭ interpolation polynomial*** on the interval $\beta \le y \le \beta + 1$ the polynomial $R(y \mid \beta]$ that is assigned to the vectors $a[\beta]$ and $a[\beta+1]$ at Step 1. Given these polynomials, we glue them together to obtain a piecewise-polynomial function that is defined on the whole integration domain. Denote the latter function by $\varphi_R(y) = R(y \mid \varphi[\beta])$.

Call $\varphi_R(y)$ the ***Ryaben'kiĭ interpolation function*** or ***Ryaben'kiĭ interpolant.*** The operator R sending $\varphi[\beta]$ to $\varphi_R(y)$ is the ***one-dimensional Ryaben'kiĭ interpolator.*** We thus settle the problem in the case of a single variable. The problem for n variables is settled by induction.

Consider $\varphi[\beta] = \varphi\big[(\beta_1, \beta_1')\big]$, a function in n variables. Interpolate it with respect to β_1, using the Ryaben'kiĭ interpolator R_1. Interpolate the resulting function $\varphi(y_1, \beta_1'] = \varphi\big(y_1, (\beta_2, \beta_2')\big]$ with respect to β_2, using the Ryaben'kiĭ interpolator R_2 in the second variable. Consecutively applying this procedure, in n steps obtain the function $\varphi(y_1, \dots, y_n) = \varphi(y)$. The whole process is described by the ***multidimensional Ryaben'kiĭ interpolator***

$$R = R_n R_{n-1} \dots R_2 R_1.$$

Establish the needed properties of the Ryaben'kiĭ interpolator.

Linearity of the interpolator R is straightforward from linearity of intermediate operators.

From the construction of $R(y \mid \varphi)$ it follows that in each mesh

$$\{y : \beta_j \le y_j \le \beta_j + 1,\ j = 1, 2, \dots, n\}$$

this interpolant is a polynomial of degree at most $(2l+1)n$.

The function $R(y \mid \varphi)$ is continuous with all derivatives up to order l on the entire $\mathbb{R}^n$. Indeed, the total operator is the product of n one-dimensional interpolators with respect to individual variables. However, each one-dimensional interpolator yields a polynomial in an interval with integer endpoints. At each integer endpoint the values of the derivatives of the interpolation polynomials to the right and to the left of the point are taken equal.

The operator $R(y \mid \varphi)$ associates with each polynomial $P[\beta]$ of degree l of a discrete variable β the polynomial of a continuous variable x with the same coefficients, i.e.,

$$R(y \mid P[\beta]) = P(x)$$

in a continuous variable x. Indeed, the Newton polynomial $Q(y \mid \beta]$ for all β agrees with the original polynomial. Moreover, for all β it satisfies the conditions (2.4), the latter written for this polynomial. By uniqueness of a polynomial meeting the conditions, the function $Q(y \mid \beta]$ is the Ryaben'kiĭ polynomial for itself.

The value of the polynomial $R(y \mid \varphi)$ at a fixed point y depends only on the values of $\varphi[\beta]$ at the points of some finite set $\Omega_{\text{infl}}(y)$ which we call the ***influence domain*** *of* y.

The set $\Omega_{\text{infl}}(y)$ for the operator $R(y \mid \varphi)$ is defined by the formula

$$\Omega_{\text{infl}}(y) = \{\beta : A \leq \beta_j - y_j \leq B,\ j = 1, 2, \dots, n\},$$

with A and B some constants.

In other words, the value of $R(y \mid \varphi)$ at y depends only on the values of $\varphi[\beta]$ in some cube of finite size independent of y and φ.

This property is obvious, since the influence domain of a point in the one-dimensional case has the form

$$\Omega_{\text{infl}}(y) = \{\beta_y, \beta_y + 1, \dots, \beta_y + l\},$$

with β_y standing for the integral part of y.

Observe that the difference $\Delta^{[\gamma]} * \varphi[\beta]$, as well as an arbitrary compactly-supported difference operator $\chi_\gamma[\beta] * \varphi[\beta]$ of order γ, may be usefully treated as a linear combination of the values of $\varphi[\varepsilon]$ at the points of some set. For $\Delta^{[\gamma]} * \varphi[\beta]$ such set is the parallelepiped

$$\{\varepsilon : \beta_j \leq \varepsilon_j \leq \beta_j + \gamma_j,\ j = 1, 2, \dots, n\},$$

which we denote by $[\beta, \beta + \gamma]$. If γ is a scalar then the same symbol stands for the cube

$$\{\varepsilon : \beta_j \leq \varepsilon_j \leq \beta_j + \gamma,\ j = 1, 2, \dots, n\}.$$

Prove that, for every γ such that $\gamma_j \leq l$, $j = 1, 2, \dots, n$, the estimate holds

$$|D^\gamma R(y \mid \varphi)| \leq K \max_{[\beta, \beta+\gamma] \subset \Omega_{\text{infl}}(y)} |\Delta^{[\gamma]} * \varphi[\beta]|. \tag{2.5}$$

Here the maximum is taken over β such that the cube $[\beta, \beta + l]$ lies in the influence domain of y.

We derive (2.5) in the case of $n = 1$. Take $y \in \mathbb{R}$ and let an integer β satisfy the condition $[\beta, \beta + l] \subset \Omega_{\text{infl}}(y)$, i.e., β is the integral part of y. Interpolating the function $\varphi[\gamma]$ from its values at the points $\gamma = \beta, \beta + 1, \dots, \beta + l + 1$ by means of the Newton polynomial of degree $l + 1$, for $\gamma = \beta, \beta + 1, \dots, \beta + l + 1$ obtain

$$\varphi[\gamma] = \sum_{k=0}^{l+1} \frac{[\gamma - \beta]^{[k]}}{k!} \Delta^{[k]} * \varphi[\beta] = Q[\gamma \mid \beta] + \frac{[\gamma - \beta]^{[l+1]}}{(l+1)!} \Delta^{[l+1]} * \varphi[\beta],$$

with $Q[\gamma \mid \beta]$ the discrete Newton polynomial of degree l. The value of $R(y \mid \varphi)$ at y is determined only by the values of $\varphi[\beta]$ at the $l+2$ points $\beta, \beta+1, \dots, \beta+l+1$. Since the Ryaben'kiĭ interpolator is identical at a polynomial of degree l; recalling also its linearity, we thus infer

$$R(y \mid \varphi) = Q(y \mid \beta] + \frac{\Delta^{[l+1]} * \varphi[\beta]}{(l+1)!} R\left(y \mid [\gamma-\beta]^{[l+1]}\right).$$

Estimate the derivatives of each summand separately.

The derivative of $Q(y \mid \beta]$ of order $k \leq l$ is a linear combination of the differences $\Delta^{[k]} * \varphi[\gamma]$ at the points γ with $[\gamma, \gamma+k] \subset [\beta, \beta+l]$. Consequently,

$$\max_{\beta \leq y \leq \beta+1} |Q^{(k)}(y \mid \beta]|$$
$$\leq K \max_{[\gamma,\gamma+k] \subset [\beta,\beta+l]} |\Delta^{[k]} * \varphi[\gamma]|.$$

The kth derivatives of $R(y \mid [\gamma-\beta]^{[l+1]})$ are obviously bounded on the interval $\beta \leq y \leq \beta+1$. Moreover, $\Delta^{[l+1]} * \varphi[\gamma]$ at $\gamma = \beta$ is also not greater than

$$K \max_{[\gamma,\gamma+k] \subset [\beta,\beta+l+1]} |\Delta^{[k]} * \varphi[\gamma]|.$$

Finally, we come to the estimate

$$\max_{\beta \leq y \leq \beta+1} |R^{(k)}(y \mid \beta]|$$

$$\leq K \max_{[\gamma,\gamma+k] \subset [\beta,\beta+l+1]} |\Delta^{[k]} * \varphi[\gamma]|,$$

which immediately entails (2.5). In the case of $n > 1$ the estimate (2.5) is readily available by induction.

Theorem 8.8 (of interpolation). *For an arbitrary function $\varphi[\beta]$ in $l_p^{(l)}$, there is an interpolant $\varphi(x)$ belonging to $L_p^{(l)}$ such that*

$$\left\|\varphi(x) \mid L_p^{(l)}\right\| \leq K \left\|\varphi[\beta] \mid l_p^{(l)}\right\| \tag{2.6}$$

with K a constant independent of $\varphi[\beta]$.

PROOF consists in checking that the Ryaben'kiĭ operator implements the desired interpolation. We are to show that the function $R(y \mid \varphi[\beta])$ is a member of $L_p^{(l)}$ and satisfies (2.6).

Estimate the $L_p^{(l)}$ norm of $R(y \mid \varphi)$. Assume given β and y such that $\beta \le y \le \beta + 1$. By the property (2.5) of the Ryaben'kiĭ operator for $|\alpha| = l$, the estimate is valid

$$|D^\alpha R(y \mid \varphi)| \le K \max_{[\gamma,\gamma+\alpha]\subset[\beta,\beta+l+1]} |\Delta^{[\alpha]} * \varphi[\gamma]|,$$

for $|\alpha| = l$, whence it follows that

$$\max_{y\in[\beta,\beta+1]} \left\{ \sum_{|\alpha|=l} \frac{l!}{\alpha!} |D^\alpha R(y \mid \varphi)|^2 \right\}^{p/2}$$

$$\le K^p \left\{ \sum_{|\alpha|=l} \frac{l!}{\alpha!} \max_{[\gamma,\gamma+\alpha]\subset[\beta,\beta+l+2]} |\Delta^{[\alpha]} * \varphi[\gamma]|^2 \right\}^{p/2}$$

$$\le C_l \left\{ \max_{|\alpha|=l} \max_{[\gamma,\gamma+\alpha]\subset[\beta,\beta+l+2]} |\Delta^{[\alpha]} * \varphi[\gamma]| \right\}^p. \tag{2.7}$$

Successively using the two routine inequalities for k nonnegative reals

$$\max(a_1, \dots, a_k) \le \sum_{j=1}^k a_j, \quad \left(\sum_{j=1}^k a_j \right)^p \le k^{p-1}(a_1^p + \dots + a_k^p),$$

conclude that the term in braces on the right side of (2.7) is dominated by the expression

$$C \sum_{|\alpha|=l} \left\{ \sum_{[\gamma,\gamma+\alpha]\subset[\beta,\beta+l+2]} |\Delta^{[\alpha]} * \varphi[\gamma]|^p \right\} \le C \sum_{\beta\le\gamma\le\beta+l+2} \left\{ \sum_{|\alpha|=l} |\Delta^{[\alpha]} * \varphi[\gamma]|^p \right\}.$$

Using the found dominant for the left side of (2.7), come to the following collection of estimates

$$\int_{\beta\le y\le\beta+1} \left\{ \sum_{|\alpha|=l} \frac{l!}{\alpha!} |D^\alpha R(y \mid \varphi)|^2 \right\}^{p/2} dy \le C \sum_{\beta\le\gamma\le\beta+l+2} \left\{ \sum_{|\alpha|=l} |\Delta^{[\alpha]} * \varphi[\gamma]|^p \right\},$$

with C independent of φ and β. Summing these estimates over all $\beta \in \mathbb{Z}^n$ and using the condition that $\varphi[\gamma] \in l_p^{(l)}$, arrive at the relation

$$\|R(y \mid \varphi[\beta]) \mid L_p^{(l)}\|^p \le K \sum_{|\alpha|=l} \sum_{\gamma} |\Delta^{[\alpha]} * \varphi[\gamma]|^p = K \|\varphi[\gamma] \mid l_p^{(l)}\|^p.$$

The proof of Theorem 8.8 is complete.

The members of $l_p^{(l)}$ and $L_p^{(l)}$ are classes of equivalent functions. Moreover, the class of polynomials of degree less than l has norm 0, i.e., is treated as zero of the corresponding space. The members of the space $W_p^{(l)}$ of functions of a continuous argument whose definition is similar to that given in § 4 of Chapter 1 in the case of $p = 2$ are individual functions rather than classes of functions. The space $W_p^{(l)}$ is constructed as the direct sum of the $L_p^{(l)}$ space and the space of polynomials $\mathbf{P}_{l-1}$, whereas the passage from $W_p^{(l)}$ to $L_p^{(l)}$ is accomplished by projection.

The same construction naturally translates to the $l_p^{(l)}$ space. An arbitrary member of the latter is a class of functions $\varphi[\beta]$ whose every two representatives differ by a polynomial of degree at most $l-1$. The new space $w_p^{(l)}$ comprises these elements we individualize. Moreover, each class may be assumed consistent of the elements $\varphi[\beta] + P[\beta]$ with $\varphi[\beta]$ some representative of the class chosen in advance and $P[\beta]$ a polynomial of degree at most $l-1$.

The choice of $P[\beta]$ is some projection π in $w_p^{(l)}$ sending $w_p^{(l)}$ to the space $\mathbf{P}_{l-1}$ comprising polynomials of degree at most $l-1$. The choice of $\varphi[\beta]$ is carried out by the projection $\widehat{\pi} = 1 - \pi$, sending $w_p^{(l)}$ to some subspace s_π isomorphic with $l_p^{(l)}$. Distinguishing s_π amounts thus to indicating the projection π or $\widehat{\pi}$.

We norm $w_p^{(l)}$ in the same manner as the $W_p^{(l)}$ by putting

$$\|\varphi \mid w_p^{(l)}\|^p = \|\pi\varphi \mid \mathbf{P}_{l-1}\|^p + \|\varphi \mid l_p^{(l)}\|^p.$$

As π we may choose, for instance, the operator

$$\pi\varphi = \sum_{|\alpha|=l-1} \left.\frac{\Delta^{[\alpha]} * \varphi}{\alpha!}\right|_{\beta=0} [\beta]^\alpha,$$

and as the norm in the space of polynomials $\mathbf{P}_{l-1}$ we may take

$$\left\| \sum_{|\alpha|\le l-1} a_\alpha[\beta]^\alpha \mid \mathbf{P}_{l-1} \right\| = \sum_{|\alpha|\le l-1} |a_\alpha|.$$

Basing on the definition of $w_p^{(l)}$ and $W_p^{(l)}$ and Theorem 8.7, conclude that the trace of each function in $W_p^{(l)}$ on Γ is a member of $w_p^{(l)}$.

Denote by $\overset{\circ}{w}{}_p^{(l)}$ the closure of the set of compactly-supported functions in $w_p^{(l)}$, and by $\overset{\circ}{l}{}_p^{(l)}$, the closure in $l_p^{(l)}$ of the set of classes each of which contains at least one compactly-supported function. We also need the subspace $\overset{\bullet}{w}{}_p^{(l)}$ that, for $n \le p$, coincides with $w_p^{(l)}$ and, for $n > p$, is defined as the set of functions $\varphi[\beta]$ in $w_p^{(l)}$ such that the difference $\Delta^{[\alpha]} * \varphi[\beta]$ of order α with $[l - n/p] + 1 \leq |\alpha| \leq l - 1$ vanishes as $|\beta| \to \infty$. Brackets stand here for the integral parts of numbers. Observe that in general the vanishing of these differences is nonuniform and understood as follows: the l_p norms of the traces of differences on the faces of the cube $\{\beta : |\beta_j| \leq N, j = 1, \dots, n\}$ vanish as $N \to \infty$.

Alongside the space $\overset{\bullet}{w}{}_p^{(l)}$ we consider an analogous space $\overset{\bullet}{W}{}_p^{(l)}$ of functions in a continuous argument x. By definition, for $n > p$ the members of $\overset{\bullet}{W}{}_p^{(l)}$ are functions of $W_p^{(l)}$ whose derivatives of order α, with $[l - n/p] + 1 \leq |\alpha| \leq l - 1$, vanish as $|x| \to \infty$. Every compactly-supported function in $W_p^{(l)}$ belongs to $\overset{\bullet}{W}{}_p^{(l)}$. Consequently, the $W_p^{(l)}$ closure of the set of compactly-supported functions, denoted by $\overset{\circ}{W}{}_p^{(l)}$, is a subspace of $\overset{\bullet}{W}{}_p^{(l)}$.

Using the above theorems of tracing and interpolation, we may extend to the discrete case the Embedding Theorem for spaces of continuous functions.

Theorem 8.9. *If $n \leq p$ then the spaces $w_p^{(l)}$ and $\overset{\circ}{w}{}_p^{(l)}$ coincide. For $n > p$ the space $w_p^{(l)}$ splits in the direct sum*

$$w_p^{(l)} = \overset{\bullet}{w}{}_p^{(l)} \oplus \mathbf{P}_{l-1}/\mathbf{P}_{[l-n/p]},$$

with $\mathbf{P}_{l-1}/\mathbf{P}_{[l-n/p]}$ standing for the factor-space $\mathbf{P}_{l-1}$ by $\mathbf{P}_{[l-n/p]}$ constructed by means of the projection π: $\mathbf{P}_{l-1} \to \mathbf{P}_{[l-n/p]}$. If $l \leq n/p$ then this factor-space coincides with $\mathbf{P}_{l-1}$.

PROOF. The Ryaben'kiĭ interpolator is a linear mapping from $w_p^{(l)}$ to the range $Rw_p^{(l)}$ which is a linear subspace in $W_p^{(l)}$. This subspace is closed, which is easy to prove by using the estimates of Theorems 8.7 and 8.8 in turn. Moreover, it contains all polynomials of degree less than l and, consequently, all members of $\mathbf{P}_{l-1}/\mathbf{P}_{[l-n/p]}$.

As is known, for $n \leq p$ the space $W_p^{(l)}$ coincides with $\overset{\circ}{W}{}_p^{(l)}$, and for $n > p$ the space $W_p^{(l)}$ may be written as the direct sum

$$W_p^{(l)} = \overset{\bullet}{W}{}_p^{(l)} \oplus \mathbf{P}_{l-1}/\mathbf{P}_{[l-n/p]}. \tag{2.8}$$

The proof of this assertion may be found, for instance, in [265, § 4 of Chapter 3]. We consider the decomposition (2.8) also in the case when $n \leq p$, presuming then that the second summand on the right side is trivial. From (2.8) it follows that $Rw_p^{(l)}$ also splits into the direct sum

$$R[w_p^{(l)}] = R[w_p^{(l)}] \cap \overset{\bullet}{W}{}_p^{(l)} \oplus \mathbf{P}_{l-1}/\mathbf{P}_{[l-n/p]}.$$

Injectivity of the mapping given by the interpolator means that the inverse image also splits

$$w_p^{(l)} = \overset{\bullet}{w}{}_p^{(l)} \oplus \mathbf{P}_{l-1}/\mathbf{P}_{[l-n/p]}.$$

The proof of Theorem 8.9 is complete.

***Theorem* 8.10** (of denseness of compactly-supported functions). *The compactly-supported functions of a discrete variable are dense in $l_p^{(l)}$. In other words, the following equality holds*

$$l_p^{(l)} = \overset{\circ}{l}{}_p^{(l)}.$$

PROOF. Take a discrete function $\varphi[\beta]$, a member of $l_p^{(l)}$. Then by Theorem 8.8 the interpolant $R(y \mid \varphi)$ belongs to $L_p^{(l)}$. As is known, the compactly-supported functions are dense in $L_p^{(l)}$. The justification of this claim may be found, for instance, in [265, §4 of Chapter 3]. Consequently, for every $\varepsilon > 0$ there is a compactly-supported function $\varphi_\varepsilon(y) \in L_p^{(l)}$ such that

$$\|R(y \mid \varphi) - \varphi_\varepsilon(y) \mid L_p^{(l)}\| \le \varepsilon.$$

The trace $\varphi[\beta]$ of $\varphi_\varepsilon(y)$ on the lattice Γ belongs to $l_p^{(l)}$ by Theorem 8.7. Moreover, the following estimate is valid

$$\|\varphi[\beta] - \varphi_\varepsilon[\beta] \mid l_p^{(l)}\| \le K\|R(y \mid \varphi) - \varphi_\varepsilon(y) \mid L_p^{(l)}\|.$$

This entails the sought equality on observing that $\varphi_\varepsilon[\beta]$ has compact support. The proof of Theorem 8.10 is complete.

§3. The Fourier Transform of a Discrete Function

We agree to consider all spaces of functions on $\mathbb{R}^n$ embedded in $\mathcal{S}'$, the dual of the well-known *Schwartz space* $\mathcal{S}$. The latter was introduced by L. Schwartz and comprises *tempered functions*, i.e., infinitely differentiable functions on $\mathbb{R}^n$ decreasing at infinity with all derivatives of every order faster than each negative power of the argument. We assume the reader to know the properties of the *Fourier transform* defined on the functions in $\mathcal{S}$ by the formula

$$(Ff)(\xi) = \tilde{f}(\xi) = \int f(x)e^{i2\pi x\xi}\,dx.$$

The inverse transform is given by equality

$$(F^{-1}\tilde{f})(x) = f(x) = \int \tilde{f}(\xi)e^{-i2\pi x\xi}\,d\xi.$$

The Fourier transform is an automorphism of $\mathscr{S}$ preserving the inner product

$$(f,g) = \int f(x)\overline{g(x)}\,dx = \int \tilde{f}(\xi)\overline{\tilde{g}(\xi)}\,d\xi = (\tilde{f},\tilde{g}).$$

At the elements l of the dual $\mathscr{S}'$ of $\mathscr{S}$ the Fourier transform $F(l) = \tilde{l}$ is defined by the equality

$$(F(l),\varphi) = (l, F^{-1}(\varphi)), \quad \varphi \in \mathscr{S}.$$

It is well known that $E(x) = e^{-|x|^2/2} \in \mathscr{S}$ and $\Phi_0(x) = \sum \delta(x-\gamma) \in \mathscr{S}'$ are invariant functions under the Fourier transform, namely,

$$\widetilde{E}(\xi) = e^{-|\xi|^2/2}, \quad \widetilde{\Phi}_0(\xi) = \Phi_0(\xi).$$

Enrich the collection of the function spaces under study by including various topological linear spaces. Introduce the relevant notations.

Let R be the set of all discrete functions, and let P be the set of *harrow-like functions*. So we call a generalized function of the shape

$$f(x) = \sum \varphi[\beta]\delta(x-\beta). \tag{3.1}$$

If the coefficients $\varphi[\beta]$ of a harrow-like function $f(x)$ grow at infinity not faster than a power of its argument then $f \in \mathscr{S}'$. We denote the operator in (3.1) by I_{RP} or by I_{PR} in accord with the implied direction of its action, namely,

$$R \underset{I_{PR}}{\overset{I_{RP}}{\rightleftarrows}} P.$$

A smooth function $\varphi(x)$ becomes a harrow-like function after multiplication by $\Phi_0(x)$, namely,

$$\varphi(x)\Phi_0(x) = \varphi(x)\sum_\beta \delta(x-\beta) = \sum_\beta \varphi[\beta]\delta(x-\beta).$$

Conversely, given a harrow-like function (3.1) and using the Ryaben'kiĭ interpolator, we may construct a smooth function $\varphi(x)$ whose trace on the lattice agrees with the coefficients $\varphi[\beta]$.

Let $\mu(x)$ be some multidimensional averaging kernel [265] possessing the additional properties $\mu \in \mathscr{S}$, $\mu(0) = 1$, and $\mu(\gamma) = 0$ for $\gamma \in \mathbb{Z}^n$, $\gamma \neq 0$. The convolution operator

$$\mu* : \sum_\beta \varphi[\beta]\delta(x-\beta) \longmapsto \sum_\beta \varphi[\beta]\mu(x-\beta)$$

transforms a harrow-like function whose coefficients grow polynomially into an infinitely differentiable function.

We denote by Π the set of periodic functions with the fundamental parallelepiped $Q = \{x : 0 \leq x_j < 1, j = 1, \dots, n\}$ and by Π^A, the set of periodic functions with the fundamental parallelepiped AQ with A some real nonsingular matrix. Distinguish periodic discrete functions with A an integer matrix; denote the set of them by $\Pi^A \cap R$. Hence, for $A = I$ the spaces Π^A and Π coincide.

In the space of variables $y \in \mathbb{R}^n$ we identify all points that differ by an integer vector, setting

$$t_j = \{y_j\} = y_j - [y_j], \quad t = V_0(y), \tag{3.2}$$

with $\{\cdot\}$ the taking of the fractional part of a number. The resulting manifold, the *torus* $\theta^n = \{t : 0 \leq t_j < 1, j = \overline{1,n}\}$, may be considered as the n-dimensional cube with identified opposite faces. Thus, the formulas (3.2) determine an infinite-to-one mapping V_0 of $\mathbb{R}^n$ onto the torus θ^n.

The torus θ^n may be transformed into a simply-connected domain Ω_0 by means of cuts. The resulting domain is called *fundamental.* We may write the condition that Ω_0 is a fundamental domain as

$$\sum_\beta \chi_{\Omega_0}(y - \beta) = 1,$$

with $\chi_{\Omega_0}(y)$ the indicator of Ω_0. We restrict our consideration to the cubic domain $\Omega_0 = Q$.

Let $\varphi(y)$ be a periodic function with period the identity matrix, i.e., $\varphi \in \Pi$. The formula $\psi(t) = \varphi(y)$, with $y \in V_0^{-1}(t)$, defines a single-valued function on the torus θ^n, since the image of $\mathbb{R}^n$ under the mapping V_0 is the whole torus θ^n, and at every point of the inverse image of t the function $\varphi(y)$ takes the same value. Under this mapping Π becomes the set T of functions on θ^n. The transformation of Π to T may be given by restricting the domain of definition of $\varphi(y)$ to the cube Q identified with the torus θ^n. Denote the corresponding operator by $\partial_Q : \Pi \to T$.

The convolution of $\Phi_0(x)$ and $\varphi(x)$ in $\mathscr{S}$, defined as

$$(\Phi_0 * \varphi)(x) = \sum_\beta \varphi(x - \beta),$$

is a periodic infinitely differentiable function of the class $\Pi \cap C^\infty$. Clearly, the convolution of φ in $\mathscr{S}$ with the functional $\Phi_0(A^{-1}x)/\det A$ possesses analogous properties for the elements of Π^A.

One of the right inverses of $\Phi_0 *$ is the operator of multiplication by a function $\lambda \in \mathscr{S}$ with the property $\sum_\beta \lambda(x-\beta) = 1$. Indeed, for every $\varphi \in \Pi \cap C^\infty$ we have

$$(\Phi_0 * (\lambda\varphi))(x) = \sum_\beta \varphi(x-\beta)\lambda(x-\beta)$$
$$= \sum_\beta \varphi(x)\lambda(x-\beta) = \varphi(x)\sum_\beta \lambda(x-\beta) = \varphi(x),$$

i.e., the operator of multiplication by $\lambda(y)$ is actually a right inverse to the operator of convolution with $\Phi_0(x)$. As a smooth factor $\lambda(x)$ we may for instance take the function

$$\lambda_n(x) = \int\limits_{Q-\mathbf{e}/2} e^{-\pi|x-y|^2}\, dy,$$

with $\mathbf{e} = (1, \dots, 1)$. Indeed, the following equality holds

$$\sum_\beta \lambda_n(x-\beta) = \prod_{j=1}^{n} \sum_{\beta_j=-\infty}^{\infty} \lambda_1(x_j - \beta_j).$$

The expression on the right side equals 1. This is easy to check by using the following equalities

$$\sum_{\beta_j=-\infty}^{\infty} \lambda_1(x_j-\beta_j) = \sum_{\beta_j=-\infty}^{+\infty} \int\limits_{-1/2}^{1/2} e^{-\pi(x_j-y_j-\beta_j)^2}\, dy_j$$
$$= \sum_{\beta_j=-\infty}^{+\infty} \int\limits_{\beta_j-1/2}^{\beta_j+1/2} e^{-\pi(x_j-\tau)^2}\, d\tau = \int e^{-\pi\tau^2}\, d\tau = 1$$

in summing the values of the function $\lambda_1(\tau)$ in a single variable.

Apply the Fourier transform to generalized functions in $P \cap \mathscr{S}'$. If $f \in P \cap \mathscr{S}'$ then $Ff \in \mathscr{S}'$, and

$$(Ff)(\xi) = \left(\sum_\beta \varphi[\beta]\delta(x-\beta), e^{i2\pi x\xi}\right) = \sum_\beta \varphi[\beta]e^{i2\pi\beta\xi},$$

i.e., $\tilde{f}(\xi)$ is a periodic function in $\Pi \cap \mathscr{S}'$.

We reveal interrelations between the above spaces with the diagram

$$\begin{array}{ccccc} R & \underset{I_{PR}}{\overset{I_{RP}}{\rightleftarrows}} & P \cap \mathscr{S}' & \underset{\Phi_0 \cdot}{\overset{(\Phi_0 \cdot)^{-1}_{pr} = \mu *}{\rightleftarrows}} & C^{(\infty)} \cap \mathscr{S}' \\ F_1 \updownarrows F_1^{-1} & & F^{-1} \updownarrows F & & F \updownarrows F^{-1} \\ T & \underset{\partial_Q}{\overset{(\partial_Q)^{-1}}{\rightleftarrows}} & \Pi \cap \mathscr{S}' & \underset{\Phi_0 *}{\overset{(\Phi_0 *)^{-1}_{pr} = \lambda \cdot}{\rightleftarrows}} & \Phi = F^{-1}(C^{(\infty)}) \cap \mathscr{S}'. \end{array}$$

Here the function $\mu(x)$ is a multidimensional averaging kernel; $\lambda(x)$ is the Fourier transform of $\mu(x)$. The operators F_1 and F_1^{-1}, defined by the relations

$$F_1 = I_{PR} \circ F^{-1} \circ \partial_Q^{-1}, \quad F_1^{-1} = \partial_Q \circ F \circ I_{RP}$$

connect an ordinary Fourier series and the sequence of its coefficients. Check this.

Given an arbitrary function φ in $\Pi \cap \mathscr{S}$, observe

$$(F^{-1}\varphi)(\xi) = \sum_\beta \int\limits_{[\beta,\beta+1]} \varphi(x) e^{-i2\pi x\xi}\, dx$$

$$= \sum_\beta \int\limits_Q \varphi(y+\beta) e^{-i2\pi y\xi}\, dy\, e^{-i2\pi\beta\xi} = \left\{ \sum_\beta e^{-i2\pi\beta\xi} \right\} \int\limits_Q \varphi(y) e^{-i2\pi y\xi}\, dy$$

$$= (F^{-1}\Phi_0)(\xi) \int\limits_Q \varphi(y) e^{-i2\pi y\xi}\, dy = \sum_\beta \int\limits_Q \varphi(y) e^{i2\pi y\xi}\, dy\, \delta(\xi - \beta).$$

Whence and from the definition of F_1 it follows that F_1 indeed assigns to $\varphi(\xi)$ in T the sequence of the Fourier coefficients of $\varphi(\xi)$. Further, if $\varphi[\beta] \in R$ then

$$F_1^{-1}(\varphi[\beta]) = \varphi(x) = \sum_\beta \varphi[\beta] e^{i2\pi\beta x}, \quad x \in Q.$$

Recall the familiar ***unitarity property*** of the operator F considered on $L_2(Q)$, i.e., the ***Parseval identity***

$$\int\limits_Q |\varphi(x)|^2\, dx = \sum_\beta |\varphi[\beta]|^2.$$

Denote by $\Pi^A \cap R$ the class of periodic discrete functions with integer period matrix A. The harrow-like functions with periodic coefficients $\varphi[\beta]$ comprise the space $C\Pi \cap P$. Denote by P^A the class of functions of the shape

$$\sum \varphi[\beta] \delta(x - A\beta).$$

Finally, denote by $C\Pi^B \cap P^A$ the set of all lattice functions in P^A with B-periodic coefficients.

Define an analog of Fourier transform on the space of discrete functions $\Pi^B \cap R$. Here B is a nonsingular matrix with integer coefficients. Let A be also a nonsingular matrix. Associate with a function $\varphi[\beta]$ in $\Pi^B \cap R$ the generalized function in $\mathscr{S}'$ that is defined by the equality

$$f(x) = \sum_\beta \varphi[\beta]\delta(x - A\beta). \tag{3.3}$$

Denote the operator (3.3) by I^A_{RP} and the inverse of it by I^A_{PR}. We display interrelations of the spaces under consideration on the following diagram

$$\begin{array}{ccc} \Pi^B \cap R & \longrightarrow & C\Pi^B \cap P^A \\ {\scriptstyle F_2}\downarrow & & \downarrow{\scriptstyle F} \\ \Pi^{B^*} \cap R & \longleftarrow & C\Pi^{B^*} \cap P^{(AB)^{*-1}}. \end{array}$$

Some explanations are now in order. For a start, prove the formula

$$F(C\Pi^B \cap P^A) = C\Pi^{B^*} \cap P^{(AB)^{*-1}}, \tag{3.4}$$

with F the Fourier transform on $\mathscr{S}'$.

Take $\varphi[\beta] \in \Pi^B \cap R$, and let the $f(x)$ be of the form (3.3). Then the equality holds

$$\tilde{f}(\xi) = \sum_\beta \varphi[\beta] e^{i2\pi A\beta\xi}. \tag{3.5}$$

Every vector $\beta \in \mathbb{Z}^n$ may be decomposed in the sum $\beta = \alpha + B\gamma$, with $\alpha, \gamma \in \mathbb{Z}^n$ and $B^{-1}\alpha \in Q$. Agree to denote the set of these α by BQ. By hypothesis, $\varphi[\beta] = \varphi[\alpha]$. With this in mind, transform (3.5) to the equivalent shape

$$\tilde{f}(\xi) = \left\{ \sum_{\alpha \in BQ} \varphi[\alpha] e^{2\pi i A\alpha\xi} \right\} \left\{ \sum_\gamma e^{i2\pi\gamma(AB)^*\xi} \right\}. \tag{3.6}$$

The second factor, as was mentioned, agrees with the function

$$\sum_\gamma \delta((AB)^*\xi - \gamma).$$

Considering this, continue (3.6) to obtain

$$\tilde{f}(\xi) = \sum_\gamma \frac{1}{\det AB} \left\{ \sum_{\alpha \in BQ} \varphi[\alpha] e^{i2\pi\alpha B^{*-1}\gamma} \right\} \delta(\xi - (AB)^{*-1}\gamma).$$

Consequently, $\tilde{f}(\xi) \in C\Pi^{B^*} \cap P^{(AB)^{*-1}}$. The reverse embedding is established similarly. The formula (3.4) is proven.

Thus, the composition $F_2 = I_{PR}^{(AB)^{*-1}} \circ F \circ I_{RP}^{A}$ of the three already-defined mappings transforms $\varphi[\beta] \in \Pi^B \cap R$ to the function

$$\psi[\beta] = (\det AB)^{-1} \sum_{\alpha \in BQ} \varphi[\alpha] e^{i2\pi\alpha(B^*)^{-1}\beta} \in \Pi^{B^*} \cap R. \tag{3.7}$$

This is a natural analog of the Fourier transform for discrete functions. The inverse operator

$$F_2^{-1} = I_{PR}^{A} \circ F^{-1} \circ I_{RP}^{(AB)^{*-1}}$$

transforms $\psi[\beta]$ to $\varphi[\beta]$ by the formula

$$\varphi[\beta] = (F_2^{-1}\psi)[\beta] = \det A \sum_{\alpha \in B^* Q} \psi[\alpha] e^{-i2\pi\alpha B^{-1}\beta}. \tag{3.8}$$

This equality results from (3.3) by replacing A with $(AB)^{*-1}$, B with B^*, and $+i$ with $-i$.

The *discrete Fourier transform* of use in number theory is the particular case of the transform F_2 defined above which results from the general formula for $A = B^{-1}$ and $B = NI$ with natural N. The formulas (3.7) and (3.8) in this specification take the form

$$\psi[\beta] = \sum_{0 \leq \alpha < N\mathbf{e}} \varphi[\alpha] e^{2\pi i\alpha\beta/N}, \quad \varphi[\beta] = N^{-n} \sum_{0 \leq \alpha < N\mathbf{e}} \psi[\alpha] e^{-i2\pi\alpha\beta/N},$$

and the Parseval identity is written down as

$$\sum_{0 \leq \beta < N\mathbf{e}} |\varphi[\beta]|^2 = N^{-2n} \sum_{0 \leq \beta < N\mathbf{e}} \sum_{0 \leq \alpha < N\mathbf{e}} \sum_{0 \leq \gamma < N\mathbf{e}} \psi[\alpha]\overline{\psi[\gamma]} e^{i2\pi(\gamma-\alpha)\beta/N}$$
$$= N^{-2n} \sum_{\alpha} \sum_{\gamma} \psi[\alpha]\overline{\psi[\gamma]} \sum_{\beta} e^{i2\pi(\gamma-\alpha)\beta/N} = N^{-n} \sum_{\alpha} |\psi[\alpha]|^2.$$

Here, $\mathbf{e}$ stands again for the vector in $\mathbb{Z}^n$ with every coordinate 1.

Chapter 9

Optimal Formulas

From the shape of the $L_2^{(m)*}$ norm of the error of a lattice cubature formula it follows that, for a fixed lattice, the norm square of the error, presenting a quadratic function in the weights $c[\beta]$, assumes a unique minimum at a specific value $c[\beta] = c_0[\beta]$. The formula with the weights $c_0[\beta]$ corresponding to this minimum is optimal.

In the current chapter we study optimal formulas with a given lattice of nodes. We show that the problem of optimizing the weights serves as an analog of the problem of finding an extension to $\mathbb{R}^n$ of some function $f(x)$ given in a domain Ω so that the norm of the extension be a minimum.

§1. Statement of the Problem of Optimal Weights

The equations for the weights of an optimal cubature formula with given nodes were derived in § 6 of Chapter 1. Recall the corresponding system

$$\sum_{\gamma} c[\gamma] G_{hH}^{(m)}[\beta - \gamma] + \sum_{|\alpha|<m} \nu_\alpha (hH\beta)^\alpha = f(hH\beta), \quad hH\beta \in \overline{\Omega}, \tag{1.1}$$

$$c[\gamma] = 0, \quad hH\beta \notin \overline{\Omega}, \tag{1.2}$$

$$h^n \sum_{\gamma} c[\gamma](hH\gamma)^\alpha = \int_{\Omega} y^\alpha \, dy, \quad |\alpha| < m. \tag{1.3}$$

Here $G_{hH}^{(m)}[\beta] = h^n G(hH\beta)$, the function $(-1)^m G(x)$ is the fundamental solution to the polyharmonic equation, H is an orthogonal matrix with $\det H = 1$, and

$$f(y) = \int_{\Omega} G(y - x) \, dx.$$

Similar equations in the continuous case where integrals replace sums are called the *Wiener–Hopf equations.*

The equations of (1.1)–(1.3) look better in convolution form

$$G_{hH}^{(m)}[\beta] * c[\beta] + P_{m-1}[\beta] = f[\beta], \quad hH\beta \in \overline{\Omega}, \tag{1.4}$$

$$c[\beta] = 0, \quad hH\beta \notin \overline{\Omega}, \tag{1.5}$$

$$h^n \sum_{\gamma} c[\gamma](hH\gamma)^\alpha = f_\alpha, \qquad |\alpha| < m. \tag{1.6}$$

Here $P_{m-1}[\beta]$ is a polynomial of degree at most $m-1$, and

$$f_\alpha = \int_\Omega y^\alpha \, dy, \quad f[\beta] = \int_\Omega G(hH\beta - y)\, dy.$$

The system of equations (1.4)–(1.6) is further called *System* B. Consider the corresponding problem.

PROBLEM B_1. *Given $f[\beta]$ and f_α, find a function $c[\beta]$ and a polynomial $P_{m-1}[\beta]$ satisfying System* B.

We call the first summand of (1.4) the ***discrete polyharmonic potential.*** This potential presents the values of the function

$$w(x) = h^n \sum_{\gamma} c[\gamma] G(x - hH\gamma) \tag{1.7}$$

at the points $x = hH\beta$. Introduce the harrow-like generalized function

$$\overline{c}(x) = h^n \sum_{\gamma} c[\gamma]\delta(x - hH\gamma). \tag{1.8}$$

Using it, write down $w(x)$ as a convolution

$$w(x) = \overline{c}(x) * G(x). \tag{1.9}$$

Thus, the discrete potential $w[\beta]$ coincides with the nodal values of the conventional polyharmonic potential with density $\overline{c}(x)$ corresponding to the weights $c[\gamma]$ by (1.8).

The representation (1.9) entails an important conclusion about the behavior at infinity of the discrete potential with compactly-supported density. Obviously, the latter admits exactly the same series expansion in polyharmonic summands in a neighborhood about the point at infinity as the ordinary polyharmonic potential. The contrast is as follows. All values of relevant functions participate in the expansion of the potential of a continuous argument, only the values of the same functions at the nodes participate in the expansion of the discrete potential.

Alongside Problem B_1, it stands to reason to consider another problem that serves as a continuous analog of the original problem.

PROBLEM A_1. *Find a function $c(x)$ of a continuous argument $x \in \mathbb{R}^n$ and a polynomial $P_{m-1}(x)$ of degree less than m satisfying the system of equations:*

$$G(x) * c(x) + P_{m-1}(x) = f(x) \quad \text{for} \quad x \in \overline{\Omega}, \tag{1.10}$$

$$c(x) = 0 \quad \text{for} \quad x \notin \overline{\Omega}, \tag{1.11}$$

$$\int c(x)x^\alpha \, dx = f_\alpha, \quad |\alpha| < m. \tag{1.12}$$

We call the system (1.10)–(1.12) *System* A. Problem A_1 is the *Wiener-Hopf problem.* Study of Problem A_1 enables us to find a clue to solution of Problem B_1 and to indicate some quantitative properties of concrete solutions.

For simplicity, assume that $f(x)$ is continuously differentiable $2m$ times and Ω is a domain with piecewise-smooth boundary. Equation (1.10) is an equation in convolutions of the first kind. Moreover, Problem A_1 requires seeking for a solution to this equation which equals 0 outside Ω and possesses given moments of order less than m.

In general, Problem A_1, as we intend to demonstrate, has no continuous or even summable solution. We establish, however, that there is a unique generalized solution to this problem serving as the derivatives of order $2m$ of some function $w \in W_2^{(m)}$.

We first establish the absence of a bounded solution in the general case. Take the left side of (1.10) as a new indeterminate

$$u(x) = G(x) * c(x) + P_{m-1}(x),$$

and designate $v(x) = G(x) * c(x)$.

In the domain Ω the function $u(x)$ assumes the given values. Everywhere it differs by $P_{m-1}(x)$ from the function $v(x)$ serving as the polyharmonic potential for the density $c(x)$ vanishing outside Ω. Consequently, $v(x)$, together with $u(x)$, is polyharmonic in the exterior of Ω. The problem of finding $u(x)$ is called *Problem* A_2.

Were there a bounded function $c(x)$ satisfying System A then the potential $v(x)$ would be continuous with all derivatives up to order $2m - 1$, which is easy to prove routinely. The function $u(x)$ would be polyharmonic in the exterior of Ω, assuming on $\partial\Omega$ the prescribed values together with all derivatives up to order $2m - 1$. However, the polyharmonic equation does not possess such solutions in general. Consequently, a solution $c(x)$ to Problem A_1 is generally unbounded.

Show now the solvability of Problem A_1 in generalized functions. Assume first that all f_α for $|\alpha| < m$ equal 0. The algorithm we suggest below enables us to write the inverse operator corresponding to System A as the superposition of the polyharmonic operator Δ^m and some operator of extension of $f(x)$ to the whole of $\mathbb{R}^n$.

Suppose that System A is solved. Using $c(x)$, construct the polyharmonic potential $v(x) = G(x) * c(x)$ and consider its behavior in a neighborhood about the point at infinity. To this end, use the routine procedure. Let $|y| < A$ and $|x| > 2A$. Then, expanding $G(x-y)$ in the Taylor series in the variable y, infer

$$v(x) = \int G(x-y)c(y)\,dy$$

$$= \sum_\alpha \frac{(-1)^\alpha}{\alpha!} D^\alpha G(x) \int c(y)y^\alpha\,dy = z(x) + w(x), \tag{1.13}$$

with

$$z(x) = \sum_{|\alpha|<m} \frac{(-1)^\alpha}{\alpha!} D^\alpha G(x) \int c(y)y^\alpha\,dy, \tag{1.14}$$

$$w(x) = \sum_{|\alpha|\geq m} \frac{(-1)^\alpha}{\alpha!} D^\alpha G(x) \int c(y)y^\alpha\,dy. \tag{1.15}$$

Each summand of the series (1.13)–(1.15) is a polyharmonic function in the neighborhood about the point at infinity which we consider. Since $c(y)$ is compactly-supported, the series (1.13) and (1.15) converge absolutely and uniformly at A sufficiently large.

Say that the function $z(x)$ is the sum of finitely many "irrelevant" terms of Problem A_2, and $w(x)$ contains only "relevant" terms. By supposition $f_\alpha = 0$ for $|\alpha| < m$ and, hence, $z(x)$ is identically 0. Thus, the expansion (1.13) of $v(x)$ consists only of relevant terms of Problem A_2. In this event, by Theorem 5.1 the lth order derivatives of $v(x)$ grow at $|x|$ sufficiently large not faster than $|x|^{m-n-l}$ in the case of n odd or $l > m-n$, whereas the derivatives grow not faster than $|x|^{m-n-l}\log|x|$ for n even, $n \leq m-l$. Thus, the potential $v(x)$ belongs to $W_2^{(m)}$ and, for $|x| > A$, satisfies the following estimate

$$|v(x) - z(x)| \leq C \begin{cases} |x|^{m-n}, & \text{if } n \text{ odd or } n > m, \\ |x|^{m-n}\log|x|, & \text{if } n \text{ even and } n \leq m. \end{cases} \tag{1.16}$$

Now take an arbitrary function $y(x)$ in $W_2^{(m)}$ that differs from $f(x)$ in Ω by a polynomial of degree $m-1$.

Assume moreover that $y(x)$ satisfies (1.16) in a neighborhood about the point at infinity and does not agree with $v(x)$. Prove that

$$\|y \mid L_2^{(m)}\| > \|v \mid L_2^{(m)}\|. \tag{1.17}$$

Observe that in Ω the functions $y(x)$ and $v(x)$ differ by another polynomial $Q_{m-1}(x)$ of degree at most $m-1$. Put $u(x) = v(x) + Q_{m-1}(x)$ and compare the norms $\|y \mid L_2^{(m)}\|$ and $\|u \mid L_2^{(m)}\|$.

Arrange the family of functions $u_\lambda(x) = u(x) + \lambda(y(x) - v(x))$. It is clear that each member of the family belongs to $W_2^{(m)}$. Consider the square of the $L_2^{(m)}$ norm of u_λ, i.e.,

$$\|u_\lambda \mid L_2^{(m)}\|^2 = \|u \mid L_2^{(m)}\|^2 + 2\lambda(u, y-u)_m + \lambda^2\|y - u \mid L_2^{(m)}\|^2.$$

The integral defining the inner product $(u, y-u)_m$ may be spread not only over the whole space but also over the complement Ω_* of Ω to $\mathbb{R}^n$, namely,

$$(u, y-u)_m = \int\limits_{\Omega_*} \sum_{|\alpha|=m} \frac{m!}{\alpha!} D^\alpha u D^\alpha (y-u)\, dx. \tag{1.18}$$

This equality is valid, since the functions $y(x)$ and $u(x)$ agree in Ω. Applying Green's identity to the right side of (1.18) and considering that all derivatives of the difference $y - u$ up to order $m-1$ equal 0 on the boundary of Ω and the function $u(x)$ is polyharmonic in the interior of Ω_*, conclude

$$(u, y-u)_m = (-1)^m \int\limits_{\Omega_*} (y-u)\Delta^m u\, dx = 0.$$

Thus, the functions $u(x)$ and $(y-u)(x)$ are orthogonal in $L_2^{(m)}$. Consequently, on the real axis of λ the only stationary point of $\|u_\lambda \mid L_2^{(m)}\|^2$ is zero. The second derivative with respect to λ of $\|u_\lambda \mid L_2^{(m)}\|^2$ at zero is strictly positive. Consequently, $\lambda = 0$ is a minimum point. In particular,

$$\|u_0 \mid L_2^{(m)}\|^2 = \|u \mid L_2^{(m)}\|^2 < \|u_1 \mid L_2^{(m)}\|^2 = \|y \mid L_2^{(m)}\|^2.$$

The sought inequality (1.17) is now available and so the polyharmonic potential $v(x)$ is an extension in $L_2^{(m)}$ of the function $f(x)$ to the whole of $\mathbb{R}^n$ which possesses a minimal norm among the norms of all extensions of the class under consideration.

The problem of extending $f(x)$ may be posed irrespectively of System A. Let us carry this out.

PROBLEM T_c. *Find a function $v(x)$ in $W_2^{(m)}$ different from $f(x)$ in Ω by a polynomial of degree $m-1$, polyharmonic in the exterior of Ω, and satisfying* (1.16) *in a neighborhood about the point at infinity.*

It is not hard to prove by the variational method that there is a unique solution to this problem.

Assume $v(x)$ to be a solution to the extension problem for $f(x)$. Then we may find the corresponding polynomial $P_{m-1}(x)$ from the condition of smoothness of $v(x)$ and its derivatives up to order $m-1$ at the points of the boundary of Ω. After that, we may put

$$c(x) = (-1)^m \Delta^m v(x) = (-1)^m \Delta^m u(x). \tag{1.19}$$

The generalized function $c(x)$ is automatically orthogonal to every polynomial of degree less than m, since $c(x)$ is compactly-supported and the potential $v(x) = G(x) * c(x)$ has no irrelevant terms in the expansion (1.13) near the point at infinity. Thus, the solvability of System A in generalized functions is proven.

The general case in which the numbers f_α in System A are other than 0 reduces to that just considered by a standard change of variables. Let $\overline{c}(x)$ be defined by (1.8), and let the weights $c[\gamma]$ satisfy (1.2) and (1.3). Then the difference $c_*(x) = c(x) - \overline{c}(x)$ solves System A with f_α zero for $|\alpha| < m$ and the first equation altered. The alteration consists only in the fact that the potential $G(x) * \overline{c}(x)$ is subtracted from $f(x)$.

Let $v(x)$ be a solution to the extension problem. Assume further that $u(x)$ results from summing $v(x)$ and the polynomial $P_{m-1}(x)$. Calculate $\Delta^m u(x)$ in the sense of generalized functions. Let $\varphi(x)$ be compactly-supported and infinitely differentiable. By definition, $\Delta^m u(x)$ satisfies the integral identity

$$\begin{aligned}\int \Delta^m u(x)\varphi(x)\,dx &= \int u(x)\Delta^m \varphi(x)\,dx \\ &= \int_{\Omega} u(x)\Delta^m \varphi(x)\,dx + \int_{\mathbb{R}^n \setminus \Omega} u(x)\Delta^m \varphi(x)\,dx.\end{aligned}$$

Integrating by parts in the classical sense each of the expressions on the rightmost side, transform the identity as follows

$$(\Delta^m u(x), \varphi(x))$$

$$= \int_{\Omega} \varphi(x)\Delta^m u(x)\,dx + \int_{\mathbb{R}^n \setminus \Omega} \varphi(x)\Delta^m u(x)\,dx + \int_{\partial\Omega} \sum_{j=0}^{m-1} \chi_j(x) \frac{\partial^j \varphi}{\partial \nu^j}(x)\,dx.$$

Here ν is an outer normal to $\partial\Omega$ and $\chi_j(x)$ are expressed through the jumps of the corresponding derivatives of $u(x)$ which arise in crossing the boundary $\partial\Omega$. The right side of the last equality is the definition of the left side. Observe also that in our case

$$\Delta^m u(x) = \overline{\Delta^m f(x)} = \begin{cases} \Delta^m f(x), & \text{if } x \in \Omega, \\ 0, & \text{if } x \notin \Omega. \end{cases}$$

Thus,

$$c(x) = (-1)^m \Delta^m u(x) = (-1)^m \overline{\Delta^m f(x)} + \sum_{j=0}^{m-1} (-1)^m \rho_j(x),$$

with $\rho_j(x)$ a generalized function supported in the boundary $\partial\Omega$. The inner product (ρ_j, ψ) for $\psi \in \overset{\circ}{C}{}^{(\infty)}(\mathbb{R}^n)$ is defined as

$$(\rho_j(x), \psi(x)) = \int_{\partial\Omega} \chi_j(x) \frac{\partial^j \psi}{\partial \nu^j}(x)\, dx,$$

i.e., as the surface potential with density $\chi_j(x)$.

The results on Problem A_1 imply some heuristic intimation that allows us to forecast the behavior of $c[\gamma]$, a solution to System B. We may expect that $c(x)$ is the limit of the function of a discrete variable $c[\gamma]$ as $h \to 0$. In this event, in the interior of Ω the function $c[\gamma]$ depends "smoothly" on the variable γ and near the boundary it oscillates, approaching the difference operator that generalizes the linear combination the normal derivatives in the expression of the generalized function $\Delta^m u(x)$.

Proceed to actual solution of System B, to which end we develop a method that is analogous to the method for solving System A.

The main idea of this solution consists in changing the unknown function. Namely, instead of $c[\beta]$ we introduce the discrete polyharmonic potential $v[\beta] = G_{hH}^{(m)}[\beta] * c[\beta]$ and the corresponding sum $u[\beta] = v[\beta] + P_{m-1}[\beta]$. The function $v[\beta]$ happens to be a solution to some extension problem which we call *Problem* B_2.

In this setting, we need to express $c[\beta]$ through $v[\beta]$ or $u[\beta]$, i.e., to find an inverse of the discrete potential operator.

§2. The Fundamental Solution to the Convolution Equation

We now turn to constructing and studying the properties of an inverse of convolution with the function $G_{hH}^{(m)}[\beta] = h^n G(hH\beta)$, i.e., the properties of a function of a discrete variable $D_{hH}^{(m)}[\beta]$ which satisfies the equality

$$D_{hH}^{(m)}[\beta] * G_{hH}^{(m)}[\beta] = \delta[\beta] \tag{2.1}$$

and decreases exponentially at infinity

$$|D_{hH}^{(m)}[\beta]| \leq C|\beta|^{1/2}e^{-s|\beta|}. \qquad (2.2)$$

This enables us to express $c[\beta]$ as

$$c[\beta] = D_{hH}^{(m)}[\beta] * v[\beta]$$

and so to translate the problem of determining the weights of optimal formulas to that of seeking for the unknown function $v[\beta]$.

To find an inverse of the discrete potential use the technique of the Fourier transform in $\mathscr{S}'$, with $\mathscr{S}$ the well-known *Schwartz space*.

From the functions of a discrete variable we pass to harrow-like generalized functions of a special shape. Put

$$\vec{D}_{hH}^{(m)}(x) = \sum_{\beta} D_{hH}^{(m)}[\beta]\delta(x - hH\beta), \qquad (2.3)$$

$$\vec{G}_{hH}^{(m)}(x) = \sum_{\beta} G_{hH}^{(m)}[\beta]\delta(x - hH\beta). \qquad (2.4)$$

If a function $\varphi(x)$ belongs to $\mathscr{S}$ then

$$(\vec{D}_{hH}^{(m)}(x), \varphi(x)) = \sum_{\beta} D_{hH}^{(m)}[\beta]\varphi(hH\beta). \qquad (2.5)$$

In view of the exponential decay of $D_{hH}^{(m)}[\beta]$ at infinity, the numerical series (2.5) converges and is continuous in φ with respect to the convergence of $\mathscr{S}$. Thus, the series $\vec{D}_{hH}^{(m)}(x)$ is a member of $\mathscr{S}'$.

The matter with the function $\vec{G}_{hH}^{(m)}(x)$ is more difficult. It is not always possible to define $\vec{G}_{hH}^{(m)}(x)$ by an equality analogous to (2.5) for an arbitrary $\varphi \in \mathscr{S}$, since the coefficient $G_{hH}^{(m)}[\beta]$ may become infinite at $\beta = 0$. Require that a test function $\varphi(x) \in \mathscr{S}$ has the zero derivatives up to order $2m - 1$ at the point $x = 0$. Denote by $\mathscr{S}(0 \mid 2m)$ the corresponding subspace of $\mathscr{S}$. Then, given $\varphi \in \mathscr{S}(0 \mid 2m)$, put

$$(\vec{G}_{hH}^{(m)}(x), \varphi(x)) = \sum_{\beta \neq 0} G_{hH}^{(m)}[\beta]\varphi(hH\beta).$$

It is clear that this equality determines a linear functional on $\mathscr{S}(0 \mid 2m)$ continuous with respect to the convergence of $\mathscr{S}$. This functional admits a continuous linear extension to the whole $\mathscr{S}$. By $\vec{G}_{hH}^{(m)}(x)$ we further mean one of these extensions. Thus, the generalized function $\vec{G}_{hH}^{(m)}(x)$ also belongs to $\mathscr{S}'$.

Alongside the problem (2.1)–(2.3), consider the following: *Find a function $\vec{D}_{hH}^{(m)}(x)$ with exponentially decreasing coefficients which solves the convolution equation*

$$\vec{D}_{hH}^{(m)}(x) * \vec{G}_{hH}^{(m)}(x) = \delta(x). \tag{2.6}$$

The problems (2.1), (2.2), and (2.6) are solvable or not simultaneously, and in the first case their solutions are related by (2.3). Prove this assertion.

Assume that to the solution $D_{hH}^{(m)}[\beta]$ to the problem (2.1) which meets (2.2) there corresponds a harrow-like function $\vec{D}_{hH}^{(m)}(x)$. Then for an arbitrary $\varphi(x) \in \mathscr{S}$ the equality is valid

$$\begin{aligned} \varphi(0) &= \sum_\alpha (D_{hH}^{(m)}[\alpha] * G_{hH}^{(m)}[\alpha])\varphi(hH\alpha) \\ &= \sum_\alpha \Big(\sum_\beta D_{hH}^{(m)}[\beta] G_{hH}^{(m)}[\alpha - \beta] \Big) \varphi(hH\alpha). \end{aligned} \tag{2.7}$$

At infinity the functions $D_{hH}^{(m)}[\beta]$ and $\varphi(hH\beta)$ decrease faster than every negative power of $|\beta|$ whereas $G_{hH}^{(m)}[\beta]$ grows as some power of $|\beta|$. Changing the order of summation over α and β in (2.7), continue (2.7) as follows

$$\begin{aligned} \varphi(0) &= \sum_\beta D_{hH}^{(m)}[\beta] \Big(\sum_\gamma G_{hH}^{(m)}[\gamma] \varphi(hH(\beta + \gamma)) \Big) \\ &= \sum_\beta D_{hH}^{(m)}[\beta] (\vec{G}_{hH}^{(m)}(y), \varphi(hH\beta + y)) \\ &= (\vec{D}_{hH}^{(m)}(x) \vec{G}_{hH}^{(m)}(y), \varphi(x + y)) = (\vec{D}_{hH}^{(m)} * \vec{G}_{hH}^{(m)}, \varphi). \end{aligned}$$

This means that the function $\vec{D}_{hH}^{(m)}(x)$ satisfies (2.6).

The converse assertion is proven similarly.

Applying the Fourier transform to both sides of (2.6), obtain the following equation

$$\Gamma_{hH}^{(m)}(p) \widetilde{\vec{G}_{hH}^{(m)}}(p) = 1. \tag{2.8}$$

If (2.8) has a solution $\Gamma_{hH}^{(m)}(p)$ not only periodic but also analytic, then the inverse Fourier transform sends $\Gamma_{hH}^{(m)}(p)$ to the harrow-like function $\vec{D}_{hH}^{(m)}(x)$ with exponentially decreasing coefficients $D_{hH}^{(m)}[\beta]$. We show this below, and now we construct a desired function $\Gamma_{hH}^{(m)}(p)$ explicitly.

Consider the following series

$$S_{hH}^{(m)}(p) = \frac{(-1)^m}{(2\pi)^{2m}} \sum_{\gamma} \frac{1}{|p - p_\gamma|^{2m}}, \tag{2.9}$$

with

$$p_\gamma = h^{-1} H^{*-1} \gamma, \quad \gamma \in \mathbb{Z}^n. \tag{2.10}$$

Suppose that $m > n/2$, with n the dimension of the space of the variables $p = (p_1, p_2, \ldots, p_n)$. This inequality, as is not hard to validate, guarantees absolute and uniform convergence of the series (2.9) for all points $p \in \mathbb{C}^n$ but those of the shape (2.10). At each of the points (2.10) the sum $S_{hH}^{(m)}(p)$ has a singularity. Put

$$\Gamma_{hH}^{(m)}(p) = 1/S_{hH}^{(m)}(p). \tag{2.11}$$

In particular, if $p = p_\gamma$ then $\Gamma_{hH}^{(m)}(p)$ is simply equal to 0.

***Lemma* 9.1.** *$\Gamma_{hH}^{(m)}(p)$ is a function periodic in p with period matrix $h^{-1}H^{*-1}$, real, and analytic for all $p \in \mathbb{R}^n$. The zeroes of this function are the points* (2.10), *and $(-1)^m \Gamma_{hH}^{(m)}(p)$ is positive at the other real p. In a neighborhood about zero $\Gamma_{hH}^{(m)}(p)$ may be written as*

$$\Gamma_{hH}^{(m)}(p) = (-1)^m (2\pi|p|)^{2m} \Big\{ 1 + \sum_{|\alpha|>0} \frac{b_\alpha}{\alpha!} (hp)^\alpha \Big\}. \tag{2.12}$$

PROOF. Assume first that $p' \in \mathbb{R}^n$ is distinct from every point p_γ. In a neighborhood about such p', the series (2.9) for $(-1)^m S_{hH}^{(m)}(p)$ converges absolutely and uniformly. This follows from the fact that its terms decrease as $|\gamma|^{-2m}$, i.e., faster than $|\gamma|^{-n}$. Since they are positive; therefore, their sum, i.e. the function $(-1)^m S_{hH}^{(m)}(p)$, is positive in a neighborhood about every p. Consequently, $(-1)^m \Gamma_{hH}^{(m)}(p)$ is analytic and positive there.

Addition to p of some summand p_γ entails only transposition of summands in the infinite series expressing $S_{hH}^{(m)}(p)$. This implies the periodicity of $S_{hH}^{(m)}(p)$, and, consequently, that of $\Gamma_{hH}^{(m)}(p)$.

Proceed to inspecting the behavior of $\Gamma_{hH}^{(m)}(p)$ in neighborhoods about the points p_γ. By periodicity, it suffices to consider one of these points, for instance, the coordinate origin. From the expression (2.9) for the function $S_{hH}^{(m)}(p)$ obtain

$$S_{hH}^{(m)}(p) = \frac{(-1)^m}{(2\pi|p|)^{2m}} \Big\{ 1 + |p|^{2m} \sum_{|\gamma|>0} \frac{1}{|p - p_\gamma|^{2m}} \Big\}.$$

In a neighborhood about the coordinate origin each term of the series in braces is an analytic function. Hence,

$$S_{hH}^{(m)}(p) = \frac{(-1)^m}{(2\pi|p|)^{2m}}\left\{1 + \sum_{|\alpha|>0} \frac{c_\alpha}{\alpha!}(hp)^\alpha\right\},$$

with the braces containing an analytic function that assumes 1 at the coordinate origin. This formula together with (2.11) entails (2.12).

The braces on the right side of (2.12) contain an analytic function that assumes 1 at the coordinate origin, i.e., $\Gamma_{hH}^{(m)}(p)$ is an analytic function in a neighborhood about this point. The proof of Lemma 9.1 is complete.

Expand the function $\Gamma_{hH}^{(m)}(p)$ in the Fourier series

$$\Gamma_{hH}^{(m)}(p) = \sum_\beta D_{hH}^{(m)}[\beta]e^{-i2\pi hpH\beta}, \tag{2.13}$$

with p ranging within the fundamental parallelepiped Ω_0 of the matrix $h^{-1}H^{*-1}$. The coefficients of (2.13) are expressed as the following integrals

$$D_{hH}^{(m)}[\beta] = \frac{1}{|\Omega_0|}\int_{\Omega_0} \Gamma_{hH}^{(m)}(p)e^{i2\pi\beta hH^*p}\,dp. \tag{2.14}$$

From the definition (2.11) of $\Gamma_{hH}^{(m)}(p)$ we readily derive the two equalities

$$\Gamma_{hH}^{(m)}(-p) = \Gamma_{hH}^{(m)}(p), \quad \Gamma_{hH}^{(m)}(p) = h^{-2m}\Gamma_H^{(m)}(hp). \tag{2.15}$$

Expanding both sides of each of them in a series of the shape (2.13) and equating the coefficients of the exponentials with the same exponents, obtain the following relations

$$D_{hH}^{(m)}[-\beta] = D_{hH}^{(m)}[\beta], \quad D_{hH}^{(m)}[\beta] = h^{-2m}D_H^{(m)}[\beta]. \tag{2.16}$$

Examine the behavior of $D_H^{(m)}[\beta]$ at infinity. We have

Lemma 9.2. *The function $D_H^{(m)}[\beta]$ decreases exponentially with $|\beta|$ growing i.e., there are positive constants C and s such that for all β the inequality holds*

$$\left|D_H^{(m)}[\beta]\right| \le C|\beta|^{1/2}e^{-s|\beta|}. \tag{2.17}$$

PROOF. Perform the change of variable $p = H^{*-1}q$ for $h = 1$ in (2.14). Then, denoting by $\Gamma^{(m)}(q)$ the function $\Gamma_H^{(m)}(H^{*-1}q)$, obtain

$$D_H^{(m)}[\beta] = \int_Q \Gamma^{(m)}(q)e^{i2\pi\beta q}\,dq. \tag{2.18}$$

As usual, here Q is the unit cube in $\mathbb{R}^n$. The function $\Gamma^{(m)}(q)$ and all its derivatives are periodic with period matrix the identity matrix. Consequently, for every multi-index α the formula holds

$$\int\limits_Q \Gamma^{(m)}(q) D^\alpha(e^{i2\pi\beta q})\, dq = (-1)^{|\alpha|} \int\limits_Q D^\alpha \Gamma^{(m)}(q) e^{i2\pi\beta q}\, dq. \tag{2.19}$$

Let β be a multi-index with nonnegative integer entries. Consider the corresponding differential operator acting on functions in the variable q by the rule

$$P_\beta(D) = \sum_{j=1}^n \beta_j D_j.$$

Here $D_j = \partial/\partial q_j$. It is not hard to see that

$$\begin{gathered} P_\beta(D) e^{i2\pi\beta q} = (i2\pi|\beta|^2) e^{i2\pi\beta q}, \\ P_\beta^k(D) = \left(\sum_{j=1}^n \beta_j D_j\right)^k = \sum_{|\alpha|=k} \frac{k!}{\alpha!} \beta^\alpha D^\alpha. \end{gathered} \tag{2.20}$$

The action of the kth power of $P_\beta(D)$ on the function $\Gamma^{(m)}(q)$ may be estimated in magnitude as follows

$$|P_\beta^k(D)\Gamma^{(m)}(q)| \le \Big(\sum_{|\alpha|=k} \frac{k!}{\alpha!}\beta^{2\alpha}\Big)^{1/2} \Big(\sum_{|\alpha|=k} \frac{k!}{\alpha!} |D^\alpha \Gamma^{(m)}(q)|^2\Big)^{1/2}. \tag{2.21}$$

Using (2.19) and (2.20) successively and applying the Hölder inequality, we readily deduce from (2.18) the following estimate valid for every natural k

$$\begin{aligned} |D_H^{(m)}[\beta]| &= \left(\frac{1}{2\pi|\beta|^2}\right)^k \left|\int\limits_Q e^{i2\pi\beta q} P_\beta^k(D)\Gamma^{(m)}(q)\, dq\right| \\ &\le \frac{1}{(2\pi|\beta|)^k} \left(\int\limits_Q \sum_{|\alpha|=k} \frac{k!}{\alpha!} |D^\alpha \Gamma^{(m)}(q)|^2\, dq\right)^{1/2}. \end{aligned} \tag{2.22}$$

Choose R so that at an arbitrary point $q_0 \in \overline{Q}$ the Taylor series for the function $\Gamma^{(m)}(q)$ converges in the *polydisk*

$$S(q_0, R) = \{q \in \mathbb{C}^n : |q_i - q_i^0| < R\}$$

with the function $\Gamma^{(m)}(q)$ continuous in the closure $\overline{S}(q_0, R)$. The existence of a positive R with this property follows from the analyticity property at the points of $\mathbb{R}^n$ of the periodic function $\Gamma^{(m)}(q)$. Then for an arbitrary point q_0 in Q the Cauchy inequality holds

$$\left|\frac{D^\alpha \Gamma^{(m)}(q_0)}{\alpha!}\right| \leq R^{-|\alpha|} \max_{q\in\overline{S}(q_0,R)} |\Gamma^{(m)}(q)|.$$

Inserting this estimate in (2.22), by easy calculations obtain

$$|D_H^{(m)}[\beta]| \leq \frac{1}{(2\pi|\beta|)^k}\left(\frac{\sqrt{n}}{R}\right)^k k! \max_{q\in S(\overline{Q},R)} |\Gamma^{(m)}(q)|. \tag{2.23}$$

The set $S(\overline{Q}, R)$ on the right side of (2.23) is the union of polydisks $S(q_0, R)$ over all $q_0 \in \overline{Q}$. The so-obtained sequence of upper bounds on the quantity $|D_H^{(m)}[\beta]|$, as it is not hard to validate, decreases for $k \leq k_0(\beta)$, with $k_0(\beta) = [2\pi|\beta|R/\sqrt{n}]$, and increases thereafter. Consequently, the minimal term of this sequence has index $k_0(\beta)$. Using the Stirling formula, we readily calculate this minimum at $|\beta|$ large. Carrying this out, obtain the sought estimate (2.17) for $s = 2\pi R/\sqrt{n}$. The proof of Lemma 9.2 is complete.

Constructing the harrow-like function $\overrightarrow{D}_{hH}^{(m)}(x)$ corresponding to $D_{hH}^{(m)}[\beta]$ by the formula (2.3), show the validity of the convolution equation (2.6). Applying the Fourier transform to both sides of (2.6), come to an equivalent equality (2.8).

Check that for an arbitrary function $\varphi(p) \in \mathscr{S}$ the integral relation holds

$$\int \Gamma_{hH}^{(m)}(p)\widetilde{\overrightarrow{G}_{hH}^{(m)}}(p)\varphi(p)\,dp = \int \varphi(p)\,dp. \tag{2.24}$$

Denote the product $\Gamma_{hH}^{(m)}(p)\varphi(p)$ by $\widetilde{\psi}(p)$. Infinite differentiability and periodicity of $\Gamma_{hH}^{(m)}(p)$ allow us to assert that $\widetilde{\psi}(p)$ still belongs to $\mathscr{S}$. Moreover, at a neighborhood about each of the points p_γ of (2.10) the function $\widetilde{\psi}(p)$ may be written as the product $(2\pi|p-p_\gamma|)^{2m}$ with some analytic factor. The inverse Fourier transform of $\widetilde{\psi}(p)$ is the function $\psi(x)$, again a member of $\mathscr{S}$, whose all derivatives up to order $2m-1$ at every point p_γ equal 0. By the definition of $\overrightarrow{G}_{hH}^{(m)}(x)$, for $\psi(x)$ the equality holds

$$\int \overrightarrow{G}_{hH}^{(m)}(x)\psi(x)\,dx = \sum_\beta G_{hH}^{(m)}[\beta]\psi(hH\beta) = (-1)^m(G(x)\Phi_{hH}(x), \psi(x)). \tag{2.25}$$

Here $\Phi_{hH}(x)$ stands for the following generalized function

$$\Phi_{hH}(x) = \sum_\beta h^n\delta(x - hH\beta) = \Phi_0(h^{-1}H^{-1}x).$$

Applying to the leftmost and rightmost sides of (2.25) the Parseval identity, obtain

$$\int \widetilde{\overline{G}_{hH}^{(m)}}(p)\widetilde{\psi}(p)\,dp = (-1)^m(\widetilde{G}(p) * \widetilde{\Phi}_{hH}(p), \widetilde{\psi}(p)). \tag{2.26}$$

The Fourier transform of $\Phi_{hH}(x)$ is known as given by the formula

$$\widetilde{\Phi}_{hH}(p) = \sum_{\beta} \delta(p - h^{-1}H^{*-1}\beta).$$

Inserting this expression in (2.26), obtain

$$\int \widetilde{\overline{G}_{hH}^{(m)}}(p)\widetilde{\psi}(p)\,dp = \sum_{\gamma}(-1)^m \int \widetilde{G}(p - p_\gamma)\widetilde{\psi}(p)\,dp. \tag{2.27}$$

Further, applying the Fourier transform to both sides of the equation

$$\Delta^m G(x) = (-1)^m \delta(x),$$

we see that at an arbitrary point $p \neq p_\gamma$ the function $\widetilde{G}(p - p_\gamma)$ is defined by the equality

$$\widetilde{G}(p - p_\gamma) = \frac{1}{(2\pi|p - p_\gamma|)^{2m}}.$$

Considering this as well as the behavior of $\widetilde{\psi}(p)$ in a neighborhood about p_γ, come to the conclusion that all integrals in (2.27) are finite and the corresponding series converges absolutely. Moreover,

$$\sum_{\gamma}(-1)^m \int \widetilde{G}(p - p_\gamma)\widetilde{\psi}(p)\,dp = \int \Big(\sum_{\gamma} \frac{(-1)^m}{(2\pi|p - p_\gamma|)^{2m}}\Big)\widetilde{\psi}(p)\,dp$$
$$= \int S_{hH}^{(m)}(p)\Gamma_{hH}^{(m)}(p)\varphi(p)\,dp = \int \varphi(p)\,dp.$$

This together with (2.27) entails the validity of (2.24) and (2.6).

Thus, $D_{hH}^{(m)}[\beta]$ is a solution to the initial problem (2.1) and (2.2).

§3. A Discrete Analog of the Polyharmonic Operator

In the space of discrete functions the convolution $c[\beta] * G_{hH}^{(m)}[\beta]$ is analogous to the polyharmonic potential. In this respect, (2.1) corresponds to the differential equation

$$\Delta^m(x) * G(x) \equiv \Delta^m G(x) = (-1)^m \delta(x).$$

Consequently, the convolution $D_{hH}^{(m)}[\beta]*$ is a discrete analog of the polyharmonic operator.

The properties of $D_{hH}^{(m)}[\beta]*$ and $\Delta^m(x)$ are similar in many aspects but distinctions are also available. For instance, the support of $\Delta^m(x)$ is a singleton, i.e., $\Delta^m(x)$ is a compactly-supported generalized function. The support of $D_{hH}^{(m)}[\beta]$ is not compact; moreover, instead of becoming 0 in a neighborhood about the point at infinity, $D_{hH}^{(m)}[\beta]$ just decreases exponentially with the growth of $|\beta|$.

In the theory of finite differences it is usual to use compactly-supported analogs of the polyharmonic operator. The function $D_{hH}^{(m)}[\beta]$ is an analog of the same operator but now constructed in another fashion and, hence, not compactly-supported. Consider some of the properties of $D_{hH}^{(m)}[\beta]$.

Theorem 9.1. *The function $D_{hH}^{(m)}[\beta]$ may be written as*

$$D_{hH}^{(m)}[\beta] = (-1)^m \sum_{|\alpha|=m} \frac{m!}{\alpha!} D_{hH}^{\alpha}[\beta] * D_{hH}^{\alpha}[-\beta], \tag{3.1}$$

where $\alpha = (\alpha_1, \dots, \alpha_n)$ is a multi-index and the convolution factors $D_{hH}^{\alpha}[\beta]$ are real functions having exponential decay with the growth of $|\beta|$.

PROOF. Alongside the function $\Gamma_{hH}^{(m)}(p)$ defined in § 2, consider the following sum

$$\Delta_{hH}^{(m)}(p) = (-1)^m \left(\frac{2}{h}\right)^{2m} \left(\sum_{i=1}^{n} \sin^2 \pi q_i\right)^m, \tag{3.2}$$

with $q = hH^*p$. Recall that H^* is an orthogonal matrix. Both functions are even, analytic in $\mathbb{R}^n$, and periodic with period matrix $h^{-1}H^{*-1}$ and have a zero of order $2m$ in a neighborhood about the coordinate origin. In regard to $\Gamma_{hH}^{(m)}(p)$, all these assertions are stated in Lemma 9.1, whereas their validity for $\Delta_{hH}^{(m)}(p)$ is obvious. The behavior of $\Delta_{hH}^{(m)}(p)$ in a neighborhood about zero becomes clear from the following equalities

$$\Delta_{hH}^{(m)}(p) = (-1)^m \left(\frac{2}{h}\right)^{2m} |\pi q|^{2m}(1 + f(p)) = (-1)^m (2\pi|p|)^{2m}(1 + f(p)). \tag{3.3}$$

Here $f(p)$ is some analytic function in a neighborhood about zero, with $f(0) = 0$.

The ratio $R(p)$ of $\Gamma_{hH}^{(m)}(p)$ to $\Delta_{hH}^{(m)}(p)$ is an analytic function strictly positive in $\mathbb{R}^n$ with period matrix $h^{-1}H^{*-1}$. If a point p differs from each of the points p_γ defined by (2.10), then the above-listed properties of $R(p)$ are easy from those of the dividend and divisor. If, on the other hand, p coincides with p_γ for some integer γ, then (2.12) and (3.3) entail all what is said about $R(p)$.

In a neighborhood about the coordinate origin, as follows from (2.12) and (3.3), the function $R(p)$ may be written as the sum

$$R(p) = 1 + \sum_{|\alpha|>0} \frac{c_\alpha}{\alpha!}(hp)^\alpha, \tag{3.4}$$

i.e., $R(p)$ is analytic and equal to 1 at $p = 0$.

These properties of $R(p)$ allow us to assert that uniformly in all real p the function $R(p)$ may be estimated from above and from below by some positive constants A_1 and A_2. In symbols,

$$0 < A_1 \le R(p) \le A_2 < \infty, \quad p \in \mathbb{R}^n. \tag{3.5}$$

Thus, the square root $F(p)$ of $R(p)$ is an analytic function in $\mathbb{R}^n$. With this in mind and using the Multinomial Theorem

$$\left(\sum_{i=1}^n \xi_i^2\right)^m = \sum_{|\alpha|=m} \frac{m!}{\alpha!}\xi^{2\alpha},$$

obtain

$$\begin{aligned} \Gamma_{hH}^{(m)}(p) &= (-1)^m \left(\frac{2}{h}\right)^{2m} \left(\sum_{i=1}^n \sin^2 \pi q_i\right)^m F^2(p) \\ &= (-1)^m \sum_{|\alpha|=m} \frac{m!}{\alpha!}\left[\Gamma_{hH}^\alpha(p)\right]^2. \end{aligned} \tag{3.6}$$

Here $\Gamma_{hH}^\alpha(p)$ stands for the following function

$$\Gamma_{hH}^\alpha(p) = \left(\frac{2}{h}\right)^m \sin^{\alpha_1} \pi q_1 \ldots \sin^{\alpha_n} \pi q_n F(p). \tag{3.7}$$

By definition, $\Gamma_{hH}^\alpha(p)$ is analytic and real. It is not hard to check also that this function satisfies the equalities

$$\begin{aligned} \Gamma_{hH}^\alpha(p + p_\gamma) &= (-1)^{\alpha\gamma}\Gamma_{hH}^\alpha(p), \quad \gamma \in \mathbb{Z}^n, \\ \Gamma_{hH}^\alpha(-p) = (-1)^m\Gamma_{hH}^\alpha(p), &\quad \Gamma_{hH}^\alpha(p) = h^{-m}\Gamma_H^\alpha(ph). \end{aligned} \tag{3.8}$$

Put

$$\Lambda_{hH}^\alpha(p) = i^m e^{i\pi phH\alpha}\Gamma_{hH}^\alpha(p). \tag{3.9}$$

The function $\Lambda^{\alpha}_{hH}(p)$ is not only analytic but also periodic with period matrix $h^{-1}H^{*-1}$. As follows from (3.4) and (3.7), in a neighborhood about the coordinate origin this function may be written as

$$\Lambda^{\alpha}_{hH}(p) = (i2\pi)^m (H^*p)^{\alpha}\left(1 + \sum_{|\gamma|>0} \frac{d_\gamma}{\gamma!}(hp)^{\gamma}\right). \tag{3.10}$$

The equalities (3.6), (3.9), and (3.10) imply that, for p real, $\Gamma^{(m)}_{hH}(p)$ decomposes in the sum

$$\Gamma^{(m)}_{hH}(p) = (-1)^m \sum_{|\alpha|=m} \frac{m!}{\alpha!} \Lambda^{\alpha}_{hH}(p)\overline{\Lambda^{\alpha}_{hH}(p)}. \tag{3.11}$$

Expand each of the factors $\Lambda^{\alpha}_{hH}(p)$ in the Fourier series

$$\Lambda^{\alpha}_{hH}(p) = \sum_{\beta} D^{\alpha}_{hH}[\beta] e^{-i2\pi hpH\beta}. \tag{3.12}$$

The coefficients $D^{\alpha}_{hH}[\beta]$ of this expansion decrease exponentially at infinity. This may be proved by a scheme analogous to that of Lemma 9.2. Expressing the coefficient $D^{\alpha}_{hH}[\beta]$ by (3.12) and using (3.9), obtain

$$D^{\alpha}_{hH}[\beta] = i^m \int_{\Omega_0} \Gamma^{\alpha}_{hH}(p) e^{i\pi phH(\alpha+2\beta)}\, dp. \tag{3.13}$$

The integrand here is a periodic function with period matrix $h^{-1}H^{*-1}$. Consequently, we may integrate in (3.13) over the parallelepiped $\Omega_{-\mathbf{e}}$ which is the translate of Ω_0 by the vector $-\mathbf{e} = (-1, \ldots, -1)$. Carrying out the change of variable $p = -p'$ in the appropriate integral, use the second of the relations (3.8) and add the result to the initial equality (3.13). Then $D^{\alpha}_{hH}[\beta]$ in represented in the case of m even as the cosine Fourier transform, and in the case of m odd as the sine Fourier transform of some real function. Consequently, the quantity $D^{\alpha}_{hH}[\beta]$ is real. Inserting the expansions (2.13) and (3.12) in (3.11), obtain the required equality (3.1). The proof of Theorem 9.1 is complete.

We list a few more properties of the functions $D^{\alpha}_{hH}[\beta]$ participating in the expansion (3.1).

Each of the functions $D^{\alpha}_{hH}[\beta]$ as a function in h maintains the relation

$$D^{\alpha}_{hH}[\beta] = h^{-m} D^{\alpha}_{H}[\beta],$$

i.e., is a homogeneous function of degree $-m$. All functions $D^{\alpha}_{hH}[\beta]$ are orthogonal to every polynomial of degree less than m, namely,

$$\left[D^{\alpha}_{hH}[\beta], [\beta]^{\gamma}\right] = 0, \quad |\gamma| < m. \tag{3.14}$$

With the monomial of degree m these functions are connected by the relations of biorthogonality

$$[D^{\alpha}_{hH}[\beta], [\beta]^m] = \gamma!\delta^m_{\alpha} \tag{3.15}$$

The homogeneity of $D^{\alpha}_{hH}[\beta]$ with respect to h may be proven by the same scheme as the homogeneity of $D^{(m)}_{hH}[\beta]$. To do so, it is of service to use the last of the equalities (3.8).

Consider the harrow-like function

$$\overline{D}^{\alpha}_{H}(x) = \sum_{\beta} D^{\alpha}_{H}[\beta]\delta(x - H\beta)$$

corresponding to $D^{\alpha}_{H}[\beta]$ and calculate the convolution of the latter with an arbitrary monomial x^{γ} of degree less than m. By definition,

$$x^{\gamma} * \overline{D}^{\alpha}_{H}(x) = \sum_{\beta} D^{\alpha}_{H}[\beta](x - H\beta)^{\gamma}. \tag{3.16}$$

It is clear that the series on the right side converges absolutely and uniformly in x in an arbitrary bounded domain in $\mathbb{R}^n$. Calculate the Fourier transform of the convolution (3.16)

$$F(x^{\gamma} * \overline{D}^{\alpha}_{H}(x))(p) = \left(\frac{1}{i2\pi}\right)^{|\gamma|} D^{\gamma}(p)\widetilde{\overline{D}^{\alpha}_{H}}(p). \tag{3.17}$$

Multiplication of a function $\varphi(p)$ by $D^{\gamma}(p)$ means the taking of the derivative of $\varphi(p)$ at the coordinate origin. Observe that the Fourier transform of $\overline{D}^{\alpha}_{H}(x)$ agrees with the function $\Lambda^{\alpha}_{H}(p)$, for which we have the expansion (3.10). Differentiating the latter γ times and letting $p = 0$, we see that, for $\gamma < m$, the Fourier transform (3.17) is identically 0. This is possible only in the case when all coefficients of the polynomial in x on the right side of (3.16) equal 0, i.e., the equalities hold

$$\sum_{\beta} D^{\alpha}_{H}[\beta](H\beta)^{\gamma} = 0, \quad |\gamma| < m. \tag{3.18}$$

The matrix H is orthogonal with determinant equal 1. Consequently, the functions of the shape $(H\beta)^{\gamma}$ constitute a basis for the space of polynomials of a discrete variable β of degree less than m. This means that the systems of equalities (3.14) and (3.18) are equivalent.

Observe that (3.1) and (3.14) imply that the function $D^{(m)}_{hH}[\beta]$ is orthogonal to every polynomial of degree less than $2m$

$$\sum_{\beta} D^{(m)}_{hH}[\beta][\beta]^{\gamma} = 0, \quad |\gamma| < 2m. \tag{3.19}$$

As we see, the properties of the operators of convolution with $D_{hH}^{(m)}[\beta]$ and $D_{hH}^{\alpha}[\beta]$ coincide with the properties of the ordinary difference operators $\Delta^{[m]}[\beta]$ and $\Delta^{[\alpha]}[\beta]$. Contrast consists only in the fact that $\Delta^{[m]}[\beta]$ and $\Delta^{[\alpha]}[\beta]$ are of compact support, whereas $D_{hH}^{(m)}[\beta]$ and $D_{hH}^{\alpha}[\beta]$ just decrease exponentially with the growth of $|\beta|$.

We may obtain a more detailed comparison between $D_{hH}^{\alpha}[\beta]*$ and the difference operator $\Delta^{[\alpha]}[\beta]*$ on using the presentation

$$D_{hH}^{\alpha}[\beta] = \Delta^{[\alpha]}[\beta] * F_{hH}^{(m)}[\beta + \alpha/2], \tag{3.20}$$

with

$$F_{hH}^{(m)}[\beta] = i^m \int_{\Omega_0} F(p) e^{i2\pi p h H^* \beta}\, dp$$

and the function $F(p)$ standing as before for the square root of the ratio of $\Gamma_{hH}^{(m)}(p)$ and $\Delta_{hH}^{(m)}(p)$. The equality (3.20) results from (3.7) by easy calculations.

Observe that the periodicity of the analytic function $F(p)$ entails the exponential decay of $F_{hH}^{(m)}[\beta]$ with the growth of $|\beta|$. This assertion may be proven by the same scheme as Lemma 9.2.

Consider the two bilinear forms defined on compactly-supported functions $u[\beta]$ and $v[\beta]$ as follows

$$\begin{aligned} [u,v]_m^D &= (-1)^m \left[D_{hH}^{(m)}[\beta] * u[\beta], v[\beta]\right] \equiv D(u,v), \\ [u,v]_m^\Delta &= \sum_{|\alpha|=m} h^{-2m} \frac{m!}{\alpha!} [\Delta^{[\alpha]}[\beta] * u[\beta], \Delta^{[\alpha]}[\beta] * v[\beta]] \equiv \Delta(u,v). \end{aligned} \tag{3.21}$$

Here $\Delta^{[\alpha]}[\beta]*$ is the operator of partial difference of order α of a function of a discrete argument.

Examine the properties of the forms $D(u,v)$ and $\Delta(u,v)$.

(a) *Given compactly-supported functions $u[\beta]$ and $v[\beta]$ and the corresponding harrow-like functions*

$$\overline{u}(x) = \sum u[\beta]\delta(x - hH\beta), \quad \overline{v}(x) = \sum v[\beta]\delta(x - hH\beta),$$

we have the relations

$$[u,v] = \frac{1}{|\Omega_0|} \int_{\Omega_0} \tilde{\overline{u}}(p) \overline{\tilde{\overline{v}}}(p)\, dp, \tag{3.22}$$

$$D(u,u) = \frac{1}{|\Omega_0|} \int_{\Omega_0} (-1)^m \Gamma_{hH}^{(m)}(p) |\tilde{\overline{v}}(p)|^2\, dp, \tag{3.23}$$

$$\Delta(u,u) = \frac{1}{|\Omega_0|} \int_{\Omega_0} (-1)^m \Delta_{hH}^{(m)}(p) |\tilde{\overline{u}}(p)|^2\, dp, \tag{3.24}$$

with Ω_0 the fundamental parallelepiped of the matrix $h^{-1}H^{*-1}$ and the functions $\Gamma^{(m)}_{hH}(p)$ and $\Delta^{(m)}_{hH}(p)$ defined by (2.11) and (3.2).

The relation (3.22) is nothing else but the Parseval identity for a trigonometric series. Applying it to the pair of the functions $D^{(m)}_{hH}[\beta] * u[\beta]$ and $u[\beta]$, derive (3.23) on involving the definition of $D^{(m)}_{hH}[\beta]$. Similar reasoning leads to (3.24). To this end, it suffices just to use the next formula

$$\Delta^{(m)}_H[\beta] = (-1)^m \sum_{|\alpha|=m} \frac{m!}{\alpha!} \Delta^{[\alpha]}[\beta] * \Delta^{[\alpha]}[-\beta], \tag{3.25}$$

an analog of (3.1) relating the Fourier coefficients $\Delta^{(m)}_H[\beta]$ of $\Delta^{(m)}_H(p)$ and the difference operators $\Delta^{[\alpha]}[\beta]*$.

(b) *There are positive constants C_1 and C_2 such that, for every compactly-supported function $u[\beta]$, the estimates hold*

$$C_1\Delta(u,u) \le D(u,u) \le C_2\Delta(u,u). \tag{3.26}$$

From (3.23) and (3.24) it follows that as such constants we may take the least upper bound and the greatest lower bound of the function $R(p)$ of (3.5).

(c) *Given every pair of compactly-supported functions $u[\beta]$ and $v[\beta]$, we have Green's identity*

$$(-1)^m \left[D^{(m)}_{hH}[\beta] * u[\beta], v[\beta]\right] = \sum_{|\alpha|=m} \frac{m!}{\alpha!} \left[D^{\alpha}_{hH}[\beta] * u[\beta], D^{\alpha}_{hH}[\beta] * v[\beta]\right]. \tag{3.27}$$

By way of demonstration, let $w[\beta]$ stand for the convolution

$$(-1)^m D^{(m)}_{hH}[\beta] * u[\beta].$$

Then the Fourier transform of the corresponding harrow-like function may be obtained by the formula

$$\tilde{\bar{w}}(p) = (-1)^m \Gamma^{(m)}_{hH}(p) \tilde{\bar{u}}(p).$$

Applying to the pair of the functions $w[\beta]$ and $u[\beta]$ a corresponding analog of (3.23), infer

$$D(u,v) = \frac{1}{|\Omega_0|} \int_{\Omega_0} (-1)^m \Gamma^{(m)}_{hH}(p) \tilde{\bar{u}}(p) \overline{\tilde{\bar{v}}}(p)\, dp.$$

Inserting here the expansion of $\Gamma^{(m)}_{hH}(p)$ in the sum (3.11), find that

$$D(u,v) = \frac{1}{|\Omega_0|} \sum_{|\alpha|=m} \frac{m!}{\alpha!} \int_{\Omega_0} \Lambda^{\alpha}_{hH}(p) \tilde{\bar{u}}(p) \overline{\Lambda^{\alpha}_{hH}(p) \tilde{\bar{v}}(p)}\, dp. \tag{3.28}$$

The Fourier coefficients of the product $\Lambda^{\alpha}_{hH}(p)\widetilde{\overline{u}}(p)$ are written down as the values of the discrete function $D^{\alpha}_{hH}[\beta] * u[\beta]$. Expanding each integral in (3.28) by (3.22), obtain the sought equality (3.27).

(d) *The bilinear form $D(u,v)$ is symmetric over real compactly-supported functions, i.e.,*

$$D(u,v) = D(v,u). \tag{3.29}$$

The quadratic form $D(u,u)$ corresponding to the bilinear form $D(u,v)$ is positive definite.

Both claims are immediate from Green's identity (3.27).

Thus, the form $D(u,v)$ possesses all properties of an inner product over the space of compactly supported functions of a discrete variable. To be more precise, if $u[\beta]$ and $v[\beta]$ are compactly-supported then

(1) $D(u,u) \geq 0$ with equality holding if and only if $u[\beta]$ is identically 0,

(2) $D(u,v)$ is linear in each of the variables, and

(3) $D(u,v)$ is symmetric.

In particular, the Cauchy–Bunyakovskiĭ–Schwarz inequality holds

$$|D(u,v)| \leq |(Du,u)|^{1/2}|(Dv,v)|^{1/2}. \tag{3.30}$$

This, together with (3.26), implies that there is a constant C such that for all compactly-supported $u[\beta]$ and $v[\beta]$ the inequality holds

$$|D(u,v)| \leq C|\Delta(u,u)|^{1/2}|\Delta(v,v)|^{1/2}. \tag{3.31}$$

The properties of $D(u,v)$ are of service further in defining a special inner product in $w_2^{(m)}$.

§4. The Weights of Optimal Formulas and the Extension Problem

Recall the definition and properties of some spaces of discrete functions.

A function $u[\beta]$ belongs to $w_2^{(m)}$ if

$$|\Delta(u,u)|^{1/2} = \left(\sum_{|\alpha|=m} \frac{m!}{\alpha!} \sum_{\beta} |\Delta^{[\alpha]}[\beta] * u[\beta]|^2 \right)^{1/2} < \infty.$$

The functional $|\Delta(u,u)|^{1/2}$ determines a seminorm on $w_2^{(m)}$. The set of functions in $w_2^{(m)}$ at which this seminorm vanishes coincides with the space $\mathbf{P}_{m-1}$ of polynomials of degree less than m. The factor space of $w_2^{(m)}$ by $\mathbf{P}_{m-1}$ is denoted by $l_2^{(m)}$ in

Chapter 8. In this event $|\Delta(u,u)|^{1/2}$ determines on $l_2^{(m)}$ the norm that is further denoted by $\|\cdot \mid l_2^{(m)}\|$.

The $l_2^{(m)}$ space is complete under the norm $\|\cdot \mid l_2^{(m)}\|$ and presents the closure of the subset of compactly-supported functions. These claims may be checked by using reciprocal embeddings between the $l_2^{(m)}$ space and the $L_2^{(m)}$ space. The pertinent theorems are given in Chapter 8.

The norm of $l_2^{(m)}$ is generated by the bilinear form $\Delta(u,v)$. Consequently, $l_2^{(m)}$ is a Hilbert space.

Assume given a projection π from $w_2^{(m)}$ to $\mathbf{P}_{m-1}$. Choosing a norm in the finite-dimensional space $\mathbf{P}_{m-1}$, endow $w_2^{(m)}$ with some norm by putting

$$\|u \mid w_2^{(m)}\| = \left(\|\pi u \mid \mathbf{P}_{m-1}\|^2 + \|u \mid l_2^{(m)}\|^2\right)^{1/2}.$$

Obviously, $w_2^{(m)}$ is complete. If the norm of $\mathbf{P}_{m-1}$ is generated by some inner product then $w_2^{(m)}$ becomes a Hilbert space.

Consider the bilinear form $D(u,v)$, defined initially on compactly-supported functions, as acting on all possible $u[\beta]$ and $v[\beta]$ in $w_2^{(m)}$. To this end, take two sequences $u_M[\beta]$ and $v_M[\beta]$ of compactly-supported functions tending in the norm $\|\cdot \mid l_2^{(m)}\|$ to $u[\beta]$ and $v[\beta]$, respectively. Existence of these sequences follows from Theorem 8.10 of denseness of compactly-supported functions in $l_2^{(m)}$. Consider the numerical sequence $D(u_M, v_M)$ and show that it converges. Applying (3.31) twice, obtain

$$\begin{aligned} |D(u_{M_1}, v_{M_1}) - D(u_{M_2}, v_{M_2})| &= |D(u_{M_1} - u_{M_2}, v_{M_1}) + D(u_{M_2}, v_{M_1} - v_{M_2})| \\ &\le K\left(\|v_{M_1} \mid l_2^{(m)}\| \, \|u_{M_1} - u_{M_2} \mid l_2^{(m)}\| + \|u_{M_2} \mid l_2^{(m)}\| \, \|v_{M_1} - v_{M_2} \mid l_2^{(m)}\|\right). \end{aligned}$$

The right side of this inequality at large M_1 and M_2 becomes however small, i.e., the sequence $D(u_M, v_M)$ converges. Consequently, we may put

$$D(u,v) = \lim_{M\to\infty} D(u_M, v_M).$$

The so-constructed bilinear form satisfies the inequality

$$|D(u,v)| \le K\|u \mid l_2^{(m)}\| \, \|v \mid l_2^{(m)}\|$$

with K a constant independent of $u[\beta]$ and $v[\beta]$. This form is thus continuous.

The properties, established in § 3 for compactly-supported functions, remain valid for the above extension $D(u,v)$. This is a consequence of the continuity of

$D(u,v)$ and the theorem of denseness of compactly-supported functions in $l_2^{(m)}$. In this event, the functional on the right side of Green's identity (3.27) must also be continuously extended to the whole of $l_2^{(m)}$. This is carried out by the same scheme as in the case of $D(u,v)$.

In particular, from (3.26) it follows that $D(u,u)$ vanishes if and only if $u[\beta] \in w_2^{(m)}$ is a polynomial of degree $m-1$. Thus, the form $D(u,v)$ determines on $l_2^{(m)}$ an inner product equivalent to the original $\Delta(u,v)$.

The operation of convolution $D_{hH}^{(m)}[\beta]*$ also extends to the entire $w_2^{(m)}$. Let $u[\beta]$ belong to $w_2^{(m)}$. Then put

$$D_{hH}^{(m)}[\beta] * u[\beta]\Big|_{\beta=\beta_0} = D(u, \delta[\beta-\beta_0]). \tag{4.1}$$

The formula on the right side of (4.1) is defined, since $\delta[\beta-\beta_0]$ is obviously a member of $w_2^{(m)}$. From the inequality

$$|D(u,\delta[\beta-\beta_0])| \le K \big\|\delta[\beta-\beta_0] \mid l_2^{(m)}\big\| \, \big\|u \mid l_2^{(m)}\big\|$$

it follows that the operator of convolution $D_{hH}^{(m)}[\beta]*$ acts continuously from $w_2^{(m)}$ to the space of bounded functions of a discrete variable.

Revert to Problem B_1 posed in § 1, and consider the case in which the parameters f_α of the hypothesis are all 0. In other words, find $c[\beta]$ and $P_{m-1}[\beta]$ satisfying the system of equalities

$$G_{hH}^{(m)}[\beta] * c[\beta] + \mathrm{P}_{m-1}[\beta] = g[\beta], \quad hH\beta \in \overline{\Omega}, \tag{4.2}$$

$$c[\beta] = 0, \quad hH\beta \notin \overline{\Omega}, \tag{4.3}$$

$$\sum_{\beta} c[\beta][\beta]^\alpha = 0, \quad |\alpha| < m. \tag{4.4}$$

We have already proven unique solvability of this system. We now consider an algorithm which allows us to express the inverse of the operator of (4.2)–(4.4) as superposition of some operator of extension of $g[\beta]$ to the whole space $\mathbb{Z}^n$ with the operator of convolution $D_{hH}^{(m)}[\beta]*$.

With a solution $c[\beta]$ of (4.2)–(4.4) associate the harrow-like function $\overline{c}(x)$ that is defined by (1.8) and the polyharmonic potential $w(x) = G(x) * \overline{c}(x)$. Then (4.4) is equivalent to following:

$$\int_{\Omega} \overline{c}(x) x^\alpha \, dx = 0 \quad \text{for} \quad |\alpha| < m. \tag{4.5}$$

Moreover, in a neighborhood about the point at infinity $w(x)$ has the expansion of the shape (1.15) containing only relevant terms and satisfies (1.16). Moreover, as mentioned in §1, $w(x)$ is contained in $W_2^{(m)}$.

The values $w(hH\beta)$ of the potential $w(x)$ at the nodes of the lattice are given as convolution $v[\beta] = G_{hH}^{(m)}[\beta] * c[\beta]$. The function $v[\beta]$ belongs to $w_2^{(m)}$ and moreover by Theorem 8.7 for $m > n/2$ it satisfies the inequality

$$\left\| v \mid l_2^{(m)} \right\| \leq K \left\| w(x) \mid L_2^{(m)} \right\|.$$

In a neighborhood about the point at infinity, $v[\beta]$ admits an expansion that is analogous to (1.13) and consists only of relevant terms. In this event, the estimate holds

$$|v[\beta]| \leq C \begin{cases} |\beta|^{m-n}, & \text{if } \ n \ \text{ odd or } \ n > m, \\ |\beta|^{m-n} \log |\beta|, & \text{if } \ n \ \text{ even and } \ m \geq n. \end{cases} \tag{4.6}$$

Assume that $y[\beta] \in w_2^{(m)}$ differs from $f[\beta]$ for $hH\beta \in \overline{\Omega}$ by a polynomial of degree $m-1$, satisfies (4.6) at $|\beta|$ large, and does not agree with $v[\beta]$. Then, the norm $\|v\|_m^D = D(v,v)^{1/2}$ is less than the norm of $\|y\|_m^D = D(y,y)^{1/2}$, namely,

$$\|v\|_m^D < \|y\|_m^D. \tag{4.7}$$

Show this by the same scheme as that for (1.17). For $hH\beta \in \overline{\Omega}$ the functions $y[\beta]$ and $v[\beta]$ differ by another polynomial $Q_{m-1}[\beta]$ of degree at most $m-1$. Put $u[\beta] = v[\beta] + Q_{m-1}[\beta]$. Then, by the properties of the operator of convolution $D_{hH}^{(m)}[\beta]*$, the norms of $\|u\|_m^D$ and $\|v\|_m^D$ coincide. Thus, it suffices to establish the validity of (4.7) with $u[\beta]$ substituted for $v[\beta]$.

Arrange the set of discrete functions depending on a real parameter λ by putting $u_\lambda[\beta] = u[\beta] + \lambda(y[\beta] - u[\beta])$. It is clear that, for all λ, the function $u_\lambda[\beta]$ is a member of $w_2^{(m)}$. Consider the norm square of $\|u_\lambda\|_m^D$ presenting a quadratic trinomial in λ, i.e.,

$$D(u_\lambda, u_\lambda) = D(u,u) + 2\lambda D(u, y-u) + \lambda^2 D(y-u, y-u).$$

Calculate the coefficient of λ in this relation. By (3.21), obtain

$$D(u, y-u) = (-1)^m \sum_\beta \left(D_{hH}^{(m)}[\beta] * u[\beta] \right) (y[\beta] - u[\beta]).$$

The convolution $D_{hH}^{(m)}[\beta] * u[\beta]$ agrees with the function $c[\beta]$ by the definition of $u[\beta]$. In particular, it equals 0 for all β such that $hH\beta \notin \overline{\Omega}$. If, on the other hand, $hH\beta \in \overline{\Omega}$; then the difference $y[\beta] - u[\beta]$ is equal to 0. The functions $u[\beta]$ and $(y-u)[\beta]$ are thus orthogonal with respect to the inner product $D(\cdot,\cdot)$.

Returning to the form $D(u_\lambda, u_\lambda)$, we see that zero is a unique stationary point of it with respect to λ. Moreover, the second derivative with respect to λ is positive at zero. Thus, the function $D(u_\lambda, u_\lambda)$ has a strict minimum at $\lambda = 0$. In particular,

$$D(u_0, u_0) = D(u, u) < D(u_1, u_1) = D(y, y).$$

The desired inequality (4.7) is established.

Consider now the following extension problem.

PROBLEM B_2. *Find $v_1[\beta]$ in $w_2^{(m)}$ satisfying* (4.6) *at infinity, different for $hH\beta \in \overline{\Omega}$ from $g[\beta]$ not more than by a polynomial summand of degree $m-1$, and possessing a minimal norm $\|\cdot\|_m^D$ in this class.*

The above reasoning shows that a sought extension is given by the function $v[\beta] = G_{hH}^{(m)}[\beta] * c[\beta]$, with $c[\beta]$ satisfying (4.2)–(4.4). A solution to Problem B_2 is unique. Indeed, if $v_1[\beta]$ and $v_2[\beta]$ are two distinct solutions then their half-sum belongs to the same convex set in $w_2^{(m)}$ as the summands $v_1[\beta]$ and $v_2[\beta]$. Moreover, the norm of the half-sum is less than the half-sum of norms, i.e.,

$$\|(v_1+v_2)/2\|_m^D < (\|v_1\|_m^D + \|v_2\|_m^D)/2.$$

The functional $\|\cdot\|_m^D$ cannot attain a minimum at $v_1[\beta]$ and $v_2[\beta]$ simultaneously.

The general case in which the numbers f_α in System B are nonzero reduces to the already-settled case by the standard change of variable $c[\beta] = c_*[\beta] + c_1[\beta]$, with $c_*[\beta]$ satisfying (4.3) and (4.4) and $c_1[\beta]$ a new unknown function of a discrete variable.

In closing, observe that the formula

$$c[\beta] = D_{hH}^{(m)}[\beta] * v[\beta]$$

may be considered as a special presentation of the operator inverse to that of (4.2)–(4.4) as superposition of the extension operator and convolution.

§5. A One-Dimensional Discrete Analog of a Derivative of Even Order

The method for constructing the weights of optimal cubature formulas we considered happens to be applicable in the case of *quadratures*, i.e., in the case of formulas approximating an integral over an interval of the real axis. We can carry out the examination of this case much further.

So, let the dimension of $\mathbb{R}^n$ equal 1. The domain Ω reduces now to the interval $[0,1]$, and the polyharmonic equation $\Delta^m u(x) = (-1)^m l(x)$ becomes an ordinary differential equation

$$\frac{d^{2m}u}{dx^{2m}} = (-1)^m l(x).$$

The norm of $L_2^{(m)}(\mathbb{R}^1)$ is defined by the equality

$$\|u \mid L_2^{(m)}(\mathbb{R}^1)\| = \left(\int \left|\frac{d^m u}{dx^m}\right|^2 dx\right)^{1/2}.$$

The homogeneous System A for functions of a single variable is written as

$$G(x) * c(x) + P_{m-1}(x) = f(x), \quad x \in [0,1], \tag{5.1}$$

$$c(x) = 0, \quad x \notin [0,1], \tag{5.2}$$

$$\int c(x) x^\alpha \, dx = 0, \quad |\alpha| < m. \tag{5.3}$$

Here the functions $G(x)$ and $f(x)$ are determined by the equalities

$$G(x) = (-1)^m \frac{|x|^{2m-1}}{2(2m-1)!}, \quad f(x) = \int_0^1 G(x-y)\, dy. \tag{5.4}$$

It is not hard to see that $f(x)$ is a polynomial of degree $2m-1$ satisfying the condition $f(x) = f(1-x)$ for $0 \le x \le 1$. Problem A_2 corresponding to (5.1)–(5.3) consists in finding a member $u(x)$ of $L_2^{(m)}(\mathbb{R}^1)$ polyharmonic in the exterior of the interval $[0,1]$ and agreeing with $f(x)$ in the open interval $(0,1)$. Thus, function $u(x)$ outside the interval $[0,1]$ is a polynomial of degree $2m-1$. In this polynomial all coefficients of the powers greater than $m-1$ vanish, or else $u(x)$ would not belong to $L_2^{(m)}(\mathbb{R}^1)$. Now we may construct the function $u(x)$ explicitly. Indeed, by the continuity of $u(x)$ with all derivatives up to order $m-1$ we have

$$u(x) = P_{m-1}^{(1)}(x) \quad \text{for} \quad x < 0, \quad u(x) = P_{m-1}^{(2)}(x) \quad \text{for} \quad x > 1, \tag{5.5}$$

with $P_{m-1}^{(1)}(x)$ and $P_{m-1}^{(2)}(x)$ known polynomials, namely,

$$P_{m-1}^{(1)}(x) = f(0) + x f'(0) + \cdots + \frac{x^{m-1}}{(m-1)!} f^{(m-1)}(0),$$

$$P_{m-1}^{(2)}(x) = f(1) + (x-1) f'(1) + \cdots + \frac{(x-1)^{m-1}}{(m-1)!} f^{(m-1)}(1).$$

Thus, $u(x)$ is a piecewise-polynomial function with nodes at the points $x = 0$ and $x = 1$.

Obviously, $u(x)$ possesses the least of the $L_2^{(m)}(\mathbb{R}^1)$ norms of all possible extensions of $f(x)$ to the whole axis. There is a unique function with this property. Indeed, were the minimum of the norm attained at two distinct functions u_1 and u_2 extending $f(x)$ to $\mathbb{R}^1$, then their half-sum would also extend $f(x)$, having the norm less than those of u_1 and u_2. This contradiction proves that $u_1 = u_2$.

Bearing in mind that the norm of $u(x)$, the solution to problem A_2, coincides with the norm of $u(1-x)$, obtain the equality

$$u(x) = u(1-x).$$

Proceed now to solving Problem A_1. Put $c(x) = (-1)^m d^{2m}u/dx^{2m}$ and observe that $c(x)$ is a generalized function, a member of $L_2^{(m)}(\mathbb{R})^*$. The support of $c(x)$ obviously lies in the unit interval. Further, the general theory of the polyharmonic equation entails existence of a polynomial $P_{m-1}(x)$ of degree $m-1$ such that

$$u(x) = G(x) * c(x) + P_{m-1}(x).$$

The first two equalities of System A are thus fulfilled.

Further, for $x < 0$ we have

$$P_{m-1}^{(1)}(x) - P_{m-1}(x) = G(x) * c(x) \equiv v(x);$$

i.e., the function $v(x)$ for x negative is a polynomial of degree at most $m-1$. On the other hand, using the familiar expansion of the convolution $G(x) * c(x)$ into a series in the derivatives of $G(x)$, obtain

$$v(x) = \int G(x-y)c(y)\,dy = \sum_{k=0}^{2m-1} \frac{(-1)^k}{k!} \frac{d^k G(x)}{dx^k} \int c(y) y^k\,dy.$$

The expression on the right side may be a polynomial of degree $m-1$ only in the case when the first m integrals equal 0. Thus, (5.3) is also fulfilled, and Problem A_1 is resolved.

The homogeneous System B may be written as

$$hG[\gamma] * c[\gamma] + P_{m-1}[\gamma] = f[\gamma], \quad 0 \le h\gamma \le 1, \tag{5.6}$$
$$c[\gamma] = 0, \quad h\gamma \notin [0,1], \tag{5.7}$$
$$\sum_\gamma c[\gamma][\gamma]^\alpha = 0, \quad 0 \le \alpha < m, \tag{5.8}$$

moreover,

$$G[\gamma] = (-1)^m \frac{h^{2m-1}}{2(2m-1)!} |\gamma|^{2m-1}, \quad f[\gamma] = \int_0^1 G(h\gamma - y)\, dy.$$

The function $f[\gamma]$ is a polynomial of degree $2m-1$.

To solve Problems B_2 and B_1 corresponding to System B we are to determine the coefficients of $D_h^{(m)}[\beta]*$. Recall that these coefficients may be found from the Fourier series for $\Gamma_h^{(m)}(p)$ satisfying the condition

$$\Gamma_h^{(m)}(p)\widetilde{\vec{G}_h^{(m)}}(p) = 1. \tag{5.9}$$

As regards the Fourier transform $\widetilde{\vec{G}_h^{(m)}}(p)$, from (2.8)–(2.11) derive the formula

$$\widetilde{\vec{G}_{hH}^{(m)}}(p) = \frac{(-1)^m h^{2m}}{(2\pi)^{2m}} \sum_{\beta=-\infty}^{\infty} \frac{1}{(ph-\beta)^{2m}} = h^{2m}\widetilde{\vec{G}_1^{(m)}}(hp).$$

Inspect $\widetilde{\vec{G}_1^{(m)}}(p)$ in more detail. As is known,

$$\sum_{\beta=-\infty}^{\infty} \frac{1}{(p-\beta)^2} = \frac{\pi^2}{\sin^2 \pi p}.$$

Differentiating this equality $2m-2$ times, obtain

$$\sum_{\beta=-\infty}^{\infty} \frac{1}{(p-\beta)^{2m}} = \frac{1}{(2m-1)!} \frac{d^{2m-2}}{dp^{2m-2}} \left\{ \frac{\pi^2}{\sin^2 \pi p} \right\}. \tag{5.10}$$

We may rearrange the expression on the right side of (5.10) on using the concept of Euler polynomial.

In line with [66], we define the *Euler polynomial* $E_k(\lambda)$ by the formula

$$E_k(\lambda) = \frac{(1-\lambda)^{k+2}}{\lambda} \left(\lambda \frac{d}{d\lambda}\right)^k \frac{\lambda}{(1-\lambda)^2}. \tag{5.11}$$

In the article [204] the polynomial $E_k(\lambda)$ is denoted by $\Pi_{k+1}(\lambda)$ and called the Euler–Frobenius polynomial. Therein the generating function is given for the sequence of the polynomials $E_k(\lambda)$ in the shape

$$\frac{\lambda - 1}{\lambda - e^z} = \sum_{n=0}^{\infty} \frac{E_{n-1}(\lambda)}{(\lambda-1)^n} \frac{z^n}{n!}.$$

The other properties of Euler polynomials are studied to some extend in the articles [256–262].

Taking the function $e^{i2\pi p}$ as λ in (5.11), come to the equality

$$\frac{d^{2m-2}}{dp^{2m-2}}\left\{\frac{\pi^2}{\sin^2 \pi p}\right\} = \frac{\pi^{2m}}{\sin^{2m} \pi p} e^{-i2(m-1)\pi p} E_{2m-2}(e^{i2\pi p}).$$

Inserting it in (5.10) and considering the function $\widetilde{\overline{G}_1^{(m)}}(p)$, obtain

$$\widetilde{\overline{G}_1^{(m)}}(p) = \frac{(-1)^m}{2^{2m}(2m-1)!} \frac{1}{\sin^{2m} \pi p} e^{-i2(m-1)\pi p} E_{2m-2}(e^{i2\pi p}).$$

So, a solution to equation (5.9) for $h = 1$ is given by the formula

$$\Gamma_1^{(m)}(p) = (2m-1)!(-2^2)^m (\sin \pi p)^{2m} \frac{e^{i(2m-2)\pi p}}{E_{2m-2}(e^{i2\pi p})}. \tag{5.12}$$

From the definition (5.11) of $E_k(\lambda)$ it is not hard to deduce the identity

$$E_k(\lambda) = \lambda^k E_k\left(\frac{1}{\lambda}\right). \tag{5.13}$$

Using (5.13) with $\lambda = e^{i2\pi p}$, rewrite the equality (5.12) in equivalent form

$$\frac{\Gamma_1^{(m)}(p)}{(2m-1)!} = (-2^2)^m (\sin \pi p)^{2m} \frac{e^{-i2\pi p(m-1)}}{E_{2m-2}(e^{-i2\pi p})}. \tag{5.14}$$

As is known, the Fourier coefficients of the product of two functions result from convoluting the Fourier coefficients of the factors. Consequently, (5.14) enables us to express $D_1^{(m)}[\beta]$ as convolution. We write down the corresponding formula.

Denote by $E_{2m-2}^{-1}[\beta]$ the following Fourier coefficients

$$E_{2m-2}^{-1}[\beta] = \int_0^1 \frac{e^{i2\pi p\beta}}{E_{2m-2}(e^{-i2\pi p})} dp, \tag{5.15}$$

Also, denote by $\Delta_2^{[m]}[\beta]$ the Fourier coefficients of $2^{2m}(-1)^m \sin^{2m} \pi p$, i.e.,

$$\Delta_2^{[m]}[\beta] = \sum_{k=-m}^{m} (-1)^{k+m} \binom{2m}{m+k} \delta[\beta - k]. \tag{5.16}$$

Recall that the convolution with $\Delta_2^{[m]}[\beta]$ means the taking of the symmetric difference of order m. Using these notations, from (5.14) obtain

$$\frac{D_1^{(m)}[\beta]}{(2m-1)!} = \Delta_2^{[m]}[\beta] * E_{2m-2}^{-1}[\beta - m + 1]. \tag{5.17}$$

The function $\Delta_2^{[m]}[\beta]$ is compactly-supported. Hence, the convolution (5.17) exists. If $B_{2m-2}[\gamma]$ are the Fourier coefficients of the product

$$e^{i2\pi p(m-1)} E_{2m-2}(e^{-i2\pi p})/(2m-1)!,$$

then (5.14) entails one more convolution relation

$$B_{2m-2}[\beta] * D_1^{(m)}[\beta] = \Delta_2^{[m]}[\beta]. \tag{5.18}$$

We now dwell in more detail on the problem of calculating the coefficients given by (5.15). To this end, we are to examine the roots of the polynomial $E_k(\lambda)$. Denote them by $\lambda_j^{(k)}$, $j = 1, 2, \ldots, k$. It is well known that all roots $\lambda_j^{(k)}$ are real, negative, and distinct [204], i.e.,

$$\lambda_1^{(k)} < \lambda_2^{(k)} < \cdots < \lambda_k^{(k)} < 0. \tag{5.19}$$

Using (5.13), we readily prove that the roots equidistant from the ends of the chain (5.19) are reciprocals, in symbols,

$$\lambda_j^{(k)} \lambda_{k+1-j}^{(k)} = 1 \quad \text{for} \quad j = 1, 2, \ldots, k. \tag{5.20}$$

Take k equal to $2m-2$ and agree to omit the superscript of the root $\lambda_j^{(2m-2)}$ up to the end of the current section. We have the following expansion in *partial fractions*

$$\frac{\lambda^{m-1}}{E_{2m-2}(\lambda)} = \sum_{j=1}^{2m-2} \frac{B_{2m-1-j}}{\lambda - \lambda_j}, \tag{5.21}$$

with

$$B_{2m-1-j} = \lambda_j^{m-1} / E'_{2m-2}(\lambda_j). \tag{5.22}$$

Differentiating (5.13) with respect to λ, putting $\lambda = \lambda_j$ in the result and using (5.20), observe that

$$B_{2m-1-j} = -\lambda_j^2 B_j.$$

Using this equality, rearrange the summands in (5.21) to come to the expansion

$$\frac{\lambda^{m-1}}{E_{2m-2}(\lambda)} = \sum_{j=1}^{m-1} B_j \left\{ \frac{1}{\lambda - \lambda_{2m-1-j}} - \frac{\lambda_j^2}{\lambda - \lambda_j} \right\}.$$

Letting $\lambda = e^{-i2\pi p}$ here and considering that, for all $j = 1, \dots, m-1$, the quantity $\mu_j = \lambda_{2m-1-j} = 1/\lambda_j$ has magnitude less than 1, find the desired Fourier series

$$\frac{e^{-i2\pi p(m-1)}}{E_{2m-2}(e^{-i2\pi p})} = \sum_{j=1}^{m-1} \frac{B_j}{\mu_j} + \sum_{n=1}^{\infty} 2 \left(\sum_{j=1}^{m-1} B_j \mu_j^{n-1} \right) \cos 2\pi n p. \tag{5.23}$$

This equality enables us to calculate the convolution factor $E_{2m-2}^{-1}[\beta - m + 1]$ in (5.17) at an arbitrary point, if only we know in advance the roots μ_j of $E_{2m-2}(\mu)$ and the quantities B_j defined as the ratio (5.22). We are thus able to calculate $D_1^{(m)}[\beta]$ using (5.17). However, with m increasing, the cost of the corresponding algorithm rises essentially. Consequently, for m large, it is desirable to find some simple asymptotic expansions of μ_j, B_j, and $E_{2m-2}^{-1}[\beta]$. In the two subsequent sections we obtain such asymptotic formulas for the roots μ_j.

§6. The Roots of the Euler Polynomial

Let the roots $\lambda_j^{(k)}$, $j = 1, 2, \dots, k$, of the Euler polynomial $E_k(\lambda)$ be ordered as prescribed in (5.19). Inspect their asymptotic behavior at large k. It is of service to consider not the roots themselves but rather the logarithms of their magnitudes

$$q_j^{(k)} = \frac{1}{\pi} \log(-\lambda_j^{(k)}).$$

Obviously, these logarithms are ordered as follows

$$-\infty < q_k^{(k)} < q_{k-1}^{(k)} < \cdots < q_2^{(k)} < q_1^{(k)}.$$

Moreover, the numbers $q_j^{(k)}$ and $q_{k+1-j}^{(k)}$ equidistant from the ends of the chain are interconnected by the equality

$$q_j^{(k)} + q_{k+1-j}^{(k)} = 0, \quad j = 1, 2, \dots, k,$$

which allows us to inspect only the positive $q_j^{(k)}$ indexed with $j = 1, 2, \dots, [k/2]$.

The numerical values of the roots of the Euler polynomials of even degree at most 12 are listed in the article [204].

Consider sequences of $q_j^{(k)}$ such that the index j is a function of the parameter k, i.e., $j = j(k)$. For these sequences we derive two asymptotic formulas.

One of them has the form

$$q_j^{(k)} \cong q_j^{\mathrm{I}(k)} = \cot \frac{(j + 1/2)\pi}{k+2}. \tag{6.1}$$

This formula is appropriate for approximation of $q_j^{(k)}$ such that the upper limit of the ratio of the modulus of $q_j^{(k)}$ to the square root of k does not exceed some $\varkappa_{\max}$, i.e.,

$$\overline{\lim_{k\to\infty}} \frac{q_j^{(k)}}{\sqrt{k}} < \varkappa_{\max} \cong 3.0215. \tag{6.2}$$

We write down the other formula in two equivalent versions:

$$q_j^{(k)} \cong q_j^{\mathrm{II}(k)} = \frac{k+1}{\pi} \log\left(1 + \frac{1}{j}\right), \tag{6.3}$$

$$\lambda_j^{(k)} \cong -\left(1 + \frac{1}{j}\right)^{k+1} \tag{6.4}$$

This formula is appropriate for approximation of $q_j^{(k)}$ such that the lower limit of the ratio of the modulus of $q_j^{(k)}$ to the square root of k is greater than some $\varkappa_{\min}$, i.e.,

$$\underline{\lim_{k\to\infty}} \frac{q_j^{(k)}}{\sqrt{k}} > \varkappa_{\min} \cong 0.3725. \tag{6.5}$$

The condition (6.5) sufficient for applying (6.3) may also be formulated in terms of the function $j = j(k)$. Namely, if

$$\overline{\lim_{k\to\infty}} \frac{j(k)}{\sqrt{k}} < \frac{1}{\pi \varkappa_{\min}}, \tag{6.6}$$

then the quantity $q_j^{\mathrm{II}(k)}$ however closely approaches $q_j^{(k)}$ at k sufficiently large. Observe that if one of the inequalities (6.5) and (6.6) is valid for the ordinary limit, then the same limit may be inserted in the other, with (6.5) and (6.6) becoming equivalent.

The results of the current section are published in brief in [262]. As regards (6.1), our result extends those by S. Kh. Sirazhdinov [219].

We first derive the asymptotic formula (6.3).

The coefficients $a_s^{(k)}$ of the Euler polynomial

$$E_k(\lambda) = \sum_{s=0}^{k} a_{k-s}^{(k)} \lambda^s$$

as was shown by Euler himself, are expressed as

$$a_s^{(k)} = \sum_{j=0}^{s} (-1)^j \binom{k+2}{j} (s+1-j)^{k+1}. \tag{6.7}$$

Comparing (6.7) with (5.1.3.–13) in [95], we easily see that $a_s^{(k)}$ is precisely $\left\langle {k+1 \atop s+1} \right\rangle$ in the notation of [95]. Observe also that [95] involves other numbers that are also attributed to Euler. Each coefficient $a_s^{(k)}$ is an elementary symmetric function in the roots $\lambda_j^{(k)}$. Write it down explicitly. To this end, enumerate by the index t, $t = 1, 2, \ldots, \binom{k}{s}$, all possible distinct combinations of s items in the pool of k without replacement. To each of the combinations there corresponds a tuple of s distinct indices $j_1(t), j_2(t), \ldots, j_s(t)$ ranging over the interval $1 \leq j \leq k$. Let

$$\nu_{s,t}^{(k)} = \left| \lambda_{j_1(t)}^{(k)} \lambda_{j_2(t)}^{(k)} \cdots \lambda_{j_s(t)}^{(k)} \right|. \tag{6.8}$$

For $0 \leq s \leq k$, the Vieta formulas and (5.19) entail

$$a_s^{(k)} = \sum_{t=1}^{\binom{k}{s}} \nu_{s,t}^{(k)}. \tag{6.9}$$

It is convenient to assume that for $s < 0$ and $s > k$ the coefficient $a_s^{(k)}$ is identically 0. Using the *squaring process* that was suggested by N. I. Lobachevskiĭ [24], arrange the polynomial

$$L_k(\mu) = b_0^{(k)} \mu^k - b_1^{(k)} \mu^{k-1} + \cdots + (-1)^k b_k^{(k)}.$$

The coefficients of $L_k(\mu)$ are expressed through $a_s^{(k)}$ by the formula

$$b_j^{(k)} = (a_j^{(k)})^2 + 2 \sum_{t=1}^{\infty} (-1)^t a_{j-t}^{(k)} a_{j+t}^{(k)}. \tag{6.10}$$

The series on the right side, obviously, terminates with finitely many summands. It is well known that the roots of $L_k(\mu)$ are the squares of the roots of $E_k(\lambda)$. Consequently,

$$b_s^{(k)} = \sum_{t=1}^{\binom{k}{s}} |\nu_{s,t}^{(k)}|^2. \tag{6.11}$$

To estimate the greatest of the numbers composing the sum (6.9), show the validity of the following

Lemma 9.3. *Let* $\nu_1 > \nu_2 \geq \nu_3 \geq \cdots \geq 0$ *and*

$$\nu_1 + \nu_2 + \cdots + \nu_k + \cdots = a < \infty, \quad \nu_1^2 + \nu_2^2 + \cdots + \nu_k^2 + \cdots = b < \infty. \tag{6.12}$$

If, moreover,

$$a^2 < 2b; \tag{6.13}$$

then the inequality holds

$$\frac{1}{2}(a + \sqrt{2b - a^2}\,) \leq \nu_1 \leq \sqrt{b}. \tag{6.14}$$

PROOF. Put $\nu_2^* = \nu_2 + \nu_3 + \cdots + \nu_k + \cdots = a - \nu_1$ and $b_* = \nu_1^2 + (\nu_2^*)^2$. Since all ν_j are positive, the quantity b_* is at least b, namely,

$$b_* \geq b, \quad b_* = \nu_1^2 + (a - \nu_1)^2 = 2(\nu_1 - a/2)^2 + a^2/2. \tag{6.15}$$

Consequently, $\nu_1 = {}^1\!/_2(a + \sqrt{2b_* - a^2}\,)$. Since $b_* \geq b$, we have the lower bound on ν_1 in (6.14). The upper bound on ν_1 is obvious from (6.12). The proof of Lemma 9.3 is complete.

Apply Lemma 9.3 to estimating the maximum of the quantities $\nu_{s,t}^{(k)}$ that are defined by (6.8). Observe that, at a fixed s, the maximal number $\nu_{s,t}^{(k)}$ appears when $j_1(t) = 1$, $j_2(t) = 2, \ldots, j_s(t) = s$. This is easy to establish by using (5.19). Consequently, (6.14) in the cases when $s = j$ and $s = j - 1$ looks like

$$\frac{1}{2}(1 + \sqrt{1 - 4\eta_j^{(k)}}\,) \leq \frac{|\lambda_1^{(k)}\lambda_2^{(k)} \ldots \lambda_j^{(k)}|}{a_j^{(k)}} \leq \sqrt{1 - 2\eta_j^{(k)}}, \tag{6.16}$$

$$\frac{1}{2}(1 + \sqrt{1 - 4\eta_{j-1}^{(k)}}\,) \leq \frac{|\lambda_1^{(k)}\lambda_2^{(k)} \ldots \lambda_{j-1}^{(k)}|}{a_{j-1}^{(k)}} \leq \sqrt{1 - 2\eta_{j-1}^{(k)}}. \tag{6.17}$$

The quantity $\eta_s^{(k)}$ in these relations may be written as the sum

$$\eta_s^{(k)} = \frac{1}{2}\left(1 - \frac{b_s^{(k)}}{|a_s^{(k)}|^2}\right) = \sum_{t=1}^{\infty}(-1)^{t+1}\frac{a_{s-t}^{(k)}a_{s+t}^{(k)}}{|a_s^{(k)}|^2}. \tag{6.18}$$

In this event the function $a_s^{(k)}$ is extended by 0 to $s < 0$ and $s > k$. Consequently, the series (6.18) terminates at a finite t. It is clear that (6.16) and (6.17) are applicable only provided that

$$\eta_j^{(k)} < 1/4, \quad \eta_{j-1}^{(k)} < 1/4. \tag{6.19}$$

Dividing (6.16) by (6.17) and taking the logarithm of the quotient, obtain

$$\frac{1}{\pi}\log\frac{1+\sqrt{1-4\eta_j^{(k)}}}{2\sqrt{1-2\eta_{j-1}^{(k)}}} \le q_j^{(k)} - \frac{1}{\pi}\log\frac{a_j^{(k)}}{a_{j-1}^{(k)}} \le \frac{1}{\pi}\log\frac{2\sqrt{1-2\eta_j^{(k)}}}{1+\sqrt{1-4\eta_{j-1}^{(k)}}}. \tag{6.20}$$

Show that, at a fixed j, the quantity $\eta_j^{(k)}$ vanishes as $k \to \infty$. Indeed, from (6.7) we have

$$a_s^{(k)} = (s+1)^{k+1}\left\{\sum_{j=0}^{s}(-1)^j\binom{k+2}{j}\left(1-\frac{j}{s+1}\right)^{k+1}\right\}. \tag{6.21}$$

At s fixed each summand in braces, from the second on, vanishes as k increases indefinitely. Consequently, for $t = 0, 1, \ldots$ and the nonzero coefficients $a_{s\pm t}^{(k)}$ we have

$$a_{s\pm t}^{(k)} = (s\pm t+1)^{k+1}\big(1+R_{s\pm t}^{(k)}\big), \quad \frac{a_{s-t}^{(k)}a_{s+t}^{(k)}}{\big|a_s^{(k)}\big|^2} = \left(1-\left(\frac{t}{s+1}\right)^2\right)^{k+1}\big(1+R_{s,t}^{(k)}\big).$$

The quantities $R_{s\pm t}^{(k)}$ and $R_{s,t}^{(k)}$ vanish as $k \to \infty$. Whence it follows that $\eta_s^{(k)}$, presented by the sum (6.18), indeed vanishes as k increases indefinitely.

Thus, at a given j, both inequalities (6.19) and (6.20) are fulfilled from some index k on. In particular, for every positive ε, the inequality holds

$$\left|q_j^{(k)} - \frac{k+1}{\pi}\log\frac{j+1}{j}\right| < \varepsilon.$$

Assume now that $j = j(k)$ grows as $\sqrt{k}$. Estimate the quality of approximating $q_j^{(k)}$ by $q_j^{\mathrm{II}(k)}$ defined in (6.3). Put

$$\varkappa_j^{\prime(k)} = \frac{\sqrt{k+1}}{\pi(j+1/2)}, \quad \varkappa' = \lim_{k\to\infty}\varkappa_j^{\prime(k)} < +\infty.$$

Obviously, we then have the equality

$$j = \frac{\sqrt{k+1}}{\pi\varkappa'}\Big(1+O(k^{-1})\Big). \tag{6.22}$$

Estimate $\eta_j^{(k)}$ from above for these js. In the formula (6.18) for $\eta_j^{(k)}$, there participate the summands $a_{j\pm t}^{(k)}$ with index t satisfying the conditions $j+t \le k$ and $j-t \ge 0$.

In this event, in particular, $j+t \le \min(k, 2j)$. Consequently, the index s of each nonzero coefficient $a_s^{(k)}$ in (6.18) at large k satisfies the inequality

$$0 \le s \le 2j = \frac{2\sqrt{k+1}}{\pi \varkappa'}\Big(1 + O(k^{-1})\Big).$$

For these s we readily prove that the ratio ρ_{t+1} of the summand in (6.21) with index $t+1$ to the preceding summand is estimated from above as follows

$$|\rho_{t+1}| \le (k+2)e^{-(k+1)/s} \le (k+2)e^{-\pi\varkappa'\sqrt{k+1}/2}, \quad t \ge 1.$$

Let $\varkappa' \ge \varkappa_0 > 0$. Then, for all nonzero $a_s^{(k)}$ in (6.18), we have the expansion

$$a_s^{(k)} = (s+1)^{k+1}\{1 + O(ke^{-\pi\varkappa_0\sqrt{k}/2})\}. \tag{6.23}$$

The quantity $O(\cdot)$ in this formula is such uniformly in all $\varkappa'$ not less than $\varkappa_0$. From (6.23) it follows that

$$\frac{|a_{j-t}^{(k)} a_{j+t}^{(k)}|}{|a_j^{(k)}|^2} \le e^{-t^2(k+1)/(j+1)^2}\Big\{1 + O(ke^{-\pi\varkappa_0\sqrt{k}/2})\Big\}.$$

Using (6.22), express the exponent of the exponential through $\varkappa'$. Then, extend the found estimate to all t. Inserting the result in (6.18), obtain

$$|\eta_j^{(k)}| \le \left(\sum_{t=1}^{\infty} e^{-t^2(\pi\varkappa')^2}\right)\Big\{1 + O(ke^{-\pi\varkappa_0\sqrt{k}/2})\Big\}.$$

Thus, we may pass to the limit in (6.18) as $k \to \infty$, obtaining

$$\lim_{k\to\infty} \eta_j^{(k)} \le \sum_{t=1}^{\infty}(-1)^{t+1}e^{-t^2(\pi\varkappa')^2} = \eta(\varkappa').$$

This is a sought upper bound on $\eta_j^{(k)}$.

We list several properties of the sum $\eta(\varkappa')$. The function $\eta(\varkappa')$ decreases from 1 to 0 for $\varkappa' \ge 0$. Table 5 gives the approximate values of $\eta(\varkappa')$ and the function

$$\omega(\eta) = \frac{1}{\pi}\log\frac{2\sqrt{1-2\eta}}{1+\sqrt{1-4\eta}}$$

related to $\eta(\varkappa')$ and accountable for the error of (6.3) in virtue of (6.20).

TABLE 5

$\varkappa'$	$\eta(\varkappa')$	$\omega(\eta)$
0.372541	0.249996	0.109050
0.380021	0.237091	0.053152
0.400035	0.204293	0.023728
0.500068	0.084696	0.001681
0.600013	0.028632	0.000147
0.660042	0.013572	0.000031

Observe also that the sum of this series is expressed through the Jacobi theta function ϑ_4 with the ratio of periods $\tau = \pi i \varkappa'^2$ [80].

Let $\varkappa_{\min}$ be the root of the equation $\eta(\varkappa') = 0.25$. Calculation shows that $\varkappa_{\min} \simeq 0.3725$. Thus, for $\varkappa > \varkappa_{\min}$, the quantity $\eta(\varkappa')$ is less than $1/4$.

If the function $j = j(k)$ is defined by (6.22) then, for k sufficiently large, the quantities $\eta_j^{(k)}$ and $\eta_{j-1}^{(k)}$ are both less than $1/4$. For these k, the inequalities (6.20) are thus satisfied. The error of (6.3) is now characterized by the maximum of the moduli of two quantities, the left and right sides of (6.20). The sequences $\eta_j^{(k)}$ and $\eta_{j-1}^{(k)}$ become however small as k increases indefinitely. Consequently, the error of (6.3) is also an infinitesimal.

Consider in more detail the question of conditions for applicability of (6.3). One of these conditions, as we have just shown, is the inequality $\varkappa' > \varkappa_{\min}$. The other two conditions are (6.5) and (6.6). Demonstrate that the latter are equivalent. Write down $q_j^{\mathrm{II}(k)}$ as

$$q_j^{\mathrm{II}(k)} = \frac{k+1}{\pi} \log\left(1 + \frac{1}{j}\right) = \frac{k+1}{\pi} \log \frac{(j+1/2)+1/2}{(j+1/2)-1/2}$$
$$= \frac{k+1}{\pi}\left(\frac{1}{j+1/2} + \frac{1}{12}\frac{1}{(j+1/2)^3} + \dots\right).$$

Recalling the denotation $\varkappa_j'^{(k)}$, obtain

$$q_j^{\mathrm{II}(k)} = \varkappa_j'^{(k)}\sqrt{k+1} + O(k^{-1/2}).$$

We see that the error $q_j^{(k)} - q_j^{\mathrm{II}(k)}$ is bounded whereas $q_j^{\mathrm{II}(k)}$ grows as $\sqrt{k}$.

Consequently, we have the following asymptotic equality

$$q_j^{(k)} = q_j^{\mathrm{II}(k)} + O(1) = q_j^{\mathrm{II}(k)}(1 + O(k^{-1/2})).$$

Hence, the following chain of equalities holds

$$\lim_{k\to\infty} \frac{q_j^{(k)}}{\sqrt{k}} = \lim_{k\to\infty} \frac{q_j^{\mathrm{II}(k)}}{\sqrt{k}} = \lim_{k\to\infty} \varkappa_j^{\prime(k)} = \varkappa'.$$

So, the inequality $\varkappa' > \varkappa_{\min}$ amounts to the condition (6.5) which is thus sufficient for (6.3) to be applicable. Bearing in mind that

$$\frac{1}{\pi\varkappa'} = \lim_{k\to\infty} \frac{j(k)}{\sqrt{k}}$$

(this equality follows from (6.22)), we see the equivalence of (6.5) and (6.6).

It is not hard to calculate that the difference between two consecutive quantities, namely, $q_j^{\mathrm{II}(k)}$ maintains the relation

$$q_{j-1}^{\mathrm{II}(k)} - q_j^{\mathrm{II}(k)} = \frac{k+1}{\pi} \frac{1+O(j^{-2})}{(j+1/2)(j-1/2)} = \pi\varkappa_j^{\prime(k)}\varkappa_{j-1}^{\prime(k)}(1+O(j^{-2})).$$

At this juncture we finish studying the asymptotic formula (6.3) for the roots of the Euler polynomial and pass to deriving (6.1).

§7. The First Asymptotic Formula

Rewrite the algebraic equation $E_k(\lambda) = 0$ in equivalent form. Introduce into consideration a new variable $q = \log(-\lambda)/\pi$, i.e., let $\lambda = -e^{\pi q}$. In the case when λ changes continuously on the interval $(-\infty, -1)$, the variable q ranges over the positive semiaxis of the real line $0 < q < \infty$. Elementary calculations make the polynomial $E_k(\lambda)$ a function of q as follows

$$E_k(\lambda) = E_k(-e^{\pi q}) = \left(\frac{2}{\pi}\right)^{k+2} \frac{\pi q}{2} \frac{d^k}{dq^k} \frac{(\pi/2)^2}{\cosh^2 \pi q/2}. \tag{7.1}$$

All $q_j^{(k)} = \log(-\lambda_j^{(k)})/\pi$ are real, and the first factors in (7.1) obviously have no real zeroes. Consequently, $q_j^{(k)}$ are the zeroes of

$$\frac{d^k}{dq^k} \frac{(\pi/2)^2}{\cosh^2 \pi q/2}$$

that lie on the real axis. Use the familiar expansion in partial fractions of the function under the differentiation sign in the preceding expression

$$\frac{\pi^2}{4\cosh^2 \pi q/2} = \frac{\pi^2}{4\sin^2(\pi(1+iq)/2)} = \sum_{\beta=-\infty}^{\infty} \frac{1}{(2\beta+1+iq)^2}.$$

Differentiating both sides of this equality k times and denoting by $S_k(q)$ the sum of the series

$$S_k(q) = \frac{1}{2} \sum_{\beta=-\infty}^{\infty} \left(\frac{i}{2\beta + 1 + iq} \right)^{k+2},$$

obtain

$$\frac{d^k}{dq^k} \frac{(\pi/2)^2}{\cosh^2 \pi q/2} = 2(-1)^{k+1}(k+1)! S_k(q).$$

The function $S_k(q)$ is real for all k. This is easy to check by replacing the sum over the negative indices β by the sum over the positive indices $\beta' = -\beta - 1$. In this event, we have thc cquality

$$\begin{aligned} 2S_k(q) &= \sum_{\beta=0}^{\infty} \left\{ \left(\frac{i}{2\beta + 1 + iq} \right)^{k+2} + \left(\frac{-i}{2\beta + 1 - iq} \right)^{k+2} \right\} \\ &= \left(\frac{i}{1 + iq} \right)^{k+2} \zeta_k(q) + \left(\frac{-i}{1 - iq} \right)^{k+2} \bar{\zeta}_k(q), \end{aligned} \tag{7.2}$$

with

$$\zeta_k(q) = 1 + \sum_{\beta=1}^{\infty} \left(\frac{1 + iq}{2\beta + 1 + iq} \right)^{k+2} \tag{7.3}$$

Denote by α the angle between the vector $1 + iq$ and the imaginary axis, i.e., let $\alpha = \arg i/(1 + iq)$, $q = \cot \alpha$, and $0 < \alpha \leq \pi/2$. Thus, the interval $0 < \alpha \leq \pi/2$ is mapped in a one-to-one fashion onto the semiaxis $q \geq 0$.

Let $q_j^{(k)} = \cot \alpha_j^{(k)}$. Then the quantities $\alpha_j^{(k)}$, as follows from (7.2), are also the roots of the equation

$$(k+2)\alpha + \arg \zeta_k(\cot \alpha) = (t + 1/2)\pi, \quad t \in \mathbb{Z}. \tag{7.4}$$

This is exactly the sought expression of the original algebraic equation $E_k(\lambda) = 0$.

Inspect (7.4) more thoroughly, assuming that

$$0 \leq \frac{q}{\sqrt{k+2}} = \frac{\cot \alpha}{\sqrt{k+2}} \leq \varkappa < +\infty. \tag{7.5}$$

The next lemma allows us to estimate the behavior of $\arg \zeta_k(\cot \alpha)$ as $k \to \infty$.

Lemma 9.4. *The series* (7.3) *converges uniformly in k with q ranging over the set defined by* (7.5).

Proof. Let $\zeta_k^N(q)$ denote a partial sum of (7.3). Then

$$|\zeta_k(q) - \zeta_k^N(q)| \le \sum_{\beta=N+1}^{\infty} \left(\frac{1+q^2}{(2\beta+1)^2+q^2}\right)^{(k+2)/2}$$

The series on the right side is an increasing function in q. Consequently, the estimate holds

$$\sup_{0\le q\le \varkappa\sqrt{k+2}} |\zeta_k(q) - \zeta_k^N(q)| \le \sum_{\beta=N+1}^{\infty} \left(\frac{1+\varkappa^2(k+2)}{(2\beta+1)^2+\varkappa^2(k+2)}\right)^{(k+2)/2} \tag{7.6}$$

Denote the term of (7.6) with a fixed index $\beta \ge N+1$ by $A_\beta(k)$. The function $A_\beta(k)$ of k decreases. The last claim is easy from the equalities

$$A_\beta(k) = \left[\left(1-\frac{1}{y(k)}\right)^{y(k)-1}\right]^{2\beta(\beta+1)/\varkappa^2} \left(\frac{y(k)}{y(k)-1}\right)^{1/2\varkappa^2},$$

$$y(k) = 1 + \frac{\varkappa^2(k+2)+1}{(2\beta+1)^2-1},$$

and the observation that $g(y) = (1-1/y)^{y-1}$ is a decreasing function for $y \ge 1$. Hence, $A_\beta(k) \le A_\beta(2)$ for $k \ge 2$. Inserting these inequalities in (7.6) and letting

$$x(\beta) = \frac{(2\beta+1)^2+4\varkappa^2}{4\beta(\beta+1)},$$

obtain the estimate

$$\sup_{0\le q\le \varkappa\sqrt{k+2}} |\zeta_k(q) - \zeta_k^N(q)| \le \sum_{\beta=N+1}^{\infty} \left(1-\frac{1}{x(\beta)}\right)^2.$$

In the right side we sum the values of a decreasing function in β. This enables us to substitute integration for summation while preserving the inequalities

$$\sup_{0\le q\le \varkappa\sqrt{k+2}} |\zeta_k(q) - \zeta_k^N(q)| \le \int_N^{\infty} \left(1-\frac{1}{x(\beta)}\right)^2 d\beta.$$

Carry out a change of variables by passing from β to $x(\beta)$. By definition, as $\beta \to \infty$ the function $x(\beta)$ decreases and tends to 1. Moreover,

$$d\left(\frac{x(\beta)}{x(\beta)-1}\right) = \frac{4(2\beta+1)d\beta}{1+4\varkappa^2}, \quad d\beta = -\frac{(1+4\varkappa^2)dx(\beta)}{4(x(\beta)-1)^{3/2}(x(\beta)+4\varkappa^2)^{1/2}}.$$

We thus obtain the estimate

$$\sup_{0\le q\le \varkappa\sqrt{k+2}} |\zeta_k(q)-\zeta_k^N(q)|$$

$$\le \frac{1+4\varkappa^2}{4}\int_1^{x(N)} \frac{(x-1)^{1/2}}{x^2(x+4\varkappa^2)^{1/2}}\,dx \le \frac{(1+4\varkappa^2)^2}{4^{5/2}}\frac{1}{N^3}. \tag{7.7}$$

Passing to the limit as $N\to\infty$, complete the proof of Lemma 9.4.

Corollary 7.1. *For every q in the interval* (7.5), *the inequality holds*

$$\lim_{k\to\infty}|\zeta_k(q)-1| \le \xi(\varkappa) = \sum_{\beta=1}^{\infty} e^{-2\beta(\beta+1)/\varkappa^2}. \tag{7.8}$$

PROOF. Using Lemma 9.4, derive the following relations

$$\lim_{k\to\infty}|\zeta_k(q)-1| \le \lim_{k\to\infty}\lim_{N\to\infty}\left|\zeta_k^N(q)-1\right| = \lim_{N\to\infty}\lim_{k\to\infty}\left|\zeta_k^N(q)-1\right|$$

$$\le \lim_{N\to\infty}\lim_{k\to\infty}\sum_{\beta=1}^{N}\left(\frac{1+\varkappa^2(k+2)}{(2\beta+1)^2+\varkappa^2(k+2)}\right)^{(k+2)/2}.$$

We may further find the limit as $k\to\infty$ of the summand with index β in the last sum by the formula

$$\lim_{k\to\infty}\left(1+\frac{1}{\varkappa^2(k+2)}\right)^{(k+2)/2} \Big/ \left(1+\frac{(2\beta+1)^2}{\varkappa^2(k+2)}\right)^{(k+2)/2} = e^{-2\beta(\beta+1)/\varkappa^2}.$$

The estimate (7.8) is thus proven.

The function $\xi(\varkappa)$ for $\varkappa>0$ decreases. Approximate values of $\xi(\varkappa)$ and $\varkappa^2\xi(\varkappa)$ are shown in Table 6.

TABLE 6

$\varkappa$	$\xi(\varkappa)$	$\varkappa^2\xi(\varkappa)$
0.650001	0.000077	0.000033
0.700021	0.000285	0.000140
1.000035	0.018327	0.018328
2.000068	0.420227	1.681805
3.000013	0.987426	8.886911
3.021542	1.000059	9.130255

We mention in passing that the function $\xi(\varkappa)$ is expressible through the Jacobi theta function ϑ_1 with the ratio of periods $\tau = 2i/\pi\varkappa^2$ [80].

Denote by $\varkappa_{max}$ the quantity for which $\xi(\varkappa_{max}) = 1$. Calculation shows that $\varkappa_{\max} \cong 3.0215$. Let $\varkappa < \varkappa_{\max}$. Then, by (7.8), $\zeta_k(q)$ lies in a disk with center 1 and radius r less than 1. Moreover, at sufficiently large k, the estimate holds

$$|\arg \zeta_k(q)| \leq h = \sin^{-1} r < \sin^{-1} \xi(\varkappa_{\max}) = \pi/2. \tag{7.9}$$

The greater the k is, the closer the h approaches $\sin^{-1}\xi(\varkappa)$. The inequality (7.9) is obvious from the geometric construction.

For $\varkappa > 0$ there is an $\varepsilon > 0$ such that for $\alpha \in (\pi/2 - \varepsilon, \pi/2]$

$$\frac{\cot\alpha}{\sqrt{k+2}} < \cot\alpha \leq \varkappa. \tag{7.10}$$

Consequently, there is a positive h such that, for all α in the above interval and all sufficiently large k, the estimate holds

$$|\arg \zeta_k(\cot\alpha)| < h < \pi/2. \tag{7.11}$$

Furthermore, at large k, we may take h equal to $\sin^{-1}\xi(\varkappa)$.

If the interval $(\pi/2 - \varepsilon, \pi/2]$ contains a root α_* of (7.4) then by (7.11) this root obviously belongs to the interval $\left(\alpha_1^{(k)}, \alpha_2^{(k)}\right)$, with

$$\alpha_1^{(k)}(t) = \frac{(t+1/2)\pi - h}{k+2}, \quad \alpha_2^{(k)}(t) = \frac{(t+1/2)\pi + h}{k+2}.$$

Let $t = 0, 1, \ldots, k+1$. Consider the points

$$\alpha_0^{(k)}(t) = \frac{\pi(t+1/2)}{(k+2)}$$

of $[0, \pi]$. Each of these points is the center of the interval $\left(\alpha_1^{(k)}(t), \alpha_2^{(k)}(t)\right)$. The distance between the right endpoint $\alpha_2^{(k)}(t)$ of the interval with index t and the left endpoint $\alpha_1^{(k)}(t+1)$ of the interval with index $t+1$ equals $2(\pi/2 - h)/(k+2)$. Since by hypothesis $h < \pi/2$, these intervals are disjoint.

If $\alpha_0^{(k)}(t)$ satisfies (7.10) and $\varphi(\alpha)$ stands for $(k+2)\alpha + \arg\zeta_k(\cot\alpha)$, then

$$\varphi\left(\alpha_1^{(k)}(t)\right) < (t+1/2)\pi < \varphi\left(\alpha_2^{(k)}(t)\right).$$

Whence it follows that the interval $\left(\alpha_1^{(k)}(t), \alpha_2^{(k)}(t)\right)$ has at least one root α_* of (7.4). Moreover,

$$\cot \alpha_2^{(k)}(t) < q_* = \cot \alpha_* < \cot \alpha_1^{(k)}(t). \tag{7.12}$$

Let $q_t^{\mathrm{I}(k)} = \cot \alpha_0^{(k)}(t)$ and $g_k = \tan(h/(k+2))$. Then, applying the routine trigonometric formula for the cotangent of the sum of two quantities, obtain

$$\cot \alpha_j^{(k)}(t) = q_t^{\mathrm{I}(k)} \pm g_k \frac{1 + \left(q_t^{\mathrm{I}(k)}\right)^2}{1 \mp g_k q_t^{\mathrm{I}(k)}}.$$

The upper sighs are chosen here for $j = 1$; whereas the lower, for $j = 2$. The inequality (7.12) is thus equivalent to the following

$$-g_k \frac{1 + \left(q_t^{\mathrm{I}(k)}\right)^2}{1 + g_k q_t^{\mathrm{I}(k)}} < q_* - q_t^{\mathrm{I}(k)} < g_k \frac{1 + \left(q_t^{\mathrm{I}(k)}\right)^2}{1 - g_k q_t^{\mathrm{I}(k)}}. \tag{7.13}$$

The sequence g_k vanishes as $k \to \infty$, for

$$\frac{\sin^{-1}\xi(\varkappa)}{k+2} + O(k^{-3}); \quad g_k = \frac{\sin^{-1}\xi(\varkappa)}{k+2} + O(k^{-3});$$

and, for the values of $\alpha_0^{(k)}(t)$ meeting (7.10), the sequence $q_t^{\mathrm{I}(k)}$ is uniformly bounded in k, i.e.,

$$\left|q_t^{\mathrm{I}(k)}\right| = \left|\cot \alpha_0^{(k)}(t)\right| \le \varkappa.$$

Thus, the error of (7.13) becomes however small at k large. Its magnitude is easy to estimate on using the positivity of g_k and $q_t^{\mathrm{I}(k)}$, which yields

$$\left|q_* - q_t^{\mathrm{I}(k)}\right| < g_k \frac{1 + \left(q_t^{\mathrm{I}(k)}\right)^2}{1 - g_k q_t^{\mathrm{I}(k)}}. \tag{7.14}$$

This inequality also characterizes the error of the asymptotic formula (6.1).

For some roots λ_* of the polynomial $E_k(\lambda)$ the quantities $q_* = \log(-\lambda_*)/\pi$ may simultaneously satisfy the two inequalities

$$\varkappa_{\min} < \varkappa' \le q^*/\sqrt{k+2} < q_*/\sqrt{k} \le \varkappa < \varkappa_{\max}. \tag{7.15}$$

Consequently, for every q_* there are two quantities $q_t^{\mathrm{I}(k)}$ and $q_j^{\mathrm{II}(k)}$ with integer t and j which are appropriate for approximation of q_*. In particular, the quantity $q_t^{\mathrm{I}(k)} = \cot \alpha_0^{(k)}(t)$ is not less than $\varkappa'\sqrt{k}$. Consequently,

$$\alpha_0^{(k)}(t) \le \operatorname{cotan}^{-1}(\varkappa'\sqrt{k}) = 1/(\varkappa'\sqrt{k}) + O(k^{-3/2}).$$

Thus, for sufficiently large k, the quantity $\alpha_0^{(k)}(t)$ becomes however small. Hence, $q_t^{\mathrm{I}(k)}$ admits the approximation

$$q_t^{\mathrm{I}(k)} = \frac{k+2}{\pi(t+1/2)} + O(k^{-1/2}). \tag{7.16}$$

From (7.15) it also follows that $q_j^{\mathrm{II}(k)} \le \varkappa\sqrt{k}$. This condition amounts to the fact that, as k increases, the index $j = j(k)$ grows slower than $C\sqrt{k}$ with C a nonzero constant. Consequently,

$$q_j^{\mathrm{II}(k)} = \frac{k+2}{\pi(j+1/2)} + O(k^{-1/2}). \tag{7.17}$$

From (7.16) and (7.17) it follows $q_t^{\mathrm{I}(k)}$ and $q_j^{\mathrm{II}(k)}$ may properly approximate the same q_* only if $j = t$; in this event, the difference between $q_j^{\mathrm{II}(k)}$ and $q_t^{\mathrm{I}(k)}$ is $O(k^{-1/2})$.

Let $q_* = q_{j_0}^{(k)}$ satisfy (7.15) for some j_0. Then, to approximate the roots $q_j^{(k)}$, $j = 1, 2, \ldots, j_0$, we use $q_j^{\mathrm{II}(k)}$. We approximate the remaining roots $q_j^{(k)}$ by $q_j^{\mathrm{I}(k)}$, $j = j_0+1, j_0+2, \ldots$. For these j, each of the numbers $q_j^{(k)}$ lies in some neighborhood about $q_j^{\mathrm{I}(k)}$, with these neighborhoods disjoint from one another. Thus, the number of the remaining roots $q_j^{(k)}$ is exactly the same as the number of $q_j^{\mathrm{I}(k)}$ for $j \ge j_0 + 1$. Consequently, in each of the intervals $\left(\cot \alpha_1^{(k)}(j), \cot \alpha_2^{(k)}(j)\right)$ with $j \ge j_0 + 1$ there is only one root q_* equal to $q_j^{(k)}$. This, together with (7.14), entails the estimate

$$\left|q_j^{(k)} - q_j^{\mathrm{I}(k)}\right| < g_k \frac{1 + \left(q_j^{\mathrm{I}(k)}\right)^2}{1 - g_k q_j^{\mathrm{I}(k)}}, \tag{7.18}$$

with $g_k = \tan h/(k+2)$ and $h = \sin^{-1}\xi(\varkappa)$.

This estimate is proven provided that $q_j^{(k)}/\sqrt{k+2} \le \varkappa$. The same must hold for the ratio $q_j^{\mathrm{I}(k)}/\sqrt{k+2}$. In other words, the inequality holds

$$\frac{\pi(j+1/2)}{k+2} \ge \operatorname{cotan}^{-1}(\varkappa\sqrt{k+2}) = \frac{1}{\varkappa\sqrt{k+2}} + O(k^{-3/2}). \tag{7.19}$$

Bearing in mind that the index j changes with the growth of k, i.e., $j = j(k)$, we pass in (7.19) to the limit as $k \to \infty$ and obtain

$$\lim_{k\to\infty} \frac{j}{\sqrt{k}} \ge \frac{1}{\varkappa\pi} > \frac{1}{\pi\varkappa_{\max}}.$$

This is a sufficient condition for the first approximate formula (6.1) to be applicable.

Compare (6.1) and (6.3) at large k. Put $\varkappa_j^{(k)} = q_j^{(k)}/\sqrt{k}$. In the interval $\varkappa_{\min} < \varkappa_j^{(k)} < \varkappa_{\max}$, as $\varkappa_j^{(k)}$ increases, the error of (6.3) grows whereas the error of (6.1) diminishes. At a point where the estimates for the errors coincide, as we see in Tables 5 and 6, the quantity $\varkappa_j^{(k)}$ is approximately 0.66. In this event we have an approximate equality

$$\left|q_j^{(k)} - q_j^{\mathrm{I}(k)}\right| \cong \left|q_j^{(k)} - q_j^{\mathrm{II}(k)}\right| \cong 0.0001.$$

The distance between two consecutive values $q_j^{(k)}$ and $q_{j-1}^{(k)}$ is approximately equal to $\pi\varkappa_j^{(k)}\varkappa_{j-1}^{(k)} \cong 1.48$. It thus happens that error is more than 10^4 times less than the distance between consecutive roots.

§8. The Weights of Optimal Quadrature Formulas

Return to the one-dimensional Problem B_2, i.e., find a function $u[\beta]$ of polynomial growth which extends

$$f[\beta] = \int_0^1 G(h\beta - y)\,dy$$

from the interval $0 \leq \beta \leq N$ to the whole set of integers $\mathbb{Z}$ and polynomials $Q_{m-1}^{(\pm)}[\beta]$ of degree at most $m-1$ which satisfy two additional conditions:

(1) the function $u[\beta]$ has the definite asymptotic expansion

$$\lim_{\beta\to\pm\infty} \left\{(-1)^m u[\beta] - Q_{m-1}^{(\pm)}[\beta] \pm Q_{2m-1}[\beta]\right\} = 0 \tag{8.1}$$

with

$$Q_{2m-1}[\beta] = \frac{h^{2m}[(N-\beta)^{2m} - \beta^{2m}]}{2(2m)!};$$

(2) the norm of $\|u\|_m^D$ is minimal among the norms of all admissible extensions $f[\beta]$ with asymptotic expansion of the shape (8.1).

The signs in (8.1) are taken either all upper or all lower simultaneously. This agreement is also effective in what follows.

Existence and uniqueness of a solution to Problem B_2 in the case of the zero polynomial $Q_{2m-1}[\beta]$ are proven in § 4 of the current chapter. The corresponding proof is easy to change so that it suit an arbitrary $Q_{2m-1}[\beta]$.

A solution to Problem B_2 may be written as the sum

$$u[\beta] = G_h^{(m)}[\beta] * c[\beta] + P_{m-1}[\beta]$$

with $c[\beta]$ and $P_{m-1}[\beta]$ satisfying (1.4)–(1.6). This implies in particular that the convolution of the discrete polyharmonic operator $D_h^{(m)}[\beta] = h^{-2m} D_1^{(m)}[\beta]$ with $u[\beta]$ vanishes at all points β outside interval $[0, N]$; i.e., for $\beta < 0$ and $\beta > N$, the equality holds

$$v[\beta] = D_1^{(m)}[\beta] * u[\beta] = 0. \tag{8.2}$$

The discrete polyharmonic operator $D_1^{(m)}[\beta]$ was written down as the convolution of two functions

$$\frac{D_1^{(m)}[\beta]}{(2m-1)!} = \Delta_2^{[m]}[\beta] * E_{2m-2}^{-1}[\beta - m + 1] \tag{8.3}$$

in § 5 of the current chapter. Here, $E_{2m-2}^{-1}[\beta]$ stands for the Fourier coefficients

$$E_{2m-2}^{-1}[\beta] = \int_0^1 e^{i2\pi p\beta} / E_{2m-2}(e^{-i2\pi p})\, dp,$$

and $E_{2m-2}(\lambda)$ is the Euler polynomial of degree $2m-2$.

The function $\Delta_2^{[m]}[\beta]$ is, obviously, compactly-supported. The convolution of each discrete function $\varphi[\beta]$ with the compactly-supported function $\Delta_2^{[m]}[\beta]$ is the taking of the mth order symmetric difference of $\varphi[\beta]$.

Denoting by $B_{2m-2}[\beta]$ the Fourier coefficients of the product

$$e^{i2\pi\beta(m+1)p} E_{2m-2}(e^{-i2\pi p})/(2m-1)!,$$

from (8.3) infer

$$B_{2m-2}[\beta] * D_1^{(m)}[\beta] = \Delta_2^{[m]}[\beta]. \tag{8.4}$$

Lemma 9.5. *Beyond the interval* $[0, N]$ *a solution* $u[\beta]$ *of Problem* B_2 *is a polynomial of degree at most* $2m-1$, *i.e.,*

$$u[\beta] = \begin{cases} P_{2m-1}^{(-)}[\beta], & \beta < 0, \\ P_{2m-1}^{(+)}[\beta], & \beta > N. \end{cases} \tag{8.5}$$

PROOF. Construct a polynomial $P_{2m-1}^{(+)}[\beta]$ that agrees with $u[\beta]$ for $\beta > N$. The definition of $B_{2m-2}[\beta]$ implies that the support of this function lies within the interval $[-m+1, m-1]$. Consequently, if the function $v[\beta] = D_1^{(m)}[\beta] * u[\beta]$ vanishes for $\beta > N$ then the convolution $B_{2m-2}[\beta] * v[\beta]$ is also equal to 0, but now for $\beta \geq N+m$. This, together with (8.2) and (8.4), implies that

$$\Delta_2^{[m]}[\beta] * u[\beta] = 0, \quad \beta \geq N+m. \tag{8.6}$$

The last relation is a linear finite-difference equation given for $\beta \geq N+m$. The dimension of its solution space equals $2m$. The polynomials $[\beta]^\alpha, \alpha = 0, 1, \ldots, 2m-1$, obviously constitute a basis for this space. Consequently, a solution $u[\beta]$ of (8.6) is a polynomial of degree $2m-1$ for $\beta \geq N+m$, namely,

$$u[\beta] = P_{2m-1}^{(+)}[\beta] = \sum_{\alpha=0}^{2m-1} c_\alpha [\beta]^\alpha, \quad \beta \geq N+m.$$

Using the asymptotic condition (8.1), find the coefficients c_α for $\alpha = m$, $m+1, \ldots, 2m-1$. We thus obtain

$$P_{2m-1}^{(+)}[\beta] = -Q_{2m-1}[\beta] + Q_{m-1}^{(+)}[\beta].$$

The coefficients of the polynomial $Q_{m-1}^{(+)}[\beta]$ of degree $m-1$, may obviously be given so that for $\beta = N+1, N+2, \ldots, N+m$ the values of $P_{2m-1}^{(+)}[\beta]$ and $u[\beta]$ coincide. In this event $u[\beta] = P_{2m-1}^{(+)}[\beta]$ for all $\beta > N$. In the case of $\beta < 0$, the equality (8.5) is deduced similarly. The proof of Lemma 9.5 is complete.

Corollary 8.1. *The polynomials $P_{2m-1}^{(\pm)}[\beta]$ giving a solution $u[\beta]$ beyond the interval $[0, N]$ may be written down as a linear combination of the polynomials $Q_{2m-1}[\beta]$ and $Q_{m-1}^{(\pm)}[\beta]$ of the asymptotic equality* (8.1):

$$P_{2m-1}^{(\pm)}[\beta] = \mp Q_{2m-1}[\beta] + Q_{m-1}^{(\pm)}[\beta].$$

We now derive formulas that allows us to find a solution to Problem B_1 granted a solution to Problem B_2. We have

Lemma 9.6. *The only solution $c[\beta]$ and $P_{m-1}[\beta]$ to* (1.4)–(1.6) *for $n = 1$ and $2m \leq N$ is connected with a solution $u[\beta]$ and $Q_{m-1}^{(\pm)}[\beta]$ to the corresponding Problem B_2 by the following relations*

$$c[\beta] = D_h^{(m)}[\beta] * u[\beta], \quad P_{m-1}[\beta] = \frac{1}{2}\left(Q_{m-1}^{(-)}[\beta] + Q_{m-1}^{(+)}[\beta]\right). \tag{8.7}$$

PROOF. The possibility of writing the weights $c[\beta]$ as the convolution $D_h^{(m)}[\beta] * u[\beta]$ is justified in § 4 of the current chapter. It thus suffices to prove the second of the equalities (8.7). Let $v[\beta] = G_h^{(m)}[\beta] * c[\beta]$, with

$$G_h^{(m)}[\gamma] = (-1)^m h^{2m} |\gamma|^{2m-1} / 2(2m-1)!.$$

Show that, for $\beta < 0$ and $\beta > N$, the function $v[\beta]$ is a polynomial of degree $2m-1$.

For $\beta < 0$, the properties of $c[\beta]$ entail the equality

$$\begin{aligned} v[\beta] &= \sum_\gamma G_h^{(m)}[\beta - \gamma] c[\gamma] = (-1)^m \frac{h}{2(2m-1)!} \sum_{\gamma=0}^{N} |\beta h - \gamma h|^{2m-1} c[\gamma] \\ &= (-1)^m \frac{h^{2m}}{2(2m-1)!} \sum_{\gamma=0}^{N} c[\gamma] \sum_{k=0}^{2m-1} \binom{2m-1}{k} (-1)^k \beta^k \gamma^{2m-1-k} \\ &= (-1)^m \frac{h^{2m}}{2(2m-1)!} \sum_{k=0}^{2m-1} \binom{2m-1}{k} (-1)^k \beta^k f_{2m-1-k} = R_{2m-1}[\beta], \end{aligned} \tag{8.8}$$

with

$$f_k = \sum_{\gamma=0}^{N} c[\gamma] \gamma^k. \tag{8.9}$$

Analogous calculation shows that, for $\beta > N$, the function $v[\beta]$ agrees with the polynomial $R_{2m-1}[\beta]$ taken with the opposite sign.

As follows from the definition of $v[\beta]$, for all $\beta \in \mathbb{Z}$ the convolution of $D_h^{(m)}[\beta]$ with the difference $u[\beta] - v[\beta]$ gives zero, namely,

$$D_h^{(m)}[\beta] * (u[\beta] - v[\beta]) = 0.$$

This is possible only on condition that $u[\beta]$ and $v[\beta]$ differ from one another by a polynomial of degree $2m-1$, i.e.,

$$u[\beta] = G_h^{(m)}[\beta] * c[\beta] + P_{2m-1}[\beta]. \tag{8.10}$$

By choice, for $0 \le \beta \le N$ the function $u[\beta]$ agrees with $f[\beta]$ and the difference $f[\beta] - v[\beta]$ is a polynomial $P_{m-1}[\beta]$. Whence and from (8.10) we conclude that the polynomials $P_{2m-1}[\beta]$ and $P_{m-1}[\beta]$ agree for $0 \le \beta \le N$. For $2m \le N$ this is possible only if $P_{2m-1}[\beta]$ and $P_{m-1}[\beta]$ agree at every point $\beta \in \mathbb{Z}$. From (8.5) and (8.10) infer now that

$$P_{2m-1}^{(\pm)}[\beta] = \mp R_{2m-1}[\beta] + P_{m-1}[\beta].$$

Inserting the decomposition of $P^{(\pm)}_{2m-1}[\beta]$ taken from Corollary 8.1 and summing the resultant equalities, arrive at (8.7). The proof of Lemma 9.6 is complete.

In virtue of uniqueness of an optimal quadrature formula, we have the following obvious equality

$$c[\beta] = c[N-\beta]. \tag{8.11}$$

Denoting the convolution $\Delta_2^{[m]}[\beta] * u[\beta]$ by $g[\beta]$ and using (8.4), obtain the equality

$$B_{2m-2}[\beta] * c[\beta] = N^{2m} g[\beta], \tag{8.12}$$

which is a linear finite-difference equation. We are interested in a solution to it such as vanishes for $\beta < 0$. Using this additional condition together with (8.11), it is not hard to obtain some recurrent relations for $c[\beta]$.

Consider one more algorithm for determining optimal weights. Assume further that the interval $[0,1]$ is transformed into $[0,N]$ and the lattice mesh-size h equals 1. In this case System B takes the form

$$\begin{aligned} G_m[\beta] * c[\beta] + P_{m-1}[\beta] &= f[\beta], \quad 0 \le \beta \le N, \\ c[\beta] &= 0, \qquad \beta \notin [0,N], \\ \sum_{\beta=0}^{N} c[\beta]\beta^\alpha &= f_\alpha, \qquad \alpha = 0,1,\ldots,m-1. \end{aligned} \tag{8.13}$$

Here

$$G_m[\beta] = G_1^{(m)}[\beta], \quad f[\beta] = \int_0^N G(\beta - y)\,dy, \quad f_\alpha = \int_0^N x^\alpha\,dx.$$

As before, let $u[\beta]$ be a solution to Problem B_2, i.e. assume that it coincides with $P_{m-1}[\beta] + G_m[\beta] * c[\beta]$. Consider the finite-difference equation

$$B_{2m-2}[\beta] * \hat{u}[\beta] = u[\beta]. \tag{8.14}$$

If $\hat{u}[\beta]$ is a solution to (8.14) then, as follows from (8.4), the weights $c[\beta]$ are tied with $\hat{u}[\beta]$ by the relation

$$c[\beta] = \Delta_2^{[m]}[\beta] * \hat{u}[\beta].$$

Thus, we may find the sought weights $c[\beta]$ by solving the equation (8.14) with the right side $u[\beta]$ given as follows

$$u[\beta] = (-1)^m \begin{cases} Q_{[m,2m-1]}[\beta] + Q^{(-)}_{m-1}[\beta], & \beta < 0, \\ [\beta^{2m} + (N-\beta)^{2m}]/2(2m)!, & 0 \le \beta \le N, \\ -Q_{[m,2m-1]}[\beta] + Q^{(+)}_{m-1}[\beta], & \beta > N. \end{cases}$$

Here $Q_{[m,2m-1]}[\beta] = \sum_{i=m}^{2m-1} q_i \beta^i$, with

$$q_i = \frac{(-1)^i N^{2m-i}}{2i!(2m-i)!}$$

and $Q_{m-1}^{(\pm)}[\beta]$ some polynomials of degree at most $m-1$.

The polynomial $Q_{[m,2m-1]}[\beta]$ may be written in a more convenient form as

$$\begin{aligned} Q_{[m,2m-1]}[\beta] &= \sum_{i=0}^{2m} q_i \beta^i - \beta^{2m}/2(2m)! - \sum_{i=0}^{m-1} q_i \beta^i \\ &= \frac{(N-\beta)^{2m} - \beta^{2m}}{2(2m)!} - P_{m-1}^{(0)}[\beta]. \end{aligned}$$

Inserting $Q_{[m,2m-1]}[\beta]$ in the preceding formula for $u[\beta]$, obtain

$$u[\beta] = \frac{(-1)^m}{2(2m)!} \begin{cases} (N-\beta)^{2m} - \beta^{2m} + P_{m-1}^{(-)}[\beta], & \beta < 0, \\ \beta^{2m} + (N-\beta)^{2m}, & 0 \le \beta \le N, \\ \beta^{2m} - (N-\beta)^{2m} + P_{m-1}^{(+)}[\beta], & \beta > N. \end{cases}$$

We do not specify the polynomials $P_{m-1}^{(\pm)}[\beta]$ since they never enter the final formula.

We may seek a solution to (8.14) in the shape

$$\hat{u}[\beta] = \hat{u}_0[\beta] + \sum_{i=1}^{m-1} a_i \lambda_i^\beta + \sum_{i=1}^{m-1} b_i \lambda_i^{N-\beta},$$

with $\hat{u}_0[\beta]$ a particular solution to (8.14) and λ_i the roots of the Euler polynomial $E_{2m-2}(\lambda)$ which lie in the interval $(-1,0)$. If, moreover, the convolution $\Delta_2^{[m]}[\beta] * \hat{u}_0[\beta]$ coincides with 1 then we have

$$c[\beta] = 1 + \sum_{i=1}^{m-1} a_i' \lambda_i^\beta + \sum_{i=1}^{m-1} b_i' \lambda_i^{N-\beta}. \tag{8.15}$$

An analogous approach was pursued by I. Schoenberg and S. Silliman in [204] who calculated the values of the weights for $m \le 7$. They also suggested that $\beta > 0$ and

$$c[\beta] = 1 + \sum_{i=1}^{m} a_i \lambda_i^\beta.$$

It turned out that the quantities a_i take values of large magnitude but with opposite signs whereas the deviations of the function $c[\beta]$ from 1 are far from being large. This hints upon the idea that at large m the decomposition (8.15) might be ineffective in calculating the weights. In connection with this, in [300] there was proposed not to seek $c[\beta]$ at large m in the form of (8.15) which reduces the problem to determining the coefficients a_i' and b_i', but start with providing a direct algorithm for calculating the two sums in (8.15). This idea was implemented in [300] for $m \leq 30$. We give a brief description for the algorithm.

Write down $B_{2m-2}[\beta]$ as the convolution

$$B_{2m-2}[\beta] = B_{m-1}^{(-)}[\beta] * B_{m-1}^{(+)}[\beta]/(2m-1)!. \tag{8.16}$$

The roots of the characteristic polynomials corresponding to the difference operators $B_{m-1}^{(-)}[\beta]$ and $B_{m-1}^{(+)}[\beta]$ coincide with the roots of the Euler polynomial $E_{2m-2}(\lambda)$ of degree $2m-2$ which lie inside and outside the interval $(-1,0)$ respectively. Let the characteristic polynomials for $B_{m-1}^{(\pm)}[\beta]$ take the form

$$E_{m-1}^{(-)}(\lambda) = \prod_{i=1}^{m-1}(\lambda - \lambda_i) = \lambda^{m-1} + k_1\lambda^{m-2} + \cdots + k_{m-1},$$

$$E_{m-1}^{(+)}(\lambda) = \prod_{i=1}^{m-1}\left(\lambda - \frac{1}{\lambda_i}\right) = \lambda^{m-1} + g_1\lambda^{m-2} + \cdots + g_{m-1}.$$

Then it is obvious that $g_i = k_{m-1-i}/k_{m-1}$ and

$$B_{m-1}^{(-)}[\beta] = \sum_{i=0}^{m-1} k_i\delta[\beta - i], \quad B_{m-1}^{(+)}[\beta] = \sum_{i=0}^{m-1} g_i\delta[\beta + m - 1 - i].$$

Let $\widehat{Q}^{(-)}[\beta]$ and $\widehat{Q}^{(+)}[\beta]$ be polynomials of degree $2m-1$ which are given on the intervals $(-\infty, m-1]$ and $[N-m+1, \infty)$ and solve the following equations

$$B_{2m-2}[\beta] * \widehat{Q}^{(\mp)}[\beta] = f[\beta] + P_{m-1}^{(\mp)}[\beta]. \tag{8.17}$$

On the interval $[-m+1, N+m-1]$ we seek a solution to the nonhomogeneous equation

$$B_{2m-2}[\beta] * \hat{u}[\beta] = f[\beta]$$

in the form of the sum

$$\hat{u}[\beta] = \hat{u}_0[\beta] + \hat{u}_-[\beta] + \hat{u}_+[\beta].$$

Here $\hat{u}_0[\beta]$ stands for some particular solution to the equation

$$B_{2m-2}[\beta] * \hat{u}_0[\beta] = f[\beta], \quad 0 < \beta < N;$$

and the functions $\hat{u}_+[\beta]$ and $\hat{u}_-[\beta]$ satisfy the following relations

$$B^{(\mp)}_{m-1}[\beta] * \hat{u}_\mp[\beta] = 0. \tag{8.18}$$

We pose the initial conditions on $\hat{u}_\mp[\beta]$ by making the functions $\hat{u}[\beta]$, $\widehat{Q}^{(-)}[\beta]$, and $\widehat{Q}^{(+)}[\beta]$ agree on the intersection of their domains of definition

$$\hat{u}_-[\beta] + \hat{u}_+[\beta] + \hat{u}_0[\beta] = \widehat{Q}^{(\mp)}[\beta]. \tag{8.19}$$

Here β ranges over the interval $[-m+1, m-1]$ if the minus sign is chosen on the right side of the formula. Otherwise, β ranges over the interval $[N-m+1, N+m-1]$. Observe that

$$\hat{u}_-[\beta] = \sum_{i=1}^{m-1} a_i \lambda_i^\beta, \quad \hat{u}_+[\beta] = \sum_{i=1}^{m-1} b_i \lambda_i^{N-\beta},$$

and take N so large that we may neglect in (8.19) either $\hat{u}_+[\beta]$ or $\hat{u}_-[\beta]$. Then (8.19) reduces to the following equalities

$$\hat{u}_-[\beta] + \hat{u}_0[\beta] = \widehat{Q}^{(-)}[\beta], \quad -m+1 \le \beta \le m-1, \tag{8.20}$$

$$\hat{u}_+[\beta] + \hat{u}_0[\beta] = \widehat{Q}^{(+)}[\beta], \quad N-m+1 \le \beta \le N+m-1. \tag{8.21}$$

The polynomials $\widehat{Q}^{(-)}[\beta]$ and $\widehat{Q}^{(+)}[\beta]$ are defined to within polynomial summands of degree at most $m-1$, since only up to such accuracy we know the right side of (8.17). Therefore, it is convenient to introduce instead of $\hat{u}_-[\beta]$ and $\hat{u}_+[\beta]$ the new unknown functions $\xi_-[\beta]$ and $\xi_+[\beta]$ that are the mth order finite differences of $\hat{u}_-[\beta]$ and $\hat{u}_+[\beta]$. Being treated so, the unknown polynomials of degree $m-1$ mentioned in the agreement conditions (8.19)–(8.21) disappear. Thus, put

$$\xi_\mp[\beta] = \Delta^{(m)}_\pm * \hat{u}_\mp[\beta],$$

with $\Delta_+[\beta] = \delta[\beta+1] - \delta[\beta]$ and $\Delta_-[\beta] = \delta[\beta] - \delta[\beta-1]$. The operator $\Delta^{(m)}_\pm[\beta]*$ signifies that we take the convolution with $\Delta_\pm[\beta]$ consecutively m times. Obviously, the function $\xi_\pm[\beta]$ satisfies (8.18). We find the initial conditions on $\xi_-[\beta]$ and $\xi_+[\beta]$ from (8.20) and (8.21), namely,

$$\begin{aligned} &\xi_-[\beta] + \tau_1[\beta] = 0, \quad -m+1 \le \beta \le -1, \\ &\xi_+[\beta] + \tau_2[\beta] = 0, \quad N+1 \le \beta \le N+m-1, \end{aligned} \tag{8.22}$$

with $\tau_j[\beta]$ standing for the following functions of a discrete variable

$$\tau_1[\beta] = \Delta_+^{(m)}[\beta] * (\hat{u}_0[\beta] - \widehat{Q}^{(-)}[\beta]), \quad \tau_2[\beta] = \Delta_-^{(m)}[\beta] * (\hat{u}_0[\beta] - \widehat{Q}^{(+)}[\beta]).$$

Rearrange (8.22) to make it more concise. To this end, first, select a solution $\hat{u}_0[\beta]$ appropriately and, second, find the coefficients of the highest powers of β in the polynomials $\widehat{Q}^{(\pm)}[\beta]$ explicitly.

Now, given $0 \leq \beta \leq N$, find a particular solution to the equation

$$B_{2m-2}[\beta] * \hat{u}[\beta] = [\beta^{2m} + (N-\beta)^{2m}]/2(2m)!. \tag{8.23}$$

Use the identity (see [95, 296])

$$\sum_{k=0}^{2m-2} a_k^{(2m-2)} \binom{x+k}{2m-1} = x^{2m-1}. \tag{8.24}$$

Here, as before, the symbol $a_s^{(2m-2)}$ stands for the coefficient (6.7) of $E_{2m-2}(\lambda)$, and the quantities in parentheses are defined as polynomials in x, namely,

$$\binom{x+k}{2m-1} = \frac{(x+k)(x+k-1)\ldots(x+k-2m+2)}{(2m-1)!}.$$

Let $x^{[k]}$ be the Newtonian power of the argument x. Then the *Worpitzky identity* (8.24) amounts to the following finite-difference relation

$$B_{2m-2}[\beta] * (\beta+m-1)^{[2m-1]} = \beta^{2m-1}, \quad \beta \in \mathbb{Z}.$$

Using the Worpitzky identity, we readily see that the integral

$$I_m[\beta] = \int_0^\beta (x+m-1)^{[2m-1]}\, dx, \tag{8.25}$$

when summed with some constant, yields a solution to the equation

$$B_{2m-2}[\beta] * v[\beta] = \frac{\beta^{2m}}{2m}.$$

The integrand in (8.25) is a polynomial $R_{2m-1}(x)$ of degree $2m-1$ with the leading coefficient 1. The function $R_{2m-1}(-x)$ is clearly a polynomial in x with the leading coefficient -1. Moreover, the zeros of $R_{2m-1}(-x)$ are the same as those of

$R_{2m-1}(x)$. Consequently, $R_{2m-1}(x)$ is an odd function of x, and so $I_m[-\beta] = I_m[\beta]$ for all $\beta \in \mathbb{Z}$. Further, the discrete function

$$\hat{u}_0[\beta] = (-1)^m \frac{I_m[\beta] + I_m[N-\beta]}{2(2m-1)!}$$

when summed with some constant, becomes a solution to the nonhomogeneous equation (8.23).

We may prove analogously that the polynomials $\widehat{Q}^{(\mp)}[\beta]$, considered on the intervals $(-\infty, m-1]$ and $[N-m+1, \infty)$, look as follows

$$\widehat{Q}^{(\mp)}[\beta] = \pm(-1)^m \frac{I_m[\beta-N] \mp I_m[\beta]}{2(2m-1)!} + \widehat{T}^{(\mp)}_{m-1}[\beta]. \tag{8.26}$$

The polynomials $\widehat{T}^{(\mp)}_{m-1}[\beta]$ of degree at most $m-1$ in this equality are unknown.

Inserting (8.26) in (8.22), translate the initial conditions on $\xi_{\mp}[\beta]$ into the desired form. We carry out the needed calculations for the left endpoint of the interval $[0, N]$. For $\beta = -m+1, \ldots, -1$, we have

$$\xi_-[\beta] = -\Delta_+^{(m)}[\beta] * \{\hat{u}_0[\beta] - \widehat{Q}^{(-)}[\beta]\} = -\frac{\Delta_+^{(m)}[\beta] * I_m[\beta]}{(2m-1)!}. \tag{8.27}$$

Clearly, for an arbitrary continuous function $f(x)$, we have

$$\Delta_+[\beta] * \int_0^\beta f(x)\,dx = \int_0^\beta \Delta_+ * f(x)\,dx + \int_0^1 f(x)\,dx,$$

with $\Delta_+ * f(x)$ standing for the difference $f(x+1) - f(x)$. This equality may be readily abstracted to the case of an arbitrary power of the operator $\Delta_+[\beta]$. We have

$$\Delta_+^{(m)}[\beta] * \int_0^\beta f(x)\,dx = \int_0^\beta \Delta_+^{(m)} * f(x)\,dx + \int_0^1 \Delta_+^{(m-1)} * f(x)\,dx.$$

Applying the last relation to the integral (8.25) and using (8.1.10), infer

$$\frac{\Delta_+^{(m)}[\beta] * I_m[\beta]}{(2m-1)!} = \int_0^\beta \frac{(x+m-1)^{[m-1]}}{(m-1)!}\,dx + \int_0^1 \frac{(x+m-1)^{[m]}}{m!}\,dx. \tag{8.28}$$

As is known, the Newtonian power of an argument x decomposes in a linear combination of the ordinary powers of x with the coefficients $s(m,k)$ the Stirling numbers of the first kind, namely,

$$x(x-1)\dots(x-m+1)=\sum_{k=1}^{m} s(m,k)x^k \equiv S_m(x),$$

$$x(x+1)\dots(x+m-1)=\sum_{k=1}^{m}(-1)^{m+k}s(m,k)x^k \equiv \overline{S}_m(x).$$

Using these notations in (8.28), rewrite the initial condition (8.27) on $\xi_-[\beta]$ as

$$\xi_-[\beta]=-\int_0^{\beta}\frac{\overline{S}_m(x)}{x(m-1)!}\,dx-\int_0^1\frac{\overline{S}_m(x)}{m!}\,dx,\quad \beta=-1,\dots,-m+1. \tag{8.29}$$

Given m, redenote $\xi_-[\beta]$ by $\xi_-^{(m)}[\beta]$. The quantity $\xi_-^{(m)}[0]$ is then given by the integral

$$\xi_-^{(m)}[0]=-\int_0^1\frac{\overline{S}_m(x)\,dx}{m!}.$$

For all $i=1,\dots,m-1$, the equality holds

$$\xi_-^{(m)}[-i]=\xi_-^{(m)}[-i+1]-\xi_-^{(m-1)}[-i+1].$$

It is not hard to deduce these recurrent relations for $\xi_-^{(m)}[\beta]$ using the formula

$$\overline{S}_m(x)=x\overline{S}_{m-1}(x)+(m-1)\overline{S}_{m-1}(x).$$

Thus, the whole set of initial data at a fixed m is easy to obtain by calculating only the quantities $\xi_-^{(0)}[0],\xi_-^{(1)}[0],\dots,\xi_-^{(m)}[0]$.

The initial conditions also possess the symmetry property or the antisymmetry property with respect to the point $-m/2$ for m even or odd, namely,

$$\xi_-^{(m)}[-i]=(-1)^m\xi_-^{(m)}[i-m],\quad 1\le i\le m-1.$$

We may summarize the above. To determine optimal weights near to 0, we are to solve the equation

$$B_{m-1}^{(-)}[\beta]*\xi_-[\beta]=0$$

with the initial data

$$\xi_-[-i] = -\int_0^\beta \frac{\overline{S}_m(x)}{x(m-1)!}\,dx - \int_0^1 \frac{\overline{S}_m(x)}{m!}\,dx, \quad 1 \le i \le m-1.$$

The sought weights are then expressed as

$$c[\beta] = 1 + \Delta_-^{(m)}[\beta] * \xi_-[\beta], \quad \beta > 0. \tag{8.30}$$

Observe also that, knowing the roots λ_j of the Euler polynomial $E_{2m-2}(\lambda)$, we may calculate the coefficients k_i of the difference operator

$$B_-^{(m-1)}[\beta] = \sum_{i=0}^{m-1} k_i \delta[\beta - i]$$

by the *Newton formula*

$$k_j = (k_{j-1}X_1 - k_{j-2}X_2 + \cdots + (-1)^{j-1}X_j)/j,$$

with $X_j = \sum_{i=1}^{m-1} \lambda_i^j$.

Our attempt at calculating the weight $c[0]$ by (8.28) fails.

To find the weight, we should use the formula

$$c[0] = \Delta_2^{[m]}[\beta] * \hat{u}[\beta]\Big|_{\beta=0}, \tag{8.31}$$

and write the operator $\Delta_2^{[m]}[\beta]$ as convolution

$$\Delta_2^{[m]}[\beta] = \Delta_-[\beta] * \Delta_-^{(2m-1)}[\beta + m]. \tag{8.32}$$

Inserting (8.32) in (8.31) and carrying out easy calculations with the representations of $\hat{u}[\beta]$ on the intervals $[-m+1, N+m-1]$ and $(-\infty, m-1]$, come to the equality

$$c[0] = \frac{1}{2} + \Delta_-^{(m-1)}[\beta] * \xi_-[\beta]\Big|_{\beta=0}.$$

The weights of optimal quadrature formulas near the right endpoint of the interval $[0, N]$ may be found by means of (8.11).

Using the just-described algorithm, F. Ya. Zagirova calculated the weights of optimal formulas for $m \le 30$. The relevant tables are published in [300]. For $m \le 7$

her results agree perfectly with the numerical values of the weights found earlier by I. Schoenberg and S. Silliman [204].

TABLE 7

Weights of $L_2^{(10)}$-Optimal Quadrature Formula

β	$d[\beta]$	β	$d[\beta]$
0	−0.230701	28	−0.092003
1	0.771127	29	0.071849
2	−1.831151	30	−0.056111
3	3.788385	31	0.043819
4	−6.363088	32	−0.034221
5	8.785364	33	0.026725
6	−10.270290	34	−0.020870
7	10.525290	35	0.016299
8	−9.792393	36	−0.012728
9	8.528898	37	0.009940
10	−7.119970	38	−0.007763
11	5.789519	39	0.006062
12	−4.632837	40	−0.004734
13	3.671494	41	0.003697
14	−2.892688	42	−0.002887
15	2.271083	43	0.002255
16	−1.779282	44	−0.001761
17	1.392206	45	0.001375
18	−1.088501	46	−0.001074
19	0.850655	47	0.000839
20	−0.664596	48	−0.000655
21	0.519146	49	0.000516
22	−0.405487	50	−0.000399
23	0.316693	51	0.000312
24	−0.247334	52	−0.000244
25	0.193161	53	0.000190
26	−0.150851	54	−0.000149
27	0.117808	55	0.000116

Table 7 provides the numerical values of the function $d[\beta]$ of a discrete argument which is connected with the weights $C_0[\beta]$ of the $L_2^{(10)}$-optimal quadrature formula by the relations

$$d[\beta] = c_0[\beta] - 1, \quad \beta > 0; \quad d[0] = c_0 - \frac{1}{2}.$$

This table is based on double precision computer calculations by F. Ya. Zagirova on a computer. We only display the values of $d[\beta]$ which exceeds 0.0001 in magnitude. For the other β, the difference between the optimal weights and unity is at most 0.0001.

It is worthy of observing that, for $m \geq 7$, there appear negative numbers among the optimal weights. Moreover, as tests show, at a fixed N and m increasing, the width of the boundary layer increases rapidly, i.e., the set widens of those β at which the optimal weights differ substantially from 1. In this layer the function $c[\beta]$ oscillates; i.e., its takes positive as well as negative values, changing sign from point to point. Of interest is the problem of analytically estimating the growth of the sum of the moduli of optimal weights as m increases.

References

1. Akopyan G. G., "Sequences of cubature formulas with a cubic lattice for domains with degenerate corners," in: Theory of Cubature Formulas and Applications of Functional Analysis to the Problems of Mathematical Physics (Trudy Sem. S. L. Soboleva) [in Russian], Inst. Mat. (Novosibirsk), Novosibirsk, 1978, No. 1, pp. 27–35.
2. Arcangeli R. and Ycurt B., "Almost sure convergence of smoothing D^m-splines for noisy date," Numer. Math., **66**, 281–294 (1993).
3. Aronszajn N. and Smith K. T., "Theory of Bessel potentials. I," Ann. Inst. Fourier, **11**, 385–475 (1961).
4. Aronszajn N., Greese T. M., and Lipkin L. J., Polyharmonic Functions, Clarendon Press, Oxford (1983).
5. Axler S., Bourdon P., and Ramey W., Harmonic Function Theory, Springer-Verlag, New York (1992).
6. Babenko K. I., "Some problems in approximation theory and numerical analysis," Uspekhi Mat. Nauk, **40**, No. 1, 3–27 (1985). English transl. in Russian Math. Surveys, **40**, No. 1, 1–30 (1985).
7. Babenko K. I., Foundations of Numerical Analysis [in Russian], Nauka, Moscow (1986).
8. Babuška I., "Optimal quadrature formulas," Dokl. Akad. Nauk SSSR, **149**, No. 2, 227–229 (1963).
9. Babuška I., "Über optimale Formeln zur numerischen Berechnung linearer Funktionale," Apl. Mat., **10**, No. 5, 441–443 (1965).
10. Babuška I., "Über universal optimale Quadraturformeln," Apl. Mat., **13**, No. 4, 304–338 and No. 5, 388–404 (1968).
11. Babuška I. and Sobolev S. L., "Optimization of numerical methods," Apl. Mat., **10**, No. 2, 96–129 (1965).
12. Bakhvalov N. S., "On approximate calculation of multiple integrals," Vestnik Moskov. Univ. Ser. I Mat. Mekh., No. 4, 3–18 (1959).

13. Bakhvalov N. S., "An estimate of the mean remainder term in quadrature formulae," Zh. Vychisl. Mat. i Mat. Fiz., **1**, No. 1, 64–77 (1961). English transl. in U.S.S.R. Comput. Math. Math. Phys., [illegible] ([illegible]).

14. Bakhvalov N. S., "On optimal properties of the Adams and Gregory formulas," in: Problems of Computational Mathematics and Computing Hardware [in Russian], Mashgiz, Moscow, 1963, pp. 9–26.

15. Bakhvalov N. S., "On optimal estimates for convergence of quadrature processes and integration methods of Monte Carlo type on function classes," in: Numerical Methods of Solution of Differential and Integral Equations and Quadrature Formulas [in Russian] (Supplement to "Zh. Vychisl. Mat. i Mat. Fiz."), No. 4, 5–63 (1964).

16. Bakhvalov N. S., "On optimum estimates for the error of integration over some classes of functions," Theor. Probab. Appl., **9**, 357–359 (1965).

17. Bakhvalov N. S., "On the optimal speed of integrating analytic functions," Zh. Vychisl. Mat. i Mat. Fiz., **7**, No. 5, 1011–1020 (1967). English transl. in U.S.S.R. Comput. Math. Math. Phys., **7**, No. 5, 63–75 (1970).

18. Bakhvalov N. S., "Statement of the problems of numerical integration," in: Problems of Numerical and Applied Mathematics [in Russian], Inst. Kibernet. i Vychisl. Tsentr Akad. Nauk UzSSR, Tashkent, 1972, No. 14, pp. 40–45.

19. Bakhvalov N. S., "A lower bound for the asymptotic characteristics of classes of functions with dominating mixed derivative," Mat. Zametki, **12**, No. 6, 655–664 (1972). English transl. in Math. Notes, **12**, 833–838 (1973).

20. Bakhvalov N. S., Numerical Methods, Mir, Moscow (1976).

21. Bakhvalov N. S. and Vasil'eva L. G., "Evaluation of the integrals of oscillating functions by interpolation at nodes of Gaussian quadratures," Zh. Vychisl. Mat. i Mat. Fiz., **8**, No. 1, 175–181 (1968). English transl. in U.S.S.R. Comput. Math. Math. Phys., **8**, No. 1, 241–249 (1970).

22. Berens H. and Hu Y., "Multivariate Gaussian cubature formulae," Arch. Math., **64**, No. 1, 26–32 (1995).

23. Berens H. and Schmid H. J., "On the number of nodes of odd degree cubature formulae for integrals with Jacobi weights on a simplex," in: Numerical Integration. Recent Developments, Software and Applications, NATO ASI Ser., Ser. C: Math. Phys. Sci., Kluwer Academic Publishers, Dordrecht, 1992, **357**, pp. 37–44.

24. Berezin I. S. and Zhidkov N. P., Computing Methods. Vol. 1 and 2, Pergamon Press, London (1965).

25. Besov O. V., "Intercellular averages and an error estimate for cubature formulas in Sobolev spaces and their generalizations," Trudy Mat. Inst. Akad. Nauk SSSR, **143**, 42–56 (1977). English transl. in Proc. Steklov Inst. Mat., **143**, 45–60 (1980).

26. Besov O. V., "Estimations of the errors of cubature formulas with respect to smoothness of functions," C. R. Acad. Bulgare Sci., **31**, No. 8, 949–952 (1978).
27. Besov O. V., "Error estimates for cubature formulas for weight integration," in: Theory of Cubature Formulas and Applications of Functional Analysis to the Problems of Mathematical Physics (Trudy Sem. S. L. Soboleva) [in Russian], Inst. Mat. (Novosibirsk), Novosibirsk, 1978, No. 1, pp. 36–45.
28. Besov O. V., "Estimates of the errors of cubature formulas in terms of smoothness of functions," Trudy Mat. Inst. Akad. Nauk SSSR, **150**, 11–23 (1979). English transl. in Proc. Steklov Inst. Mat., **150**, 11–23 (1981).
29. Björck Å. and Pereyra V., "Solution of Vandermonde systems of equations," Math. Comp., **24**, 893–904 (1970).
30. Blinov N. I., "Approximate calculation of double integrals," in: Algorithms and Programs [in Russian], VNTITsentr, Moscow, 1974, No. 2, p. 3.
31. Blinov N. I., "Approximate calculation of n-dimensional integrals," in: Algorithms and Programs [in Russian], VNTITsentr, Moscow, 1974, No. 3, p. 17.
32. Blinov N. I., "An algorithm for calculation of multiple integrals from functions with singularities," in: Algorithms and Programs [in Russian], VNTITsentr, Moscow, 1977, No. 2 (16), p. 9.
33. Blinov N. I., "On piecewise-lattice cubature formulas having optimal mode of convergence for some classes of functions with singularities," in: Theory of Cubature Formulas and Applications of Functional Analysis to the Problems of Mathematical Physics (Trudy Sem. S. L. Soboleva) [in Russian], Inst. Mat. (Novosibirsk), Novosibirsk, 1977, No. 1, pp. 25–42.
34. Blinov N. I. and Voĭtishek L. V., "Cubature formulas with regular boundary layer for a cube," in: Problems of Numerical and Applied Mathematics [in Russian], Inst. Kibernet. i Vychisl. Tsentr Akad. Nauk UzSSR, Tashkent, 1970, No. 38, pp. 8–15.
35. Blinov N. I. and Voĭtishek L. V., "Construction of cubature formulas with regular boundary layer for rational polyhedra," in: Theory of Cubature Formulas and Applications of Functional Analysis to the Problems of Mathematical Physics (Trudy Sem. S. L. Soboleva) [in Russian], Inst. Mat. (Novosibirsk), Novosibirsk, 1979, No. 1, pp. 5–15.
36. Bojanov B. D. and Nikolov G., "Comparison of Birkhoff type quadrature formulae," Math. Comp., **54**, 627–648 (1990).
37. Bojanov B. D., Hakopian H., and Sahakian A., Spline Functions and Multivariate Interpolation, Kluwer Academic Publishers, Dordrecht (1993).
38. Brass H., Quadraturverfahren, Vanderhoeck & Ruprecht, Göttingen (1977).
39. Brass H., "Error bounds based on approximation theory," in: Numerical Integration. Recent Developments, Software and Applications, NATO ASI

Ser., Ser. C: Math. Phys. Sci., Kluwer Academic Publishers, Dordrecht, 1992, **357**, pp. 147–163.

40. Brass H., "Asymptotically optimal error bounds for quadrature rules of given degree," Math. Comp., **61**, 785–798 (1993).

41. Burenkov V. I., "Embedding and extension theorems for classes of differentiable functions in many variables over the whole space," in: Itogi Nauki. Matematicheskiĭ Analiz, 1965, Inst. Nauchnoĭ Informatsii, Moscow, 1966, pp. 71–155.

42. Burenkov V. I., "Sobolev's integral representation and Taylor's formula," Trudy Mat. Inst. Akad. Nauk SSSR, **131**, 33–38 (1974). English transl. in Proc. Steklov Inst. Mat., **131**, 33–38 (1975).

43. Butler R., "On the evaluation of $\int_0^\infty (\sin t)^m / t^m dt$ by the trapezoidal rule," Amer. Math. Monthly, **67**, No. 6, 566–569 (1960).

44. Carlitz L., "Eulerian numbers and polynomials," Math. Mag., **32**, No. 5, 247–260 (1959).

45. Chevalley C., "Invariants of finite groups generated by reflections," Amer. J. Math., **77**, No. 4, 778–782 (1955).

46. Collatz L., Functionalanalysis und Numerische Mathematik, Springer-Verlag, Berlin (1964).

47. Cools R., "A survey of methods for constructing cubature formulae," in: Numerical Integration. Recent Developments, Software and Applications, NATO ASI Ser., Ser. C: Math. Phys. Sci., Kluwer Academic Publishers, Dordrecht, 1992, **357**, pp. 1–24.

48. Cools R. and Haegemans A., "A lower bound for the number of function evaluations in an error estimate for numerical integration," Constr. Approx., **6**, 353–361 (1990).

49. Cools R. and Rabinowitz P., "Monomial cubature rules since 'Stroud': A compilation," J. Comput. Appl. Math., **48**, 309–326 (1993).

50. Davis P. J., "On the numerical integration of periodic analytic functions," in: On Numerical Approximation: Proceedings of a Symposium, Wisconsin Univ. Press, Madison, 1959, pp. 45–59.

51. Davis P. J. and Rabinowitz P., Methods of Numerical Integration, Academic Press, London (1984).

52. Didur L. I., "Weighted Hermite cubature formulas on periodic functions," in: Theory of Cubature Formulas and Applications of Functional Analysis to the Problems of Mathematical Physics (Trudy Sem. S. L. Soboleva) [in Russian], Inst. Mat. (Novosibirsk), Novosibirsk, 1981, No. 1, pp. 48–56.

53. Dimitrov D. K., "Integration of polyharmonic functions," Math. Comp., **65**, 1269–1281 (1996).

54. Disney S. and Sloan I. H., "Error bounds for the method of good lattice points," Math. Comp., **56**, 257-266 (1991).

55. Ditkin V. A. and Lyusternik L. A., "On one method of practical harmonic analysis on the sphere," in: Computational Mathematics and Computing Hardware [in Russian], Mashgiz, Moscow, 1953, No. 1, pp. 3–13.

56. Ditkin V. A. and Prudnikov A. P., Transformations Intégrales et Calcul Opérationnel, Mir, Moscow (1978).

57. De Doncker E., "New Euler–Maclaurin expansions and their application to quadrature over the s-dimensional simplex," Math. Comp., **33**, 1003–1018 (1979).

58. Dubinskiĭ Yu. A., "Some questions of the theory of Sobolev spaces of infinite order and nonlinear equations," in: Differential Equations with Partial Derivatives [in Russian], Nauka, Novosibirsk, 1980, pp. 75–80.

59. Dubrovin B. A., Novikov S. P., and Fomenko A T., Géométrie Contemporaine. Méthodes et Applications. I. Géométrie des Surfaces, des Groupes de Transformations et des Champs, Mir, Moscow (1982).

60. Duchon J., "Splines minimizing rotation-invariant seminorms in Sobolev spaces," Lecture Notes in Math., **571**, 1977, pp. 85–100.

61. Elepov B. S., Kronberg A. A., Mikhaĭlov G. A., and Sabel'fel'd K. K., Solution of Boundary Value Problems by the Monte Carlo Method [in Russian], Nauka, Novosibirsk (1980).

62. Engels H., Numerical Quadrature and Cubature, Academic Press, London (1980).

63. Ermakov M., Monte Carlo Method and Related Problems [in Russian], Nauka, Moscow (1975).

64. Erugin N. P. and Sobolev S. L., "Approximate integration of some oscillating functions," Appl. Math. Mech., **14**, No. 2, 193–196 (1950).

65. Evans G., Practical Numerical Integration, Wiley, Chichester (1993).

66. Frobenius G., "Uber die Bernoullischen und Eulerschen Polynome," Sitzungsber. Preussische Akad. Wiss., 809–847 (1910).

67. Gatermann K., "Linear representations of finite groups and the ideal theoretical construction of G-invariant cubature formulas," in: Numerical Integration. Recent Developments, Software and Applications, NATO ASI Ser., Ser. C: Math. Phys. Sci., Kluwer Academic Publishers, Dordrecht, 1992, **357**, pp. 25–35.

08. Gautschi W., "A survey of Gauss–Christoffel quadrature formulae," in: Elwin Bruno Christoffel, Birkhäuser, Basel, 1981, pp. 72–147.

69. Gautschi W., "Gauss–Kronrod quadrature — a survey," in: Numerical Methods and Approximation Theory III, Faculty of Electronic Engineering, Niš Univ., Niš, 1988, pp. 39–66.

70. Gautschi W., "Remainder estimates for analytic functions," in: Numerical Integration. Recent Developments, Software and Applications, NATO ASI Ser., Ser. C: Math. Phys. Sci., Kluwer Academic Publishers, Dordrecht, 1992, **357**, pp. 133–145.

71. Gautschi W., "Spline approximation and quadrature formulae," Atti Sem. Mat. Fis. Univ. Modena, **40**, No. 1, 1992, pp. 169–182.

72. Gel′fand I. M. and Chentsov N. N., "On numerical calculation of continual integrals," Zh. Èksper. Teoret. Fiz., **31**, No. 6, 1106 (1956).

73. Gel′fand I. M., Frolov A. S., and Chentsov N. N., "Calculation of continual integrals by the Monte Carlo method," Izv. Vyssh. Uchebn. Zaved. Mat., No. 5, 32–45 (1958).

74. Gel′fond A. O., Calcul des Différences Finies. Collection Universitaire de Mathématiques, XII. Dunod, Paris (1963).

75. Ghizzetti A. and Ossicini A., Quadrature Formulae, Birkhäuser, Basel and Stuttgart (1970).

76. Girschovich J., "Extremal properties of Euler–Maclaurin and Gregory quadrature formulas," Izv. Akad. Nauk ÈstSSR Ser. Fiz.-Mat., **27**, No. 3, 259–265 (1978).

77. Godunov S. K., Spinor Theory and Representations of the Rotation Group [in Russian], Novosibirsk Univ., Novosibirsk (1978).

78. Gokhberg I. Ts. and Kreĭn M. G., Introduction to the Theory of Linear Nonselfadjoint Operators, Transl. of Math. Monographs, **18**, Amer. Math. Soc., Providence (1969).

79. Golomb M. and Weinberger H. F., "Optimal approximation and error bounds," in: On Numerical Approximation: Proceedings of a Symposium, Wisconsin Univ. Press, Madison, 1959, pp. 117–190.

80. Gradshteyn I. S. and Ryzhik I. M., Tables of Integrals, Series, and Products, Academic Press, New York and London (1965).

81. Grundmann K. and Möller H. M., "Invariant integration formulas for the n-simplex by combinatorical methods," SIAM J. Numer. Anal., **15**, 282–290 (1978).

82. Haegemans A. and Verlinden P., "Construction of fully symmetrical cubature rules of very high degree for the square," in: Numerical Integration IV: Proceedings of the Conference, Birkhäuser, Basel, 1993, pp. 167–174.

83. Hammer P. C. and Stroud A. H., "Numerical evaluation of multiple integrals. II," Math. Tables Aids Comput., **12**, 272–280 (1958).

84. Ignatov M. I. and Pevnyĭ A. B., Natural Splines of Several Variables [in Russian], Nauka, Leningrad (1991).

85. Israilov M. I. and Èndibaev K., "The Fourier series for the Epstein zeta function," in: Problems of Numerical and Applied Mathematics [in Russian], Inst.

Kibernet. i Vychisl. Tsentr Akad. Nauk UzSSR, Tashkent, 1975, No. 34, pp. 15–38.

86. Israilov M. I., Shushbaev S., and Èndibaev K., "On an algorithm for calculating the Epstein zeta function," in: Problems of Numerical and Applied Mathematics [in Russian], Inst. Kibernet. i Vychisl. Tsentr Akad. Nauk UzSSR, Tashkent, 1970, No. 38, pp. 36–41.

87. Kalmykov V. D., "Asymptotically optimal quadrature formulae in the space," in: Problems of Numerical and Applied Mathematics [in Russian], Inst. Kibernet. i Vychisl. Tsentr Akad. Nauk UzSSR, Tashkent, 1972, No. 14, pp. 24–29.

88. Kantorovich L. V. and Akilov G. P., Functional Analysis, Pergamon Press, Oxford (1982).

89. Kazakov A. N. and Lebedev V. I., "Gauss-type quadrature formulas invariant under the dihedral group on a sphere," Trudy Mat. Inst. Ros. Akad. Nauk, **203**, 100–112 (1994).

90. Keast P., "Software for integration over triangles and general simplices," in: Numerical Integration. Recent Developments, Software and Applications, NATO ASI Ser., Ser. C: Math. Phys. Sci., Kluwer Academic Publishers, Dordrecht, 1992, **357**, pp. 283–294.

91. Khaitov T. I., "Cubature formula with given derivatives," Dokl. Akad. Nauk TadzhSSR, **12**, No. 10 (1969).

92. Khamaev E. A. and Shoĭnzhurov Ts. B., "Some problems of the theory of cubature formulas in the $L_p^{\mu}(E^n)$ space," in: Theory of Cubature Formulas and Applications of Functional Analysis to the Problems of Mathematical Physics (Trudy Sem. S. L. Soboleva) [in Russian], Inst. Mat. (Novosibirsk), Novosibirsk, 1980, No. 1, pp. 132–148.

93. Khoroshikh A. N., "Precise asymptotic expansions for the norm of error functionals of Gregory formulas in $L_2^{(m)}[a,b]^*$," in: Problems in Computational and Applied Mathematics [in Russian], Inst. Kibernet. i Vychisl. Tsentr Akad. Nauk UzSSR, Tashkent, 1976, No. 40, pp. 3–8.

94. Knuth D. E., The Art of Computer Programming. Vol. 1. Fundamental Algorithms, Addison-Wesley, Reading etc. (1968).

95. Knuth D. E., The Art of Computer Programming. Vol. 3. Sorting and Searching, Addison-Wesley, Reading etc. (1973).

96. Konyaev S. I., "Quadrature formulas of degree 25–29 for the sphere," in: Problems of Numerical and Applied Mathematics [in Russian], Inst. Kibernet. i Vychisl. Tsentr Akad. Nauk UzSSR, Tashkent, 1975, No. 32, pp. 69–76.

97. Konyaev S. I., "Gaussian quadratures for the sphere that are invariant with respect to the icosahedral group with inversion," Mat. Zametki, **25**, No. 4, 629–634 (1979). English transl. in Math. Notes, **25**, 326–329 (1979).

98. Konyaev S. I., "Formulas for numerical integration on the sphere," in: Embedding Theorems and Their Applications (Trudy Sem. S. L. Soboleva) [in Russian], Inst. Mat. (Novosibirsk), Novosibirsk, 1982, No. 1, pp. 75–82.

99. Konyaev S. I., "On invariant quadrature formulae for a sphere," Russian J. Numer. Anal. Math. Modelling, **324**, No. 1, 41–47 (1995).

100. Korobov N. M., Number-Theoretic Methods of Approximate Analysis [in Russian], Fizmatgiz, Moscow (1963).

101. Korneĭchuk N. P., Extremal Problems of Approximation Theory [in Russian], Nauka, Moscow (1976).

102. Kronrod A. S., Nodes and Weights of Quadrature Formulas, Consultants Burean, New York (1965).

103. Krylov V. I., Approximate Calculation of Integrals, Macmillan, New York (1962).

104. Ladyzhenskaya O. A., Boundary Value Problems of Mathematical Physics [in Russian], Nauka, Moscow (1973).

105. Lebedev V. I., "Spherical quadrature formulas of highest algebraic degree," in: Theory of Cubature Formulas and Applications of Functional Analysis to Some Problems of Mathematical Physics [in Russian], Nauka, Novosibirsk, 1973, pp. 31–35.

106. Lebedev V. I., "Calculation of quadratures of Gauss–Markov type for domains and weights invariant under the octahedron group with inversion," in: Problems of Numerical and Applied Mathematics [in Russian], Inst. Kibernet. i Vychisl. Tsentr Akad. Nauk UzSSR, Tashkent, 1975, No. 32, pp. 77–84.

107. Lebedev V. I., "Values of the nodes and weights of ninth to seventeenth order Gauss–Markov quadrature formulae for the sphere invariant under the octahedron group with inversion," Zh. Vychisl. Mat. i Mat. Fiz., **15**, No. 1, 48–54 (1975). English transl. in U.S.S.R. Comput. Math. Math. Phys., **15**, No. 1, 44–51 (1976).

108. Lebedev V. I., "Quadratures on a sphere," Zh. Vychisl. Mat. i Mat. Fiz., **16**, No. 2, 293–306 (1976). English transl. in U.S.S.R. Comput. Math. Math. Phys., **16**, No. 2, 10–24 (1977).

109. Lebedev V. I., "Spherical quadrature formulas exact to orders 25–29," Sibirsk. Mat. Zh., **18**, No. 1, 132–141 (1977). English transl. in Siberian Math. J., **18**, 99–107 (1977).

110. Lebedev V. I., "Quadrature formula of degree 35 for the sphere," in: Theory of Cubature Formulas and Numerical Mathematics [in Russian], Nauka, Novosibirsk, 1980, pp. 110–114.

111. Lebedev V. I., "A quadrature formula of degree 59 on a sphere," Dokl. Akad. Nauk, **338**, No. 4, 454–456 (1994).

112. Lebedev V. I. and Skorokhodov A. L., "Quadrature formulas for a sphere of orders 41, 47, and 53," Dokl. Akad. Nauk, **324**, No. 3, 519–524 (1992). English transl. in Russian Acad. Sci. Dokl. Math., **45**, No. 3, 587–592 (1993).

113. Lo Y. T., Lee S. W., and Sun B., "On Davis method of estimating quadrature errors," Math. Comp., **19**, 133–138 (1965).

114. Lyness J. N., "Quadrature over a simplex," SIAM J. Numer. Anal., **15**, 122–133 and 870–877 (1978).

115. Lyness J. N., "The canonical forms of a lattice rule," in: Numerical Integration IV: Proceedings of the Conference, Birkhäuser, Basel, 1993, pp. 225–240.

116. Lyness J. N. and McHugh B. J. J., "On the remainder term in the N-dimensional Euler–Maclaurin expansion," Numer. Math., **15**, 333–344 (1970).

117. Lyness J. N. and Puri K. K., "The Euler–Maclaurin expansion for the simplex," Math. Comp., **27**, 273–293 (1973).

118. Lyness J. N. and Jespersen D., "Moderate degree symmetric quadrature rules for the triangle," J. Inst. Math. Appl., **15**, 19–32 (1975).

119. Lyness J. N. and de Doncker E., "Quadrature error expansions. II. The full corner singularity," Numer. Math., **64**, 355–370 (1993).

120. Lyness J. N. and Cools R., "A survey of numerical cubature over triangles," in: Mathematics of Computation 1943–1993: A Half-Century of Computational Mathematics, Proceedings of Symposia in Applied Mathematics, **48**, Amer. Math. Soc., Providence, 127–150 (1994).

121. Maergoĭz L. S., Elements of the Theory of Growth of Convex and Entire Functions in Many Variables [in Russian], Krasnoyarsk. Univ., Krasnoyarsk (1976).

122. Mandelbrojt S., Quasianalytic Classes of Functions [in Russian], ONTI, Moscow and Leningrad (1937).

123. Mandelbrojt S., Series Adhérentes. Régularisation des Suites. Applications, Paris (1952).

124. Maxwell J. C., "On approximate multiple integration between limits by summation," Math. Proc. Cambridge Philos. Soc., **3**, 39–47 (1877).

125. McLaren A. D., "Optimal numerical integration on a sphere," Math. Comp., **17**, 361–383 (1963).

126. Meyers L. F. and Sard A., "Best approximate integration formulas," J. Math. Phys., **29**, 118–123 (1950).

127. Moan T., "Experiences with orthogonal polynomials and 'best' numerical integration formulas on a triangle; with particular reference to finite element approximations," Z. Angew. Math. Mech., **54**, 501–508 (1974).

128. Möller H. M., "Kubaturformeln mit minimaler Knotenzahl," Numer. Math., **25**, 185–200 (1976).

129. Monegato G., "Stieltjes polynomials and related quadrature rules," SIAM Review, **24**, No. 2, 137–158 (1982).

130. Mynbaev K. T., "Cubature formulas for functions in a weighted anisotrophic Sobolev–Liouville space," Dokl. Akad. Nauk SSSR, **310**, No. 6, 1300–1302 (1990). English transl. in Soviet Math. Dokl., **41**, 182–185 (1990).

131. Mynbaev K. T., "Optimal quadrature formulas for functions from weight class," Mat. Zametki, **49**, No. 2, 133–140 (1991). English transl. in Math. Notes, **49**, 207–214 (1991).

132. Mysovskikh I. P., "On invariant cubature formulas," in: Theory of Cubature Formulas and Applications of Functional Analysis to the Problems of Mathematical Physics (Trudy Sem. S. L. Soboleva) [in Russian], Inst. Mat. (Novosibirsk), Novosibirsk, 1978, No. 1, pp. 69–78.

133. Mysovskikh I. P., Interpolatory Type Cubature Formulas [in Russian], Nauka, Moscow (1981).

134. Mysovskikh I. P., "On cubature formulas invariant under the octahedral group," in: Cubature Formulas and Their Applications. Reports of a Seminar-Conference [in Russian], Ufa and Krasnoyarsk, 1996, pp. 41–51.

135. Nikol′skiĭ S. M., Approximation of Functions of Several Variables and Imbedding Theorems, Die Grundlehren der Mathematischen Wissenschaften, Band 205, Springer-Verlag, New York (1975).

136. Nikol′skiĭ S. M., Quadrature Formulas [in Russian], Nauka, Moscow (1979).

137. Noskov M. V., "Asymptotically optimal cubature formulas on lattice surfaces," in: Application of Functional Analysis to Partial Differential Equations (Trudy Sem. S. L. Soboleva) [in Russian], Inst. Mat. (Novosibirsk), Novosibirsk, 1983, No. 2, pp. 103–112.

138. Noskov M. V., "Approximate integration of functions periodic in some variables," in: Embedding Theorems and Their Applications (Trudy Sem. S. L. Soboleva) [in Russian], Inst. Mat. (Novosibirsk), Novosibirsk, 1982, No. 1, pp. 83–102.

139. Noskov M. V., "On cartesian products of cubature formulas," in: Theory of Cubature Formulas and Numerical Mathematics [in Russian], Nauka, Novosibirsk, 1980, pp. 114–116.

140. Patterson T. N. L., "Modified optimal quadrature extensions," Numer. Math., **64**, No. 4, 511–520 (1993).

141. Peherstofer F., "On Stieltjes polynomials and Gauss–Kronrod quadrature," Math. Comp., **55**, 649–664 (1990).

142. Polovinkin V. I., "Weighted cubature formulas for periodic functions," Mat. Zametki, **3**, No. 3, 319–326 (1968). English transl. in Math. Notes, **3**, 201–205 (1968).

143. Polovinkin V. I., "Weighted cubature formulas," Dokl. Akad. Nauk SSSR, **179**, No. 3, 542–544 (1968). English transl. in Soviet Math. Dokl., **9**, 417–419 (1968).
144. Polovinkin V. I., "Some estimates of norms of error functionals of cubature formulas," Mat. Zametki, **5**, No. 3, 317–322 (1969). English transl. in Math. Notes, **5**, 192–195 (1969).
145. Polovinkin V. I., "Asymptotically $L_2^{(m)}(\Omega)$-optimal cubature formulas," in: Problems of Numerical and Applied Mathematics [in Russian], Inst. Kibernet. i Vychisl. Tsentr Akad. Nauk UzSSR, Tashkent, 1970, No. 38, pp. 88–92.
146. Polovinkin V. I., "Cubature formulas in $L_2^{(m)}(\Omega)$," Dokl. Akad. Nauk SSSR, **190**, No. 1, 42–44 (1970). English transl. in Soviet Math. Dokl., **11**, 38–40 (1970).
147. Polovinkin V. I., "Questions in the theory of weight cubature formulas," Sibirsk. Mat. Zh., **12**, No. 1, 177–196 (1971). English transl. in Siberian Math. J., **12**, 128–142 (1971).
148. Polovinkin V. I., "Composite cubature formulas," in: Problems of Numerical and Applied Mathematics [in Russian], Inst. Kibernet. i Vychisl. Tsentr Akad. Nauk UzSSR, Tashkent, 1972, No. 14, pp. 53–62.
149. Polovinkin V. I., "On cubature formulas with regular boundary layer," Sibirsk. Mat. Zh., **13**, No. 4, 951–954 (1972). English transl. in Siberian Math. J., **13**, 663–665 (1973).
150. Polovinkin V. I., "Generalization of Sobolev's theory to cubature formulas exact for polynomials of degree not exceeding $m - [(n+1)/2]$," Sibirsk. Mat. Zh., **14**, No. 5, 1144–1148 (1973). English transl. in Siberian Math. J., **14**, 800–803 (1973).
151. Polovinkin V. I., "Sequences of functionals with a boundary layer," Sibirsk. Mat. Zh., **15**, No. 2, 413–429 (1974). English transl. in Siberian Math. J., **15**, 296–308 (1974).
152. Polovinkin V. I., "Asymptotic optimality of sequences of formulas with a regular boundary layer for odd m," Sibirsk. Mat. Zh., **16**, No. 2, 328–335 (1975). English transl. in Siberian Math. J., **16**, 253–258 (1975).
153. Polovinkin V. I., "Asymptotically best sequences of cubature formulas," Sibirsk. Mat. Zh., **16**, No. 6, 1255–1262 (1975). English transl. in Siberian Math. J., **16**, 963–968 (1976).
154. Polovinkin V. I., "Sequences of quadrature formulas with boundary layer," in: Theory of Cubature Formulas and Applications of Functional Analysis to the Problems of Mathematical Physics (Trudy Sem. S. L. Soboleva) [in Russian], Inst. Mat. (Novosibirsk), Novosibirsk, 1977, No. 1, pp. 149–158.
155. Polovinkin V. I., "Lattice cubature formulas with positive coefficients," in: Theory of Cubature Formulas and Applications of Functional Analysis to the

Problems of Mathematical Physics (Trudy Sem. S. L. Soboleva) [in Russian], Inst. Mat. (Novosibirsk), Novosibirsk, 1978, No. 1, pp. 79–87.

156. Polovinkin V. I., "Convergence of sequences of quadrature formulas with boundary layer on individual functions," in: Mathematical Analysis and Related Problems of Mathematics [in Russian], Nauka, Novosibirsk, 1978, pp. 183–191.

157. Polovinkin V. I., "Asymptotically best sequences of cubature and quadrature formulas," in: Theory of Cubature Formulas and Numerical Mathematics [in Russian], Nauka, Novosibirsk, 1980, pp. 116–118.

158. Polovinkin V. I., "Sequences of quadrature formulas with boundary layer and sequences of Gregory type formulas," in: Quadrature and Cubature Formulas. Solution of Functional Equations [in Russian], Leningrad Univ., Leningrad, 1981, pp. 7–25.

159. Polovinkin V. I., "Sequences of cubature formulas with periodic systems of nodes," Sibirsk. Mat. Zh., **22**, No. 3, 147–155 (1981). English transl. in Siberian Math. J., **22**, 442–448 (1982).

160. Polovinkin V. I., "Weighted cubature formulas corresponding to interpolation operators with boundary layer," in: Embedding Theorems and Their Applications (Trudy Sem. S. L. Soboleva) [in Russian], Inst. Mat. (Novosibirsk), Novosibirsk, 1982, No. 1, pp. 103–115.

161. Polovinkin V. I., "On existence of best cubature formulas," in: Problems of Numerical and Applied Mathematics [in Russian], Inst. Kibernet. i Vychisl. Tsentr Akad. Nauk UzSSR, Tashkent, 1982, No. 69, pp. 145–153.

162. Polovinkin V. I., "Weighted cubature formulas with arbitrary nodes," in: Application of Functional Analysis to Partial Differential Equations (Trudy Sem. S. L. Soboleva) [in Russian], Inst. Mat. (Novosibirsk), Novosibirsk, 1983, No. 2, pp. 113–128.

163. Polovinkin V. I., "On realization of the error of a cubature formula in L_p^m-type spaces," in: Boundary Value Problems for Partial Differential Equations [in Russian], Inst. Mat. (Novosibirsk), Novosibirsk, 1988, pp. 125–136.

164. Polovinkin V. I., "Asymptotically optimal cubature weight formulas," Sibirsk. Mat. Zh., **30**, No. 2, 151–160 (1989). English transl. in Siberian Math. J., **30**, 289–297 (1989).

165. Polovinkin V. I., "Cubature formulas with regular boundary layer and universal asymptotic optimality," in: Optimal Methods of Calculation and Their Application to Data Processing [in Russian], Politekhn. Inst., Penza, 1991, pp. 36–41.

166. Polovinkin V. I., "Asymptotically optimal sequences of errors in $L_p^m[a, b]$ at large m," in: Cubature Formulas and Their Applications. Reports of a Seminar-Conference [in Russian], Ufa and Krasnoyarsk, 1996, pp. 64–69.

167. Polovinkin V. I. and Didur L. I., "Asymptotically optimal sequences of Hermitian cubature formulas," Sibirsk. Mat. Zh., **19**, No. 3, 663–669 (1978). English transl. in Siberian Math. J., **19**, 467–479 (1979).

168. Radon J., "Zur Mechanischen Kubatur," Monatsh. Math., **52**, No. 4, 286–300 (1948).

169. Rahman Q. I. and Schmeisser G., "Quadrature formulae and functions of exponential type," Math. Comp., **54**, 245–270 (1990).

170. Ramazanov M. D., "Construction of an asymptotically optimal cubature formulas in $\widetilde{W}_2^m(\Omega)$," Dokl. Akad. Nauk SSSR, **202**, No. 2, 290–293 (1972). English transl. in Soviet Math. Dokl., **13**, 98–102 (1972).

171. Ramazanov M. D., "An optimal error functional over periodic functions in $\widetilde{H}_p^\mu$," Sibirsk. Mat. Zh., **13**, No. 1, 225–229 (1972). English transl. in Siberian Math. J., **13**, 162–164 (1972).

172. Ramazanov M. D., "Optimal error functionals over periodic functions from Banach spaces," Sibirsk. Mat. Zh., **13**, No. 2, 481–484 (1972). English transl. in Siberian Math. J., **13**, 334–336 (1972).

173. Ramazanov M. D., "Asymptotically optimal errors on an anisotropic Hilbert space," in: Problems of Numerical and Applied Mathematics [in Russian], Inst. Kibernet. i Vychisl. Tsentr Akad. Nauk UzSSR, Tashkent, 1972, No. 14, pp. 72–82.

174. Ramazanov M. D., Lectures on the Theory of Approximate Integration [in Russian], Bashkirsk. Univ., Ufa (1973).

175. Ramazanov M. D., "Asymptotic optimality of grid cubature formulas on Hilbert spaces," Dokl. Akad. Nauk SSSR, **216**, No. 1, 44–45 (1974). English transl. in Soviet Math. Dokl., **15**, 755–757 (1975).

176. Ramazanov M. D., "Asymptotic optimality of grid cubature formulas on spaces of continuously differentiable functions," Dokl. Akad. Nauk SSSR, **216**, No. 2, 270–273 (1974). English transl. in Soviet Math. Dokl., **15**, 789–793 (1975).

177. Ramazanov M. D., "Cube formulas on spaces of continuously differentiable functions," Sibirsk. Mat. Zh., **16**, No. 6, 1263–1285 (1975). English transl. in Siberian Math. J., **16**, 969–985 (1976).

178. Ramazanov M. D., "On the degree of convergence of lattice cubature formulas in spaces with dominant derivative," Dokl. Akad. Nauk SSSR, **226**, No. 3, 551–553 (1984). English transl. in Soviet Math. Dokl., **30**, 135–137 (1984).

179. Ramazanov M. D., "Universal optimality of lattice cubature formulas," Dokl. Akad. Nauk, **324**, No. 5, 933–937 (1992). English transl. in Russian Acad. Sci. Dokl. Math., **45**, No. 3, 665–669 (1992).

180. Ramazanov M. D., "Functions of a boundary layer of asymptotically optimal lattice cubature formulas," Dokl. Akad. Nauk, **335**, No. 1, 25–26 (1994).

181. Ramazanov M. D., "Optimizing the location of the nodes of lattice cubature formulas," Dokl. Akad. Nauk, **343**, No. 4, 456–457 (1995).

182. Ramazanov M. D., Blinov N. I., et al., "Development of the Sobolev theory of cubature formulas," in: Theory of Cubature Formulas and Computational Mathematics [in Russian], Nauka, Novosibirsk, 1980, pp. 118–126.

183. Ramazanov M. D. and Umarkhanov I., "A quadrature formula with a simple weight function," Dokl. Akad. Nauk UzSSR, No. 5, 4–7 (1982).

184. Rasputin G. G., "On exact estimates for the errors of quadrature formulas," in: Theory of Cubature Formulas and Applications of Functional Analysis to the Problems of Mathematical Physics (Trudy Sem. S. L. Soboleva) [in Russian], Inst. Mat. (Novosibirsk), Novosibirsk, 1978, No. 1, pp. 110–115.

185. Rasputin G. G., "On construction of interpolation cubature formulas of Gauss type for approximate calculation of double integrals," in: Embedding Theorems and Their Applications (Trudy Sem. S. L. Soboleva) [in Russian], Inst. Mat. (Novosibirsk), Novosibirsk, 1982, No. 1, pp. 116–122.

186. Reihel L. and Opfer G., "Chebyshev–Vandermonde systems," Math. Comp., **57**, 703–721 (1991).

187. Rogers C., Packings and Coverings, Cambridge University Press, Cambridge (1964).

188. Rozenfel'd B. A., Many-Dimensional Spaces [in Russian], Nauka, Moscow (1966).

189. Ryaben'kiĭ V. S., "Local splines," Comput. Methods Appl. Mech. Engrg., **5**, No. 2, 211–225 (1975).

190. Ryaben'kiĭ V. S., The Method of Difference Potentials for Some Problems of Continuum Mechanics [in Russian], Nauka, Moscow (1987).

191. Ryshkov S., "On the question of final ζ-optimality of lattices providing the closest lattice packing of n-dimensional spheres," Sibirsk. Mat. Zh., **14**, No. 5, 1065–1075 (1973). English transl. in Siberian Math. J., **14**, 743–750 (1974).

192. Salikhov G. N., "On cubature formulas on the 4-dimensional sphere, invariant under finite rotation groups," Izv. Akad. Nauk UzSSR Ser. Fiz.-Mat., No. 6, 25–29 (1963).

193. Salikhov G. N., "On some cubature formulas on hyperspheres," Izv. Akad. Nauk UzSSR Ser. Fiz.-Mat., No. 2, 37–40 (1966).

194. Salikhov G. N., "On a method of improving efficiency of the cubature formulas of S. L. Sobolev on the sphere," in: Problems of Numerical and Applied Mathematics [in Russian], Inst. Kibernet. i Vychisl. Tsentr Akad. Nauk UzSSR, Tashkent, 1971, No. 8, pp. 3–9.

195. Salikhov G. N., "To the theory of polyhedral groups," Dokl. Akad. Nauk SSSR, **205**, No. 1, 33–35 (1972).

196. Salikhov G. N., "Polyhedral groups and cubature formulas on hyperspheres," in: Problems of Numerical and Applied Mathematics [in Russian], Inst. Kibernet. i Vychisl. Tsentr Akad. Nauk UzSSR, Tashkent, 1972, No. 14, pp. 139–147.

197. Salikhov G. N., "To the theory of cubature formulas on the sphere," in: Theory of Cubature Formulas and Applications of Functional Analysis to Some Problems of Mathematical Physics [in Russian], Nauka, Novosibirsk, 1973, pp. 22–27.

198. Salikhov G. N., "The 600-hedral group and cubature formulas of highest degree on the hypersphere," in: Problems of Numerical and Applied Mathematics [in Russian], Inst. Kibernet. i Vychisl. Tsentr Akad. Nauk UzSSR, Tashkent, 1975, No. 32, pp. 139–154.

199. Salikhov G. N., "Cubature formulas for the hypersphere invariant under the 600-hedral group," Dokl. Akad. Nauk SSSR, **223**, No. 5, 1075–1078 (1975).

200. Salikhov G. N., Cubature Formulas for Multidimensional Spheres [in Russian], Fan, Tashkent (1985).

201. Sard A., "Best approximate integration formulas, best approximation formulas," Amer. J. Math., **71**, 80–91 (1949).

202. Schmid H. J., "On lower bounds for the number of nodes in cubature formulae," in: Cubature Formulas and Their Applications. Reports of a Seminar-Conference, Ufa and Krasnoyarsk, 1995, p. 34.

203. Schoenberg I. J., "Spline interpolation and best quadrature formulae," Bull. Amer. Math. Soc., **70**, 143–148 (1964).

204. Schoenberg I. J. and Silliman S. D., "On semicardinal quadrature formulae," Math. Comp., **28**, 483–497 (1974).

205. Schoenberg I. J. and Sharma A., "The interpolatory background of the Euler–Maclaurin quadrature formulas," Bull. Amer. Math. Soc., **77**, 1034–1038 (1971).

206. Secrest D. H., "Best approximate integration formulas and best error bounds," Math. Comp., **19**, No. 89, 79–83 (1965).

207. Shadimetov Kh. M., "Optimal quadrature formulas in the spaces $L_2^{(m)}(\Omega)$ and $L_2^{(m)}(\mathbb{R}^1)$," Dokl. Akad. Nauk UzSSR, No. 3, 5–8 (1983).

208. Shanks D., "Calculation and applications of Epstein zeta functions," Math. Comp., **29**, 271–287 (1975).

209. Sharygin M. F., "A lower bound on the error of a quadrature formula on a function class," Zh. Vychisl. Mat. i Mat. Fiz., **3**, No. 2, 370–376 (1963).

210. Sharipov T. Kh., "An upper bound for the norm of the error of a cubature formula with boundary layer regular in the S. L. Sobolev sense in $H_p^\mu(\Omega)$," Dokl. Akad. Nauk SSSR, **202**, No. 1, 51–53 (1972).

211. Shilov G. E., Mathematical Analysis. A Special Course, Pergamon Press, Qxford (1965).

212. Shilov G. E., Mathematical Analysis. Second Special Course [in Russian], Nauka, Moscow (1965).

213. Shoĭnzhurov Ts. B., "Some problems of the theory of cubature formulas in the space $W_p^{(m)}$," Sibirsk. Mat. Zh., **7**, No. 2, 433–446 (1967).

214. Shoĭnzhurov Ts. B., "Some problems of the theory of cubature formulas in anisotropic Sobolev spaces," Dokl. Akad. Nauk SSSR, **209**, No. 5, 1036–1038 (1973).

215. Shoĭnzhurov Ts. B., "On the approximate integration of functions in Sobolev spaces with a weight," Dokl. Akad. Nauk SSSR, **218**, No. 6, 1036–1038 (1974). English transl. in Soviet Math. Dokl., **15**, 1429–1432 (1975).

216. Shoĭnzhurov Ts. B., "Asymptotically optimal cubature formulas with nodes at the points of a curved lattice," in: Theory of Cubature Formulas and Applications of Functional Analysis to the Problems of Mathematical Physics (Trudy Sem. S. L. Soboleva) [in Russian], Inst. Mat. (Novosibirsk), Novosibirsk, 1976, No. 1, pp. 157–164.

217. Shoĭnzhurov Ts. B., "The norm of an error in the space $W_p^{(m)}(E_n)$," in: Cubature Formulas and Their Applications. Reports of a Seminar-Conference [in Russian], Ufa and Krasnoyarsk, 1996, pp. 123–127.

218. Silvester P., "Symmetric quadrature formulae for simplexes," Math. Comp., **24**, 95–100 (1970).

219. Sirazhdinov S. Kh., "On the limit distribution of the roots of the Euler polynomials," Dokl. Akad. Nauk SSSR, **239**, No. 1, 56–59 (1978).

220. Sloan I. H., "Lattice methods for multiple integration," J. Comput. Appl. Math., **12–13**, 131–143 (1985).

221. Sloan I. H., "Superconvergence," in: Mathematical Concepts and Methods in Science and Engineering, Plenum Press, New York, 1990, pp. 35–70.

222. Sloan I. H., "Numerical integration in high dimensions — the lattice rule approach," in: Numerical Integration. Recent Developments, Software and Applications, NATO ASI Ser., Ser. C: Math. Phys. Sci., Kluwer Academic Publishers, Dordrecht, 1992, **357**, pp. 55–69.

223. Sloan I. H. and Joe S., Lattice Methods for Multiple Integration, Oxford University Press, Oxford (1995).

224. Sobolev S. L., "Formulas for mechanical cubatures in n-dimensional space," Dokl. Akad. Nauk SSSR, **137**, No. 3, 527–530 (1961). English transl. in Soviet Math. Dokl., **2**, 317–320 (1961).

225. Sobolev S. L., "On the interpolation of functions of n variables" Dokl. Akad. Nauk SSSR, **137**, No. 4, 778–781 (1961). English transl. in Soviet Math. Dokl., **2**, 343–346 (1962).

226. Sobolev S. L., "Various types of convergence of cubature and quadrature formulas," Dokl. Akad. Nauk SSSR, **146**, No. 1, 41–42 (1962). English transl. in Soviet Math. Dokl., **3**, 1246–1248 (1963).

227. Sobolev S. L., "Cubature formulas on the sphere invariant under finite groups of rotations," Dokl. Akad. Nauk SSSR, **146**, No. 2, 310–313 (1962). English transl. in Soviet Math. Dokl., **3**, 1307–1310 (1963).

228. Sobolev S. L., "The number of nodes in cubature formulas on the sphere," Dokl. Akad. Nauk SSSR, **146**, No. 4, 770–774 (1962). English transl. in Soviet Math. Dokl., **3**, 1391–1394 (1963).

229. Sobolev S. L., "On the formulas of mechanical cubatures on the sphere," Sibirsk. Mat. Zh., **3**, No. 5, 769–796 (1962).

230. Sobolev S. L., "Some questions of the theory of cubature formulas," in: Outlines of the Joint Soviet-American Symposium on Partial Differential Equations, Sibirsk. Otdel. Akad. Nauk SSSR, Moscow, 1963, pp. 241–245.

231. Sobolev S. L., "A method for calculating the coefficients in mechanical cubature formula," Dokl. Akad. Nauk SSSR, **150**, No. 6, 1238–1241 (1963). English transl. in Soviet Math. Dokl., **4**, 878–882 (1963).

232. Sobolev S. L., "On cubature formulas," Studia Math. (Ser. Specjalna), **1**, 117–118 (1963).

233. Sobolev S. L., "Some new problems in the theory of partial differential equations," in: Differential Equations and Their Applications, Proceedings of the Conference (Prague, 1962), 1963, Academic Press, New York, pp. 167–171.

234. Sobolev S. L., Lectures on the Theory of Cubature Formulas. Part 1 [in Russian], Novosibirsk Univ., Novosibirsk (1964).

235. Sobolev S. L., Lectures on the Theory of Cubature Formulas. Part 2 [in Russian], Novosibirsk Univ., Novosibirsk (1965).

236. Sobolev S. L., "On the oder of convergence of cubature formulas," Dokl. Akad. Nauk SSSR, **162**, No. 5, 1005–1008 (1965). English transl. in Soviet Math. Dokl., **6**, 808–811 (1965).

237. Sobolev S. L., "Convergence of approximate integration formulas for functions from $L_2^{(m)}$," Dokl. Akad. Nauk SSSR, **162**, No. 6, 1259–1261 (1965). English transl. in Soviet Math. Dokl., **6**, 865–867 (1965).

238. Sobolev S. L., "Evaluation of integrals of infinitely differentiable functions," Dokl. Akad. Nauk SSSR, **163**, No. 1, 33–35 (1965).

239. Sobolev S. L., "Cubature formulas with regular boundary layer," Dokl. Akad. Nauk SSSR, **163**, No. 3, 587–590 (1965).

240. Sobolev S. L., "A difference analogue of the polyharmonic equation," Dokl. Akad. Nauk SSSR, **164**, No. 1, 54–57 (1965). English transl. in Soviet Math. Dokl., **6**, 1174–1178 (1966).

241. Sobolev S. L., "Optimal mechanical cubature formulas with interpolation points on a regular grid," Dokl. Akad. Nauk SSSR, **164**, No. 2, 281–284 (1965). English transl. in Soviet Math. Dokl., **6**, 1226–1229 (1966).

242. Sobolev S. L., "The theory of cubature formulas," Wiss. Z. Hochschule Architektur Bauwesen Weimar., **12**, No. 5–6, 537–546 (1965).

243. Sobolev S. L., "Representation of periodic analytic functions by a sum of squares," Dokl. Akad. Nauk SSSR, **165**, No. 1, 40–43 (1965). English transl. in Soviet Math. Dokl., **6**, 1412–1415 (1966).

244. Sobolev S. L., "Construction of cubature formulas with a regular boundary layer," Dokl. Akad. Nauk SSSR, **166**, No. 2, 295–297 (1966). English transl. in Soviet Math. Dokl., **7**, 80–82 (1966).

245. Sobolev S. L., "Density of finite functions in l_p^m," Dokl. Akad. Nauk SSSR, **167**, No. 3, 516–518 (1966). English transl. in Soviet Math. Dokl., **7**, 421–423 (1966).

246. Sobolev S. L., "Sur une classe des fonctions de plusieurs variables indépendantes," Atti Accad. Naz. Lincei Rend. Cl. Sci. Fis. Mat. Natur. (8) Ser. 42, 133–137 (1967).

247. Sobolev S. L., "Theory of approximation of integrals of functions of several variables," in: Proceedings of the International Congress of Mathematicians, Nauka, Moscow, 1966, pp. 659–663 (1968). English transl. in Amer. Math. Soc., translat, II Ser. **70**, 216–221 (1968).

248. Sobolev S. L., "Some problems of the theory of functions of several discrete variables," in: Proceedings of the International Conference on Functional Analysis and Related Topics (Tokyo, 1969), Tokyo Univ. Press, Tokyo, 1970, pp. 148–160.

249. Sobolev S. L., "On optimal cubature formulas in a finite domain," in: Theory of Cubature Formulas and Applications of Functional Analysis to Some Problems of Mathematical Physics [in Russian], Nauka, Novosibirsk, 1973, pp. 5–21.

250. Sobolev S. L., Introduction to the Theory of Cubature Formulas [in Russian], Nauka, Moscow (1974). An abridged English transl.: Cubature Formulas and Modern Analysis: An Introduction, Gordon and Breach, Philadelphia (1992).

251. Sobolev S. L., "Convergence of cubature formula on infinitely differentiable functions," Dokl. Akad. Nauk SSSR, **223**, No. 4, 793–796 (1975). English transl. in Soviet Math. Dokl., **16**, 991–995 (1976).

252. Sobolev S. L., "Les formules optimales pour l'intégration des fonctions de plusieurs variables," in: Metodi Valutativi Nella Fisica-Matematica (Conf. Inter., 1972, Roma), Roma, 1975, pp. 423–441.

253. Sobolev S. L., "The convergence of cubature formulas on the elements $\widetilde{L}_2^{(m)}$," Dokl. Akad. Nauk SSSR, **228**, No. 1, 45–47 (1976). English transl. in Soviet Math. Dokl., **17**, 660–663 (1976).

254. Sobolev S. L., "Convergence of cubature formulas on various classes of periodic functions," in: Theory of Cubature Formulas and Applications of Functional Analysis to the Problems of Mathematical Physics (Trudy Sem. S. L. Soboleva) [in Russian], Inst. Mat. (Novosibirsk), Novosibirsk, 1976, No. 1, pp. 122–140.

255. Sobolev S. L., "Theory of cubature formulas," in: Fundamental Studies. Fiz. Mat. Tekhn. Nauk [in Russian], Nauka, Novosibirsk, 1977, pp. 5–8.

256. Sobolev S. L., "On the roots of Euler polynomials," Dokl. Akad. Nauk SSSR, **235**, No. 2, 277–280 (1977). English transl. in Soviet Math. Dokl., **18**, 935–938 (1978).

257. Sobolev S. L., "The coefficients of optimal quadrature formulas," Dokl. Akad. Nauk SSSR, **235**, No. 1, 34–37 (1977). English transl. in Soviet Math. Dokl., **18**, 896–900 (1977).

258. Sobolev S. L., "On extreme roots of Euler polynomials," Dokl. Akad. Nauk SSSR, **242**, No. 5, 1016–1018 (1978). English transl. in Soviet Math. Dokl., **19**, 1253–1257 (1978).

259. Sobolev S. L., "Les coefficients optimaux des formules d'intégration approximative," Ann. Scuola Norm. Sup. Pisa Cl. Sci. (4), **5**, No. 3, 455–469 (1978).

260. Sobolev S. L., "On the asymptotics of the roots of the Euler polynomials," Dokl. Akad. Nauk SSSR, **245**, No. 2, 304–308 (1979). English transl. in Soviet Math. Dokl., **20**, 306–310 (1979).

261. Sobolev S. L., "More on the zeros of Euler polynomials," Dokl. Akad. Nauk SSSR, **245**, No. 4, 801–804 (1979). English transl. in Soviet Math. Dokl., **20**, 364–367 (1979).

262. Sobolev S. L., "On the roots of Euler polynomials," in: Theory of Cubature Formulas and Numerical Mathematics, Nauka, Novosibirsk, 1980, pp. 125–141. English transl. in: Differential Equations and Numerical Mathematics (Sel. Pap. Nat. Conf., 1978, Novosibirsk), 1982, pp. 49–68.

263. Sobolev S. L., "Comportement asymptotique des racines des polynômes d'Euler," Rend. Sem. Mat. Fis. Milano, **52**, 221–243 (1985).

264. Sobolev S. L., "On the order of formulas for approximate integration," in: Partial Differential Equations, Proceedings of the International Conference, 1983, Novosibirsk [in Russian], Nauka, Novosibirsk, 1986, pp. 4–11.

265. Sobolev S. L., Selected Questions of the Theory of Functional Spaces and Generalized Functions [in Russian], Nauka, Moscow (1989).

266. Sobolev S. L., Some Applications of Functional Analysis in Mathematical Physics, Transl. of Math. Monographs, **90**, Amer. Math. Soc., Providence (1991).

267. Solak W. and Szydelko Z., "Quadrature rules with Gregory–Laplace end corrections," J. Comput. Appl. Math., **36**, No. 2, 251–253 (1991).

268. Stroud A. H., "A bibliography on approximate integration," Math. Comp., **15**, No. 73, 52–80 (1961).

269. Stroud A. H., Approximate Calculation of Multiple Integrals, Prentice-Hall, Englewood Cliffs (1971).

270. Stroud A. H. and Secrest D., Gaussian Quadrature Formulas, Prentice-Hall, Englewood Cliffs (1966).

271. Taylor M. A., "Cubature for the sphere and the discrete spherical harmonic transform," SIAM J. Numer. Anal., **32**, 667–670 (1995).

272. Temlyakov V. N., "A new way of obtaining a lower bound on errors in quadrature formulas," Mat. Sb., **181**, No. 10, 1403–1413 (1990).

273. Temlyakov V. N., "On universal cubature formulas," Dokl. Akad. Nauk SSSR, **316**, No. 1, 44–47 (1991).

274. Triebel H., Interpolation Theory, Function Spaces, Differential Operators, North-Holland Mathematical Library, **18**, Amsterdam (1978).

275. Umarkhanov I., "Describing lattice cubature formula for many-dimensional domains with smooth boundaries," in: Problems of Numerical and Applied Mathematics [in Russian], Inst. Kibernet. i Vychisl. Tsentr Akad. Nauk UzSSR, Tashkent, 1979, No. 58, pp. 6–13.

276. Umarkhanov I., "A method for numerical integration of multiple integrals," in: Algorithms [in Russian], Tashkent, 1980, No. 41, pp. 75–82.

277. Umarkhanov I., "Cubature formulas with simple weight function," Dokl. Akad. Nauk UzSSR, No. 12, 5 (1983).

278. Vaskevich V. L., "On the norm in $L_2^{-m}[0,1]$ of an error with regular boundary layer," in: Theory of Cubature Formulas and Applications of Functional Analysis to the Problems of Mathematical Physics (Trudy Sem. S. L. Soboleva) [in Russian], Inst. Mat. (Novosibirsk), Novosibirsk, 1980, No. 1, pp. 41–59.

279. Vaskevich V. L., "On the convergence of Euler–Maclaurin quadrature formulas on a class of smooth functions," Dokl. Akad. Nauk SSSR, **260**, No. 1, 1040–1043 (1981). English transl. in Soviet Math. Dokl., **24**, 360–363 (1981).

280. Vaskevich V. L., "On the convergence of Gregory's quadrature formulas," Dokl. Akad. Nauk SSSR, **261**, No. 5, 1041–1043 (1981). English transl. in Soviet Math. Dokl., **24**, 634–637 (1981).

281. Vaskevich V. L., On One Problem of the Theory of Quadrature Formulas [Preprint, No. 3] [in Russian], Inst. Mat. (Novosibirsk), Novosibirsk (1982).

282. Vaskevich V. L., "Classes of infinitely differentiable functions and approximate integration," in: Some Applications of Functional Analysis to the Problems of Mathematical Physics (Trudy Sem. S. L. Soboleva) [in Russian], Inst. Mat. (Novosibirsk), Novosibirsk, 1984, No. 2, pp. 52–64.
283. Vaskevich V. L., "Approximate calculation of the Dirichlet integral of a harmonic function," in: Differential Equations with Partial Derivatives [in Russian], Inst. Mat. (Novosibirsk), Novosibirsk, 1991, pp. 64–83.
284. Vaskevich V. L., "On approximate solutions of the Dirichlet problem in composite spatial domains," Siberian Adv. Math., **1**, No. 2, 146–184 (1991).
285. Vaskevich V. L., "Approximate computation of some quadratic forms of integral type," Siberian Adv. Math., **4**, No. 3, 109–150 (1994).
286. Vaskevich V. L., "Cubature formulas in harmonic spaces of Bergman type," in: Cubature Formulas and Their Applications. Reports of a Seminar-Conference [in Russian], Ufa and Krasnoyarsk, 1995, p. 10.
287. Vaskevich V. L., "Cubature formulas of total precision on spaces of harmonic functions and their applications," in: Abstracts: International Conference AMCA-95, Novosibirsk, NCC Publisher, 1995, p. 340.
288. Vaskevich V. L., "Analogs of Hermite cubature formulas for the Dirichlet integral of harmonic functions," Siberian Adv. Math., **6**, No. 1, 105–150 (1996).
289. Vaskevich V. L., "Cubature formulas in harmonic spaces of Bergman–Polovinkin type," in: Cubature Formulas and Their Applications. Reports of a Seminar-Conference, Ufa and Krasnoyarsk, 1996, pp. 4–13. English transl. in: Siberian Adv. Math., **7** (1997) (to appear).
290. De Villiers J. M., "A nodal spline interpolant for the Gregory rule of even order," Numer. Math., **66**, 123–137 (1993).
291. Voĭtishek L. V., "On calculation of the coefficients of the formulas of mechanical cubatures," Apl. Mat., **10**, No. 3, 312–315 (1965).
292. Voĭtishek L. V., "A particular case of the construction of cubature formulas with a boundary layer," Zh. Vychisl. Mat. i Mat. Fiz., **9**, No. 2, 417–419 (1969). English transl. in U.S.S.R. Comput. Math. Math. Phys., **9**, No. 2, 213–216 (1971).
293. Voĭtishek L. V., "Calculation of the Epstein zeta functions," in: Problems of Numerical and Applied Mathematics [in Russian], Inst. Kibernet. i Vychisl. Tsentr Akad. Nauk UzSSR, Tashkent, 1970, No. 38, pp. 16–21.
294. Voĭtishek L. V., "On the choice of lattices for integration by Sobolev's formulas," Sibirsk. Mat. Zh., **17**, No. 4, 774–781 (1976). English transl. in Siberian Math. J., **17**, 584–589 (1977).
295. Voĭtishek L. V., "Construction of cubature formulas with varying mesh-size," in: Theory of Cubature Formulas and Applications of Functional Analysis

to the Problems of Mathematical Physics (Trudy Sem. S. L. Soboleva) [in Russian], Inst. Mat. (Novosibirsk), Novosibirsk, 1978, No. 1, pp. 46–53.

296. Worpitzky J., "Studien über die Bernoullischen und Eulerschen Zahlen," J. Reine Angew. Math., **94**, 203–232 (1883).

297. Xu Y., "Gaussian cubature and bivariate polynomial interpolation," Math. Comp., **59**, 547–555 (1992).

298. Yanovich L. A., Approximate Calculation of Continual Integrals by Gaussian Measures [in Russian], Nauka i Tekhnika, Minsk (1976).

299. Zagirova F. Ya., "An estimate of the remainder in a cubic formula with a regular boundary layer," Sibirsk. Mat. Zh., **15**, No. 6, 1262–1275 (1974). English transl. in Siberian Math. J., **15**, 888–898 (1975).

300. Zagirova F. Ya., "On expressing the weights of optimal quadrature formulas," in: Application of Functional Analysis to Partial Differential Equations (Trudy Sem. S. L. Soboleva) [in Russian], Inst. Mat. (Novosibirsk), Novosibirsk, 1983, No. 2, pp. 71–86.

301. Zagirova F. Ya., On Constructing Optimal Quadrature Formulas with Equidistant Nodes [Preprint, No. 25] [in Russian], Inst. Mat. (Novosibirsk), Novosibirsk (1982).

302. Zhamalov Z. Zh., "Construction of Green's functions for determining the extremal function of a cubature formula in $L_2^{(m)}(\Omega)$," in: Problems of Numerical and Applied Mathematics [in Russian], Inst. Kibernet. i Vychisl. Tsentr Akad. Nauk UzSSR, Tashkent, 1970, No. 38, pp. 29–35.

303. Zhamalov Z. Zh., "On an extremal problem of quadrature formulas with given derivatives," in: Boundary Value Problems for Differential Equations [in Russian], Fan, Tashkent (1973).

304. Zhensykbaev A. A., "Monosplines with a minimal norm and quadrature formulas," Uspekhi Mat. Nauk, **36**, No. 4, 171–196 (1981).

305. Zhensykbaev A. A., "Characteristic properties of best quadrature formulas," Sibirsk. Mat. Zh., **20**, No. 1, 49–68 (1979). English transl. in Siberian Math. J., **20**, 34–49 (1979).

306. Zhensykbaev A. A., "On optimal methods for reconstructing an integral," in: Theory of Cubature Formulas and Numerical Mathematics [in Russian], Nauka, Novosibirsk, 1980, pp. 104–107.

307. Zhileĭkin Ya. M., "On quadrature formulas on classes of functions," Zh. Vychisl. Mat. i Mat. Fiz., **8**, No. 3, 507–516 (1968).

Notation Index

Subject Index

Other *Mathematics and Its Applications* titles of interest:

W.L. Miranker: *Numerical Methods for Stiff Equations and Singular Perturbation Problems*. 1980, 220 pp. ISBN 90-277-1107-0

K. Rektorys: *The Method of Discretization in Time and Partial Differential Equations*. 1982, 470 pp. ISBN 90-277-1342-1

L. Ixary: *Numerical Methods and Differential Equations and Applications*. 1984, 360 pp. ISBN 90-277-1597-1

B.S. Razumikhin: *Physical Models and Equilibrium Methods in Programming and Economics*. 1984, 368 pp. ISBN 90-277-1644-7

A. Marciniak: *Numerical Solutions of the N-Body Problem*. 1985, 256 pp. ISBN 90-277-2058-4

Y. Cherruault: *Mathematical Modelling in Biomedicine*. 1986, 276 pp. ISBN 90-277-2149-1

C. Cuvelier, A. Segal and A.A. van Steenhoven: *Finite Element Methods and Navier-Stokes Equations*. 1986, 500 pp. ISBN 90-277-2148-3

A. Cuyt (ed.): *Nonlinear Numerical Methods and Rational Approximation*. 1988, 480 pp. ISBN 90-277-2669-8

L. Keviczky, M. Hilger and J. Kolostori: *Mathematics and Control Engineering of Grinding Technology. Ball Mill Grinding*. 1989, 188 pp. ISBN 0-7923-0051-3

N. Bakhvalov and G. Panasenko: *Homogenisation: Averaging Processes in Periodic Media. Mathematical Problems in the Mechanics of Composite Materials*. 1989, 404 pp. ISBN 0-7923-0049-1

R. Spigler (ed.): *Applied and Industrial Mathematics*. Venice-1, 1989. 1991, 388 pp. ISBN 0-7923-0521-3

C.A. Marinov and P. Neittaanmaki: *Mathematical Models in Electrical Circuits. Theory and Applications*. 1991, 160 pp. ISBN 0-7923-1155-8

Z. Zlatev: *Iterative Improvement of Direct Solutions of Large and Sparse Problems*. 1991, 328 pp. ISBN 0-7923-1154-X

M. Vajtersic: *Algorithms for Elliptic Problems*. 1992, 352 pp. ISBN 0-7923-1918-4

V. Kolmanovskii and A. Myshkis: *Applied Theory of Functional Differential Equations*. 1992, 223 pp. ISBN 0-7923-2013-1

A.D. Egorov, P.I. Sobolevsky and L.A. Yanovich: *Functional Integrals: Approximate Evaluation and Applications*. 1993, 426 pp. ISBN 0-7923-2193-6

G.I. Marchuk: *Adjoint Equations and Analysis of Complex Systems*. 1994 ISBN 0-7923-3013-7

Other *Mathematics and Its Applications* titles of interest:

A. Bakushinsky and A. Goncharsky: *Ill-Posed Problems: Theory and Applications*. 1994, 256 pp. ISBN 0-7923-3073-0

G.N. Milstein: *Numerical Integration of Stochastic Differential Equations*. 1995, 170 pp. ISBN 0-7923-3213-X

V.V. Shaidurov: *Multigrid Methods for Finite Elements*. 1995, 331 pp. ISBN 0-7923-3290-3

A.N. Tikhonov, A.V. Goncharsky, V.V. Stepanov and A.G. Yagola: *Numerical Methods for the Solution of Ill-Posed Problems*. 1995, 264 pp. ISBN 0-7923-3583-X

H.W. Engl, M. Hanke and A. Neubauer: *Regularization of Inverse Problems*. 1996, 328 pp. ISBN 0-7923-4157-0

G.I. Marchuk: *Mathematical Modelling of Immune Response in Infectious Diseases*. 1997, 358 pp. ISBN 0-7923-4528-2

V. Feinberg, A. Levin and E. Rabinovich: *VLSI Planarization*. 1997, 190 pp. ISBN 0-7923-4510-X

S.L. Sobolev† and V.L. Vaskevich: *The Theory of Cubature Formulas*. 1997, 438 pp. ISBN 0-7923-4631-9

Made in the USA
Las Vegas, NV
13 November 2024

11636299R00247